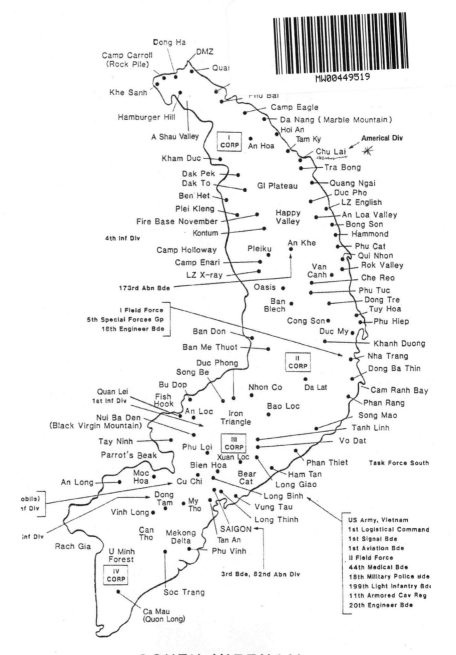

SOUTH VIETNAM

I'M READY TO TALK

This book is dedicated with respect and profound gratitude to all those who honorably answered our country's call to duty during the Vietnam War, and all other wars our nation has fought. It is also dedicated to the Family members and friends who lived the experience with us and who want to learn more from the point of view of those of us who participated in Vietnam and other wars — past, present, future.

ACKNOWLEDGMENTS

First, I thank the Vietnam veterans and wives who have written their stories to include in this book. You have far exceeded my expectations, both in number of stories to include and in the quality of the way you have expressed your honest memories of what to most of us is the most memorable time of our lives.

I thank my team at Deeds Publishing for making sure this book is as good as it can be. Jan quickly sold me on the title and cover picture, Mark used his design talents to make it into a cover that will catch anyone's eye, and Mark and Matt for their layout work on such a large and demanding book.

My editors—Norman Zoller, Mark Walker, Vince Corica teamed their editor's eye with the experiences they had during the Vietnam war to make sure this book is as real and accurate as it can be.

And I thank all who encouraged me to make this project happen. This is a legacy that all who participated are leaving for generations of Americans to come. I've never had a person tell me they knew too much about the military service of their Family member. This is our gift to our Family, our military unit, and to American history. No political agendas are included in these stories, just our personal experiences.

FOREWORD

When Bob Babcock and I began discussing the idea of asking our AVVBA members to share their stories for a book about our collective Vietnam experiences, we were thinking such a book would make a nice gift for our grandchildren.

However, as we started reading the stories our members submitted, we began to realize that what we were reading was evolving into far more than just a nice group of individual stories for our grandchildren. It was becoming a powerful piece of combined personal history, told by those who were actually in the middle of it—a historically accurate view of the entire Vietnam war!

The more stories we read, the more intense our invitation to our members to participate became. The more our members responded, the stronger was the realization by everyone that this project was far more important than we ever imagined it would be.

Every phase of the war was represented, from advisory to withdrawal. Every type of service was included, from support to battlefield. Every branch of service was represented, from the relatively few who served in the Coast Guard to the Army in which the largest number served.

Bob knew that each of us had at least one interesting and meaningful story to tell. What we came to discover is that the collection of all of our stories would provide *A Story of Vietnam*, end to end, front to back, top to bottom.

We still believe our children and grandchildren (and as yet unborn great-grandchildren) will enjoy reading our stories. But we are now convinced that every person wanting a clear and unbiased view of the Vietnam War will gain important historical insight by reading this book.

John Butler, President, Atlanta Vietnam Veterans Business Association

PREFACE

In 1998, when our WWII fathers were roughly the same age we Vietnam vets are now, I realized that far too many of those "Greatest Generation" veterans were dying and taking their stories to their graves with them. As the newly elected president of the National 4th Infantry Division Association, I assigned myself the project of collecting as many of the WWII stories as I could. I had good success—I preserved 325 stories from WWII vets who had served with the 4th Infantry Division and fought in Europe from D-Day to VE Day.

Today, all but a few of those WWII vets are gone as we are commemorating the 75th anniversary of their fight across Europe to defeat the savagery of Adolf Hitler... but their stories live on for future generations to value and learn from. They were in their 70s when they decided, with my prodding, that "I'm Ready to Talk."

This project has similar timing and the same objective for this unique organization of veterans in the Atlanta Vietnam Veterans Business Association. My fellow Vietnam vets, with my prodding, have agreed, it's time for them to agree—"I'm Ready to Talk." And in another 25 years, when the 75th anniversary of Vietnam is being commemorated, even if we are gone, our stories will be preserved in this book for younger generations to enjoy reading and to learn from.

I give a heartfelt "thank you" to the 115+ members and wives of the Atlanta Vietnam Veterans Business Association who have shared their personal experiences in this book. It will be a gift to our Families, our military units' legacy, and American history for generations to come.

By now you have figured out that this is a BIG book. While we have about 160 stories, versus the 325 that were in my WWII book, the WWII stories averaged a half page or less; our stories here are typically multiple pages long.

It shows the higher level of education the Vietnam generation had over our depression era fathers, and it also shows the passion in stories that have been held back as we Vietnam vets are just now able to tell the true feelings and emotions that we experienced in a war that most Americans wanted to sweep under the rug. Now, read on and you'll get quite an education as the unfiltered experiences you read come out. Our words express our deeds that most didn't want to hear about when we got home. There are others who call themselves "experts" who have tried to tell our story. In this book we are sharing the many pieces of the tapestry that made up the Vietnam War from what we personally lived and experienced. Put each thread together and you will find a complete tapestry of the real truth of the Vietnam War from the personal side that we lived.

VIETNAM WAS NOT A 16-YEAR WAR...

...it was a one-year war fought 16 times, by new troops every year.

Unlike World War II where people were drafted for the "duration of the war plus six months," the time period each person served in Vietnam was a 12-month tour of duty for the Army (which had the most personnel in Vietnam), 13-months for Marines, and varying times for Air Force, Navy, and Coast Guard. Career military personnel typically served multiple tours in Vietnam. Those who served in the military only two or three years usually had only one tour of duty.

No one can tell you, "This is how it was in Vietnam." Everyone's experiences were different. There are close to three million different experiences to be told from those who served in Vietnam, less the 58,000+ who didn't make it home alive. Sadly, many Vietnam veterans are now deceased and took their stories to their graves with them, never to be preserved for posterity.

Think of a tapestry hanging on a wall. If all the threads are still there, you have a beautiful piece of art. If you have many threads missing, the tapestry doesn't tell the whole story that needs to be told; it is incomplete. The same holds true of veterans telling their stories. If each veteran doesn't take the time to include his thread in the tapestry of any specific part of history, he or she leaves a hole that detracts from the overall story. In this book, we have a strong cross-section of individual stories from veterans who served in Vietnam, the threads that make up the history of the Vietnam War. We thank the veterans who decided, "I'm Ready to Talk." We also have some stories from wives who sweated through the war on the home front. These are our stories...

There were many diverse factors that shaped each veteran's memories from the Vietnam War. Among them are:

Branch of Service — Army, Marines, Air Force, Navy, Coast Guard, Civilian

Rank of the Person — Private, Specialist, Sergeant, Warrant Officer, Lieutenant, Captain, Major, Colonel, General (and equivalent ranks from other branches of service)

Region of Country — I Corps (far north, along border with North Vietnam and Laos), II Corps (center of country, going from South China Sea across to Central Highlands and Cambodia border), III Corps (area around Saigon, going from South China Sea to Cambodia border), IV Corps (farthest south, delta area of Mekong River), Off-shore (Navy and Coast Guard in South China Sea, Air Force flying from Thailand, Guam, Philippines, US, etc.)

Time Period Served —
Advisor Phase (1959-1964)
Buildup Phase (1965-1967)
Tet and Major Battles (1968-1969)
Withdrawal Phase (1970-1972)
Vietnamization Advisory Phase (1973-1975)
(Stories are arranged in this book by the time period the story refers to).

Public Opinion — Ignorance and Apathy to Mild Support, Disillusionment, Opposition, Protests and Riots, Relief and Indifference

Within the pages of this book are stories and experiences from 115+ Vietnam veterans who served in Vietnam. None of us will say, "This is the way it was"—rather, we are simply relating our stories to be left long after we are gone to be read, enjoyed, learned from by our Family members and others who want to read first-hand experiences from those who served in America's most unpopular war. None of us claim to be experts other than in what we personally experienced during our time in Vietnam.

And a quick side note, I understand "Any battle is a big battle... if you are in it." The time period above showing Tet and Major Battles is not to take away from the major battles fought in other periods of the war—Ia Drang Valley in November 1965, Dak To in November 1967, NVA Easter Offensive in 1972 are three big battles, along with others, that immediately come to mind. Tet 1968 started a series of big battles and escalation of troop strength to the highest level of the war during 1968 and 1969.

We believe it is our responsibility as American veterans to share our stories. To not do so is doing a disservice to our Families (current and future generations), current and future members of the branch of service and unit we served with, and to American history.

If Vietnam veterans let those who did not serve in Vietnam tell our story (with an agenda of their own to "prove"), we run the risk of perpetuating the myth that the Vietnam veterans were drug addicts, baby killers, and fought in a war that could not be won. After reading the stories in this book, I'll bet your view of Vietnam veterans changes for the better. We, like American veterans in wars before us, answered our country's call to duty and did our jobs, whatever that job was, to the best of our abilities.

When you have read our stories, we encourage you to ask veterans among your family and friends to tell you their stories. They will all be different—and absolutely true for that veteran, based on the factors shown above.

Steadfast and Loyal—Deeds not Words!
Proud, Professional, Patriotic
Robert O. "Bob" Babcock, B/1-22 IN, 4ID, Vietnam 1966-1967
Book Released on 3 December 2019

ABOUT THE AUTHOR

You may wonder, who is the guy who put this book together and why did he do it—what makes him qualified to take on this job. I won't answer that question, I'll leave it to my wife to tell my story from her point of view...

JAN BABCOCK
Wife of Bob Babcock

I must warn you, this story is quite different from the others in this book. It is a peek into the past of a high school girl who knew nothing about the Vietnam War. It didn't make the cut! After all, I had important things to do... cheerleading practice, writing notes to my friends, staring into the eyes of my boyfriend, etc. You get the picture.

Now fast forward to July, 1981; the day I married the love of my life, Bob Babcock. This was a second marriage for both of us. We were blissfully happy but extremely poor. Life was good for us as we tried to merge two families together. We had a child together and got into our separate roles as wife, mother, and IBM person for me, and head of household for Bob. As an IBM executive, Bob travelled a great deal. I still recall his coming home from a meeting in the winter of 1983 and telling me about this guy that he met and got into a very deep conversation about time spent in Vietnam, both with the 4th Infantry Division, but at different times. I should have known to nip that in the bud—just kidding!

In the summer of 1984, Bob came home telling me his new-found friend had invited him to go to Washington, DC on Veterans Day to attend the dedication of some kind of monument at the "Wall" that had been dedicated in 1982. I knew little about it but Bob asked me to go with him. I decided, it can't hurt anything. DC should be interesting and maybe I can learn more about Bob's Vietnam experiences, so I agreed to go.

Bob and I together attended our first Veterans' event, the dedication of the 3-Man Statue near the Vietnam Memorial Wall in Washington, DC. I just remember thinking what kind of group is this? Men telling stories, crying, laughing, and drinking lots of beer. I listened to them talk and was shaken by some of the things that I heard. You did what??? At one point there was a man, a Vietnam Marine (my father was a WWII Marine — Semper Fi!) that was truly struggling with his past. Somehow, I got in the middle of his anguish and he grabbed me and wouldn't let go. I was terrified by what I saw in his eyes. It took several veterans to remove me from his grasp. They did it gently, understanding from personal experience what his mind was flashing back to. It scared me; I didn't belong there; this wasn't for me. But being the supportive, loving wife, I encouraged Bob to continue with his newfound "hobby."

From that first experience, Bob started talking about Army reunions and how he wanted to go. "Fine," I said, "you go, and I'll stay with the children." At that time, little did I know that this would become a way of life for us. (His first National 4th Infantry Division Association reunion was in 1991 when the 4IDA was made up of 90% WWII veterans. He has only missed one since then. Vietnam veterans were just beginning to join). But I understood; I certainly had things that I loved to do so it seemed natural that he would as well. His involvement in the reunions grew as he became a part of the leadership and executive boards. Bob is a big-picture kind of guy and had huge plans for expanding his Associations' reach to other misplaced Veterans. He kept saying that he wanted me to attend these events. Naturally I had a great excuse; who would take care of the children??

In life, if we are blessed with the ability to follow our passion, we are indeed fortunate. That is exactly what Bob has done; not just for the past 30 plus years but for the majority of his life. As a boy, Bob constantly played Army. In college he took ROTC and received his commission. That was the springboard for the most memorable year in his life — serving as a rifle platoon leader leading men in combat in the central highlands of Vietnam with his beloved 22nd Infantry Regiment of the 4th Infantry Division. He always said he made more important decisions as a 23-year old lieutenant in Vietnam than he ever made as an executive at IBM.

From there, life took over and it resulted in a 34-year career with IBM. Remember when I told you about Bob's conversation with a fellow IBMer who also had been in Vietnam? That was 1983 when his passion was ignited, yet carefully tucked away.

As time passed, Bob began attending more and more reunions, he loved talking with WWII vets and hearing their stories. As he looked forward to hearing more each year, he quickly realized that those Vets were dying, along with their stories. In an effort to preserve this precious history, Bob took it upon himself to put together a book of 450 stories which included 325 WWII accounts, 25 Cold War, and 100 Vietnam. Although the vast majority of these WWII and Cold War men are now deceased; their stories live on.

Although I didn't attend many of these reunions with Bob, I understood that this was his love and passion. I had no problem with it as I was busy taking care of our children. In 2003 when the war in Iraq began, Bob got heavily involved in communicating via a daily email to family members of the 4th Infantry Division, letting them know what was happening. It was all consuming; sometimes taking 12-18 hours each day and reached 35,000 people. Now Bob's passion was starting to impact me. At first, I was resentful of the time spent away from our children and me. It was taking a huge emotional toll on Bob; there was nothing left for us.

I'll never forget the day in April 2004 that we were in Ft. Hood at the welcome home for the 4th Infantry Division after their first year in Iraq. Bob was recognized by Maj. General Ray Odierno, the Commanding General of the 4ID (later Chief of Staff of the Army), in front of the large assembly of Soldiers, Family, and friends, with a Commander's Award for Public Service. Bob has said he cherishes that award above any he has ever earned, other than his Combat Infantryman's Badge earned in Vietnam.

After the ceremony, as we were walking to the area where he would be meeting people, we saw a huge line of people and I wondered what it was for. As we got closer, Bob was engulfed by the crowd. They were there to see him; they wanted to meet this man who was their lifeline to their loved one in Iraq. It was the turning point for me. I finally understand that this was more than a passion; it was what God had whispered in his ear.

I stood back and watched as Mothers, Fathers, and wives hugged my husband in gratitude for what he had done. I finally understood the importance of his passion. He continued doing that mission for the division's next three Iraq deployments, through 2011.

Because of Bob's extensive knowledge, he is constantly asked to speak to different groups. I never tire of hearing him talk. There is not a week that goes by that someone doesn't ask him to help them find out information about a loved one's service record. He has helped me see the importance of love for our country and the willingness to lay down one's life to keep all of us safe.

So I want to thank you, Bob, for your service and for teaching our children and me the importance of standing up for what you believe in. I am very proud of you and for what you have accomplished. You exemplify Deeds Not Words! I love you dearly.

Now read on, you will read stories from many more veterans who are as passionate as Bob is about their patriotism and pride from answering their country's call to serve during the Vietnam War.

A COMMENT FROM THE AUTHOR

With over 115 authors contributing stories to this book, we (the editing team) have agreed to keep the style and voice of each author in his/her story. Thus, you will see inconsistencies in use of capitalization, abbreviations, military jargon, and style. We considered putting a glossary of military terms in the book, but it is already longer than we thought it would be (thank goodness we got so many stories sent my way). We were adamant that we preserved each author's story in their unique style and manner. Therefore, don't be an English teacher critic if you find some things that don't suit you. Our focus is the story, not making it so vanilla and proper that the story's intent is lost.

If you want to understand a term we use, do like most people do in this day and age, go to your computer and Google it—or, better yet, ask a veteran. We all speak the same language and would love to explain it to you, and maybe tell you another one of our stories because you showed an interest in us. Remember, we are reaching the age, "I am Ready to Talk." Thus, that is why we titled the book *I'm Ready to Talk*. Show some interest, ask your favorite veteran a few questions, and I bet he or she will tell you more than you expected to hear.

Also, you will see some profanity sprinkled through the stories. That has been a way of life for all wars since the beginning of time. It's hard to express yourself with a, "Gee whiz, that was bad…" Instead, it takes some profanity to best express yourself in these toughest and most memorable times of our lives. If you are offended, it is not our intent to offend but to tell our stories in the best way we know how to tell them.

DAVE HAMBRICK

USMC – Combat Engineer and Administrative Legal/Chief
Dates of Military Service (Active Duty and Reserves Combined): 1964 – 1985
Unit Served with in Vietnam: "A" Company, 7th Engineer Battalion, 1st Mar.
Div; "A" Company, 1st Tank Battalion, 3rd Mar. Div; HML 367- 3rd Mar. Air.
Wing
Dates in Vietnam: Aug 1965 to Oct 1966; Mar 1968 to Jan 1969; Apr1975
Highest Rank Held: Captain
Place of Birth and Year: Maynard, AR – 1947

We will start our stories with one from a Marine who was there close to the beginning, in the middle, and at the end of the war. His story comes from all three times he served in Vietnam.

"THE BEGINNING, MIDDLE, AND END"

It has been 55 years now, and who would have thought it, and, I rarely think of my tours in Vietnam unless some veteran mentions it, or if there is a sound, a smell, a flash of light, or a glimmer of movement in the trees behind our back yard. Instead I see small vignettes, some are vivid but most are mundane. Recently I was asked by my VA PTSD Counselor what are some of my first, or most vivid memories that affected my senses — some are:

The smell of *nuoc mam*. The heat, humidity, and dust. The blue exhaust of cycles clogging the dirt roads, streets and villages/hamlets. Local villagers, including women, relieving themselves along the dusty roads and trails. Curious eyes behind the questionable smiles of the villagers. Naked children in the villages, the sight of young women with their flowing *Ao Dais*. Water buffalos moving gracefully, for their enormous size, down a rice patty dike with a young small boy in charge. The smells of rotten vegetation. Monsoon rains that seemingly would never end and the smell of your own feet that never seemed to get dry, and the daily, if not hourly, sounds of the helicopters, trucks and jeeps.

So, from the cotton and soybean fields of northeast Arkansans and southeast Missouri who would have known that the young 17-years old high school drop-out would be in a war, so far from home in a matter of just a short year. I joined the Marine Corps in September 1964 and by the spring of 1965 I had completed boot camp and infantry training, and then joined Company "A", 7th Engineer Battalion located at Camp Pendleton California for Combat Engineer Training. It was the lead company of the engineer battalion assigned to the Regimental Landing Team-7 of the 1st Marine Division. Along with this unit, we boarded the USNS General Blatchford transport ship in San Diego harbor in June 1965 and headed for "destination to be determined." Twenty-two (22) days later we arrived on the island of Okinawa and completed jungle training. Afterwards, we boarded the LST 1166 Washtenaw County and headed for Vietnam (even though we did not know it at the time) arriving at the Da Nang pier after seven more days at sea on August 16th.

Throughout the next year, many large and small events and operations occurred, some that I recall are Operations "Starlight" and "Harvest Moon." Teams sweeping several roadways, trails, and villages for mines and booby traps. Not only did we sweep roadways for mines, etc. we used heavy equipment for building roadways, our construction workers built hard back tents, buildings, bunkers, and sandbagged those bunkers. Like others I pulled guard duty. I sat on a listening post (LP) in a garbage dump infested with rats. Movement to the left or movement to the right, quietly report it up the chain of command.

More thoughts: 122mm rockets and/or B-40 rockets launched down the hill from the listening post LP headed for the Da Nang airstrip. Death along the roadway. Shrapnel in the knee and ankle. Tracers bouncing through the compound. Spider bite on the thigh that had to be drained by a medical corpsman daily for a week or so. Pain in the right ankle—broken. Hill 41 and Hill 55 under repeated nightly insurgent attacks, small arms and mortars.

Other thoughts: seeing Ann Margaret, Joey Heatherton, and Johnny Rivers over at Freedom Hill. October 1966, this extended tour of 14 months is over and now to the Freedom Bird over at the Da Nang airstrip. But instead of returning stateside like so many combat vets, I was sent to the Marine Corps Air Station, Iwakuni, Japan for a year. After Japan, I was sent to train/teach the Marine reservists in a Communications Battalion in Cincinnati, OH, on what to expect should they be called up to go to Vietnam. During that tour here are the various campaigns we were involved with in country:

Vietnam Defense Campaign (8 March 1965 — 24 December 1965)

Vietnamese Counter-Offensive Campaign (25 December 1965 — 30 June 1966)

Vietnamese Counter-Offensive Campaign Phase II (1 July 1966 — 31 May 1967)

I was only in the states for a few months and received orders to go to Vietnam for a second tour because my MOS was short. This tour was from March 1968 until January 1969. I flew this time, instead of ship, and when I reported into the Division Headquarters, I told them that I was already a one-tour Vietnam vet and I wanted to go as far north in "I Corps" as I could. I was assigned to Company "A", 1st Tank Battalion located at the Gia Le Combat Base southwest of Hue City and Phu Bai at the base of the Ashau Valley. I got there in time for the backside of the Tet offensive and the company I was assigned to had multiple Silver Star and Bronze Star Medal recipients. I was happy to be in a unit that was very much involved with the support not only of many Marine Corps infantry units but also supporting the 101st Airborne Army units located nearby at Camp Eagle.

By this time, I was no longer a Combat Engineer having been injured in the first tour I was assigned as the Administrative/Legal Chief and supported all the various operations. There were casualties, both WIA and KIA. One Tank Commander (TC) a sergeant was shot in the back, through-and-through by an outlying sniper. Routinely received H&I (harassment and interdiction) fire within the compound. Major wire breach and gunships overhead — hot brass falling down our body armor-flak jackets. About a dozen NVA bodies in the morning at first light; they were already bloating.

Time to leave this wretched, but beautiful, country — it was taking its toll on me and I didn't know at the time it would be called PTSD sometime later. Our company left the Hue City area by way of the Navy taking us down the Perfume River out to the South China Sea and back down to the Da Nang area — all in I Corps. During that tour here are the various campaigns we were involved in and some of the operations from that tour that I can still remember were Operations "Nevada Eagle," "Houston" and "Owen Mesa" and the campaigns were:

Tet Counter Offensive (30 January 1968 — 1 April 1968)
Vietnam Counter Offensive Phase IV (2 April 1968 — 30 June 1968)
Vietnam Counter Offensive Phase V (1 July 1968 — 1 November 1968)
Vietnam Counter Offensive Phase VI (2 November 1968 — 22 February 1969)
Tet 1969 Counter Offensive (23 February 1969 — 8 June 1969)

One combat tour is certainly enough and two can be considered too much, but several years later I would be serving my third overseas tour, this time back in Okinawa. I was assigned as the Squadron Administrator/Legal Chief. Seemingly just getting there the squadron (HML-367-Scarface and later Hover Cover) was chosen, along with many others to go afloat the USS Blue Ridge and USS Hancock with the challenge and opportunity to help perform the extraction and evacuation of Americans, South Vietnamese, and Cambodians from the American embassies in Saigon (RVN) and Phnom Penh (Cambodia) in Operations "Frequent Wind" and "Eagle Pull." After this tour of duty, I was reassigned to Camp Pendleton, CA, where we assisted in billeting many of the refugee "boat people" from the South Vietnam area.

So, literally I was in Vietnam at the beginning (1965-1966), the middle (1968-1969), and at the very end (1975).

Later, with my combat assignments behind me, I was chosen to go to officer's school in Quantico, VA. Even though I was the second oldest and not in as good as shape as the youngsters, I graduated in the top 2% in a class of 248 candidates and was placed on the Honors List . I was both surprised and honored, and later I was sent to the Administrative Officer's Course in Parris Island, SC, graduating 2nd in the class. Later I was sent to the Naval Justice Course (Nonlawyer) and, even though a student, was asked to teach the Courts-Martial prep phase. As an officer I had several exciting assignments at Parris Island, SC; Camp Pendleton, CA; Okinawa (again); Naval Air Station Memphis, TN; and, as Inspector-Instructor Staff on the Georgia Institute of Technology campus when I retired.

So here I am—55 years ago, a young 17-year-old left the cotton and soybean fields as a high school dropout. Joined the Marine Corps. Went to sea. Went to war a few times. Got a lot of experience and got hurt. That was a long 55 years ago. Along the way I got formally educated—high school completed, bachelor's degree achieved (after 12 years of night school), got an MBA degree, and pursued a second Masters' degree (Operations Management) until I began instructing at the Embry-Riddle Aeronautical University as an adjunct professor/instructor of Management. I rose in the enlisted ranks from buck private (E-1) to Gunnery Sergeant (E-7), performed in the warrant officer ranks through Chief Warrant Officer-3, then as a commissioned officer ranks through Captain. But, more importantly, along the way while on Recruiting Duty stationed in Atlanta in the early 1970's, I met the absolute love of my life and was fortunate enough to marry Gail Donald of Atlanta, Georgia, 47 years ago.

A great career of over 21 years I must say. It has been a great ride—Semper Fi to all.

CONTENTS

ADVISOR PHASE

(1959 TO 1964)

July 8, 1959 — First two Americans killed in Vietnam.

November 30, 1961 — President Kennedy expanded American involvement in Vietnam.

January 12, 1962 — Operation Ranch Hand Began (spraying of Agent Orange in Vietnam).

February 20, 1962 — John Glenn orbited the Earth.

October 14, 1962 — Cuban missile crisis began.

January 2, 1963 — Battle of Ap Bac (VC guerillas defeated a much larger South Vietnam force, American advisors were killed).

January 11, 1963 — Buddhist monk burned himself to death to protest President Diem.

August 28, 1963 — Martin Luther King, Jr gave his "I Have a Dream" speech at Lincoln Monument in Washington, DC.

November 2, 1963 — President Diem, president of South Vietnam was assassinated.

November 22, 1963 — President Kennedy was assassinated.

March 24, 1964—COL Floyd Thompson captured by VC, longest held POW in American History.

July 2, 1964—Civil Rights Act was signed.

August 2, 1964—Gulf of Tonkin incident—the Gulf of Tonkin Resolution by Congress gave the President unprecedented power to commit US Forces without a declaration of war.

August 5, 1964—LT Everett Alvarez, US Navy, shot down—became first pilot detained in Vietnam, was released in 1973.

Source: www.vvmf.org

RICHARD HAMIL

US Army — Aviator

Dates of Service: 1959 to 1965

Unit Served with in Vietnam: UTT (Utility Tactical Transport)

Dates Served in Vietnam: Nov 1962 to Nov 1963

Highest Rank: Captain

Place of Birth and Date: Villa Rica, GA — 1937

UTILITY TACTICAL TRANSPORT COMPANY

The UTT arrived in-country in the fall of 1962. They had been stationed on Okinawa and were brought to Vietnam on TDY orders by way of Thailand! The company had been in the planning stage for months and finally was able to enter the country equipped with rigged "Rube Goldberg" machine guns and rockets mounted on UH-1A helicopters. All the armament was "homemade "and scrounged from multiple sources. Rocket tubes from the Air Force, electronics from the Navy, and machine guns from the Army. Company personnel designed, fabricated, and installed the systems. The machine guns and rockets were bore-sighted. There was no sight system, so we used tracer ammo in the machine guns to access the target and then fired the rockets. Pilots would use a grease pencil to mark their sighting preferences on the wind shield. A bit primitive, but it worked and was very accurate.

The first duty performed by the UTT was transport helicopter escort. Before we arrived, the H-21 transport helicopters had been subject to deadly ground fire due to the incompatibility of Air Force fixed wing aircraft to support helicopters. The T-28 was not able or accurate enough to provide the coverage needed. So after many meetings, on-the-ground discussions, and Army/ Air Force politics, the following deal was struck. The Air Force T-28 aircraft would escort until two minutes from the LZ and then would climb to altitude and loiter while the UTT gunships swept the LZ and broke left and right and formed a "daisy chain" support. This tactic worked very well as we were able to keep guns on target with three aircraft on each side of the LZ. Additionally, the crew chief and a door gunner provided more fire power out of the two rear cargo doors. One of these gunners fabricated a "monkey strap" from a parachute harness, thus giving the door gunners the ability to get on the skids and fire below the aircraft.

We were attached to MACV and our mentor was BG Joe Stilwell. He loved the armed helicopter concept and helped us get any tools we needed to perform our mission. Almost daily, he could be found in the cargo door of one of our aircraft working as a door gunner. He was a Soldier's Soldier! Loved the troops and wanted, no, demanded, to be in the middle of the action.

In mid-November, the UTT received 10 brand new UH-1B helicopters with Emerson XM 6 quad M60 machine gun kits with flex sight gun controls operated by the co-pilot. I was part of this group. We were sent to Ft Rucker for a UH-1B model check and a gunnery course. Then we went on to pick up the aircraft at the Bell plant in Ft Worth. We flew the Bs to Ft. Sill to test fire the guns and back to Texas for loading on Air Force C-124s for transport to Vietnam. A few days later, we arrived in Vietnam and became part of the UTT.

Major Ivan Slavich had just taken command and was preparing to take the concept of armed helicopters to the next level. Known to his men as "Drivin' Ivan," behind his back of course. The Major was no nonsense, a competent, savvy, and fearless leader. Never would he ask his men to do anything he had not done or that he would not do. He worked hard and played hard. He was well-connected with the brass as well as the press. The UTT was a magnet for major news outlet reporters. They flew with us almost daily. Dave Halberstan, Neil Sheehan, Dick Tregaskis, correspondents from all news outlets, as well as other writers. Our Operations Center was always open for them to write and file their stories. This was the way Slavich got our story out to the world. He was a publicity master!

Major Slavic was always thinking of new tactics to employ against Charlie. We survived the Battle at Ap Bac, losing an aircraft to ground fire while attempting to rescue downed H-21 crews. One of our crew chiefs was killed. The other aircraft on the mission received damage, but were flyable. We immediately did an internal review to determine what we could have done differently. The following day we returned to retrieve our aircraft and fly cover for the brass on the ground. We located the main Ap Bac VC force but our request to engage was denied. The VC were traveling by sampan down a canal and presented a beautiful target. The whole Ap Bac engagement had been a nightmare and a lot of folks were really upset. Never really understood the politics!

Slavich worked with American advisors to develop a concept they named "Eagle Flight." This involved two or three armed aircraft searching at low level for suspicious activity. When found, troop carrying helicopters would land and put troops on the ground to engage the target. The interesting part was the unknown size of the enemy, anywhere from a few to company or larger force!

There were several interesting encounters. On one Eagle, there was garbled radio contact with the ground unit that had been deployed and our Platoon Leader landed to establish contact. A prisoner had been captured and it was decided to return him to the staging area on UTT aircraft. Our Crew Chief and ground troops searched him, bound his arms and placed him on the helicopter. On board was Dick Tragaskis, a freelance writer, as well as our crew. When the flight got back to the staging area and turned the prisoner over to Intelligence for questioning, a live hand grenade was found strapped to his leg! (The complete story can be found in Vietnam Diary by Dick Tregaskis).

On another Eagle flight, the action was slow. so Major Slavich decided to do a low-level recon and spotted a VC lying under water, breathing through a reed. He blew the reed away and the VC surfaced and was immediately captured. Then the remainder of the flight used this tactic and captured several more VC. Boys will be boys! Maybe not smart but effective.

Another thing we did was deploy one platoon to Da Nang to fly support for the Marines. Evidently they had heard the loss of the Army's troop-carrying H-21helicopters had dramatically dropped since the UTT's introduction to the battle field. I do not know how this deal was brokered but later we heard the Navy Commander in the Pacific was not happy but we continued the support for many months as the Marine unit said they would not fly without gunships. We stayed with the Marines until they received UH-1 armed helicopters and were trained to fly their own support.

We also introduced the "Hog" which was a rocket system that deployed 24 rockets on each side of the Huey. This was quite effective in preparing an LZ or providing close support without having to waste time re-arming. On an early test, our Armament Officer was test firing the system and all 48 rockets fired at once; he said he felt as if the aircraft had stopped in mid-air!

The UTT was the first armed helicopter unit deployed into combat. We developed tactics and proved the armed helicopter could deliver fire support for troop transport, supply missions, and deliver fire power in support of engaged ground forces. Best of all, we proved the helicopter could survive on the battlefield. Many attempts had been made to arm helicopters but the UTT HU-1 A models were truly the first in combat. From UH-1A came the future generation of gunship Hueys and finally the Cobra, all serving in Vietnam. The UTT motto was "First With Guns!"

EDWARD ETTEL

U.S. Navy

Dates of Military Service (Active and Reserve): 1963 to 1989

Unit Served within Vietnam: USS Hollister (DD-788)

Dates you were in Vietnam: 1964

Highest Rank Held: Navy Captain (O-6)

Place of Birth and Year: Corpus Christi, TX – 1940

TONKIN GULF YACHT CLUB

The Tonkin Gulf Yacht Club was a tongue-in-cheek nickname for the

United States Seventh Fleet during the Vietnam War. Throughout the war in Vietnam, the Seventh Fleet engaged in combat operations against enemy forces through attack carrier air strikes, naval gunfire support, amphibious operations, patrol and reconnaissance operations and mine warfare.

In July 1964, my ship, the USS *Hollister (DD-788)*, was part of an anti-submarine group (Task Force 77) based in Yokosuka, Japan. As head of the Communications Division, among other things, I was Communications Officer and Crypto Officer. In early August, we were scheduled to stay in port in Yokosuka, Japan for two weeks.

Meanwhile, down in Southeast Asia, part of the U.S. Seventh Fleet was deployed off the South Vietnamese coast to show U.S. determination to preserve South Vietnam. At least one carrier task group steamed at the soon-to-be famous Yankee Station, the operational staging area off Vietnam's coast in the South China Sea. The mission was to conduct low-level aerial reconnaissance of suspected Communist infiltration routes in eastern and southern Laos. No combat missions had started, but the Soviets had assisted the North Vietnamese in the construction of more sophisticated anti-aircraft installations, so the USS *Maddox* (DD-731) was dispatched to conduct electronic warfare "DeSoto missions" to identify these installations. This set up the famous "Tonkin Gulf Incident."

On 4 August 1964, the American destroyers USS *Maddox* and the USS *Turner Joy* (DD-951) evaded what lookouts and sonar identified as torpedoes, and fired on contacts visually identified by the Turner Joy crewmen as P-4 motor torpedo boats, which then headed for the USS *Ticonderoga* (CV-14) carrier task group steaming around the entrance to the Tonkin Gulf. President Johnson went on national TV at mid-day on Tuesday, 4 August (0100 5 August in Japan) to announce the attacks on the *Maddox* and *Turner Joy* in international waters. President Johnson ordered U.S. naval forces to prepare for a retaliatory air strike against North Vietnam to be carried out at 0800 local time on 5 August. A few days later, the U.S. Congress passed the Tonkin Gulf Resolution, which gave the Government authorization for what eventually became a full-scale war in Southeast Asia.

Back in Japan on 4 August, our task force was preparing for upcoming anti-submarine exercises in Japanese waters several days hence. However, because of the new events in Vietnam, our Captain, CDR Robert F. Stanton, called the officers together and read our orders to get underway with the task force for Vietnam waters at 0500 the next morning, 5 August. The weather started out well enough, but as we passed Taiwan, it was overcast and the winds and sea state began to increase. Super Typhoon Ida (Seniang), a category 4 super typhoon (SSHS) had intensified to a 155 mph super typhoon on August 6 and struck northeastern Luzon at that intensity that night. After weakening over the island, Ida turned to the northwest, blocking our path through the Bashi Channel, so our task force slowly circled, waiting for the storm to clear. You don't want to pass through a typhoon if you don't have to, and even aircraft carriers knew to avoid it.

The next day the typhoon had barely moved, and because its wind velocities were increasing, and because it was imperative that we reach the Tonkin Gulf, the decision was made to go through part of the typhoon. As we approached the storm, we could see tops of the waves being blown off by the wind, so we knew we were already in gale force winds. As we approached closer to the typhoon, the weather began to deteriorate further. We had to pass by the typhoon in the shortest possible route to the Tonkin Gulf, which meant we would pass through its front-right quadrant, where the maximum effects of a hurricane are usually felt. Here the winds are usually strongest, storm surge is highest, and the possibility of tornadoes is greatest. This also meant that as we approached, swells came from our port bow and then our port beam. It is usually difficult to steer a steady course with a following sea, but you sure don't want waves coming from your beam. The winds crept up over 80 knots, with higher waves and swells, and I saw my first green waves hitting the bridge windows. Of course, everything was tied down topside (outside), secured inside, and no one was allowed topside.

The combination of wind and sea would shove the *Hollister* over on her side, and hold her there until we would wonder if she would right herself. That afternoon we took several severe rolls, up to 47 degrees. Even though things were stowed well, we heard crashes all over the ship, with things falling everywhere, including loud crashes of china from the galley. I remember holding on as best I could in one big roll, as the ship rolled and seemed to keep leaning further and further. It was a terrible feeling because it didn't seem like it was going to roll back, but then we bounced back like a cork, the way a destroyer is supposed to do. With these swells and winds, the Hollister was acting more like a cruise ship, with long exaggerated rolls, rolling over, then rolling a little more, and then continuing the roll a little more. Visions came to me of a World War I destroyer that rolled over upside down, until a wave helped knock it back upright, thanks to the good righting arms that destroyers are designed to have.

When it began to settle down, I decided I needed a photo. It wasn't possible to take one from inside, because the windows were awash with spray, so I ventured out on the starboard side of the wardroom on the main deck, with a rope tied to my waist. I had very little time, so took only three photos looking astern. Later, the winds shifted to our port quarter and then from our stern, as we continued south into the Tonkin Gulf. I don't remember if we had chow during the storm. Sleep was difficult, and poor Ensign Queeney (with an obvious nickname) was only able to sleep on the evaporator in Engine Room Number One, amidships close to the ship's center of gravity, thereby the most stationary spot on the ship. He subsisted only on crackers and water that his enlisted men would bring him.

Thus began our time in the South China Sea, continually serving as part of the screen around the aircraft carrier, or serving as 'plane guard', positioned to rescue pilots if one of her birds went into the drink. In addition to our regular daytime work hours, we stood four-hour duty watches, normally eight hours off watch and four on watch, and sometimes 'port and starboard,' which means four hours on, and four hours off watch.

One five-inch dual gun-mount was manned 24 hours a day, but we also had dawn and dusk General Quarters every day for an hour or so, because these are crucial hours at sea, when visual and radar (I think it is a word now, not an acronym we old farts used) difficulties make it hard to pick out low-flyers or small boats. Officers attended 'Don McNeill Breakfast Club' meetings in the Wardroom each morning with the Captain. (Referring to a long-run morning variety radio show). We would follow this routine at sea for 30 days or more, and then pull into Subic Bay, Philippines, and become lost in a fog of Planter's Punches at the Subic Officer's Club (only $0.15 during Happy Hour). And try to stay away from the raunchy places in Olongapo City outside the base gates. But that is another story.

The ship also visited Hong Kong, which was a treat. However, we were warned never to go to the top of the tall apartment buildings, because of the opium dens. But our Hong Kong experiences are another story. Interestingly, we have fewer memories of visiting Sasebo, on the southern Japanese island of Kyushu.

Even without a storm, some of the crew were frequently seasick. I found I had a continuous dull headache as my inner ears kept pinging on my brain that things were tipsy. LTJG Cleary had arranged his lower bunk mattress and cruise box in the forward junior officer stateroom so he could sleep with his head in the sink. He remembers his very first Office-of-the-Deck (OOD) watch on a mid-watch (midnight to 0400) trying to stay out of the way of USS *Constellation* (CV-64), which had launched retaliatory strikes into North Vietnam. He said he would never forget the ominous call on the radio — "Unknown Navy destroyer — this is War Chief — I am conducting flight operations — stand clear" or something to that effect. He didn't bother to wake Captain Stanton, he just stood clear.

On 19 September, I assumed new duty as Combat Information Center (CIC) Officer. As such I was responsible for the Center, which brings together and manages information on the warship's status and its surroundings, supplying recommendations to the captain, who would generally be present on the nearby bridge. CIC, or 'Combat', is the tactical center of a warship that funnels communications, radar, sonar, electronic emissions, and other data received over multiple channels, which is then organized, evaluated, weighted and arranged to provide ordered timely information flow to the battle command staff under the control of the CIC officer and his deputies.

CIC is usually depicted in movies and television as being filled with large maps, numerous computer consoles and radar/sonar repeater displays, as well as the ubiquitous grease-pencil annotated plots on vertical edge-lighted transparent plotting boards. The room was noisy due to humming electronic equipment and background static and traffic from multiple tactical radio speakers. As in most ship spaces, there was the ever-present large coffee urn, the life-blood of the ship, with strong coffee that most Navy men drink straight with no cream or sugar. The coffee became stronger with time because new coffee was often brewed over standing coffee and grounds, until the urn had to be emptied of several inches of bottom sludge.

I also continued as Crypto Officer, which became a burden due to heavy FLASH secret and top-secret cryptographic message traffic, the codes of which had to be broken manually. There were other officers on the Crypto Board, (a group of personnel qualified to break codes), but because several members of the board had left ship with no chance to train others, we were very short. So I was awakened at all hours of the night, often every one or two hours, to break critical in-coming or out-going messages in cryptographic code.

I would just doze off, and feel my bunk jiggle, and hear a soft voice, "Mr. Ettel, you have a Flash Secret message to break." Of course to save time I slept in my khaki pants and T-shirt, so I would make my way to the COMM room, open a second locked combination door to my little five by five-foot crypto space I knew as my CC, 'cold coffin', and close the spring safe door behind me. Besides performing the precise, unforgiving code-breaking procedures while very sleepy; because of high security, there was only one vent in the room, and it could not be closed off. The room temperature was between a refrigerator and a freezer and the only vent was a three-inch vertical tube that came out of the ceiling and blew right down my neck. I attached all types of devices to limit the flow, to no avail. At least it helped keep me awake. I dared not mention how cold it was to the crew, because few enlisted spaces had any air conditioning.

For our patrol services off Vietnam, we were eligible for the Armed Forces Expeditionary Medal. Military personnel who served later in Vietnam were awarded the Vietnam Service Medal, and we were granted the option to exchange the Armed Forces Expeditionary Medal for the Vietnam Service Medal.

P.S. Historians conclude that the first PT boat attack on the USS *Maddox* was real, but the second "attack" on the *Maddox* and the *Turner Joy* was not.

NORMAN E. ZOLLER

US Army — Field Artillery Officer (later Judge Advocate General Officer)
Dates of Military Service (Active and Reserve): 1962 to 1993
Units Served with in Vietnam: Detachment B-130, US Special Forces; 3rd Bde, 82nd Airborne Div.
Dates in Vietnam: Sep 1964 to Mar 1965; Mar 1968 to Feb 1969
Highest Rank Held: Lt. Colonel
Place and Date of Birth: Cincinnati, OH — 1940

DUTY WITH SPECIAL FORCES

Norman Zoller served almost seven years on active duty in the Army as a field artillery officer, including two tours of duty in Vietnam, first with Special Forces (Detachment B-130; 1964-65) and then with the 3rd Brigade of the 82nd Airborne Division in response to the Tet Offensive in 1968-69. He served 15 additional years in the National Guard and the Army Reserves as a judge advocate general officer for a total of 22 years, retiring as a Lieutenant Colonel.

When Zoller first deployed to Vietnam in September 1964, he joined about 17,000 military service members then in-country. In those days for Special Forces, the principal missions were pacification, security, and counter-insurgency, and his B Team located initially at Can Tho in the Mekong River Delta region supervised and supported eight A Teams on the Cambodian and Laotian borders. Objectives varied: protection of village inhabitants and village leaders, conduct of reconnaissance and appropriate offensive and defensive operations, delivery of public education, public utility, public welfare, and refugee and local government support.

After about two months in-country at Can Tho, it was determined for several reasons that the B Team needed to be closer geographically to the A Teams. As a result, the B Team built a new camp and moved midway from Can Tho to near the village of Long Xuyen, which was about 45 kilometers from the border regions.

From this location, another civil affairs and military government officer and I made and maintained contacts with, and provided support and resources to, village chiefs, not only at Long Xuyen but also at the eight Special Forces A Team locations: Don Phuoc, Ha Tien, Du Tho, Ap Bac, Long Khat, Moc Hoa, Tien Binh, An Long, Tri Ton, and Tan Chau. In thinking today about those locations, the names seem foreign and unfamiliar. But in the mid-60s, those places are where we worked every day seeking to create and maintain healthy and trustworthy relations with village people, their leaders, and a safe village environment: digging wells to provide potable water; building schools, providing school supplies (many of which came from CARE USA), and other Special Forces support.

On reflection, I will always carry with me the comforting realization that of the 30 members on our B Team, although some sustained wounds and injuries, all returned to our home base in Okinawa safely and no one was killed. Sadly, however, one of our team members subsequently became Missing In Action on his sixth tour of duty.

These were committed, courageous, and specially-trained soldiers, and I was and am proud to have served with them.

ROBERT LANDIN

United States Army – Infantry Officer, Chemical Officer

Dates of Military Service: 1960 – 1981; 2003 – Present (Georgia State Defense Force)

Unit Served with in Vietnam: 10th ARVN Division Advisory Detachment, Senior Adviser 4th Battalion, 48th Infantry Regiment

Highest Rank Held: Colonel

Place of Birth and Year: Albia, IA – 1937

MIDNIGHT (?) REQUISITION

All military are familiar with the use of the "Midnight Requisition" to make up shortages or get needed equipment or supplies not available through channels.

In Vietnam, a twist on this method was used by my assistant. I was leading a five-man advisory team with a Vietnamese infantry battalion in an area north of Ben Hoa and adjacent to the Dong Nai River in an area the Viet Cong called War Zone C. The battalion base camp was on a small hill that overlooked the river. The Advisory Team lived in a large rectangular hole on the south side of the camp. Dunnage lumber from the docks was used to make a roof and then covered with sand bags. Within the hole each man had a canvas cot and a footlocker positioned next to the wall and about three feet below the ground line. At one end of the hole at ground level was a covered shelter with a handmade table and stool for each man. We called it the Team Hut. Air circulation in the hole was minimal at the best of times, so sleeping was often difficult.

The team had become acquainted with several members of an Air Force Medcat team from the base at Ben Hoa when they had come to a village in the area. A result of this is that we were invited to visit them at the base. My assistant was the first to do so. When he stopped at the base on his return from a bartering trip (captured weapons for US food). While there he noticed some men unloading boxed furniture into their barracks and when he asked was told where the issue point was located. As an enterprising Lieutenant, he drove to it to see if anything might be available. They were issuing bunks, wall lockers, and mattresses for the barracks. He asked to get some and was told to show his requisition. His explanation was that he did not have one and was just doing a favor for the Headquarters Company since he had a trailer to haul stuff in so if there was a problem he would just leave and return to his unit.

The end result was they loaded his trailer with six double wall lockers, five metal cots, and five mattresses. He immediately headed out of the main gate and directly to the base camp, in case the issue point decided to check. Thanks to the Air Force generosity, that night each team member had a new double wall locker to store gear and a bunk with a mattress to sleep on. Each bunk was supplied with a set of extensions for conversion into bunk beds. These were attached to the legs and placed on blocks to raise the level of the beds to ground height to get any breeze available. The extra wall locker was set up in the Team Hut to store our food. This illustrates that for an enterprising individual, the midnight requisitions need not be done at night.

EDWARD J. NIX

U S Army — Infantry Officer
Dates of Military Service: 1951 to 1972 (All Active)
Unit Served with in Vietnam: MACV — 21st Infantry Division, South Vietnam Army; MACV HQ
Years Served in Vietnam: Aug 1964 to Aug 1965; Jul 1971 to Jul 1972
Highest Rank Held: Colonel
Date and Place of Birth: Cleveland, GA — 1930

AIR MOBILE OPERATION — SOUTH VIETNAM

When I arrived in South Vietnam in August 1964, I was assigned as the G-3 (Operations) Advisor to the 21st Infantry Division, South Vietnam Army. The Division's Area of Operations was the five southern provinces of South Vietnam. The Advisory Group and the Division were conducting all larger operations as Air Mobile Operations. We would develop targets, stage troops and helicopters at one of the province air fields, and conduct the operation from there.

In the spring of 1965, we received very reliable intelligence that indicated heavy VC activity in Can Tho Province—just south of the Basac River. We developed the intel and agreed that the intel was correct.

My counterpart (Division G-3) and I developed the plan of attack over the next 24 hours. We staged the entire Division (three Regiments, plus two Ranger Battalions), 45 lift helicopters, and five gun platoons at a nearby airfield. The LZ was an open rice field, with rice cut and stacked in shocks. The troops were to land in the rice field and attack the tree line on the enemy side of the field. The "gun-ships" fired on the tree line before the ground troops were landed in the LZ.

As the first assault wave of "slicks" were landing, the rice shocks were knocked down and we faced machine guns and recoilless rifles. We found ourselves in a big battle and changed our strategy quickly. We used the gun ships to destroy the machine guns and recoilless rifles and started landing the ground troops closer to the trees to reduce their exposure time.

We used two Command and Control choppers to direct each air mobile operation. One included my Senior Advisor (Colonel) and the Division Commander (Major General). The second included me and my Vietnamese counterpart (both Majors). One of the "C & C's" were over the operation at all times. This gave us the capability to talk to the US Advisor on the ground, helicopter commander (Pilot) and the Vietnamese Commanders. The operation lasted well into the night when the enemy broke contact.

During one of my flights in the afternoon, our chopper took a hit and dropped about fifty feet. We all checked our body parts and found no injuries—the helicopter was still flying and we were heavily involved in the operation. So, we flew another two hours before refueling. When we landed, the crew chief checked the plane for damage and found a large hole in the main rotor blade. He took out his green tape and covered the hole—top and bottom—and we flew another eight hours that day. The next day we extracted the ground troops with captured weapons, prisoners, wounded, etc. The prisoners were in khaki uniforms (shorts). This was the first time we had made contact with VC not in black uniforms. Our intelligence people determined that the unit was part of the North Vietnamese U Minh III Regiment.

When MACV Headquarters learned that we had had an operation against a unit in other than black uniforms, they requested that we send someone to brief at the 5 PM Press Briefing. My Senior Advisor boss said to me, "You planed this operation, you go brief them." I took the C&C Pilot, one US ground Advisor, and maps of the operation and flew to Saigon for the briefing.

There were about 25 press people present—UPI, AP, NY Times, Stars and Stripes, etc. I gave the overall briefing with maps, followed by US ground advisor and C&C pilot. During my briefing, various press members tried to change the meaning of my comments to fit their thinking. I had to restate my correct comments several times.

The next morning at breakfast, I read the front-page article of the briefing In the Stars & Stripes and learned that the reporter had taken editorial liberties with my comments—splitting them and inserting them as he saw fit. I started complaining out loud to the Advisory Group about the article. My Senior Advisor (Boss) said, "Now you know why I didn't go do the briefing."

As an irony to this experience with the press, later the Army sent me to graduate school (UGA) to get a Masters' Degree in Journalism/Public Relations. I spent four of my last five years on active duty in this field.

BUILDUP PHASE

(1965-1967)

March 2, 1965—Rolling Thunder bombing campaign of North Vietnam began.

March 8, 1965—First brigade sized US force of Marines landed in Da Nang, first large scale ground force deployed to Vietnam.

August 18, 1965—Operation Starlite began, first major US battle of the war, 5,000 Marines against a stronghold of VC. Two Marines earned the Medal of Honor.

November 14, 1965—Battle of Ia Drang Valley started, first major battle between US and NVA forces.

November 27, 1965—March on Washington for Peace in Vietnam, 25,000 protestors.

March 8, 1966—Australia sent first troops to Vietnam, one of five allies to send troops to fight alongside US.

August 6, 1966—President Johnson increased troops deployed to 292,000, which would later peak at more than a half million deployed *(Author's note: I landed in Vietnam on this date—will never forget it, as will all our vets remember when they arrived).*

October 21, 1967—Over 50,000 protestors marched on the Pentagon in protest against the war.

November 30, 1967—American deaths in Vietnam hit the 15,000 mark.

Source: www.vvmf.org/VietnamWar/Timeline

JOHN "MARK" WALKER

United States Navy — Supply Officer

Dates of Military Service (Active and Reserve): 1963 to 1992

Unit Served with in Vietnam: USS Wrangell (AE-12)

Dates Served in Vietnam: Nov 1965 to May 1966

Highest Rank Held: Captain (O-6)

Place and Date of Birth: Memphis, TN — 1941

LET'S TRADE: MY SPARE TIRE FOR YOUR PAINT

In October of 1965, our ship, the USS Wrangell (AE-12), departed our home port of Charleston, SC, for the South China Sea in support of the newly initiated war in Vietnam. Our trip over took about six weeks. We transited the Panama Canal, entering the Pacific Ocean, and were detached from the Service Force, Atlantic, or SERVLANT and attached to the Service Force, Pacific or SERVPAC. The next stop was Pearl Harbor, Hawaii for three days to load more materials. We were so loaded up that our decks were covered with crates of bomb fins, and 2 x 12 boards had to be laid atop the crates to allow the crew to walk fore and aft on the decks. Normally we worked Monday thru Saturday on this part of our cruise. But we crossed the International Date Line on a Saturday, giving us two Saturdays in a row, making that work-week 7 days. Arriving in Subic Bay, Philippine Islands, our deck-load was removed and we loaded up for our first trip to support the combatant ships in the South China Sea at Yankee or Dixie Station.

Our typical schedule was two weeks at sea replenishing the bomb and ammunition loads of the ships we served, and one week back in Subic to reload. While in the Operating Area we replenished ships virtually every day by conducting "UNREPS" or Underway Replenishments. The supply ship would choose a steady course into the wind. The large combatants like aircraft carriers and cruisers would take station (come along side and steam adjacent to us) on our port side; the smaller ships, on our starboard side, for the duration of the UNREP. It was not unusual for us to replenish two vessels simultaneously, one on each side. It was quite an operation. Sometimes we replenished ships at night. Because white light destroys night vision the whole night replenishment operation was conducted with red lights, which do not impact one's vision. The Supply Department's Commissary Division (food service) would provide "mid-rats" (midnight rations) for the crew when these operations were conducted. Baloney sandwiches were the most popular mid-rats. It was common for the aircraft carriers to launch and recover aircraft (even in the dark) while we were sending over 500 pound bombs and other ammunition or supplies. We could go up to the flying bridge, which was the highest point on our ship, and still not be able to see onto the flight deck of a Carrier. But we could sure hear the noise!

I was fortunate to have another officer as my Assistant Supply Officer, plus a Chief Store Keeper. They were both very competent and pleasant people, as were most of the people in the Supply Department. As I recall we had about 33 enlisted members and our department served a crew of approximately 235. Our work included storage, inventory control, and ordering all spare parts; ship's services such as laundry, ship's retail store and barber shop; food service for the crew, except for the officers, who had their own cook and galley (kitchen); and disbursing of funds. The most important part of disbursing was payday for the crew! As the senior of the two Supply Officers (by this time I was a Lieutenant Junior Grade (LTJG) or O-2), I chose parts inventory and storage, and food service. The Assistant Supply Officer handled Disbursing and the Ship's Services. I was the only Reserve Officer Department Head. The other three, Deck Force, Operations and Engineering were "Navy Mustangs," or former enlisted men who had become commissioned officers. All of them were Lieutenants, which is the Navy's rank of O-3, equivalent to an Army, Marine Corps or Air Force Captain. Without exception, they were very good at their jobs, having been E-7s or higher in one of the professional jobs which was included in their departments. I took a certain amount of grief from these very senior men. They were all 15 or 20 years older than me.

When the ship was underway (at sea) there were six "watches" or duty rotations each day of four hours each, which required four officers in each duty period. In addition, the enlisted members of each department had "watch" duty assignments. We had 24 officers besides me, the XO (Executive Officer) and the CO (Commanding Officer). So the "watch bill" could be filled without me, and for this cruise I was not required to stand underway watches. I was assigned to a duty section while in port.

The ordering, handling, storage, and inventory of bombs, ammunition, and associated materials was handled by the Deck Force, managed by the First Lieutenant, a former Senior Chief Gunners Mate (E-8). This was a position, not a rank, even though our First Lieutenant was of the LT rank. Everybody worked hard throughout the crew, but the Deck Force was really under pressure when we were on station. A typical work day was 12-18 hours or more. However, when in port, we had a normal working day, and the resupply of bombs, etc., was handled by civilians. In port, "Liberty" started after the work day for all hands (everyone) except those in the "duty section," whose members stayed aboard in case of emergency, and to insure necessary services were performed. The crew was usually divided into three duty sections, so one "had the duty" every third day. For most of the officers, being in port meant a much lighter workload. But the Supply Department was working the same schedule as if we were at sea. For example, three meals a day are served whether at sea or in port. Also, I spent many days at the Subic Bay Supply Depot trying to track down parts for our ancient equipment.

One of the most difficult supply items to procure through the Navy supply system was non-skid paint. Our decks were used all day long by fork trucks carrying heavy loads, and the paint wore off. It was a safety issue because the steel decks became as slick as ice when wet, so they really needed to be covered by non-skid paint. The aircraft carriers had priority for non-skid paint because their flight decks had to remain "non-slip." No matter what I did, I could not persuade anyone in the supply chain to make an exception for Wrangell. In the spring-time of 1966, we were preparing for the 6-week return to the States. We would have plenty of time to paint the decks if I could only get that non-skid paint.

On one of my trips to the Supply Depot, I was chatting with a First Class Aviation Storekeeper (E-6) sailor. I asked him if his carrier had trouble getting non-skid paint. He said, "Heck, no! We have skid loads that we'll never use! But I cannot get a spare tire for our Commanding Officer's vehicle." The bottom line was that, with the Engineering Department's permission, (they were in charge of our ship's vehicles) I traded the brand new spare tire and wheel from our pick-up truck for a skid-load of non-skid paint. That is what is called "comshaw" in the Navy; trading something you don't want or need for something that you do want or need. I ordered a new wheel and tire for the truck, and they were waiting for us when we got back to Charleston.

We had four short visits to "liberty ports" during our six months, each lasting three or four days. Two in Hong Kong; two in Kaohsiung, Taiwan. I loved Hong Kong because of all the great buys. I had two business suits made for $30 each and six tailor-made shirts for $2.00 each. They came in handy in my subsequent civilian life.

During this cruise, my wife and I communicated by letter and audio tape. Our first child was born on April 1st of 1966, a little girl. At first I thought the congratulatory message was an April Fool's joke, but it was for real! Suddenly I was a "daddy." My wife and I have often said that nine months of letters and tapes helped us get to know each other better than if I had never deployed.

In early May of 1966, we were detached from SERVPAC to return to Charleston, and our Commanding Officer decided to complete an around-the-world cruise. We headed south, crossed the Equator, making us all "shellbacks," and stopped in four ports of call on the way home: Singapore; Bombay, India; Beirut, Lebanon; and Barcelona, Spain. When we transited the Suez Canal to enter the Mediterranean Sea, the daytime temperatures reached 100+ degrees. We started ship's work at 5:30 AM and stopped at 2:00 PM. A Navy ship in that kind of heat is unbelievably hot. The only air conditioned spaces were the Crew's Mess and sleeping quarters, Chief's Mess and sleeping quarters, the Wardroom, and the Captain's spaces. As I remember, the Officers' sleeping area was poorly air-conditioned, and often uncomfortable. We only suffered that for four or five days.

We got home in mid-June. The Charleston Navy League was proud of their first ship to serve "in Vietnam," so they had a Navy Band on the pier to welcome us, and held a luncheon for all the officers in downtown Charleston. That was very nice, but I had been away from my bride for nine months, so I just wanted to get home. I did not hang around after the lunch was over. I went back to our apartment and got reacquainted with my wife and met my new, 2 and 1/2 month-old daughter. She was so beautiful! (Still is!)

Because Wrangell served in two combat zones, our ship and crew was awarded the Vietnam Service Medal with two stars. That has allowed me to fellowship with an incredible group of people, the Atlanta Vietnam Veterans Business Association.

In recent years, I have come to realize that the U. S. Navy has been a very important part of my life, particularly because I met my wife while serving. I am proud of my Navy service and especially proud that I was able to contribute in a small way to the success of our efforts in the Vietnam War.

JUDY C. WALKER, PH. D.

Wife of Mark Walker, US Navy

MY EVENTFUL 1965 AND 1966 AS A NEW NAVY WIFE

My husband, Mark Walker, and I had only been married a short time when the ship in which he served as Supply Officer left Charleston, SC for Subic Bay, PI in October of 1965. I was alone for the first time in my life, having lived at home most of the time I attended the University of Georgia, which is in my hometown of Athens. Mark and I met when he was a student at the "Navy School" in Athens, where he learned to be a ship-board Supply Officer. Mark had rented a furnished duplex apartment in Charleston and I was looking forward to being independent and learning the ropes of a Navy wife whose husband was deployed.

Captain Homer Doran was Commanding Officer of the ship, USS Wrangell (AE-12). Mrs. Doran did a great job of keeping the officers' wives together, having coffees and other social gatherings. As a result, I made friends with several wives and we still keep in touch every Christmas. Mrs. Doran was really good at keeping our morale up during the 9-month deployment.

During the deployment I was pregnant with our first child. There was no technology at the time to tell us the sex of the baby, so I bought a little book that chronicled the life of a sexless baby, Egbert. So we called our baby "Egbert" until birth.

Mark and I wrote a lot of letters. We also sent audio tapes back and forth. Tape cassettes had not yet made it to market, so we purchased two inch reel-to-reel tapes at the Navy Exchange, which came with little mailers. We each had a recorder/player and we sent tapes back and forth regularly. Somewhere I still have most of those tapes stored away in a box. I wrote to Mark several times each week, keeping him up to date on Egbert's progress. One time I sent him a profile photograph of my pregnant body in a two-piece bathing suit. I thought that would help him feel a part of the process, but he said that he was shocked when he saw the picture. I looked like a toothpick with an olive on it.

During the third month of his deployment, the neighbor in our little duplex apartment reminded me that our rental agreement specified that the apartment was for "two people." Of course, that would not work when the baby was born. I could have waited until Mark returned to find another place, but, being excited about my first project as a wife, I started looking for an apartment that had no such restriction. I found a furnished one-bedroom apartment right outside the "hospital gate" to the Navy Base. It was bigger and more secure than the duplex, and my Mom and Dad came over to help me "move." Because we were just beginning our lives together, we had almost no furniture, but we had purchased a large piece of plywood to put under the sagging mattress in the duplex. I remember my Dad tying that big plywood piece on top of his car and driving down the section of Interstate highway under construction, holding on to it through the window with his left hand. A kind Highway Patrolman stopped him. After explaining that the minimum speed limit was 40, not 20, he had Dad follow him to the new apartment. I will always appreciate that kindness.

Our baby, Egbert, turned out to be Cheryl, our only daughter. (We later added three sons.) She was born on April 1, 1966. Mark said that when he got the radio message, he thought the Communications Officer was playing an April Fool's joke on him. Not only was it April 1, but his first child was supposed to be a son! When Cheryl was only about four weeks old, I sent Mark a tape which included her crying. She was such a good baby, that I had to pinch her (Can you believe it?) to start her crying. There must have been three minutes of that crying jag, and Mark later told me it was "way too long."

It turned out that Wrangell was the first Charleston-based ship to serve the US forces in Vietnam, so the Navy League organized a welcome reception for the ship. In late June, 1966, the ship returned, and as she approached the pier, with over a hundred family members waiting for their sailor husbands and fathers, a Navy Band began to play. After a luncheon for the ship's officers, Mark and I had a wonderful afternoon getting reacquainted and having him meet his almost three-month-old daughter for the first time. Five months later, Mark's three year and four month obligation to active Navy service was over, and we said good-bye to Charleston and our Navy friends.

We often talk about how that nine-month deployment really helped us get to know each other in ways that being together for that time would not have accomplished. As of this writing we have celebrated 54 years as husband and wife.

CLINT JOHNSON

US Navy — Aviator

Dates of Military Service (Active and Reserve): 1954 to 1990

Unit Served with in Vietnam: VA-25 Attack Squadron 25 "The Fist of the Fleet"

USS Midway CVA-41 and USS Coral Sea CVA-43

Dates in Vietnam: Mar 1965 to Nov 1965; Jun 1966 to Nov 1966

Highest Rank Held: Captain USN (O-6)

Place and Date of Birth: Detroit, MI — 1937

SKYRAIDER VS MIG-17

Frustration and fatigue were starting to simultaneously set in on me on 20 June 1965. We were 30 days into our third at-sea period, and the ops tempo was intense. Ten days prior we had our first loss, one of our nuggets, Carl Doughtie (Nugget = A freshly minted Naval Aviator with no operational squadron experience). The last four days we had not been especially successful. During those four days I had flown 21 hours on an Alfa strike (a maximum effort full airgroup attack on a target assigned by the Joint Chiefs of Staff in DC), two road recces (an armed combat mission for targets of opportunity), and a seven and one half hour RES-CAP (Rescue Combat Air Patrol — a mission to cover downed pilot and escort for rescue helos). The strike was marginally successful with 40 percent BDA, the RESCAP was not. We had to leave the downed pilot when it got dark. One road recce was nothing more than harassment. The other I scored one truck, but someone almost scored me while I was executing a life-saving pullout just short of bending the prop. I logged two nice round holes in the aft fuselage.

The day began normally with the starboard catapult crashing into the water-brake outside my door, acting as my alarm clock. It was supposed to be a stand-down day, but by noon we were suiting up for an emergency RESCAP. An Air Force photo-recon pilot had been shot down very deep into the northwest corner of North Vietnam. There were already RESCAP aircraft over the downed pilot, but they were running low on fuel. We were needed for backup coverage. We manned up, started and were told to shut down. Someone else had covered the pilot, and they did not need us. We unmanned and returned to the ready room and waited. Two hours later we got the call again. We manned up, but did not get started again before we were again put on hold. By the time we got to the ready room, we were told to man up again. By now we were fast becoming the leaders in the squadron sweat stain contest.

The sweat stain contest was unique to Skyraider squadrons. The winner was the pilot who could merge the salty white left and right armpit stains in the center of his flight suit first. This contest was made possible by the USS MIDWAY (CVA-41) laundry and morale officer who would accept only one flight suit per week per pilot from us. At any rate we were hot, sweaty, and beginning to worry that this man up was going to mean no dinner. This time, however, we started, were told that we were a go mission and began our taxi forward to the catapults. At the last minute my Plane Captain, AN Halcomb, gave me a slush filled thermos and a hopeful look (hopeful that he would not have to do a fourth preflight on old 577). I gave him thumbs up and taxied forward to the starboard catapult. It was almost 1800. I spread and locked the wings, got thumbs up from the final checker, and agreed with the flight deck officer on a 21,300 pound launch weight.

As I felt the Skyraider settle into the catapult holdback, I released the brakes, added full power, and scanned the engine instruments. Everything looked good and with the canopy open everything sounded good — well at least loud. I returned the cat officer's salute and waited. I saw my flight leader go off the port cat and turn right for our standard starboard side rendezvous. The humidity was so high that his flap tips left contrails and my prop was making corkscrew contrails as the carrier moved through the sultry gulf air. The cat shot killed my radio. We rendezvoused 1,000 feet on the starboard side of MIDWAY and headed west. After reforming in a finger four formation I tried to get my radio working. As the second element leader I had a "Middleman" aircraft. My airplane had two radios with a relay control box that could be switched so that the low aircraft covering the downed pilot could transmit through my aircraft to the ship using my aircraft at a higher altitude as an antenna relay. I was able to get the number two radio working, but continued to fiddle with number one so that I could act as relay. I got it working and checked in on tactical frequency as we went feet dry. Then it failed again.

Feet dry at 12,000 feet heading northwest we were passing north of Thanh Hoa. LCDR Ed Greathouse was in the lead. On his port wing was LTJG Jim LYNNE. I was on his starboard wing with Charlie Hartmann on my starboard. We all had the standard RESCAP load: two 150 gallon drop-tanks on the stub racks, four LAU-3 pods with 19 2.75 inch rockets apiece and 800 rounds of 20mm for the four wing cannons. We were flying steadily toward the downed pilot while I navigated, searched for active low frequency ADF stations (Until September 1965 the North Vietnamese MiGs used the ADFs listed in our 1964 navigation supplements) and considered what the situation ahead might be.

Suddenly Ed Greathouse rolled inverted into a near vertical dive with Jim Lynne following. I rolled and followed him down. I was concerned that I had not heard anything and that we were only 70 miles inland, at least 80 miles from our RESCAP point. A quick radio check confirmed that my radio was dead. I had missed the buildup to the run-in with the USS STRAUSS (DE-408) alerting us to MiGs in the area. The MiG pilots were on an intercept for two Skyraiders south of us, but missed and were coming around for another intercept when they spotted us. STRAUSS was keeping Ed Greathouse updated, and when it was apparent that we were the target, Ed took us down. At 12,000 feet and 170 knots we looked like Tweetybird to Sylvester the Cat. Our only hope was to get down low and try to out turn the MiGs. Ed was doing just that. Our split-S got us some speed and reversed our course toward the ship. I figured that any time my nose was pointed at the ground my ordnance should be armed. I armed the guns and set up the rockets.

About that time I saw a large unguided rocket go past downward. My first inclination was that it was a SAM, but SAMs generally go up. A second rocket hit the ground near Ed and Jim. There was no doubt we were under attack by MiGs. This was confirmed when a silver MiG-17 with red marking on wings and tail streaked by Charlie and me heading for Ed. Tracers from behind and a jet intake growing larger in my mirror were a signal to start pulling and turning. As I put g's on the Skyraider, I could see the two distinct sizes of tracers falling away. (The MiG-17 had two 23mm and one 37mm cannon in the nose.) He stayed with us throughout the turn, firing all the way. Fortunately, he was unable to stay inside our turn and overshot. As he pulled up, Charlie got a quick shot at him but caused no apparent damage. He climbed to a perch position and stayed there.

Our turning had separated us from Ed and Jim. Now that we were no longer under attack, my main concern was to rejoin the flight. I caught a glimpse of the leader and his wingman and headed for them. As we had been flying at treetop level in and out of small valleys, we had to fly around a small hill to get to them. Coming around the hill we saw Ed Greathouse and Jim LYNNE low with the MiG lined up behind them. I fired a short burst and missed, but got his attention. He turned hard into us to make a head-on pass. Charlie and I fired simultaneously as he passed so close that Charlie thought that I had hit his vertical stabilizer with the tip of my tail hook and Charlie flew through his wake. Both of us fired all four guns. Charlie's rounds appeared to go down the intake and into the wing root and mine along the top of the fuselage and through the canopy. He never returned our fire, rolled inverted and hit a small hill, exploding and burning in a farm field. Charlie and I circled the wreckage while I switched back to number two radio. We briefly considered trying to cut off the other MiG, but were dissuaded by the voice of Ed Greathouse asking what we thought we were doing staying in the area when STRAUSS was reporting numerous bogeys inbound to our position. We took the hint and headed out low level to the Tonkin Gulf where we rejoined with our flight leader.

By now the sun was setting, guaranteeing a night arrested landing back at MIDWAY. Our radio report was misunderstood by MIDWAY CIC which believed that one of us had been shot down. It took some effort for Ed Greathouse to convince them that we were OK and the North Vietnamese were minus one. Rarely does a night carrier landing evoke as little response from a pilot as ours did. We were so pumped up that we hardly noticed it.

After debriefs all around, the politics started. Charlie and I were informed that we would get no recognition or awards for our MiG kill. SECNAV had been aboard three days earlier when VF-21 F-4 pilots had bagged the first kills of the war. Their awards were being held until SECNAV could get to Washington, announce it to the President and present it to Congress with the plea for more funds for F-4 Phantoms to fight the air war.

Obviously, the success of primitive Skyraiders would undermine his plans. Unfortunately, someone had included our kill in the daily action report to MACV where it was read by COMSEVENFLT DET "C" who thought that it would be an excellent opportunity for Navy public relations. Indirectly Ngyuen Cao Ky, the new Premier of South Vietnam, and a Skyraider pilot, heard of it and recognized Ed Greathouse's name as one of the Skyraider instructors from the RAG. He then demanded our appearance for Vietnamese awards.

The next day we flew to Saigon for the Five O'clock Follies and were instant celebrities, since the news media did not yet know about the F-4 kills. They assumed that we were the first which made an even better story. We stayed at the Majestic Hotel in Saigon where we thoroughly enjoyed the lack of water hours and the availability of our favorite beverages. The next day we were guests of Premier Ky at the palace were we were awarded Air Gallantry Medals and honorary commissions in the South Vietnamese Air Force. After the awards ceremony we sat down to tea with Premier Ky and some of his young hot pilots and traded war stories. He told us that the Skyraider MiG kill had boosted morale tremendously in the VNAF Skyraider squadrons.

Upon arrival back at MIDWAY, we were surprised to learn that there had been a change of heart and we would be recognized at the same ceremony as the F-4 pilots. Since they had already been recommended for Silver Stars, Charlie and I go the same while Ed and Jim got Distinguished Flying Crosses. Due to slow processing of earlier awards

Charlie and I wore the Silver Star and one foreign decoration for about a month as our only medals. Nothing like starting from the top.

A few days later the carrier went to Yokosuka where Japanese reporters were very interested. We even became the subject of an article in a boy's adventure comic book. There was a lot of hometown interest also with reporters looking up our wives and parents for comments. This caused me a problem because I had not told my mother that I was flying combat to avoid worrying her.

Needless to say, the VA-25 pilots were not about to let the slack-jawed beady-eyed jet pilots (Ed Greathouse's description) forget our success. The squawk box in the fighter ready rooms got plenty of incoming from our ready room. There was much frustration in the swept wing tail hook community as the next two kills went to the Air Force in July. Then the North Vietnamese pulled the MiGs for more pilot training. The only kill between July 1965 and April 1966 was a single Navy kill in October 1965. We maintained that we embarrassed them into pulling the MiGs.

A combat action happens fast and it is difficult to include all the influences that affect the outcome, but some sidelights are of interest. The day of the shoot down was the first that gun camera film was not loaded in our planes. Charlie fired 75 rounds and I fired 52. We both thought we had fired more. I had considered firing rockets to ensure a kill, but was afraid that the widespread pattern of the LAU-3s would also hit Ed or Jim. Three of our aircraft suffered engine failures in the near future. There were no fighters airborne at the time and they missed a great opportunity for the bogeys launched after the shoot down. Two years later I was invited to Miramar to brief the people setting up "TOP GUN." My briefer said, "Well, you were flying the F-4?" "No." "Oh, the F-8?" "No." "The A-4?" "No." "A-7?" "No." "Well, what the hell were you flying?" "The Skyraider." Then his jaw went slack and his eyes got beady. They're all the same. (See editorial comments below.) Our squadron, VA-25, "The Fist of the Fleet," was the last operational Skyraider attack squadron in the Navy. We were flying a 20-year-old design that had been perfected about as far as the engineers could take it. Everyone thought that our time was over as front-line attack. What everyone forgot was that Ed Heinemann had mandated that the Skyraider not only had to be able to carry that 2,000 pound bomb a thousand miles to Tokyo and return to the ship, but that it also had to be able to defend itself against air attack. We never forgot. Unfortunately, even Ed Heinemann could not foresee SAMs. The Skyraider just did not have the top end speed to evade them. In April 1968, VA-25 retired the Skyraider in favor of the A-7 Corsair II. The aircraft and pilot, Ted Hill, that made the last combat carrier landing led four A-7s in a flyby, broke off to the east and disappeared out of our sight, but not our hearts. Ted flew it to Pensacola where it resides in the National Museum of Naval Aviation in our squadron colors. I flew six combat missions in that aircraft.

I flew as many hours in the A-4 Skyhawk as I did in the Skyraider and later flew the A-7. I truly enjoyed my A-4 time and it became my favorite. However, the Skyraider was something special. Even though my right leg has shrunken to the same size as my left leg, the carbon monoxide is cleared from my blood and the stack gas from my lungs, there is still that feeling that the Skyraider was where I was meant to be.

One final note. The first flight of the Skyraider was on 18 March 1945, my eighth birthday.

Editor: When news of the MiG shoot down arrived in VA-122, we fired off a message to our sister RAG squadrons at Miramar—offering "our assistance in improving their air-combat training." Another MiG shoot down by VA-176 on October 9, 1966 proved the ACM skill of SPAD pilots was not a fluke. Shortly, we heard that Miramar would be the home of the new TOP GUN School. What SPAD pilots had known all along really was important in combat. (Source: midwaysailor.com)

MARILYN WEITZEL

Army Wife

A SOMETIMES PAINFUL MEMORY

As a young Army bride in December of 1961, I had joined my husband at his duty station in Fort Benning, GA.

Soon thereafter I was hired as the secretary to the Lawson Army Airfield commander, a continuation of my service as a DA civilian which had begun a few years earlier at an Army Ordnance depot in Ohio.

Three years later I was with child, concurrent with the founding of the 11th Air Assault Division which would later deploy to Vietnam as the 1st Air Cavalry Division. My pilot husband had volunteered to accompany the advance Caribou unit; however, the battalion commander denied the request as he realized I was due to deliver within the month.

Two weeks later I went into labor. Alas, my husband was on a field exercise and a neighbor delivered me to Martin Army Hospital where I gave birth to my first son with no new father in attendance. Welcome to the Army, Miss Ohio Farm transplant.

Nine months later, my husband received orders to Nam. I elected to remain at Benning as I had returned to my job and preferred being in my adopted home to continue my civilian career rather than to return to my rural roots.

When the first large wave of 11th Air Assault orders came down, including my husband's, we were given little notice to vacate government quarters. Moving vans could be seen on post sometimes as late as 3 AM as wives and children scrambled to move. Many returned to their home towns; I found an apartment just a few miles off post where my baby son and I would reside for the next duration. The wives who also elected to stay in the military community formed support groups to ease the burdens they were now forced to endure.

The airfield was abuzz with the latest war happenings as more units were mobilized and dispatched. I had to cease watching Walter Cronkite because I was having nightmares about the chaplain making that fateful knock on my door.

A year later, my husband returned from the war front, but alas our marriage disintegrated due to his persistent bouts of alcoholism. A divorce ensued; sadly, he was KIA during the first few months on his second tour.

I remained in the rented digs, eventually remarrying another pilot at Fort Benning and began an odyssey of eight PCSs (permanent change of station) in seven years. Tucked into these moves was another tour in Vietnam when I gave birth to my second child. Thanks to the communications of those times, it was an entire week before I heard from the new father and six months before he met his new daughter.

More changes of address ensued before I finally succumbed to an abusive situation and fled to Atlanta, no job, no relatives, and no home, now with three kids in tow, two of whom were preschoolers. Fortunately, I was able to be hired at Fort McPherson and resume my status within the civil service ranks.

My Vietnam Era children are grown, educated, and tax paying citizens. However, the blight of the inhumanity of the treatment of the Vietnam veterans remains unspeakable.

But a little known offset was to affect our small family later. My oldest son, whose father had made the ultimate sacrifice, was the undeserving victim. At the end of his senior year, just before he was to enter college, Jimmy Carter passed a rider on a bill denying the children of deceased Vietnam veterans the social security benefits they had been promised should they continue to higher education.

Had my son's father elected to defect to Canada rather than give his life serving his country, he would have been able to support his son's college years. However, because I was struggling on a lower government salary with two other children approaching college age, sometimes unable to buy milk for them, I was unable to provide minimal financial aid for him, thus forcing him to work two and three jobs to get through UGA.

Despite the ugliness of our countrymen's treatment of the Vietnam vets, the camaraderie of the Army has prevailed in my life. Many of the early 11th Air Assault aviators I knew all those years ago became airline pilots. They found me in Atlanta and looked after me and the children during layovers. They later formed a reunion which I have attended many times over the years, often at their expense. There has been no end to their generosity and love.

My life continues with Vietnam as just a sometimes painful memory. Over the years and even now, I continue to volunteer for veterans' causes which serves as a connection to my many Army friends and, in small measure, my way of giving back.

DON PLUNKETT

US Army — Engineer Officer
Dates of Military Service: 1964 to 1966
Unit Served with in Vietnam: 169th Engineer Battalion, 18th Engineer Brigade
Dates in Vietnam: Jan to Sep 1966
Highest Rank Held: 1LT
Place of Birth: Atlanta, GA — 1942

"ENLISTED OR DRAFTED?"

Upon graduating from The Citadel in 1964, I was commissioned in the U.S. Army. My branch assignment was 'Corps of Engineers.' After the Engineer Branch School (EOBC) at Ft. Belvoir, VA, my first unit assignment was the 2nd Engineer Battalion of the 2nd Infantry Division, based at Ft. Benning, GA. I was a platoon leader in Company E, the bridge company. We conducted "assault river crossings" on the Chattahoochee River in support of the Infantry Battalions of the 2nd Infantry Division (Second to None!). Shortly thereafter, the 2nd ID was merged with the 11th Airborne Division (Provisional) to form the 1st Calvary Division (Air Mobile).

This resulted in my new assignment to the 169th Engineer Battalion (Construction) at Ft. Stewart, GA. We were soon notified of our unit deployment orders, first to Okinawa, with final destination Vietnam. The advance party flew out of Hunter Air Force Base and the soldiers and equipment followed by train, plane, and boat.

After a stay in Okinawa where the 169th prepared some marshaling areas for more combat equipment bound for RVN, we soon thereafter departed. Again, I was in the 'advance party' group flying into Bien Hoa in C-130s while the remaining officers and men, along with our equipment were treated to an exciting ride from the Ryukyu Islands to the Republic of Vietnam aboard the U.S. Navy's finest LSTs. I can still remember the "LST soldiers" describing the voyage. Needless to say, almost all comments are still unprintable!

Upon the arrival of our soldiers and equipment, we began our numerous projects. One of the first was establishing water purification facilities, then building base areas for the new units arriving in country. We would normally build a battalion headquarters, dining hall, latrines, and a pad or platform for a GP tent for sleeping. Most of our customers were the brave infantry soldiers and they were very happy customers!

Other projects included, a barge off-loading facility on the Saigon River, field hospitals, highway construction and maintenance and bridge building, and some demo in support of the infantry.

In summary, serving my country in Vietnam is one of the highlights of my life. I served with young and older troops who gave their best efforts to accomplish whatever mission we were given. Some of them were volunteers and some were drafted, but they all were amazing. Once when a VIP was driven to one of our work sites, I was asked by him which of the men working on the project were enlisted (RA) and which were drafted (US). My truthful reply was I did not know and asked him if he could tell them apart. He looked at me as though he was not pleased with my answer, but I know he got my drift.

At 23 years of age, I was a company commander with 140 enlisted, NCOs, and Warrant Officers. We also had about $25 million worth of equipment. We were today's "mid-sized construction" company. The lasting feelings and memories of working with such great human beings is priceless!

My identical twin was also a company commander in Vietnam. Growing up, we were often asked, "What is it like to be an identical twin?" My reply became, "What is it like not to be an identical twin?" Today, when I am occasionally asked about serving my country in Vietnam, my thoughts are along the same lines, "What is it like not to have served in Vietnam."

ARTHUR "ART" KATZ

US Coast Guard—Captain of CG Cutter, Point Cypress
Dates of Military Service (Active Duty and Reserves Combined): 1959 to 1967
Unit Served with in Vietnam: Coast Guard Squadron Zone, Division 13, patrolling the Mekong River Delta and Coastal III Corps, South Vietnam
Dates in Vietnam: Jan 1966 to Oct 1966
Highest Rank Held: Lieutenant (O3)
Place of Birth and Year: New York, NY—1942

VIETNAM VIGNETTES

Submitted by Arthur "Art" Katz, Commanding Officer of the US Coast Guard Cutter, Point Cypress, patrolling the Mekong River Delta area of the coast of Vietnam during most of 1966.

1. Maria Clara—She stands regally across the room, staring directly at me as I sit at my home office desk. She maintains her svelte figure, just as she has for these past 50 plus years. She is roughly three feet high, and was hand carved of monkeypod wood in a small village in the Philippine highlands near Baguio City. She embodies the traditional Filipino feminine ideal.

I purchased her in January, 1966, as a gift for my pregnant wife. That is when Maria's (we'll call her Maria) journey begins. A week later, she was wrapped in bubble pack, and stored below the water line (where she was less likely to be shattered by enemy gunfire) of the Coast Guard gunboat I commanded, as we sailed from the Philippines (where each gunboat was outfitted for war) to the Mekong River Delta area Vietnam. We were part of Division 13, Coast Guard Squadron One. Once in the war zone, we saw multiple battle actions. Although our floating "home," the US Coast Guard Cutter Point Cypress, was hit multiple times by Viet Cong bullets, Maria remained safe in her below waterline shelter.

As part of Cypress' duties, we sometimes escorted merchant supply ships up the unsafe portion of the Saigon River. During one of the evolutions, I chanced to talk with one of the merchant seaman. It turned out that we both lived in New Jersey, about 150 miles apart. I learned that he would return home within a month. On a whim, I asked if he would consider delivering my gift (Maria) to my wife. He readily agreed. I quickly wrote a note to my wife, which would accompany Maria, and parted with her, knowing only that "Bob" said he would deliver her.

A month or so later, Bob showed up at my home, with Maria Clara tucked under his arm. Unbeknownst to me, he did so wearing his Merchant Marine uniform. When my pregnant wife, Carol, saw a stranger in military looking uniform through the curtains, she thought her worst fears for my safe return home were realized, and she refused to open the door and confront a possible dreaded reality.

Fortunately, my mother-in-law was staying with Carol, and she bravely opened the door. Of course, instant relief followed, and Maria continues to be part of our home. To this day, the only thing I know about "Bob" is his first name.

2. Playboy Power—Part of Cypress' duty was to investigate any unidentified vessels in our patrol area. The purpose was to interdict any North Vietnamese waterborne attempts to supply the Viet Cong. On one particular patrol, we noted a suspicious radar contact, and proceeded to intercept. It turned out to be an innocent Chinese Nationalist fishing trawler, just outside Vietnamese territorial water. As we approached, the trawler swung a very large net full of the seas' bounty onto their deck for processing.

It occurred to me that fresh seafood would be a wonderful break for the crew from the monotony of C rations. In spite of sign language attempts, the trawler showed no interest in sharing, or even understanding, our seafood request.

In a moment of command brilliance, I directed one of the crew below decks, instructing him to return with one of the Playboy magazines the crew had on board. He did so, and I immediately opened to the centerfold, displayed it to our Chinese counterparts, and made motions that said, in essence, we get fish in exchange for magazine.

Shortly thereafter their net came up from the bottom, filled to the brim with fish. They swung the net over the open back of our boat, and dropped the living contents on our deck. The ensuing scramble by a well-trained and disciplined crew to capture some of this bounty before it flipped back into the seas, was chaotic at best, and occurred while their Commanding Officer (me) was shouting "get the lobster first."

It wasn't pretty, it wasn't well organized, but we ate well (we had a small refrigerator) for several days thereafter... and the Playboy... it was last seen in the fishing boat's net. Perhaps it was their best catch of the season.

3. It is all about the delivery—Patrolling the coastal waters of South Vietnam was tedious, often dangerous (because of the enemy and because of the often terrible sea conditions) duty. One of the things that brightened our day was receiving an airdrop from a Navy patrol plane. We never knew when these might happen (every month or so, but with no predictability). Imagine a low flying US Navy patrol plane passing close overhead to get our attention, and then making a second pass, during which it dropped a two foot long, round, metal, artillery shell canister for us. The canister would hit the waves at 150—250 miles per hour, and bounce several times, with Cypress in hot pursuit (some canisters blew apart because of the impact).

Once we retrieved the canister, we would take it down to our small kitchen to open on our 3x3 foot table. The canisters were typically packed with magazines and cookies, baked lovingly by any number of organizations "back home."

One crew member was assigned to do the unpacking, with one assistant. The key was to remove each magazine slowly and carefully, so that NO cookie crumbs (the cookies were blasted to crumbs upon the canister impact with the water) escaped the table top. This then involved paging through each magazine to release any captured cookie crumbs. Once the unpacking was completed, the cookie crumbs were gathered into a mound in the middle of the table.

Then, our carefully recorded rotation began, with each crew member in turn, taking a spoonful of cookie crumbs, until all were consumed. The magazines were treasured for their content, and for the occasional crumb that had avoided prior discovery. We never failed to enjoy this process, and to this day I remain thankful for those back home who supported us in so many ways.

CAROL KATZ

Wife of Art Katz

MEMORIES OF A COAST GUARD WIFE

In Fall 1965, my husband Art received orders to Vietnam to be Commanding Officer of an 82 foot (gunboat) cutter. We had about two weeks' notice before he was to report to Alameda, California. At the time, Art was a First Lt. in the US Coast Guard, and the Executive Officer on a Coast Guard AIDS to Navigation (AtoN) 180 foot ship, based in New York Harbor. We lived on Governors Island, off the tip of Manhattan, in base housing.

We married immediately after his graduation from the Coast Guard Academy in June 1963, and in the ensuing two+ years, moved from Connecticut to Virginia to New York City with Uncle Sam. I learned that I was pregnant just before Art left for Alameda and eventually Vietnam. It was a difficult time for a young couple to be faced with a prolonged separation. Long distance international communications, particularly into and out of a war zone, were by snail mail and radio telephone (of intermittent reliability).

Further, because Art was part of the first Coast Guard task force (Coast Guard Squadron One) that was going to Vietnam to blockade the coast of South Vietnam, we had no idea what danger he would be sailing into.

In short order, we vacated our government housing, with me going back to my parents' home on nearby Long Island, and Art departing. While pregnant and living at home with my folks, I got a job in order to earn my keep and be useful and busy. When the time came for Art to sign up for a week away from the war, we had the opportunity to meet in Hawaii. I found myself in an extremely difficult position.

My pregnant body and I desperately wanted to see my husband, but knew that meant I would also have to say goodbye to him (as he returned to the dangers of war) a second time. It is difficult to express the emotional stress associated with the thought of having to make that decision. Ultimately, I felt it would be less stressful and traumatic to not have to say goodbye a second time, and we chose to wait until Art completed his tour.

When he did return home, safe and sound, it was to me and his oldest child, Lisa. I would never have chosen giving birth and going through the first few months as a Mom without a husband to share it all with. Looking back some 53 years, it all seems forever ago. Fortunately, Lisa and her two younger sisters have grown to be wonderful women in their own right, married, and we've been blessed to have seven grandchildren grow up and living within a few miles.

Thank you for the opportunity to think and reflect back those many years, and try to capture in writing those thoughts of long ago.

ED WOODS

US Navy—Seabees

Dates of Military Service: 1964 to 1970

Unit Served with in Vietnam: MCB 10

Dates Served in Vietnam: Jun 1966 to Feb 1967; Sep 1967 to Dec 1967

Highest Rank Held: Petty Officer 2nd Class

Place and Date of Birth: Atlanta, GA—1942

SEABEES: "WE BUILD, WE FIGHT"

In January 1966, I went on active duty. I became 6953894, a service number; a number that is never forgotten. I flew from Atlanta to New York to Providence, RI. This was the first time I have ever flown. I flew with Doug Middleton, still a friend of mine today. My first duty station was 'A' School in Davisville, RI. This was a three-month school in my rate, which was Builder. This is where I learned to be a Builder. There I met a lot of different people, some that I would serve in Vietnam with and still stay connected with. I was assigned to NMCB 10, a west coast battalion. I came home from 'A' school April 8th to start a 30-day leave.

MCB 10 left for Vietnam on May 11, 1966. It took eight days and a total of six C-141s to move the entire battalion. We had just arrived in California and I was scheduled to leave with them. *Then* they realized I had not received any combat or rifle training. I stayed back for another month with some other guys. After being taught how to use the M-14 weapon, I was off to Vietnam. We were issued the M16 at some time during our tour in Vietnam. We flew from Port Hueneme, to San Francisco, Hawaii, Wake Island, Guam, and on to Da Nang in a four engine Navy prop plane, possibly a P3. We flew into Da Nang at night over a firefight. This was the first of many that we would see. We left the airport and went to our base, Camp Hoover, in what I would call a cattle car. It's a trailer that is commonly used today to transport cows and hogs to slaughter. It was dark and very hot.

During this time I made friends with a guy running the base radio. One night he was able to patch me through to Carol at the bank. Patching means to bounce radio signal from one radio tower to another. We were not able to talk long. He told me he would have to break it off if we got hit. Just my luck, we were hit in about three minutes. This would be the only time I would talk to her until January the next year. This was long before the internet and cell phones. After about a month of guard duty, I was assigned to Charlie Company.

No matter where you were in Vietnam you could always hear small arms fire, rocket fire, mortar fire, and larger artillery fire. At night, flares were always going off with their little parachutes so they would stay up a while and just later float down. Helicopters were always flying, 24 hours a day. No let up! Always moving men, both dead and alive, supplies, and providing fire power.

Unlike the guys that lived in the 'field,' we had dry sleeping quarters and except when we were out all day at an outpost, we had three good meals a day. I ate a lot of things and sometimes I had no idea what it was. The drink was always 'Kool-Aid'. Every day was a different flavor. We did have coffee as well, but in that temperature not many of the younger men drank it. When we were in the field during a meal we would eat 'C- Rations'.

NMCB 10 started leaving Vietnam January 27, 1967 to return home. As we were leaving, another battalion was replacing us. It took about two weeks to complete the exchange. We landed at Mugu Naval Air Station in California. Mugu is on the coast, south of and not far from Oxnard. Carol met me there. I would return to Vietnam in September 1967.

During home deployment we trained by building a bridge over land. After building it, we had to drive over it to prove its worthiness. We went to classes and built more huts. We also spent two weeks at Camp Pendleton for more combat training. We also had weekend combat training at a place called Conejo Grade. Conejo Grade is a mountain range near Thousand Oaks, CA. Home duty was good. Often we could leave the base early. I would go home and spend time at the pool and wait for Carol to come home.

In 1967, we were the *Pacific Alert Battalion* meaning we had to be prepared at a moment's notice to be deployed. In August 1967, MCB 10 was deployed to Okinawa. We were stationed at Camp Kinser, a Marine camp on the west side of the island. This was a typical mount out. It could take as many as ten C-141 planes to move a battalion from home base to its final destination. We stayed at Kinser for about five weeks waiting on assignment to Vietnam. We did a lot of work at the Marine Base, repairs, upgrades and etc., as well as, spent a lot of nights on 'BC' (Bring Cash) Street. Okinawa was full of service personal (Marines, Soldiers, Seabees, and Airmen) waiting on deployment to Vietnam. The whole island catered to the military.

When we left Okinawa we boarded C-130s for the trip to Vietnam. On August 29, 1967, we left Okinawa for South Vietnam. We landed at Gia Le Combat Base. From there we were airlifted to Dong Ha. Men from MCB 10 were spread all over the I Corps area; Quang Tri, Camp Carroll, Qua Viet, Phu Bai, Cam Lo and Khe Sanh. Charlie Co., my company, was assigned to Detail Golf and was sent to Camp Carroll. This was the first time a battalion was so widely dispersed over such a large area.

When we landed my sense of smell brought back memories of my first tour. The aroma was almost overwhelming. This was a combat zone in a third world country. There was the smell of death associated with gun powder, napalm, Agent Orange, and human excrement. A Porta Potty would be a luxury to the Vietnamese.

Camp Carroll was the command post of the 3rd Marine Regiment and was located on a ridge just south of the DMZ. The Marines and Seabees had the outer perimeter. The Army had an artillery base in the center.

The Army had 175 mm artillery which could fire a projectile about 33,000 meters or 21 miles. The Marines had smaller artillery, 105 mm and 155 mm. We were tasked with building 300 huts, three galleys capable of handling 1,250 men, five clubs (enlisted and officer), twelve heads, and three shower facilities. There was constant artillery firing, either in support of a nearby base or into North Vietnam. We were working twelve-hour days, seven days a week. During the fall, it constantly rained. No matter what the conditions were, rain or shine, the work had to be done. *The war never stopped.* This was not like the Da Nang tour where we got Sunday afternoons off. The area around Carroll was very hostile. We seldom left the base. There were occasions we left to go to other fire bases, such as The Rock Pile or Khe Sanh, to do one-day jobs.

I did have to drive to Quang Tri one day in a pickup with a shotgun rider. I cannot remember why I was sent. We got back just before dark. There were incidents in my Vietnam experience when I was afraid, however, driving to Quang Tri was my most frightening moment. Often it was raining and the roads mud covered and narrow.

We could see North Vietnam and the South China Sea from camp. It was during this time, Con Thien, a Marine combat base, about three miles away and very visible to us at Carroll, was being blasted 24 hours a day. Often at night we could see the Navy ships shooting from the Indian Ocean in support of Con Thien or to a target in North Vietnam or the DMZ. We could also see B-52 raids in the DMZ. This is the most awesome display of American fire power any one could possibly see. One night there was a B-52 raid between us and Con Thien. Not only did it wake us up, the whole hut was shaking. Between what the Army and the Marines had there, there was always someone firing into North Vietnam.

On occasion we were attacked, but nothing like Con Thien. Often the firebase west of us, Khe Sanh, would fire artillery in support of Con Thien. Sometimes a short round would fall on us. When short rounds were coming in, or mortar fire, we would run to the mortar trenches. Often there was water in them but they were much safer than being on the outside.

On December 6, our camp was hit by three mortar rounds. Damage was minimal, but a couple of unfinished huts were damaged. The VC come in fast, set off a few mortar rounds and leave, however, mostly for harassment purposes. Over 1,400 men were killed on their last day in country. When it's close to your time to leave, all kind of thoughts go through your mind. I left Camp Carroll on December 9, 1967. I had to stay at a Marine exit camp for a couple more days, finally flying out of Da Nang on Dec. 13. One month later, the North would launch the Tet

Offensive. The North Vietnamese would start the biggest offensive push into South Vietnam of the war. I was very lucky, because I left Vietnam in December and on January 31, 1968, 245 men were killed there in the worst day of the war.

On my last trip home, I was one of the last to board the commercial (Continental) plane, called 'Freedom Bird', therefore, I flew in the first class section. Regardless of rank, all men were boarded from rear to front of the plane. We flew into Norton Air Force Base where Carol met me. She took me straight to the motel so that I could quickly get out of that uniform, and she had champagne waiting as well. She already had everything packed and was ready to leave for Atlanta. When I landed, it was snowing. Supposedly this was the first snow in the Los Angeles area in over thirty years.

I exited active duty at the Naval Air Station in Marietta, GA on 23 Dec 1967. I returned to my reserve unit in January 1968 where I was promoted to Petty Officer 2nd Class on March 5, 1968. I stayed in the reserve unit for two additional years, attending monthly meetings and doing a two week tour in Gulfport, MS in 1969. At that time, I learned Carol was pregnant with Scarlett, who was born September 5, 1969. On January 8, 1970, I completed my military requirements and was honorably discharged. I am proud and will never regret my military service.

CAROL WOODS

Wife of Ed Woods

SPOUSE, US NAVY SEABEES

I married Ed in April 1965. We knew he had to go on active duty January 1966. In January 1966, he left for Davisville, RI, to attend Navy 'A' school. I stayed in Atlanta until he finished 'A' school.

In April he returned to Atlanta and we prepared to leave for California. He had been assigned to NMCB 10, a West Coast Seabee Battalion located in Port Hueneme, CA. We did not know the status of this battalion. We bought a 1965 Volkswagen and left Atlanta in May. He reported to the base and found out his battalion was mounting out to Vietnam within the next two weeks. We as young kids foolishly thought we may be able to spend our active duty time in California. He was not allowed to leave with the battalion because he did not have any combat training. He did about three weeks of training and was off to Vietnam. He was in the Da Nang area and I was assured it was a fairly safe area.

We found a small, affordable, I think $65 per month, two-room apartment not far from the base. This would be my home for the next six months. Ed left for Vietnam in the middle of June. I really was by myself.

As soon as Ed left, I had to open a checking account at the bank. All of his pay would be direct pay. I went to Bank of America in Ventura, opened the checking account, and applied for a job. Surprisingly I got the job. I say that because they knew I was a military wife and was subject to relocation at any time. Little did I realize that I would retire from Bank of America in 2014.

I did not know anyone except my co-workers. Later I met a wife, Sue Bailey, of another Seabee that was in Vietnam. We spent time together, but mostly we were in our own little world. Working was helpful and allowed me to interact with other people. I filled the time window writing to Ed daily and baking at least once a week to make sure he had fresh cookies and an occasional cake. I also met another military wife with two children that was having financial issues. On occasion I was able to help her.

About four weeks after Ed left, I got a phone call from him at the bank. We only talked a short time, but that would be the only time I would talk to him while he was gone. A couple of months later he sent me a dozen roses. What I do remember is how beautiful they were and how good they smelled.

I spent most of my spare time writing letters to Ed and also to his mother. I also spent my weekends at the complex pool. The bank needed people to go to San Luis Obispo to help open a new branch on the college campus. I went. This was fun and a new experience.

While Ed was gone I moved into a nicer apartment in nearby Ventura. A one-bedroom but with a separate living room and kitchen.

Ed returned in February the next year. We had fun living in southern California. We went to Dodger Stadium when the Braves played, did the 'Homes of The Stars' tour (still have the map), went to San Francisco, Santa Barbara, Disneyland, and sometimes just drove to L.A. and Hollywood just to drive around. We tried to take full advantage of all free time. Sometimes he had weekend duty but that was rare.

Ed was scheduled to return to Vietnam in August. In August, he received orders to go to Okinawa as part of a standby battalion for possible return to Vietnam. On September 29, he left Okinawa for Camp Carroll, Vietnam. That was not a safe area.

I knew he would leave active duty in December and we would be headed back to Atlanta. I stayed as busy as possible. Sometimes not a lot to do. He wrote and said he would be leaving Vietnam on or about December 12. He left Vietnam and landed at Norton Air Force Base on the evening of December 13. I already had everything packed and I met him when he landed. We left the next day for Atlanta.

CHARLES TUTT

USMC — Infantry Officer, F-4 pilot
Dates of Military Service (Active and Reserves): 1965 to 1992
Unit Served with in Vietnam: 3d Battalion, 9th Marines, 3d Mar Div; VMFA-115, Marine Aircraft Group 13, 1st Marine Aircraft Wing
Dates in Vietnam: Feb 1966 to Jan 1967; Oct 1969 to Sep 1970
Highest Rank Held: Colonel
Place of Birth and Year: Birmingham, AL — 1943

THE MATURING OF A MARINE

I grew up in Atlanta, Georgia in a middle class family. Both of my parents worked to provide for our family. The 1950s and 60s were in many respects a relaxed and calmer time than today. It was also a time when the Civil Rights Movement was moving forward. My family was apolitical and not involved either for or against the Movement. On June 12, 1965, I graduated from Georgia Tech in the morning, was commissioned a 2d Lieutenant in the United States Marine Corps in the afternoon and married Susan Christopher that night.

After a honeymoon, we reported to The Basic School (TBS) in Quantico, Virginia. TBS is where new USMC 2d LTs are trained as new Marine officers. I graduated in December and was assigned an 0-3 MOS, which is an infantry officer. I was also given orders to proceed and report to the 3d Marine Division in Vietnam. Susan and I traveled back to Atlanta to get her settled as that is where she would live while I was overseas. While in Atlanta, my orders were modified to report to Camp Pendleton, California, to become part of a group of Marines that would fly out together.

Prior to flying to California, I had never been farther west than New Orleans, farther north than Washington, DC, or farther south than Miami. At Camp Pendleton I was in charge of a group of Marines with many skills and MOSs that were being prepared to go to Vietnam. Our time was spent in physical conditioning, weapons refresher training, and administrative requirements.

After a couple of weeks at Camp Pendleton, we boarded a USAF C-135 and flew to Okinawa with a stop on Wake Island for refueling. I spent a couple of days on Okinawa for more processing and then was flown to Da Nang, South Vietnam. When we got off the aircraft, we were taken into a building beside the runway and told to wait until someone came to get us. It was after a number of hours and well after dark before my name was called and I was taken to the headquarters area for the 9th Marine Regiment.

I reported to the S-1 officer. His office was a general purpose tent with a light bulb hanging from the top of the tent. There were bugs flying throughout the tent and every surface, including his desk, was covered with bugs. I grew up in the South and was used to bugs, but I remember thinking this is a different place than anything I have ever seen.

By the time that night was over, I had become the 3d Platoon Commander for Mike Company, 3d Battalion, 9th Marines. The previous Platoon Commander had been Killed In Action several months earlier, and I was the first officer assigned to the platoon since his death. The Marines in our platoon were outstanding, particularly the non-commissioned officers (NCOs). The Marines in my platoon were from every geographic area of the country and ethnically diverse as well.

My platoon sergeant, SSgt Hayes, and one squad leader were African American. One squad leader was Hispanic and my radio operator was an Irish kid from Boston. It is said that NCOs are the backbone of the Marine Corps and that was certainly true in my experience. Here I was in charge at 22 years old with really no worldly experience. The NCOs had more time in the Corps, more combat experience, and more time together in the platoon, yet they supported and guided me without hesitation.

For the first week that I was in Vietnam, we manned the southeast perimeter of the Da Nang Air Base, providing security. After about a week, we moved to an area just south of Marble Mountain which is in the Da Nang area and along the coast of the South China Sea. Each platoon was assigned an Area of Operation and had minimal contact with the rest of the company. Our contact with the VC was low key, consisting mostly of sniper fire and mines and booby traps.

While I did not fully appreciate it at the time, SSgt Hayes was educating and guiding me as a platoon leader. He was certainly helping me mature as an officer and a person. After a few months, our battalion was moved to the An Hoa area which is southwest of Da Nang and an area that had long been controlled by the VC and NVA. The battalion conducted a ground operation to get to An Hoa and preceded us by several days. Our company was flown by helicopter to An Hoa.

As soon as we landed, we were told that there were a couple of companies of NVA several kilometers on the other side of the Thu Bon River. We were to board AmTracs (we traveled on top as the Marines had learned that the location of the fuels cells made the inside of the AmTrac a death zone if the AmTrac struck a mine), cross the river, and try and make contact with the NVA troops.

We crossed the Thu Bon River and landed on the opposite bank in the vicinity of a geographic feature called Football Island. We were dismounting from the AmTracs when the NVA opened fire on us. They had not been several kilometers inland but were waiting for us in the hedgerows at the top of the river bank. This was my first major engagement, and it lasted most of the day, involving artillery and air support as well as two more infantry companies coming to our aid. The area where this engagement took place was later named the Arizona Territory.

Unfortunately, SSgt Hayes was shot in the leg and evacuated. He was a fine Marine and helped me mature immeasurably. I was fortunate that one of my Squad Leaders, Sgt Harvey, was also a fine NCO and was able to assume the position of Platoon Sergeant. Like SSgt Hays, he did an outstanding job in that position.

In our new Operating Area, we worked in company units and the contact was more intense. While we still encountered snipers and mines, we also engaged larger VC units and also received mortar fire. Not too long after reaching the An Hoa area, my Company Commander rotated back to the States. His replacement was Captain Gary Cooper. Captain Cooper was the first African American Marine to lead a Marine rifle company in combat. He was from Mobile, Alabama and a Notre Dame graduate.

So here we were, two Marine officers, a white kid from Georgia learning to be a leader and a black leader from Alabama teaching him, while at the same time making a statement about ability and equality. We were half-way around the world, fighting a war, leading a geographic and ethnic diverse group of Marines. Our Marines performed exceptionally well. They conducted themselves in a professional manner—attacking the VC fiercely and treating the civilians with care and compassion.

At this point, I would like to devote the remainder of this article to the Marines that helped me mature as a Marine Officer. First, I want to acknowledge Captain Bill Northcut, my last Company Commander. He, like Captain Cooper, proved strong leadership as well as mentoring me. I will be forever in his debt. As I noted earlier, I grew up in the South during the Civil Rights Movement and was apolitical. I thought folks, no matter their skin color, should have equal rights, but I did nothing to support them. However, two Marines, SSgt Hays and Captain Cooper, were instrumental in helping me mature and develop as a Marine officer.

I think two examples illustrate their willingness to educate me. The first one involves SSgt Hayes. SSgt Hayes normally assigned replacement Marines to a squad. One day we received two replacement Marines and SSgt Hays was not available. I assigned the two Marines, one to the First Squad and one to the Second Squad. When SSgt Hays got back, I told him about the two replacement Marines. He went about his business but shortly came back and said we need to reassign the two Marines. The two of us had a normal give and take in most matters and unless I was doing something really stupid he would not persist in the matter.

However, he continued to tell me we needed to reassign the Marines. Finally I asked why, should have earlier, and he said the Marine I assigned to the first squad was African American and that made too many African Americans in that squad and the Marine assigned to the Second Squad was white and that made too many white Marines in the Second Squad. I told him to assign the Marines as he recommended.

His point was that if we kept things balanced, we would not have a problem, and we never did. I learned to ask him, "Why?" sooner.

The second example involves Captain Cooper. In the field we were all business, but occasionally the company would get to a rear area for some much needed relaxation. It would have been easy for Captain Cooper to spend the time alone or with other officers of his rank, but frequently he would spend time with me. On several occasions he acquired a bottle of whiskey which he shared with me. We had wide ranging discussions, many of which involved our experiences in the States and how we were treated. I learned what his life was like and he mine.

I learned to appreciate the Civil Rights Movement and also how each of us can make a difference in improving the lives of others. A great example of a teaching moment was when we discussed that we were fighting together for our country and would die for each other, but in our home states we could not eat together at the same restaurant.

I was proud to have served and proud of the Marines I served with.

I have been asked if I would serve again under the same circumstances. The answer is yes and I hope I would be a better leader.

Let me close by saying that many Marines helped me mature. They taught me that hard work and perseverance pays off. By being associated with Marines of different backgrounds and ethnicity, I learned that if we treat each other fairly and with respect, we can build great teams that can accomplish even the most difficult tasks.

JOHN FRASER

US Army — Chaplain

Dates served: 1953 — 1983 Texas National Guard (Enlisted) 1953-1956; Army Ready Reserve (Enlisted) 1956-1960; Army Active Duty (Commissioned) 1963-1983.

Units Served with in Vietnam: 2/14 Inf and 1/27 Inf, 25th Inf Div; 8 RRFS.

Dates Served in Vietnam: Apr 1966 to Apr 1967; Nov 1969 to Nov 1970

Highest Rank Held: Lt. Colonel

Place of Birth and Year: Greenville, TX — 1936

BEST COMPLIMENT I EVER RECEIVED

I arrived at Schofield Barracks in Hawaii in February 1966, just a few weeks before the 2/14 Infantry Battalion shipped out of Pearl Harbor on the USNS Walker for Vietnam. These few weeks gave me the opportunity to get to know the officers and enlisted men with whom I would serve in Vietnam. I spent a lot of time on the firing ranges and on training exercises in the Kahuka Mountains training area. This story comes from a day on the .45 pistol range with an infantry company which was getting familiarization training on the .45 pistol.

The Range Officer was a first lieutenant, the Executive Officer of the infantry company that was on the range. He was very proficient in his job, but he had impressed me as kind of a wise guy who was a little too big for his pants. Someone had told me that when the company had finished firing, that lieutenant might challenge me to a contest with the

.45 and a wager on the outcome of a one-magazine shooting match. And he was considered very proficient with a .45. Sure enough, when the company had finished firing and the troops were seated in the bleachers waiting for transportation, the lieutenant looked at me and, still using the speaker system, asked if I would like to fire the .45. Well now, all the soldiers knew that chaplains don't carry weapons and aren't trained in their use. Some of the troops were looking at me and shaking their heads. What no one knew was that I was raised in a county seat town in Texas where there wasn't much to do except go to a movie with a pretty girl or walk out to the woods with a friend to go rabbit hunting or just plinking at tin cans with whatever firearm happened to be taken along that day. And I had continued to enjoy shooting every now and then and was the owner of long guns and hand guns. Add to that I had joined the Texas National Guard when I was seventeen, had transferred to the Ready Army Reserve about three years later, and after eight years had been discharged as an engineer sergeant.

The lieutenant showed me how to aim and squeeze a trigger. He fired one magazine at the paper target, and by golly he was pretty good. Then he handed me the pistol and also a loaded magazine. I very carefully tried to insert the magazine backwards. He kindly took the pistol from me, inserted the magazine and handed it back. I innocently asked, "Is this what you do next?" as I chambered a round.

Then he asked the magic question! "Would you like to wager a lunch at the officers club?" Some of the troops in the bleachers started shaking their heads again. "Well, I guess I can afford that," I said. I hope it is not too immodest to say that I shot the pants off him. When I cleared the pistol and handed it back to him, he announced on the speaker system, "This is the first time I've ever been hustled by a chaplain." I thought that I would never hear anything about this little episode, but I didn't take into consideration how stories get around.

Fast forward several weeks. No longer were we in beautiful Hawaii. We were at the hot, dusty 25th Infantry Division base camp in Vietnam near Cu Chi. After a few days, the Battalion had been involved in a few minor missions, had taken a very few casualties while we were getting up to speed in the area. A few of the platoon leaders (2nd lieutenants all) had started carrying only a .45 pistol on short missions such as sweeping around part of the base camp perimeter where no contact with the Viet Cong was expected.

Soon the Battalion Commander had an Officers Call. The subject was the M16 rifle. At the end of the lecture, he declared, "ALL officers will be armed with an M16 rifle! There are only three people in this Battalion who can effectively engage the enemy with a .45! I'm one of them," he said, "the other two are the Armorer and the Chaplain, if he wants one!" And I think that's the best compliment I've ever received.

GEORGE MURRAY

US Army — Armor Officer / Aviator
Dates of Military Service: 1964 – 1968
Unit Served with in Vietnam: A/82d Airmobile Light, re-designated 335th Assault
Helicopter Company
Dates in Vietnam: Feb 1966 to Feb 1967
Highest Rank Held: Captain
Place of Birth and Year: Grenada, MS — 1942

SOME FOOL MADE A SOLDIER OUT OF ME, CAN I GET A PURPLE HEART?

My birthday is seven days away. I will be 24 and am two years out of college and a fraternity house at Mississippi State. As I write years later, I know I was a kid with no experience and carrying responsibility beyond my years. But now, at age 76, I have read a lot and know that kids fight all wars. I know the title of my article is more appropriate than I first imagined.

It is July 1966; I am wearing U. S. Army Aviator Wings. I have just been cleared by the flight surgeon, Doc Hammergren, to fly beyond the Army's fatigue limit of 80 hours for the month — a mere formality in our unit, the 335th Assault Helicopter Company "Cowboys." Doc simply asked me how I felt. In my best macho manner I replied, "Great!"

Doc loves to fly with us and often jumps in to ride in the door gunner's spot. If he is crazy enough to fly there, he must have confidence that the pilots are not too tired to be safe. Doc also loves to fire the M60 machine gun, but a "real live gunner" has to be right there with Doc or he will fire the machine gun until the barrel melts down. He has no concept of fire control; we all figure he practices medicine with the same level of "throw caution to the wind."

Doc is not with us today, because we are going to fly a long day. 1st Battalion, 503rd Infantry Regiment, 173rd Airborne Brigade, "Airborne, All the Way, Sir" is moving around the Vo Dat area looking for VC tax collection crews.

These bandits simply steal the harvested rice from the peasants who have worked to grow their meager food supply. Amazingly, the battalion has managed to interdict some of the VC and has killed a few and captured some. "Charlie" is definitely upset and has fought back some.

We are briefed on a lift of five slicks and a pair of guns to move Alpha Company several miles to a new LZ that is expected by the S-2 to be hot. Intelligence has information that several of the tax collection/extortion teams are on the move and have come to rest near a small stream. The plan calls for a pickup of the company from the area they are working now and a quick movement to intercept the bad guys.

As a 1st Lieutenant, I am aircraft commander (AC) of a UH-1 D helicopter. My co-pilot today will be the unit operations officer, a Major, who only flies enough to meet the four-hour monthly minimum for flight pay. He has no command function over a super-confident, experienced AC and is comfortable being in MY helicopter. I have a crew chief on one machine gun and a door gunner on the other.

We have two senior operations officers of the 173d as passengers who will do a reconnaissance of the area the troops will be assaulting to make contact with the VC. We have climbed out to about 1,500 feet, all engine instruments are in the green, and I am flying a course that will take us over the VC area for a high recon.

With no warning, my windshield explodes and a projectile strikes me in center chest. My helmet plastic shield that covers my eyes and nose is covered with blood; wind is tearing through the cockpit. Both passengers are apoplectic. My co-pilot grabs the controls and throws the helicopter into a high speed diving left turn as he goes into autorotation. (Autorotation is an emergency solution to engine failure and will get us from 1,500 feet to a rice paddy in seconds).

I know I am alive, but blood is everywhere. I am still on the flight controls trying to stop the fall from 1,500 feet; not realizing my co-pilot has put us in autorotation. In my most macho manner, I am screaming repeatedly into my intercom mike, "I have the airplane, I have the airplane!"

Finally our free fall recovers at several hundred feet over bad guy country. I can see now and notice an unusual aspect of the bloody mess on my face and helmet shield. Mixed with the blood are copious quantities of feathers. Resting in my lap is a very dead, large DUCK.

My door gunner and crew chief recover their sphincter muscles and are laughing hilariously, in spite of knowing they will soon be cleaning the mess in their helicopter. My copilot logs four hours in the 45 minutes we flew and goes back to his paper work. Our windshield is quickly replaced and put in place temporarily by "100 mile/hour tape" aka, duct tape.

All I want to know as we fly the rest of the day with a new co-pilot is, "Do I get a Purple Heart?" NO!

BRUCE BURGEE GEIBEL

US Navy—Seabees and Civil Engineer

Dates of Military Service (Active and Reserve): 1962 to 1991

Unit Served with in Vietnam: Naval Construction Battalion Eleven (NMCB-11)

Dates in Vietnam: Jun 1966 to Sep 1968 (three tours to Vietnam/Thailand)

Highest Rank Held: Naval Captain (O-6)

Place of Birth and Year: Washington, DC—1942

FIRST TIME UNDER FIRE—JULY 23, 1966

Oh, for the memories: The good, the bad and the ugly. Vietnam—Saturday, July 23, 1966. This was the date of my first wartime experience. I was an Ensign in the Civil Engineer Corps, U.S. Naval Reserve, and had been with the Seabees of Naval Mobile Construction Battalion ELEVEN (NMCB-11) at Camp Adenir, DaNang, Vietnam a little less than a month, since June 26, 1966. I had gotten settled into my job as Assistant Operations Officer (S-3A). I was living in a tent top with three other officers with wide-open sides in the blowing sand almost around the clock. I knew where the fox holes and sandbagged bunkers were located near my living hut and my office space. We trained in peace to be prepared for war. The Marine saying posted on a sign at Camp Pendleton Marine Corps Base in California, where we did two weeks weapons firing and a field exercise to practice our combat skills and techniques before deployment, "The more you sweat in peace, the less you bleed in war!" sure rang true this date—Saturday, July 23, 1966.

We had various fighting posts dug in the sand around our perimeter overlooking all four sides of our campsite. Observation Point/Post No. 8 kept watch over the rice paddies and riverfront. Post No. 9 observed the sand dunes area to our north and Post No. 10 was the back-up position for Bravo Company, Marine Military Police Battalion, with their tank or Amtrak vehicle gate located on the highway passing the front of our camp site. We manned these and other posts around our perimeter 24 hours a day. We had an M60 machine gun at each of our posts and several 81 mm mortar pits strategically located throughout the Seabee camp that were manned much of the day and every night to provide fire support should the enemy attack the camp. Seabees manned all sites and were outfitted with a .45 cal. pistol and/or an M-14 or M16 rifle. We were ready for anything or anybody coming our way—friendly or not.

On the evening of July 23, 1966, there were approximately 300 men playing bingo in the Enlisted Men's Club—The Grass Shack. Other Seabees were manning their fighting positions, some getting some early sleep since they would be manning the bunkers on the next shift and others just relaxing in their huts cleaning their weapons and/or reading or whatever suited their fancy at night.

At 8:01 p.m. that evening NMCB-11 and surrounding military components including, NSA Hospital and MAG-16 Helicopter Squadron adjacent to our camp, we came under 81 mm mortar attack by the Viet Cong enemy forces. The first barrage hit with a heavy concentration of rounds landing near the galley, just 25 yards from the Enlisted Men's Club. The Alpha Company (equipment company) compound and our Supply open storage area also received hits. The attack lasted 20 minutes with 19 rounds landing within our perimeter. Within minutes after the barrage lifted, the Seabees gathered their weapons and combat gear and were fully manning their fighting positions and/or in their respective company command posts. I personally relocated to the battalion command post where I was assigned to be during any enemy attack on our camp site.

Four duds were found following the attack and were removed by an EOD Team from MAG-16 located across the highway from our base camp. Other installations hit during the attack were MAG-16, NSA Hospital, Force Reconnaissance and the Army Special Forces camp sites.

Light casualties were inflicted on battalion Seabees with five men receiving shrapnel wounds and four of these men being medevaced. They were: Thomas Burkhart, CN, D-Company, who was medevaced by air to the Naval Hospital, Yokosuka, Japan; Fred Copsen, SWE3, sent to Clark Air Force Base, Philippines, for further evacuation to the States; Charles Long, BUH3, D-Company, who was sent to Charlie Med, DaNang, for treatment; Leroy Miller, CN, A-Company, sent to Charlie Med, and transferred to NSA Hospital, DaNang; and Gary Sylte, EOH3, A-Company, who was treated and released from our own sickbay in camp. Copsen's and Burkhart's injuries were later reported as light and Sylte received superficial wounds.

Our Commanding Office, Cdr. William L. Wilson, CEC, USN, later commented to the battalion after the mortar attack:

"I would like to say a few words concerning the Viet Cong mortar attack Saturday evening 23 July on our camp and the adjacent Marble Mountain facilities. How you all reacted was an example of how the many long hours of training at Port Hueneme and Camp Pendleton paid off. You were all professionals—You reacted quickly, knowing where to go and what to do. Because of this, our casualties were light.

"Our lines were manned and we had excellent fire discipline. I have heard many favorable comments concerning your actions and once again you have upheld the tradition of the Seabees as a fighting unit. Well Done." Signed: William L. Wilson, CDR, CEC, USN, Commanding Officer.

So, there you have it—my first experience of an enemy attack on our campsite in Vietnam. My overall reaction at the time: scared—no shit; a wake up call; remembering to hit the deck or find a mortar hole; lay low until the incoming barrage ceased; head post haste to my command post bunker when the incoming ceased; be prepared to do anything the command ordered; praise the Lord I wasn't injured. I certainly didn't experience anything like our Marine and Army guys in the jungle or my Seabee brothers in Dong Ha in 1967 being pounded by the Viet Cong almost every day while I was safe in Thailand with my Seabee Team 1109. They are the true heroes. But, this was my wake up in the war zone. A big Welcome Home to all of my Seabee Brother's from Vietnam. We went over as strangers and came home as Brothers. God bless the Seabees and may God bless America.

EDWARD ETTEL

US Navy — Line Officer

Dates of military service (Active duty and Reserves): 1963 – 1989

Unit served with: Military Sealift Command Office, Manila

Dates Served in Vietnam (Philippines): 1965 to 1966

Highest Rank Held: Navy Captain (O-6)

Place of Birth and Date: Corpus Christi, TX — 1940

BOOZE AND CHOPPED LIVER: OCEAN LOGISTICS SUPPORT OF THE VIETNAM WAR

Between mid-1965 and late 1966, approximately 90% of all Vietnam War cargo was supplied by ship. Early in the Vietnam War, troop ships carried two-thirds of U.S. troops to Vietnam; but after the troopships were retired, airlift was used to move the great majority of troops. Priority cargo was also airlifted, but it accounted for only a small fraction of the total tonnage moved.

As a LTJG, (O2), after serving aboard the USS Hollister (DD-788) in the Pacific and South China Sea in 1963-64, I was ordered to the Military Sealift Command Manila (MSCO Manila).

My initial duties were Assistant Operations Officer, Passenger Officer, and Boarding Officer. My office was in the TRANSCO terminal building on a pier near downtown Manila, so my wife, Mary, and I moved into an apartment in downtown Manila, living on the economy instead of a military base. The nearest military base was Naval Station Sangley Point, but because the base was twenty-two driving miles on difficult roads around Manila Bay, it was too far away for a daily commute to the office.

Prior to the Vietnam War, Military Sea Command (MSC) operations in the Philippines were relatively calm and easily managed, which was the operational state when I reported for duty on 12 January 1965. That pace changed with the introduction of U.S. combat troops to Vietnam beginning on March 8, 1965. Soon the office was handling as much cargo tonnage in a month than in the entire year in 1964. The Vietnam

MSC operation could not wait for the eventual increase of MSC's total complement of men, so our Executive Officer (a Lieutenant Commander) was temporarily transferred to Vietnam to set up an MSC office there. Because our office was without its Executive Officer, Watch Officer duties, previously split three ways between the Executive Officer, Operations Officer, and Assistant Operations Officer, were now split just two ways between the Operations Officer and Assistant Operations Officer. The Commanding Officer did not assume Watch Officer duties.

MSC is responsible for ocean shipping of all U.S. Government cargoes throughout the world and was already the largest ocean shipping operator in the world before the escalation. MSC cargo included munitions, marine and aviation fuels, vehicles, weapons, base supplies, construction materials and equipment, household effects and personal items. Nonmilitary cargoes included items in support of various U.S. government agencies such as the U.S. Embassy, USAID programs, and the Peace Corps.

Soon Vietnam supply requirements grew such that too many ships were anchored off Vung Tau, Qui Nhon, Da Nang, and Cam Ranh Bay. This was such a security, economic, and readiness problem that a smart decision was made to stage the ships in the Philippines instead of waiting offshore Vietnam. In Philippine anchorages the ships were secure, making hazardous duty pay unnecessary. Also, ships' provisions could be maintained, fuel tanks topped off, and maintenance performed. In addition, the crews were happier because they could have great liberty. Soon there were over 300 ships pre-positioned in the Philippines, with only a 36-hour cruise over to Vietnam. Obviously, we were very busy maintaining control and support of this volume of ships. At this time, my fellow officer LTJG Mattox, received retirement orders, and so I assumed additional duties as Operations Officer and Cargo Officer.

Early every morning, I would board the Philippine Pilot boat and board newly-arrived ships at anchor in Manila Bay, meeting the captain, providing port status and information, giving ships orders and satisfying requests. It was a civilian ship custom for the captain to offer a beverage as he and I conducted business. This was usually some type of alcoholic drink; scotch, bourbon, whisky, or, in the case of Japanese captains, sake. This protocol could be a problem after visiting over eight ships, especially on hot days climbing ladderways, so because I had to do it, I started taking snacks with me to help absorb the alcohol.

The Philippine Quarantine, Customs, and Immigration officials also boarded the ships, but used their special converted tugboat that was much more stable and had a lot more space. They needed more space for their booty, because another 'custom of the port' was for every captain to give each official a bottle or two of booze. At the end of the day, instead of going back to their dock, the officials would motor up the Pasag River and unload their booty into their private warehouse. Of course, I had no such privilege.

Sometimes I would come back to the dock late at night. One time I was walking down an isolated pier, a week ahead of Philippine elections, when automatic weapons opened up and I heard bullets whizzing around the pier, and along the wall of the warehouse. I ran toward the side of the warehouse and jumped into a large pile of trash. I was wearing my tropical white uniform, and Mary was not happy when I came home very dirty and smelly, but at least in one piece. The Philippines could be dangerous around election time, with rival factions resorting to violence. They were not targeting me, but were shooting up the place for some reason I never knew.

My favorite duty was Passenger Officer, which involved passenger operations responsibility for Southeast Asia. MSC was responsible for transport of U.S. government personnel (military and civilian) designated for ocean travel who could not fly due to medical or other reasons. These included high ranking officers and officials being transferred in and out of Southeast Asia. I billeted passengers and troops for onward passage to Guam, Okinawa, Yokohama, Honolulu, and San Francisco; and for passenger cruise lines for onward passage to Hong Kong, Tokyo, Honolulu, and San Francisco. One MSC troop transport called on Manila and Subic Bay once a month, and many military personnel would be given orders to return to the States via MSC troop transports, which included five non-air-conditioned ships (the USNS Marshall, USNS

Mitchell, etc.) and three air-conditioned ships (the USNS Darby, USNS Barrett and USNS Geiger, which were acquired from American President Lines (APL).

The air-conditioned ships were greatly favored by the troops, and space available people as well. The Barrett was nicknamed the 'Grin and Barrett', which I thought was unfair because it was air-conditioned. I also booked lucky personnel and their families on American President Lines (APL) ships; the SS President Roosevelt, the SS President Wilson and the SS President Cleveland. All were a combination cruise-ship and cargo ship, and it was a special prize to be able to be billeted on one, especially for the 23-day voyage back to San Francisco via Hong Kong, Tokyo, and Honolulu, considered as 'Proceed Time' and not vacation time.

I would receive long booking lists from all the South China Sea area ports, and then assign specific cabins or troop spaces based on many factors, too complex to mention here.

Wearing my Service Dress Whites, I would catch the Philippine Pilot boat very early in the morning before sunrise, and as the cruise-ship would pass Corregidor Island and enter Manila Bay, it would slow to about eight knots. When the ship was about a thousand yards away, my pilot boat would do an intercept in order to come along the port side of the ship. The ship would go as slow as she could, however my putt-putt pilot boat could not go very fast, so we only had about 30 seconds for the pilot and me to climb about twenty feet up a Jacobs ladder, a flexible hanging ladder consisting of vertical ropes supporting horizontal rope rungs. Pilot ladders are often incorrectly referred to as Jacobs ladders, with specific regulations on step size, spacing and the use of spreaders, long treads that extend well past the vertical ropes to stop the ladder from twisting on its long axis, usually when the ship rolls and the ladder is no longer in contact with the ship's side. If this happened, the climber would be trapped between the ship's side and the ladder. In our case, I am pretty sure the ladders we used were not as fancy.

While the pilot was climbing up the ladder, the ship's crew would throw me a line and I would tie a knot around the handles of my ugly brown antique leather satchel, and they would haul the satchel up, which contained all the tickets and other papers. Simultaneously timing my jump with the waves, I would jump on the swaying rope ladder and start climbing as the ladder bounced off the hull, and avoiding the Pilot's heels as I did so. It was not easy to climb the moving ladder, and although I knew there was no margin for error, I thought it kinda fun.

After several months of doing this three times a month, I asked my captain why the Quarantine, Customs, and Immigration officials boarded after the ship was docked, and not do as the pilot and I did. It would speed up the whole process. I was told that a year ago two Customs officials were killed when they fell from the ladder, and because they had to fall right next to the ship, they were sucked into the propellers and ended up resembling chopped liver. After that incident, the officials refused to climb the Jacobs ladder. I then realized why so many people were hanging over the ship watching us climb the ladder, and I was much more careful from then on, but continued two more years without incident. It was always a thrill with the ship underway and all, but this was probably the most dangerous thing I did on active duty.

The best part was that once aboard, I was a hero. Not because I avoided the ship's propellers, but because I held the golden tickets and paperwork for families that would be boarding the ship when the ship tied up to the pier. I would then meet each family as they boarded, describe the process, give them their tickets and lead them to their cabin. They were always very, very excited, and thought I was the cat's meow. It was like they thought I had paid for their 23-day transpacific cruise myself. I felt like Santa Claus, and they couldn't stop thanking me, and usually had someone take our group photos. However, I was not very smart, because this was the time to trade business cards and say, "Have a wonderful cruise. Please keep my card and someday we might cross paths." Because in June of 1967, I was retiring from active duty, did not have a job, and could have used a favor or two.

JOHN FRASER

US Army — Chaplain

Dates served: 1953 — 1983 Texas National Guard (Enlisted) 1953-1956; Army Ready Reserve (Enlisted) 1956-1960; Army Active Duty (Commissioned) 1963-1983.

Units Served with in Vietnam: 2/14 Inf and 1/27 Inf, 25th Inf Div; 8 RRFS.

Dates Served in Vietnam: Apr 1966 to Apr 1967; Nov 1969 to Nov 1970

Highest Rank Held: Lt. Colonel

Place of Birth and Year: Greenville, TX — 1936

A WAR STORY: A DOG AND A CHAPLAIN

This is a war story, but there's no war in it. Not a shot was fired this day in 1966. I was the Battalion Chaplain for the 2/14 Infantry Battalion, 1st Brigade, 25th Infantry Division at Cu Chi, Vietnam. We had come in from the field where we had inflicted very substantial damage on the Viet Cong, but not without losing several men killed and a good many others wounded. One of our companies had just received a new company commander, and he was about to take his new company on a sweep around a part of the division base camp perimeter. I decided that I would spend the day with that company.

Just before we walked beyond the bunker line, a young soldier with a Scout Dog reported to the company commander. To this day, I have no idea why he was there. I had never seen a Scout Dog and Handler before and never did see one again. The Scout Dog was a beautiful German shepherd. I just love dogs and I hadn't been close to one in a long time, so I decided to stay more forward than usual so I could enjoy being close to the dog.

Because we were going out for just part of the day, I carried only two canteens of water and a small can of C-ration turkey loaf. We were supposed to be back in time for supper, so only lunch needed to be taken on this short mission. Of course we could have gotten into a fight which would change our plans, but I knew that if we got into trouble the helicopters would show up with a resupply of ammo, water, and food. If things got bad enough that the choppers couldn't resupply us, I wouldn't have much appetite anyway. So that one little can of turkey loaf would be enough.

We walked through the woods, sweating and getting thirsty, until about noon. I had watched that dog all morning. The soldier and the dog seemed to move with a single mind, and their affection for each other was wonderful. When we stopped to eat, I sat on a log next to the Handler and complimented him on his dog. I was opening my turkey loaf when it occurred to me that the soldier was taking care of his dog before he made any motion to take care of himself. The dog was given a drink of water from the soldier's canteen, and a doggie burger was opened and put on the ground in front of the dog, and only then did the soldier move to open his own C-ration and get a drink of water.

I sat there for a while, looking without any enthusiasm at my open can of turkey loaf. "How many of these things have I eaten?" I wondered. "I just can't eat another can of C-ration turkey." Then I noticed that the dog was looking at his doggie burger about the same way I was looking at that can of dead turkey. "Your dog is about as interested in his lunch as I am in this C-ration," I said. "Would it be okay to let him have this turkey?" "Yes sir, that would be fine," he replied.

I put the can on the ground in front of the dog. The dog sniffed it carefully, looked up suspiciously at me, and ate the doggie burger.

C-ration turkey loaf never tasted very good after that.

JACK HORVATH

Branch of Service and Job: US Army—Armor and Transportation Corps

Dates of Military Service (Active Duty and Reserves Combined): 1961—1988

Unit Served with in Vietnam: 64TH Transportation Co, 8th TC Group; 54th TC Bn, and 124th TC Bn, 8th TC Group

Dates in Vietnam: Jul 1966 to May 1967; Jul 1969 to Jul 1970

Highest Rank Held: Lt. Colonel

Place of Birth and Year: Cleveland, OH—1939

TWO MILLION MILES OF BAD ROAD

Vietnam Central Highlands, 1966-1967

In May of 1966, I assumed command of the 64th Tractor/Trailer Company at Ft Bragg. We deployed in July, 1966, with 60 tractors, 120 12-ton flatbed trailers, 20 headquarters vehicles, and 186 officers and men. We were able to take our company mascot, a German shepherd mix named Huntz. As our tour neared its end, we paid to have Huntz returned to Fayetteville, NC, where he lived with his master, Sam Hovey and his family, until Huntz passed away in 1975.

Leaving Ft Bragg had challenges. After we were alerted, the post management took away our maintenance building and asphalt parking area. We moved our supplies and vehicles to an open sandy field surrounded by barbed wire. We later had a better maintenance area in Vietnam. We were beginning to feel a bit paranoid. Then, post management told us that they planned to keep our tractors. We took all of the tires off our tractors and put these good tires on our trailers. After this job, DA in DC overruled Ft. Bragg and told us that we were going to take our tractors. You guessed it. We returned all of those best tires back to our tractors. We obtained 2x4 lumber and nailed a double floor of lumber on to the bed of every trailer. We would later use this lumber to reinforce our medium tents in our new area. We came by 26 large steel shipping containers. We loaded them with everything from refrigerators to portable electric saws, cement, plywood, machine guns, all kinds of supplies, and tires.

We were then told that we would be flying west by commercial air. We were to prepare our 190 weapons to move as air cargo. We used dozens of wooden crates, tubs of grease, and rolls of waxed paper to crate our weapons. When we finished, we were notified that we would travel by charter air, and we would carry our weapons. We were indeed convinced that our being paranoid was well justified. We cleaned our greasy weapons and had a 2 ½ ton truck load of greasy paper and greasy wood boxes for the post dump.

We had borrowed a few vehicles, since our own vehicles had headed overseas. The post dump gate dispatcher told our supply truck driver to back up to the cliff-like dumping point to dump our load. Our greasy trash was shoved out to flow down the steep hill behind our borrowed truck. So far, so good. But, there was a fire burning at the bottom of the hill. This fire quickly climbed up our trash stream, into the truck, and burned off all of the paint, wooden bows, canvas cushions, wood troop seating, tarpaulin, canvas cab cover, and ruined the instrument panel. The Ft Bragg paperwork demanding $1,200 for the new instrument panel followed us to Vietnam. My reply ended the matter. I wrote that the whole problem was caused by the original directions which were given to our truck driver by the disposal yard gate dispatcher.

When we arrived at the port of Qui Nhon, our advance party greeted us. Buses took us to our company area, about ten miles inland, on Highway 1, near the village of Phu Tai. We were assigned to the 27th Truck Battalion. Our daily convoy route ran west from Qui Nhon 110 miles to Pleiku on Highway 19, where the varied logistic locations were our destinations. We would run this round-trip every day.

With the addition of the tractor trailer rigs from the two other tractor trailer companies in the 27th Transportation Battalion, these daily convoys were usually made up of approximately one hundred cargo vehicles.

After nine months, our dispatch logs showed that we had run two million miles over our route which had 35 bridge bypasses, hundreds of axel-breaking potholes, and two serious mountain passes.

I shared the daily convoy commander duties with the platoon leaders of the company. When on convoy duty, I woke up at 0200, then went to our company operations tent to learn how many tractors and which assistant convoy commander would be with me. I reported to the battalion operations tent at 0230. The sergeant there would give me our daily trailer pickup locations. Our drivers woke up at 0330, and at 0400 ate breakfast in our award-winning company mess hall. At 0430 I would announce where each driver would pick up his trailer at the supply locations. We were not able to send assistant drivers.

At 0630 our drivers with loaded trailers would assemble along the right side of Highway 19, about 15 miles west of Qui Nhon, in the Cha Rang valley, near the location of the 54th Truck Battalion. At 0645, I would give our outbound briefing, and make the final list of our convoy information. We ran with a trail party of a mechanic driving a bobtail tractor, with breakdown supplies, and a second bobtail tractor. The convoy lead driver was always a dependable driver who would keep a reasonable pace. The assistant convoy commander jeep and my jeep would move up and down the convoy, with one keeping at the trail at all times. At 0700 the convoy departed.

By 0915 we arrived at An Khe after the slow low range first gear grind drive up the steep winding An Khe pass. Next came the steep winding Mang Yang pass, each climb usually took our convoys an hour to reach the top. We would marshal just beyond An Khe to filter out any stray traffic. At about 0945 we would be released from the An Khe checkpoint. At about 1200 we arrived at the Pleiku highlands marshalling area. Our drivers would disperse to their trailer drop locations, drop their loaded trailer, pick up an empty trailer, and return to the marshalling area. By 1400 we would depart from Pleiku on our return run. If we delayed beyond that time, the Military Police would be forced to hold the convoy overnight since the bridges closed just before dusk.

Thankfully there was not much danger during this period before the serious enemy danger on September 2, 1967. That convoy ambush killed and wounded many soldiers and destroyed and damaged many vehicles.

Daily convoy operations were never interrupted on our Highway 19. By 1830 our convoy rolled back into our Phu Tai company area. Since

I had arranged that our mess hall served 24 hours every day, our drivers could eat supper before or after doing their daily truck maintenance. By 2200 hours, drivers could enjoy a beer or a soft drink at our company club, watch an outdoor movie from a wooden bench, shower, and get ready for the next day on the road. At 0230, every day, the whole operation would begin once again. The maintenance section was divided into two crews, a day crew and a night crew. The night owls worked all night to return as many tractors as possible to the platoons for the next day's convoy. Tractors which could be fixed by our day crew company mechanics were set up for day repair. Tractors which needed to go to ordnance for repair had all services brought up to date before turn-in to the Ordnance Battalion.

My 64th Truck Company suffered its only fatality of that first year in the dark of an early August morning when driver Kenneth Tierney was crushed between two vehicles in the trailer transfer point. Creeping fatigue made accidents a constant problem. If a person had not experienced what it meant to be out on this road, he could have no understanding of what our drivers were going through every day. We put sandbags on the driver cab floors and armor plating on the doors. This plating reached up halfway above each door on the way to the canvas cab top. This trapped even more heat in the cab.

With helmets and flak jackets always worn by those out on the convoy, it all made for a long day. Both the An Khe and the Mang Yang passes were a series of zig zag steep climbing turns. One turn in the An Khe pass was so much of a switchback that it became known as the "Devil's Hairpin." During one climb up the An Khe pass, I spotted one of my drivers who was driving while "frozen," with the dashboard engine throttle handle pulled wide open, but he was sound asleep behind the wheel. His hands, wrists, and arms were stiffly locked in place. I loosened my canteen from my belt, told my jeep driver what I was about to do, and got out of the jeep, ran between the jeep and the tractor, and jumped up on to the running board. I grabbed the steering wheel, and splashed some water into the face of the astonished driver.

Our Thanksgiving Day became known as 'Black Thursday" when

three of our tractors were wrecked in accidents. Even considering the accidents and the maintenance challenges, it was the road itself which caused us the most trouble. Many of the pot holes were so wide and so deep that our trucks left the road and made a bumpy dirt bypass off the road and back on to the road again. Our trailers were rated for 24 tons for highway use, but no one knew exactly how much weight we carried. We began our tour in August 1966 with 60 tractors, but by December 1967 we were down to 36 tractors. Most of these losses were due to cracked frames caused by the road conditions.

Coming in from the road and pulling maintenance did not give the drivers much sleep. They took naps in their truck cabs whenever they could, waiting in the morning lineup area or in the marshalling area in Pleiku. We were able to name our vehicles on the armor plate. My jeep was the 'Patmobile', named after my wife, Patricia. Drivers named their tractors. Our daily runs gave us two climates, one was hot with dust and one was monsoon rain with red staining mud. In the monsoon rain, the drops would come down so hard, large, and fast that even on a slope the water would build up to over an inch deep and just stay there on the slope because the rain just could not run down fast enough. During the dry season, many drivers used disposable surgical masks from the clinics to fight for air in the thick dust.

Our diesel fuel came through the quartermaster supply system and somehow our diesel was found to contain a serious amount of sand. Replacement fuel filters were not available. On one day I saw a used fuel filter sitting on a large piece of cardboard and oozing raw sand out of the bottom of the filter. We finally used our parts cleaning machine to flush each of these filters with clean fuel and then put them back into place. The rear-view tractor mirrors were also a concern. As the miles piled up, mirrors became broken and could not be replaced. Eventually each driver fixed a single mirror to his driver side with wing nuts and took his mirror with him when he left his tractor.

We avoided overnighting at Pleiku as much as possible. When I found that an engineer depot crew would not unload our trailers which were full of 55 gallon tar drums, I talked to them but this did not help with their delay. Their delay was causing our drivers to stay overnight. I solved the problem by cutting the tie down bands, and having each driver drive in tight circles. After once having the barrels spread around their yard, the engineer crew changed their unloading operations so that our drivers could join the returning convoy.

We had been issued arctic sleeping bags prior to leaving Ft Bragg, and we put them to good use during cold highland nights. We also had been issued rubber galoshes, they were very valuable in the thick monsoon mud. When my Vietnam tour neared the end, I noticed that my M151 jeep, the Patmobile, had logged 20,000 convoy escort miles.

If you google "twomillionmilesofbadroad," you will come to a movie in which I put together narration, pictures, and super 8 film covering our unit move to Vietnam and our long haul trucking.

Our 64th Truck Company lost a total of seven soldiers during convoy contact with the enemy, with five non-combat deaths, during the five years before being inactivated in April of 1972.

JAMES B. "JIM" STAPLETON, JR

US Army — Infantry Officer
Dates of military service: 1964 to 1994
Units served with: 1/22 Inf, 2nd Brigade, 4th ID; G-3 4th ID; Aide to CG, First Field Force Vietnam; HQ; Advisor, Vietnam Airborne Division (Team 162)
Dates in Vietnam: Aug 66 to Dec 68 (28 months)
Highest Rank Held: Colonel
Place of Birth: Fort McPherson, GA — 1942

ENGINEER SHOWER SUPPORT FOR INFANTRY

In late August 1966, after arriving as a unit in Vietnam on 6 Aug 66, our 1st Bn 22nd Infantry went out immediately from our base in Pleiku to secure the famous Highway 1 (QL1 — the Street Without Joy)* in the vicinity of Tuy Hoa and south along the South China Sea. The Battalion Commander told our ever-overachiever Support Platoon Leader that he wanted a first-class shower unit for our Infantry Company Grunts as they rotated in from our "Search and Destroy" operations in the mountains, rice paddies, and jungles surrounding the Highway.

LT Bill S found a Corps Support Engineer unit doing Highway repair along QL1 that had a first class shower unit—certainly nothing they needed nor deserved. The Support Platoon NCO drove up to the Engineer CP at dark and told them he and his truck were there to pick up the shower unit to take it back for scheduled maintenance. It was dutifully loaded on the truck and headed back to Tuy Hoa and the beach.

We maintained the shower unit at peak performance for the rest of our mission through mid-October when we moved back to Pleiku and participation in Operation Paul Revere IV, along the Cambodian border. While our battalion HQ was on the beach south of Tuy Hoa during the time the Air Force runway was being constructed, our Grunts took "engineer showers" at every opportunity. The Engineers never asked how long the maintenance would take. Thanks to our Engineers, we had Joy in their Shower Unit!

*The title of the book *Street Without Joy* was taken from the name given by French soldiers to a stretch of Route 1 which had been fortified by their enemies, the Viet Minh, running along the central coast of Vietnam. In French it was called *La Rue Sans Joie* which translated means *Street without Joy*.

ROBERT O. "BOB" BABCOCK

US Army — Infantry Officer

Dates of Military Service (Active Duty and Reserves Combined): 1965 to 1974

Unit Served With in Vietnam: Bravo Company, 1st Bn, 22nd Inf Regt, 4th Inf. Div.

Dates Served in Vietnam: Jul 1966 to Jul 1967

Highest Rank Held: Captain

Place of Birth: Heavener, OK — 1943

THEY SAID WHAT?

3 November 1966 — It was our first night in the jungles along the Cambodian border. Even though we had been in country for three months, we felt our inexperience in this new environment. Our previous work had been against the Viet Cong along the coast south of Tuy Hoa, now we would be facing the North Vietnamese Army (NVA) regular troops who had traveled down the Ho Chi Minh trail. We knew they were better trained, better disciplined, and better equipped than the VC. Not a person in our company had served in Vietnam before.

To heighten our concern on our first night of Operation Paul Revere IV, we had been working with a scout dog that had kept alerting all afternoon. There had to be something out there the dog was sensing. Our movement had been slowed as we checked out the terrain each time the dog alerted. By the time we finally stopped for the night, evening was approaching and we had not had as much time to build good, strong bunkers and clear fields of fire as was to become the norm for us.

We kept working to improve our positions as the day's light began to disappear. As dusk was about to turn into darkness, Captain Fiacco called us on the radio. "Oscar, this is Oscar 6. Make sure everyone is in his bunker. I'm about to adjust in the def-cons."

We all knew that "def-cons" was short for "defensive concentrations" of artillery. By preregistering the def-cons, the artillery knew where to fire if we were attacked. Def-cons filled all the likely avenues of attack and also covered the low spots in the terrain where the enemy could hide and not be hit by direct rifle or machinegun fire.

Methodically, he and his artillery and mortar forward observers adjusted the 105mm artillery fire and 4.2 inch and 81mm mortar fire until it fell in a nice iron curtain around our perimeter. This became an every night ritual throughout the remainder of Paul Revere IV.

"Oscar, this is Oscar 6. All's clear. Make sure everyone eats and then let me know when your listening posts are going out."

Most of the men had already eaten their C-rations while on break from their bunker digging chores. Hacking through the jungle, carrying all our worldly possessions on our backs was back breaking work. Add digging foxholes and chopping down trees to make strong bunkers and you have a recipe for ravenous troops. Even C-rations had the flair of a seven course dinner.

Several of the men moved forward of the bunker line to set out claymore mines. Like the def-cons, this became a nightly ritual. We always set up claymores in depth. There were three claymores in front of each bunker, staggered at distances from approximately five yards on out to twenty-five yards in front of our positions. Trip flares were rigged on some to alert us if the NVA tried to crawl up and turn the claymores around to fire at us rather than toward the enemy approaches.

The men took great care in making sure that everything was set just right before they returned to the bunkers. We were not in VC territory any longer, we were in NVA territory and nothing was left to chance.

As full darkness settled in, we sent our listening posts out. Listening posts (LP) consisted of three or four men with a radio and several claymores. They went out in front of our perimeter along the most likely avenues of approach to listen for the enemy and give us early warning if anyone tried to sneak up on us during the night. Each platoon typically sent out one LP in front of its defensive sector. It was not good duty but everyone had to do it when his turn came. It gave a great sense of security to know someone was out front to give advance warning of an NVA attack.

When a listening post confirmed they heard NVA, they could blow their claymore mines and high tail it back to the perimeter, or lay low and hope they were not seen or heard. The decision was up to the LP leader.

All was quiet for about an hour when the jungle night was shattered with a loud, "F**K YOU." Everyone immediately tensed up and scrambled for their bunkers, not believing what we had just heard.

My listening post was the first one to break radio silence, "Oscar 61, this is Oscar 61 Bravo, did you hear what we just heard? Over."

"I sure did. Stay quiet and be alert. Let us know if you hear any movement. Out." I had already heard the other platoons call Captain Fiacco on the company radio net to report what they had heard.

"F**K YOU," came the sound again. My adrenalin was pumping full speed. I knew the NVA were better than the VC but I did not think they were so good they would stand out in the jungle and taunt us in English.

"Oscar 61 Bravo, this is Oscar 61. Do you hear any movement? Over."

"Negative. Where are they yelling from? Can we come in? Over."

"This is Oscar 61 — stay where you are. I will advise. Out."

Rather than add to the already high level of radio traffic, I crawled out of my bunker and half ran, half crawled to Captain Fiacco's command bunker 25 yards away. "What in the world do you think they are doing?" I asked Sandy, hoping he could shed some light on what was scaring my men and me to death. He was as bewildered and concerned as we were as we heard another, "F**K YOU," come out of the jungle.

"Babcock, bring your listening post in. Harry, call the other platoons and tell them to bring in their LP's. Let me know when everyone is in. We're going to put some firepower out there. And do not let anyone fire his weapon or give away his position. Keep everyone alert and in his bunker."

I scrambled back to my bunker and called my listening post in. It took no time for them to come thrashing up through the jungle thickets. As they neared the perimeter, you could hear a loud whisper with the challenge, "Green!" "Bean," came the whispered password as the men crashed through the perimeter and flung themselves into the bunkers along the platoon front. I reported to Captain Fiacco that they were safely in and soon heard reports the other LP's were also in.

I could hear Captain Fiacco calling the battalion fire base. "This is Oscar 6. We have bad guys around us. Request a fire mission, over."

"How many and where are they?" came the reply.

"I don't know how many or where they are, they're hollering at us. I need a fire mission, over."

"What are they saying? Over." came the reply. "They said 'F**k You.' Now quit asking questions and get me some fire out here, over."

"Say again. They said what?" was the astonished reply.

"F**k you!. Damn it, get some artillery fire going out here! NOW! Fire the def-cons! Out."

Soon the welcome sound of artillery shells whistling through the air and exploding around us filled the jungle. We all stayed hunkered down in our bunkers and peered out into the darkness waiting for the inevitable attack. The safeties were taken off the claymore charger handles. Rifles and machine guns were checked to make sure rounds were in the chambers and ready to fire. A curtain of steel was exploding around our perimeter as we continued to wait in anxious anticipation.

Finally, the artillery fire stopped and silence engulfed the jungle night once more. The NVA still had not attacked. Maybe the artillery fire had worked. It should have. It was hard to see how anyone or anything could have lived through that bombardment.

After what seemed like an eternity peering out into the darkness, listening for any sound of movement, I once again crawled up to the command post bunker to talk to Captain Fiacco.

"Keep your troops on 50% alert. Send your listening posts back out. Let me know if you hear anything," were Sandy's instructions to the platoon leaders.

Keeping our troops on 50% alert was not a problem. Getting the listening posts to go back out could have been. But, the troops responded like the true professionals they were and moved back out into the darkness.

If any one of the more than 120 men we had with us that night got more than fifteen minutes sleep, it was highly unusual. We lay awake all night long, looking, listening, and wondering what kind of monsters the NVA were that would scream at us saying, "F**K YOU." They had to be some kind of tough fighting men!

Only men who have lived through combat can understand how happy we were to see daylight. The listening posts came in at first light. Each platoon sent a squad sized patrol out to check the area in front of its position. None of the patrols found anything except a lot of leaves and branches that had been blown down by the intense artillery firepower.

We kept outposts outside the perimeter as we broke camp and prepared to move out on our next day's trek through the jungle. Caution was our byword as we cleared out of our night defensive position and started plowing through virgin jungle forests again.

After moving all morning, we were more than happy to stop for a lunch and rest break. After security was posted, the rest of the troops leaned up against trees or plopped down on the ground to catch a few winks of sleep to make up for the sleepless night. As we rested there in the jungle, we heard it again. "F**K YOU."

This time we found the culprit immediately. Perched on the side of a tree next to one of our men was a little green lizard. From then on we ignored the sound of the Vietnamese F**k You lizard. We heard its cry many times in the weeks and months ahead. Never again did it bring fear into the hearts of so many men as it did that first night in the jungle.

Note... we were a new unit who had come by ship from Fort Lewis, Washington to Vietnam. We didn't have a Vietnam experienced person in our company. From then on we were jungle veterans. As we heard the lizards, we assured any new guys who had come in as replacements what it was they were hearing. They were never scared by lizards, nor did we waste so many artillery and mortar shells as we did that first night in the jungle.

RONNIE RONDEM

US Army — Infantry Officer

Dates of Military Service (Active Duty and Reserves Combined): 1961 to 1996

Units Served With in Vietnam: D and C Companies, 2/1 Infantry, 196th Light Infantry Brigade; HQ and B Companies, 2/506 ABN Infantry, 101st Abn Div

Dates Served in Vietnam: Aug 1966 to Aug 1967; Dec 1968 to Jan 1970

Highest Rank Held: Colonel

Place of Birth (and year of birth): Oslo, Norway — 1943

THE LONG JOURNEY (THE FIRST TOUR)

After being commissioned as a second lieutenant, I was assigned to Company D, 2d Bn, 196th Infantry Brigade (Sep) at Ft Devens, MA. That was a "train and retain" unit where the soldiers reported to the Brigade from the reception station. The unit conducted Basic Training, Advanced Individual Training (AIT), and small unit training at Ft Devens and at Otis AFB. That all occurred before I was assigned in April, 1966. To complete the platoon, company and battalion training, the Brigade went by Motor March to Camp Drum, NY (now Ft Drum), where there was more maneuver space.

I had been assigned to the Anti -Tank Platoon in Company D. That was six each 106 mm Recoilless Rifles mounted on a M151 chassis, i.e. a jeep. The whole Brigade had been formed with a TO&E suited to specifically perform in Santo Domingo in relief of the Brigade from the 82d Airborne. The final exercise at Camp Drum was a Field Training Exercise (FTX) where the whole Brigade was rated by umpires as to how well we performed. Much of the FTX was written to use the mobility of the helicopters and that played into the exercise.

This was in the first week of May and overnight we were surprised by 6-10 inches of new snow. Not really conducive to the Dominican Republic weather (or to Vietnam). But we did not know we were heading there. One of the moments that has stuck in my head was when we were conducting road clearing and road security for a supply convoy. We came upon an opposing force (OPFOR) that was heading toward the supply convoy and we engaged them. We believed that we had stopped the OPFOR quite well, until an umpire stepped out of the woods and told us that we could not shoot at them as the ¾ ton truck they were riding in was "actually" a UH-1 helicopter and the 2 ½ truck behind them was "actually" a CH-47 Chinook Helicopter. So much for the reality of the FTX.

After the FTX we had green beer provided by Genesee at the O-Club, the NCO Club and the Enlisted Club to send us on our way. In June, our Brigade Commander reported to the 82d Brigade Commander in Santo Domingo for coordination of the changeover. On the way back, he was told to report to DA in Washington. There he received a change of orders for the Brigade to deploy to Vietnam.

The Brigade was organized for the Main Body, which deployed by boats (or ships) from Boston harbor on 15 July (my birthday), and a Rear Detachment/Advance Party, that deployed by C-141s on 8 August 1966. As the term of overseas service began on the date the unit, or individual, departed CONUS, the 30 days at sea counted for the Main Body and the 8th of August was the start date for the Rear Detachment /Advance party. I was assigned to the latter group.

We departed from Westover AFB in Massachusetts with the first stop in Alaska at Elmendorf AFB. There were three C-141s for the unit. We landed around 4 AM, Alaska time and were told to stretch our legs a bit. Then we were told that they needed to correct some mechanical issues and that it might take a couple of hours. The OIC of our group, Major Snow, somehow found the Manager of the Officers Club and persuaded him to open the Club at 5:30 AM. He also persuaded the same manager to allow all our troops to come in as well. And, he found some of the staff. Short answer, by the time the planes were ready to fly, there were quite a few that crawled back onto the planes. And, did I mention that the planes were fitted for cargo transport and there were no seats?

Next stop, Yokota AFB, Japan. "Same, same, GI." Mechanical issues and a delay. There it was mid-afternoon and all the clubs were open. Gambling was allowed there and our guys tasted that to various degrees.

Two of the guys boarded the planes with approximately $1,000.00 more than they came with.

Next stop Tan Son Nhut AFB, Saigon, Vietnam. The temperature and humidity hit you like a wall when the doors and ramps opened. Due to our two delays en route, we could not continue on to our destination, Tay Ninh, that evening. We also had to change planes to C-130s in order to land at Tay Ninh. We were transported to what I believe was called Camp Red Ball in the vicinity of Tan Son Nhut.

There were no barracks or obvious places to bed down. But there was one large building that was a storage site for coffins. The coffins were empty and 6-10 men slept their first night in Vietnam in open coffins. NOT ME!!

Next stop Tay Ninh. Our base camp will be there, said the Major. Our Main Body arrived in country at Vung Tau on the 15th of August and arrived at Tay Ninh over several days. Medium and large tents were erected by our advance party, allowing everyone to throw something on the ground to sleep on. With most of the Brigade in tents of some type, the building of the base camp began. Not too many water points. All of the men, from Colonel to Privates, learned how to shower.

When you see an afternoon shower forming in the distance, work stopped. We ran inside to fetch our bars of soap and our towels. We got out in the open and waited for the tropical rain to begin and then got to clean up some. The times that I wish I had a camera was when the cloud turned before it reached the desired area. The picture would have been the dejected look of the naked men walking back to their tents with towels over their shoulders and a bar of soap in their hand.

During the course of our tour in Vietnam, we spent some rowdy time in the Officers Club. It was there where Army officers and Marine aviators worked hard to impress and get the best of each other. Of course, there were rules for the "contests" that happened almost every night. Here are two sets of rules that have survived these many years…

CARRIER QUALIFICATION FOR THE ARMY (BY THE MARINE AVIATORS)

1. The candidates had to be ARMY Grunts
2. The candidates had to have imbibed many, MANY, drinks or beers.
3. The Carrier deck had to be marked by bottles or cans, on the wet application (of beer or similar) onto the wooden floor of the Officers Club tent, in the shape of a carrier deck. That included the beginning, sides, and end of the deck.
4. The candidate would start at the incoming lane of the aircraft, and simulate landing on the deck with his arms out on each side, as the wings. He could not disturb the beginning of the deck nor go past the end of the deck, in order to be certified as a qualified carrier aviator. Not an easy task.

BAYONET QUALIFICATION FOR THE MARINE AVIATORS (BY THE ARMY GRUNTS)

1. The candidates had to be Marine Aviators.
2. The candidates had to have imbibed many, MANY, drinks or beers.
3. The bayonet attack course was marked at the beginning and end of the Officers Club tent while the attack lane was lined by ARMY grunts and bayonet-qualified Marine Aviators.
4. Provide a bayonet (a broomstick) to the candidate.
5. Have the candidate grab the bayonet and hold it straight up to a marked point in the ceiling.
6. Assist the candidate by spinning and turning them ten times while the candidate held the bayonet strongly against the point in the ceiling.
7. Place a cardboard box with a center hole cut out at the far end of the bayonet lane.
8. Insure that the lane guides were properly positioned to assist (read physically assist) the bayonet qualifier candidates to proceed at a rapid pace toward his target.
9. Help the poor soul up at the end of the course.

There are many other stories in the Big City and these were some of them.

MIKE HAMER

US Marine Corps – Squad Leader; US Army – Infantry Officer
Dates of Military Service: 1956 to 2011
Unit Served with in Vietnam: B Company, 3rd Battalion, 12th Infantry Regiment, 4th Inf Div; Ops Advisor to I Corps Vietnamese Training Center
Dates Served in Vietnam: Sep 1966 to Sep 1967; Dec 1968 to Dec 1969
Highest rank: Major
Place and Date of Birth: Chicago, IL – 1938

"I HAD YOUR JOB IN WORLD WAR II"

I deployed to Vietnam in September 1966 from Fort Lewis, WA. I was the Battalion S2 in 3rd of the 12th Infantry Regiment, 1st Brigade, 4th Infantry Division. In early 1967, I received a letter from a Jake Hay saying, "I had your job in WWII." He asked about what I was doing as S2 and what I thought of what was happening back in the US. I gave him a "candid" opinion of what I thought about the protests.

A short time later I received another letter from him, telling me he was a columnist for the Baltimore Sun and had included my letter in his paper! It's attached here. I thought I'd be put in the brig for what I told him.

He lived in Linwood, Maryland, a small community in Eastern Maryland and drove to Baltimore to work. We corresponded for several months; I took command of a company in March 1967. He arranged for the ladies of Lynwood to send cookies and cakes to my company. Of course, most were pilfered at mail stops along the way, but a few got through. I reported to Fort Benning for the Infantry Career Course in Sept 1967. We had a week break for Thanksgiving; my wife and I drove to Linwood and spent three days with Jake and his lovely wife Joy, an English woman he married over there before coming home from

WWII. I discovered Jake was a martini lover! We shared many while telling war stories.

I learned he could speak French. When 3/12 Infantry of 4ID entered Paris as one of the first American units, he was on a jeep announcing to the French they were being liberated; he said it was unreal, everyone throwing flowers and kisses.

We continued corresponding over the years; however, Jake passed away in the early 80s; I never saw him again.

I like to say, "I had his job in Vietnam!"

THE NEWSPAPER ARTICLE FROM JAKE:

The Kooks Aren't in Charge Yet!

Editor's Note: In the thick of the Vietnam fighting is Capt. Martin E. (Mike) Hamer, son of Mr. and Mrs. C. O. Hamer of Muncie.

The Evening News today reprints a remarkable article by Joseph Hey, Linwood (Mass.) newspaper columnist, based on correspondence with Mike, whom he has never seen. Some background on Capt. Hamer follows the Hay article.

By JACOB HAY

More years ago than I care to think about, I held down a temporary job that involved a good deal of walking; and a couple of weeks ago, purely on an impulse, I dropped a line to the guy who has the job now, not really expecting a reply.

But I got one.

From Capt. Mike Hamer, Commander of B Company, 3rd Battalion, 12th Infantry Regiment, 4th Infantry Division, USA, APO SF 96262, somewhere in Vietnam. (And if you have a minute, you could do worse with your time than to drop the captain a brief note, for reasons I'll explain. Or rather, he'll explain...)

What touched me, grabbed me hard, was that here is a young officer taking time to reply to what was doubtless a maudlin letter on my part. I felt that I had to tell him how much I appreciated the job he and his men were doing for me while I sit here, fat, dumb, and happy in Linwood.

If you have never lived a 24-hour day in the knowledge that the next tree could hide the sniper who will kill you, you might find my attitude hard to understand.

What shook me was that Capt. Hamer took very seriously something we here in the States tend to regard as a kind of adolescent, academic acne—the draft card burners, the hot-eyed, fanatic pacifists who would just as soon kill you rather than have you approve of our operation in Vietnam.

"The biggest problem," Capt. Hamer wrote, "is the knowledge of mixed emotions and non-support at home—viewed with distaste by the troops."

Distaste?

The Army's gone to hell again—when an infantry captain uses so mild a word as "distaste." I have an idea of the word Capt. Hamer really had in mind—but was too polite to write.

What scares me is the knowledge that a fighting man half a world away is deeply concerned by the yells and screams of the kook fringe.

I happen to believe, with all my heart, that we are doing the best we can with a ghastly mess, and I am horrified that the morale of the men in the line should be threatened by the loonies who scream bloody murder if a bomb drops on a Hanoi resident and clam up when Viet Cong terrorists heave a grenade in an American barracks. These are the same vicious nuts who phone servicemen's families and hiss obscenities at them.

Capt. Hamer's letter brought me to the realization that our men in Vietnam are fighting under a handicap I never knew, my generation never knew, thank God. War itself is hell enough, but to die a thousand deaths a day in the knowledge that the grand folks back home couldn't give a tinker's damn one way or the other, that some of them think you're a mercenary murderer... that's asking more of a man than can be tolerated.

So I wrote a reply to the captain, and while it will win no Pulitzer nominations, I beg the liberty to reproduce a part of it here, and I would like to think that a few of you who read this may be prompted to do likewise.

"Captain," I wrote, "if this letter does nothing else, I do hope it will reassure you and your men that a whale of a lot of us back here, snug, secure, smug, you name it, do give a damn. Our plight is that we can do so little about this whole shebang.

"Your war is in our living room, and every Huntley-Brinkley report brings it to us, and there's not a bloody thing we can do to make your situation any better. But the kooks are not in charge yet, and I doubt if they'll ever make it. Don't let the odd-balls discourage you or your men; odd-balls we've always had."

Maybe it will cheer Mike Hamer up. And I hope his section sergeant has got the battlefield promotion to second lieutenant that Mike says he has recommended.

I hope they're both still alive to go on risking their lives, 24 hours a day, for the bleeding hearts and lily-livered bleating of imperialism. The gutless-generation, if you will. I will.

Capt. Mike Hamer is the reason I will not join a "peace vigil" in front of a Westminster church, will not carry a peace sign, will not stab a fighting man in the back at a safe, comfy remove where my ideals aren't about to get muddied up by rice paddies, and my nostrils will never again be abused by the aroma of a good friend two days dead in the sun, with a booby-trap on his corpse.

Some fine day, if we're both lucky, I hope to see Capt. Hamer drive up here to Linwood, alive and whole, for the infamous martini I have promised him; all gin and forget the vermouth.

Maybe then I will be able to explain to him that most of us know and devoutly appreciate the job he has performed and that the whimpers came from the kind of people he would have heaved out of his intelligence section at first glance.

I can sit here writing this because Mike Hamer is out there for all of us.

The Hamer family moved in 1958 to Muncie, where Capt. Hamer's parents live at 2615 Ball. His brother is a Ball State graduate and assistant director of bands at Elwood High School.

Mike is a graduate of Henderson State Teachers College in Arkansas, where he had a four-year football scholarship. His wife and 2-year-old son live in Pine Bluff, Ark., her parents' hometown.

Mike served a two-year hitch in the Marines after graduating high school in Columbus, Ohio, in 1958. He served as an Army security officer in Germany following officer training at Fort Benning, Ga., and three years as a training officer at Ft. Ord, Calif. In September last year, he was transferred to Vietnam.

ROBERT O. "BOB" BABCOCK

US Army—Infantry Officer
Dates of Military Service (Active Duty and Reserves Combined): 1965 to 1974
Unit Served With in Vietnam: Bravo Company, 1st Bn, 22nd Inf Regt, 4th Inf. Div.
Dates Served in Vietnam: Jul 1966 to Jul 1967
Highest Rank Held: Captain
Place of Birth: Heavener, OK—1943

THE OLD MAN AND OLD WOMAN

For the second week, our search and destroy mission moved through the dense, triple canopied Vietnamese jungle. The area had been declared a "free fire zone" by the South Vietnamese government. Several months earlier, they had warned all their citizens to get out of this area. Anyone left was to be considered enemy.

Cautiously, we moved down a slope, toward one of the many small streams we crossed each day. The small trail we were following made us stay more alert than if we had been moving through untracked jungle. My platoon was the point platoon and, as always, I was positioned behind my lead squad.

Three rapid rifle shots from Ernie Redin, my point man, ruptured the jungle calm. Without thinking, our instincts drove us to the ground as we quickly scanned the jungle around us. Was this going to be our first fight with the NVA? Half crawling, half crouching, I moved to the front of the column to see what was happening. Quiet again engulfed the jungle.

Ernie was pointing his rifle at an old Montagnard man and woman standing by the stream with their hands in the air. Sergeant Benge had moved the rest of his squad up to form a hasty skirmish line on either side of Ernie.

"What's up?" I asked Ernie.

"A Montagnard boy ran up that hill when he saw me. I fired at him but don't think I hit him," was his reply.

"How big was he? Did you see anyone else?" I asked.

"I would guess he was about twelve and I didn't see anyone else."

As I radioed the information back to Captain Fiacco, I sent the point squad across the stream and up the hill to see what they could find. They cautiously worked their way up the hill, alert to any sign of an enemy ambush. I led the old man and old woman back to where the rest of my platoon had formed a defensive perimeter. We waited to see if our squad found the boy, or anything else.

The squad soon returned, having found nothing. In the meantime, our Vietnamese interpreter, Sergeant Quann, was trying to question the captives. As Montagnards, they had their own language and did not speak Vietnamese. (The nomadic Montagnard tribes, or Hmongs, roamed the central highlands of Vietnam, living a very primitive, almost prehistoric existence. They were frequently discriminated against by the Vietnamese). None of us, including Sergeant Quann, understood their jabbering, punctuated with a lot of pointing in the way the boy had run. Looks of anguish clouded their faces. The old woman's teeth were stained from the beetle nuts that were chewed by so many of the Montagnards. (It had a numbing effect on the gums and helped ease the pain from bad teeth that had never seen a toothbrush or dentist).

They were a pathetic site, old and weak, scared to death, and obviously not a threat to anyone. We were not so sure about the boy who had run away, however.

They caused us a real dilemma. It was mid-afternoon and we needed to get moving to find a defensive position for the night. The man and woman were too old and slow to be able to move with us and keep up. The thick, triple canopied jungle precluded us getting a helicopter in to take them out. And, they had been with a fighting aged boy who ran away when he saw us coming. What were our options?

Though Captain Fiacco had the ultimate decision, I talked our options over with him. The obvious thing was to shoot them and move on. The Vietnamese government had said that anything in this area could be considered to be enemy. They had been with a healthy boy, old enough to lead enemy troops to us, who had run away. That option was immediately dismissed. They were, at worst case, now our prisoners of war and protected by the Geneva Convention.

We also could have moved them with us, knowing that it would slow us down. If they were innocent civilians, we would have left the young boy out in the jungle to fend for himself and that option did not appeal to us. Since the jungle was so thick, cutting a helicopter landing zone would have been impossible. Our only other choice was to let them go.

Sandy and I concluded they were innocent civilians and, hopefully, were not going to hurt anyone. And, we really did not like the idea of leaving that boy out there in the jungle by himself. So, we decided to let them go.

As I gave the command to saddle up and move out, they looked very confused when we started to move out and left them standing there. We veered off from the trail and took a different route towards what looked like a good hill to set up on for the night. All our senses were alert as we resumed our movement through the jungle.

A squad from the second platoon stayed a little further behind than rear security would normally stay, just to make sure nothing moved in behind us. They radioed, "The old man and old woman are following us." Sandy radioed back, "Fire a few warning shots into the air and try to scare them off." The warning shots worked, and we moved on out of their sight.

As we moved up the hill to prepare our nightly defensive position, we posted security outposts on all sides. Our nightly artillery and mortar defensive concentrations were registered especially close to our perimeter.

It was a restless night; none of us were comfortable we were not still being followed. Staying awake was easy when it was your turn to be on guard duty.

As the last shadows of night disappeared, and the first waves of daylight started to creep into our perimeter, we breathed a collective sigh of relief. Before we started breaking down our defensive positions, patrols swept the surrounding area in search of ambushes that might have been set up around us during the night. Nothing was found. Fully aware that a unit is most vulnerable when it first starts moving, we moved out with extra caution. (The troops tend to be clustered together and confusion seems to be the norm).

We continued with our habit of filling in the bunkers so they could not be used by the enemy at a later time. That was good for the future, but it did not give us any retreat options if we were hit as we moved out.

By noon, we were convinced nothing was going to come from our encounter with the old man and old woman. We had put quite a distance between us and the spot where we had left them. We started to breathe easier.

I have often thought about that incident. It never crossed my mind to shoot them. I'm sure Captain Fiacco felt the same way. When someone was armed and capable of hurting us, it was a different story. But, when we had taken away the means to fight and had them as a captive, my heart always went out to them with compassion as a human being. I could only guess how frightened they must have been.

And, I also wonder how I would have handled it later in life if either Sandy or I had decided to shoot them. I am sure there are veterans of all wars, not just Vietnam, who have had to live with decisions made under adverse circumstances.

Much has been written about the atrocities, real or alleged, that were committed by our servicemen during the Vietnam war. Fortunately, I was never a party to, nor a witness of, any of the things that the press seemed to think made good news stories. I firmly believe there are more stories similar to the one I have written than there are of the headline grabbing spectacles that the press corps so frequently pounced on. The way we conducted ourselves on this occasion makes me very proud of Sandy Fiacco and of myself. I am sure most of our fellow Soldiers would have made the same decision. It would have been difficult to live with any decision other than the one we made that day.

GEORGE MURRAY

US Army — Armor Officer / Aviator
Dates of Military Service: 1964 to 1968
Unit Served with in Vietnam: A/82d Airmobile Light, re-designated 335th Assault
Helicopter Company
Dates in Vietnam: Feb 1966 to Feb 1967
Highest Rank Held: Captain
Place of Birth and Year: Grenada, MS — 1942

LZ STUMP

Christmas was rapidly approaching in Vietnam in 1966. Our unit, the 335th Assault Helicopter Company (Cowboys) attached to the 173rd Airborne Brigade, was operating in III Corps near Bien Hoa. I had been in the country nine months, been on R&R to Hong Kong, but flying most of the time. Vance Gammons* was the platoon leader of the Gun Platoon which flew under the call sign of "Falcons." I was a "slick" pilot flying for 1st platoon, under radio call sign Cowboy, followed by a 3 digit tail number of the helicopter I was flying that day.

We had been operating south of Bien Hoa in an area near Bear Cat. Action had drawn focus to a small landing zone (LZ) located at YS308857 on the map. Viet Cong activity had been noted in the area and the 173rd Airborne was planning to position a battalion known as 4th of the 503rd Infantry in the landing zone. The "landing area" was approximately the size of a football field, surrounded by 200-foot tall trees. On the ground, trees of various heights were scattered about with a thick carpet of undergrowth covering fallen trees, stumps, ditches, and pits. Running from north to south in the landing zone was a belt of trees and brush approximately 10 meters wide that bisect the LZ. This LZ was used as an emergency extraction of two elements of a long-range reconnaissance patrol (LRRP) on 16 December 1966. The intelligence gathered by this LRRP indicated heavy enemy activity in the area.

Using ten UH-1 D helicopters in two flights of five, because the LZ was so small, paratroopers of the 4th Battalion were flown into the LZ. On the first flight in, the trail aircraft in the second group of five lost a tail rotor due to a strike on a tree. Maintenance aircraft landed in the area and began repair at the same time Viet Cong fire was opened across the landing zone. As we delivered troops to the area throughout the afternoon until 570 paratroopers were on the ground, we received continuous automatic weapons fire.

At this point, Captain Vance Gammons' platoon of Falcons gunships were delivering supporting fire to the landing "slicks." Artillery and airstrikes were conducted in the immediate area, causing tremendous radio traffic and heavy congestion in and out of the landing zone.

The infantry maintained contact with the Viet Cong until the brilliant political decision of 23 December to remove the troops from contact. This halt in operations was ordered so President Johnson could brag about a Christmas truce. With this mission in mind, the Cowboys, using a flight of ten UH1Ds and four gunships, arrived at LZ Stump at approximately 1300 hours to extract the 700+ troops on the ground. Again, entering the landing zone took each pilot's maximum skill to maneuver through the trees, stumps, and other obstructions on the landing zone.

In order to pick up the troops, it was necessary to land. As each flight of five came in, we were dodging trees and keeping our tail rotors off stumps and obstructions while watching tracers crisscross the landing zone. On departure from the landing zone, each slick was carrying five infantrymen and struggling to get off the ground because of the weight. Having reduced our fuel loads and dealing with the high humidity and high heat, the aircraft simply were not powerful enough to carry full loads and be agile enough to dodge the obstructions.

Each helicopter having FM and UHF radios were monitoring all channels and the radio traffic was tremendous. Crews were calling "fire" on takeoff and landing and trying to direct the gunships onto the targets. At the same time, command and control aircraft were calling for artillery and airstrikes. To make it more fun, Air Force F-105s were delivering napalm right along our flight path; no 4th of July fireworks display can match napalm striking the bright green jungle.

Note that Vance Gammons and his lead gunship known as Falcon 86 was running a tape-recorder picking up this traffic. At this time in the war, the enemy use of .50 caliber anti-aircraft was not common in this part of Vietnam. However, my door gunner was screaming that he was firing back at a .50 cal coming off the landing zone. His observation was verified at the end of this mission when one of these ships was found to have a large hole in a rotor blade caused by an obvious .50 caliber strike.

All of us being military knew what a .50 caliber round looked like and the damage it could do to an individual or to aircraft controls. In Vance's audio of the action at LZ Stump, there is mention of the .50 caliber action several times, mostly in loud screams from aircraft commanders calling ".50 caliber fire."

Adding to the fun of artillery rounds falling, fighter aircraft dropping napalm, and the general confusion around the landing zone, rain starting falling in the area. And, as troop strength dwindled on the LZ, it became imperative that the final lifts get every last trooper off the landing zone. Knowing that each aircraft was only able to struggle with five troops aboard, the ground commander reported 60 troops still on the LZ. The Cowboy company commander, Cowboy 6, radioed all ten aircraft, asking if anyone was not able to carry six on the final lift. All the aircraft commanders were silent. It was obvious that maximum effort would be made to extract everyone on this last lift in a rain storm with darkness approaching.

The maximum effort was effective as the Viet Cong put up a last blast of fire and received, in turn, helicopter gunships, tac air, and artillery on their positions. All 700+ airborne troops were lifted from the final pick-up zone and the President got his stupid Christmas cease fire.

As our flight of ten slicks and four gunnies made the approximately 20-minute trip back to our base camp, aircraft were calling damage reports from bullet strikes. To top the day off, the rain was closing in fast as we ducked through an opening in the storm and made it back in time for supper and a cold beer.

*Vance Gammons arrived at the Cowboys/Falcons a couple of months after I joined the unit. As a senior Captain, he has extensive helicopter flying time. In 1978, I was transferred to Georgia and reconnected with Falcon 86. Today we both are members of Atlanta Vietnam Veterans Business Association and Mt Bethel United Methodist Church, members of the monthly meeting Vets Ministry of the church, and play golf together occasionally—small world!

ED FELL

US Army — Armor Officer, Aviator (Huey Pilot)

Dates of Military Service: 1965 to 1968

Unit Served with in Vietnam: 68th Assault Helicopter Company

Dates in Vietnam: Nov 1966 to Nov 1967

Highest Rank Held: Captain

Place of Birth and Year: Stockport, OH — 1943

THE BATTLE OF LZ GOLD — AN ASSAULT INTO HELL

From November, 1966 to November, 1967, I had the honor and privilege to serve with the 68th Assault Helicopter Company (the Top Tigers) based in Bien Hoa, an Air Force Base located about 20 miles north of Saigon in the III Corps area. Our mission was to fly combat assault and support missions for the various military ground units in that area. During that year, I logged 1,325 flight hours and was shot down twice. There are so many stories to relate about that year, but the one that I want to highlight is what happened on Palm Sunday, March 19, 1967.

On that day, the 68th had just completed a three-week field assignment in Tay Ninh in support of Operation Junction City. After spending three weeks in tents and living in nasty conditions, we were ready to get back to the relative cleanliness of our home base. Instead, on March 19th, we were diverted to a staging area near the small village of Soui Da near Nui Ba Den (the Black Virgin Mountain). Our mission, along with the 118th Assault Helicopter Company (the Thunderbirds), was to insert troops from the 1st Brigade of the 4th Infantry Division into a small landing zone designated LZ Gold. The 118th was down a ship due to maintenance so we loaned our spare aircraft and crew to them. Everyone knew that this entire area was saturated with enemy forces, but very little reconnaissance had been done in this area, so we were unaware that the 271st North Vietnamese Regiment with over 2,000 soldiers was lying in wait.

After a 30-minute artillery prep, we began our assault. An assault helicopter company normally conducts LZ landings with ten aircraft; but because of the size of the LZ, each company divided up into two flights of five. My company, the 68th, was the first company in. I was flying ship number two in the second lift. Each lift made it in OK, but we did receive automatic weapons fire on the way out. The 118th followed us into the LZ. By that time the North Vietnamese were ready, and all hell broke loose. Automatic weapons fire was heavy and land mine explosions were going off everywhere. Two helicopters were completely destroyed, one of which was our ship that was loaned to the 118th. Three of the four crewmembers were killed, along with all of the troops on board. Five other ships were seriously damaged.

Realizing the need to reinforce the troops that were already on the ground and being ambushed, the two companies returned to the pick-up point for more troops and a return into the LZ. The 118th was so badly crippled that they were down to five aircraft. During the next assault, the lead ship and crew received damage and injuries and had to discontinue the mission. That left me in charge of the mission for the rest of the day. We flew five more assaults into that LZ on that day. Automatic weapons fire was intense, and at the end of the day, there wasn't an aircraft that did not receive a hit. Heroism was the norm rather than the exception on that day. The battle on the ground continued for two more days, and, in the end, enemy body count exceeded 600 and we had lost 79. It was the largest III Corps battle to date (and largest single day body count of the war).

Fifty one years had gone by in 2018 since that fateful day, and through those years I was bothered by so many things; why hadn't we been given better intelligence on the enemy situation? Why did so many young men have to lose their lives? Why did three members of my own company have to die? Why was I awarded the Distinguished Flying Cross when others gave their all? What was it like now in Vietnam at the site of LZ Gold? There was only one way to come to closure with all of these questions, and that was to go back.

So, in August, 2018, I returned to Vietnam with my wife. We saw many interesting places and things, and I was amazed at how the country had changed over those 51 years. But the primary purpose of my visit was to find LZ Gold and stand in that very spot to come to closure with all of those memories that still lingered. It was not an easy task to pinpoint where the LZ was, but with the help of the military coordinates that I had kept over the years, Google Earth, and an app on my smartphone that would accept military coordinates, I found it! We got as close to the site as possible by road, and then trudged through rice paddies and sugar cane fields for about 300 meters until the app indicated that we were there.

My first impression was of amazement. There I stood, looking over peaceful and serene rice paddies and sugar cane fields—there was no indication whatsoever that one of the fiercest battles of the war had been fought there. A woman working the rice paddies spotted us and approached us. Through our interpreter, we learned that she was 14 years old when the battle occurred and remembered it well. She said that they were still finding war artifacts in the fields. She took us over to a small clearing carved out of a sugar cane field and told us that this was where over 400 North Vietnamese troops were buried. Accounts that I have read about the battle refer to bulldozers coming in, digging a huge hole, and burying enemy bodies, so her account seemed credible.

I stayed there a while longer, just looking out over this once chaotic and now peaceful spot. I was able to shed all of the anger, and came to the realization that all of those men hadn't died in vain. They died while supporting and defending their brothers in arms. There can be no nobler way to die.

VANCE S. GAMMONS

U S Army- Infantry Officer/AviatorDates of military service: 1959 to 1982
Units Served with in Vietnam: 117th, 119th, 335th Assault Helicopter Co; 1st Bde, 25th Inf. Div.
Dates Served in Vietnam: Jan 1964 to Dec 1964; Jun 1966 to Oct 1967; Jun 1969 to May 1970
Highest Rank Held: Lt. Colonel
Place and Date of Birth: Cowan, TN — 1936

BULLET THROUGH THE CYCLIC STICK

Back in 1967, I was assigned to the 335th AHC and my brother was in an Air Force FAC/ALO outfit named SIGMA. He and I extended our tours in order to get free 30 day leaves and an airplane ticket to anyplace we wanted to go. We selected Melbourne, Australia and then paid Pan American the difference to go by way of Bangkok, Thailand, New Deli, India (for a 10 day tiger hunt), Singapore, then Sydney. We enjoyed the hunting but never got past Sydney.

On returning I was reassigned to the 1st Brigade, 25th Infantry Division at Cu Chi as the Aviation Officer. We had six H-23 D helicopters. Bubble type — not unlike the H-13 of MASH fame — but flown from the center with a passenger seat on either side. We flew liaison, recon, artillery adjustment, commander command and control, and whatever we were assigned. For me personally, there were two that were memorable.

Both were during a Brigade operation in The Iron Triangle area of NW III Corps in late September. Our HQ was located at Trung Lap.

One afternoon we were advised that a patrol had reported seeing an airplane canopy. The Air Force Major we had with the HQ requested that he be flown out to view and identify it. I cranked up with him in the left seat and we found it in a few minutes and he wanted to circle it. We were less than 1000 feet above the ground and we descended some as we circled. The second time around he yelled, "That guy is shooting at us!" I took a very quick look—leaned back behind the Major—and we accelerated out of the area. No hits—but happy elsewhere. It was most unusual for me—as a pilot—to look at a person who has a rifle and is shooting at me.

The Brigade had a daily late afternoon briefing on the operation. An officer from each of our three battalions attended. We picked them up and flew them back. This particular afternoon there was active Air Force and artillery so it was necessary to fly low level and know how not to interfere with either. I had two young Lieutenants and had dropped one off and was en route with the other when we received fire and one round came through the floor, cutting the tail rotor control cable (anti-torque) and damaging my cyclic control stick and my right hand.

I retained thumb and first finger control and was flying slightly nose low and a bit sideways, but with full power. We were within 10 minutes of the Trung Lap grass strip, and I recalled the emergency procedure was to approach long and low and attempt to do a running landing so I did pretty well till just before ground contact I slowed too much and the chopper started to spin. I immediately cut the power, which stopped the spin, then did a hovering autorotation. We were not hurt nor was the helicopter further damaged. I went to the hospital and a few days later was talking to one of my fellow pilots who had witnessed the landing. I asked him if he liked my landing and he replied, "Which one, you bounced that thing two or three times."

When I opened my B4 bag at home, I found the cyclic stick with the bullet hole mounted for a trophy. My thanks to the maintenance crew who saved that for me. And, I've always wished I could hear the 2LT's version of that ride.

CARY KING

US Army — Artillery and Infantry Officer
Dates of Military Service (Active and Reserve): 1963 to 1987
Unit Served with in Vietnam: 2nd Bn, 28th Infantry Regt, 2nd Bde, 1st Inf.
Div.;1st Bn, 7th Field Artillery Bn; HHB, Division Artillery
Dates in Vietnam: Apr 1967 to Oct 1968
Highest Rank Held: Lt. Colonel
Place of Birth and Year: Atlanta, GA — 1941

ARTILLERY TO INFANTRY AND BACK AGAIN...

I am an Atlanta boy, born and bred, and was commissioned as a Second Lieutenant through the college Army ROTC program in August 1963. My original branch assignment, Military Occupational Specialty (MOS), was in the Artillery and that is where I began my military career at Ft. Sill, Oklahoma, in the Artillery Officers Basic Course.

Upon completion of Ft. Sill, I was assigned to Ft. Benning, Georgia, and then received orders for Germany with 2d Battalion, 83rd Artillery, an 8" Self-Propelled Howitzer unit tasked with providing artillery supporting fire to U.S. and NATO units protecting West Germany from a threatened or potential invasion by the Russians or the East Germans. It was there that I decided I wanted to make the Army a career. I served there for two and a half years at several different levels of command and left in late 1966 with orders for Vietnam, based on my request for reassignment to what had now become a major military escalation.

While I was in Germany, I attended a school where I was trained to be an Aerial Observer which qualified me to do target detection from the air, call in artillery strikes on an enemy and in some cases, direct air strikes when needed.

However, when I reached the U.S., my orders had been changed from

Vietnam to Ft. Leonard Wood, Missouri. When I reached Ft. Leonard Wood, I immediately volunteered again for Vietnam and several months later, received my second set of orders for Vietnam with an assignment to the 1st Infantry Division. This time, I went to Vietnam. My daughter was born just weeks before my departure for Vietnam. All the way over to Vietnam, I realized, maybe for the first time, that I had just left my wife and daughter, my parents, and my brother, whom I might never see again.

I arrived in-country in late Spring 1967, reported in to the 1st Infantry Division and was waiting briefly for more specific orders to get assigned to what I expected was an artillery unit command of some kind, based on my experience. Instead, I was advised only 16 hours in country, that we, the 1st Infantry Division, (also called the 1st I.D.) were short Infantry Captains and over on Artillery Captains and therefore I was being branch-transferred to the Infantry and assigned to a line Infantry unit.

What they failed to tell me is that I was on my way to a horrible base camp/airfield in a highly hostile village called Loc Ninh in the northern part of our area of operations ("AO") where a six-day battle was still raging. So, before I had sweated through my first set of jungle fatigues and boots, I would see more than I was prepared for, no matter how many times I envisioned it in my head. I would also experience first-hand the incredible quality and dedication and bravery of the men in my unit.

I stayed with that unit until I was told I was being assigned as the Assistant S-2, Intelligence Officer of 2nd Brigade, 1st I.D. Within a few weeks after my assignment, the S-2, my boss, was killed and I took over as Brigade S-2. As Brigade S-2, I advised the Brigade commander of intel reports of enemy movements and where we should put our resources, i.e., our three Infantry Battalions.

At that time our Brigade conducted numerous Search and Seal Operations where we would seal off a village in the middle of the night with an Infantry company based on Intel reports our team had collected from prisoner interrogations, captured documents, etc., that indicated the Viet Cong ("VC") were infiltrating or controlled the village. During these operations, we frequently captured VC from these operations or a fire fight would break out as we started to move through the village. These Search and Seal operations were done numerous times, and I would be present for all or most all of them, along with my Military Intelligence interrogation team.

In late December 1967 and early January 1968, we started to get reports of large enemy movements in our area but could never pin down any specific units or catch them. Several of our prisoners talked about a possible huge clandestine offensive, and I reported it but there wasn't much to corroborate those reports. In spite of the lack of corroboration, I passed the info to the Brigade CO at night briefings and passed it up the chain of command to Division Headquarters and on to 2nd Field Force Headquarters in Long Binh.

In early January 1968, my Brigade CO, in spite of his skepticism, ordered me to start flying missions at night to see if we could confirm any of these reported enemy movements. If we spotted movements in our AO, we could hit them with air strikes and/or artillery or bring in one of our Infantry companies to engage. During these nighttime aerial missions, I did spot some enemy units moving of various sizes, and we had some success hitting them and neutralizing their effectiveness.

When the 1968 Tet Offensive hit, in spite of that, we were caught by surprise as to the size and scope of it and the fact that they had essentially snuck by us by merging with the civilian population. We later determined that they had been dressing like peasants while they were infiltrating the cities and smuggling in weapons. The fighting that started then was the most intense of the war and was a coordinated nationwide massive attack by the North Vietnamese and the Viet Cong, the local guerillas.

In very early February 1968, my brigade was sent to prevent an attack on a water purification facility in Thu Duc, north of Saigon with a reinforced company of either the 1st or 2d Battalion of the 18th Infantry Regiment, or, the 2n Battalion of the 16th Infantry Regiment, I'm not sure which. I was sent by the Brigade CO with the unit defending as the

Brigade's eyes and ears and to call in additional artillery or air support if needed. We got hit that night but held on to the Water Plant.

The weeks went by in a blur after that, and we were involved in so many operations over the next two to three months. I can't remember every specific battle or engagement we had where I was present or where I briefed the CO, but we were enormously successful in rooting out the enemy and inflicting huge casualties on them. We were bringing in huge numbers of prisoners, to include at least one North Vietnamese Lt. Colonel whom we interrogated at length.

Overall, 2d Brigade casualties were relatively moderate considering the size of the offensive and the enemy casualties were huge. Later we heard that the enemy lost 50% of their men, a statistic I have seen verified many times since. The primary Viet Cong enemy unit we were up against since I had been there was the 273rd VC Regiment, and it was essentially decimated during the Tet Offensive of 1968. I was involved in providing intel support and operational targeting information during March and April 1968 to Task Force Ware, led by Maj Gen. Keith Ware, Division Deputy CG and later our Division CG, who was killed in a helicopter that either was shot down or crashed in September 1968.

From the start of the Tet Offensive until late April 1968, I just remember there were weeks and weeks without mail and often weeks without hot chow or even knowing what was going on except right in front of you. In late February 1968, I attended a briefing in Saigon at some old French building and remember all of the bodies all over the streets on the way there. Now as I think back over that time, I realize how numb I had become to seeing death and accepted it without even a thought. Maybe that was my way of mentally dealing with it.

By late April and early May 1968, the Tet Offensive was pretty much over, and the Brigade CO, Colonel Allen asked me what I would like to do. I told him that I missed the artillery and wanted to command an artillery unit in combat. One week later I was assigned as Battery Commander of "C" Battery, 1st Battalion, 7th Artillery Regiment. From May 1968 through August 1968, my unit participated in and supported almost every major battle and campaign in the 2d Brigade area, and we supported all of my old Infantry brothers. Charlie Battery, I am proud to say, had some of the most significant unit awards of any artillery unit in the 1st I.D.

During one of our campaigns in June and July 1968, where we were air-lifted into an area where there had been an old French house in the middle of a heavily jungled area called War Zone D, we had heavy casualties but successfully continued support for our Infantry and Armored Cav brothers in spite of our losses and the loss of one of our howitzers. The loss of the brave men in those campaigns is a memory I will never forget.

Ultimately, I was asked to take over Headquarters Battery Division Artillery and move the Division Artillery Headquarters from Phu Loi to Lai Khe with the rest of the division headquarters elements. I believe this was based on my experience at this point and the Division Artillery CO believing I had already spent many months in combat, giving me a "safe" last 30-45 day job.

My time in Vietnam shaped the rest of my life in so many ways. I went there a somewhat naïve 25-year-old and came home changed forever. I will always love and respect the men I served with and the incredible sacrifices of every service member, then and now. I will always be proud of my service in Vietnam and in the Army.

MIKE HAMER

US Marine Corps — Squad Leader; US Army — Infantry Officer
Dates of Military Service: 1956 to 2011
Unit Served with in Vietnam: B Company, 3rd Battalion, 12th Infantry Regiment, 4th Inf Div; Ops Advisor to I Corps Vietnamese Training Center
Dates Served in Vietnam: Sep 1966 to Sep 1967; Dec 1968 to Dec 1969
Highest rank: Major
Place and Date of Birth: Chicago, IL — 1938

NINE DAYS IN MAY

In Vietnam in March 1967, I took command of B Company, 3rd Battalion, 12th Infantry Regiment of the 4th Infantry Division. I will not relate the battles fought on 18-27 May 1967; for those please see the book entitled *Nine Days in May* written by Warren Wilkins. Briefly we had two companies in a perimeter about 5,000 meters from 1st Battalion 8th Infantry Regiment who had been in contact with NVA units since 18 May. We were to reinforce that unit but needed resupply on 22 May. We were attacked at 0710 and fought for five hours.

We had five artillery batteries firing for us. We even had close air support from B-52s! When the B-52 pilot called me on FM radio he asked for grid squares in which to drop his ordnance. As the NVA were dragging their dead and wounded toward the Se San River which was the Cambodian border and the safety of that country, I sent coordinates for 8 by 2 grids from 1000 meters from our location to the river, knowing the last two squares were in Cambodia which was where their huge camp was located. They always camped near water. I heard the navigator chuckle. When he plotted the target and the captain said we're supporting a unit in combat, I'm putting it right on your target. Three aircraft with full loads of 750 pound bombs! Too bad we could not go look at the damage; it had to be tremendous.

I'll tell you of a most memorable moment during the fighting. SP4 Mike Horan from Staten Island was blown out of his foxhole, probably by an RPG. He managed to roll over behind the rubble of his foxhole.

He had splinters and shrapnel wounds to his head, back, and legs and a partially collapsed lung. A medic happened by and Horan asked for morphine; the medic said he could not give it to him with a head wound. Horan said, "Aw shucks, first time in my life I could get high and he won't let me!"

After four hours of fighting, we lost 70% of our company but beat back several attempts to penetrate our perimeter. We had only a one helicopter LZ so evacuating wounded and dead took all day. The intelligence folks told us because Budda's birthday was today or tomorrow there was another regiment of NVA close by. Alpha Company was lifted to the fire base for security and C Company was brought out to continue operations. Our battalion commander brought the SGM and all the clerks and bottle washers from headquarters out and set up a perimeter for the entire battalion, which was about 240 soldiers. I was happy because the LTC was in charge; I laid down on the ground near my foxhole and did not stir for 10 hours!

Since no NVA appeared overnight on 23 May, we evacuated weapons and equipment all day. On 24 May 1967, it was raining and murky while finishing evacuations. I was told to take B Company, which had 50 soldiers, 800 meters up a hill and set 1/3 of a perimeter on an east west; C Company would come up later to fill out the other 2/3s of the perimeter. I had one lieutenant (Pestikas) and one E7 (Tharp) who positioned our foxholes on the point of the finger.

We had been there 20 minutes when we heard the mortars dropping in the tubes and knew we were under attack. It was a reinforced company of NVA that had a camp nearby and thought we knew they were there so they decided to hit us. Fortunately, they attacked right up the hill where we had our foxholes. Had they come from any other direction, the first person they would see was me!!

For a reason I cannot recall, and I've thought about it for 52 years, I had our 50-man unit bring four M60 machine guns with 3000 rounds each with us. That was devastating for the enemy and he did not penetrate our 1/3 perimeter. C Company ran the 800 meters to our position and arrived in time to knock out snipers in trees and save our butts; we were low on ammunition and probably could not have continued much longer.

B Company had 17 soldiers still standing and fighting after three days of intense combat. The American citizen soldier was at his best in Nine Days in May. I see many of them at reunions and at the Vietnam Memorial on Veterans Day every year. We are family.

Author's note: This nine-day fight was going on not far (about 25 kilometers) from where my company was located. In fact, our company, Bravo, 1-22 Infantry of 4ID was alerted to go reinforce the 1st Brigade as the battle was winding down. Fortunately, Mike and his fellow Soldiers ran the NVA off and we didn't get into that fight. I highly recommend the book, Nine Days in May by Warren Wilkins. He did extensive research on it. Three Medals of Honor were earned during that series. of major battles in May 1967. Mike Hamer is extensively quoted in the book. What Mike wrote above is just a small piece of his part in the battle.

ANN HAMER
Wife of Mike Hamer

THE STORY OF MY HERO

It was 1958 and I was a senior education major at a small college in Arkansas when I met this handsome blonde, blue-eyed freshman named Martin Hamer, later to be known as Mike by the Henderson State football team (and so to everyone else). We talked in the Student Union, walked to class together, studied in the library, and went for cocoa at a local hangout close to campus. I found out that he had joined the Marine Corps right out of high school, that his parents lived in Indiana, and that he too was an education major—coaching was his goal.

One thing led to another and by spring we talked marriage. After my graduation and teaching grade school one year in a school 75 miles from our college, we were married in June 1960. I would travel to every game during football season, and he would come with me to my folk's farm on other weekends. At a meeting with his advisor at the beginning if his junior year, Mike was told how many credits he was lacking and that was when he decided ROTC was the answer to graduating! So he then also explained to me that he would be obligated to the US Army for two years after being commissioned a second lieutenant.

OK, I thought, we would just delay our riding off into the sunset teaching school together. After Mike's graduation and Infantry School at Fort Benning, Georgia, we were stationed at Fort Ord, CA. I had visited California but never thought I'd be living there. I applied for a teaching job and got a second grade assignment with no problem. My salary tripled from the Arkansas one and we were ecstatic!

By the way, we soon found out we were in a very special part of California—Carmel, Monterey, and Pacific Grove were just minutes from post and San Francisco was one hour away. We had a great time at Fort Ord and met wonderful people. Just before Mike's obligation to the Army was coming to an end, a Colonel named Beightler came to me at a party in the officer's club one night and told me what a wonderful life he had as a career officer, and he thought Mike would make a good leader blah, blah, blah… Well, Mike and I talked it over at home and decided to stay in the Army, and the rest is history!

Our first son Michael was born in April 1965. In October that year Mike got orders for a three-year tour in Germany. We were excited to be going to a special assignment right away. Mike had a great job there—no field duty, home on weekends, and even a car to drive. We traveled in Germany a little but were saving our money for nearby countries. Then in May 1966, our lives changed once more. Mike was on the golf course when a call came for him. I loaded Michael in the car and went to find Mike (remember no cell phone days).

When Mike returned the call, the message was that he was to be reassigned to Fort Lewis for reassignment to a classified area overseas… we all knew that meant Vietnam. Mike started making arrangements for our return to the States. We first went to Arkansas to say good-bye to my parents and then to Indiana to visit with his folks where Mike left for Fort Lewis and then Vietnam. Before leaving Arkansas, we found a two bedroom duplex in the town where I began my teaching career. Michael and I flew back there to begin our 12 months without Mike.

The first month was hard because I had no mail from Mike or address to write him. After Mike's first letter I could write every day, send pictures of Michael, send favorite magazines, Louisiana hot sauce, and various food items that would travel well. This helped me feel "needed" in some way. I began watching the news about the war in Vietnam until Mike wrote and said I should not watch….same with reading the newspaper.

I grew up in Arkansas and have lots of relatives there; visiting them kept us busy and helped Michael learn about his kinfolks. We also flew to Indiana once a month to visit Mike's folks and relatives nearby. I felt that my job as an Army wife was to keep the home for our son and myself as if nothing had changed—Daddy's pictures all around us and talking about him every day. I would read the parts of Mike's letters to Michael where he would say, "kiss Michael for me." Then I would kiss him. This may not have helped a two-year-old but it sure helped me.

After twelve months, Mike finally came home. He flew into Little Rock, Arkansas where Michael and I met him. We stayed in Arkansas a few days and then drove to Indiana to see Mike's parents. Unbeknown to all of us, in May of 1967, Mike and the company he commanded was in the battle to end all battles — one that would stay with him the rest of his life — one where he was wounded but stayed with his men and kept fighting. He told his parents and me about those nine days and then asked us not to talk about it again.

Fifty years later, a book was written about this battle—a book called Nine Days in May. Martin "Mike" Hamer was my hero on the football field, my hero in the US Army, and is still my hero today.

HUBERT "HUGH" BELL

US Army – Aviator

Dates of Military Service (Active Duty and Reserves Combined): 1966 to 1997

Unit Served with in Vietnam: Co A, 25 AVN BN, 25 Inf Div (first tour); 73rd
Surveillance Airplane Co (second tour)

Dates in Vietnam: Mar 1967 to Mar 1978; Jul 1971 to May 1972

Highest Rank Held: Lt. Colonel

Place of Birth and Year: Elberton, GA – 1942

FIREFIGHTS CURE TOO MUCH CHAMPAGNE

In 1967, I served in Vietnam with Company A, 25th Aviation Battalion, 25th Infantry Division, the Little Bears. B company was the battalion's attack helicopter company, call sign "Diamond Head." In late 1967, and I don't remember the date or month, I was grounded for three days because I had logged more than 140 hours in the preceding 30 days. This happened to me almost monthly because the recurring mission which I flew was a very high-time mission; in fact, I logged 1,327 hours in UH-1D and UH1-H helicopters in approximately 11 months of flying service.

I had not been assigned any extra duties for that day of not flying. Therefore, I hopped on the Little Bear ash-and-trash mission helicopter which every day flew a pattern all around the 25th Division AO to haul people, cargo, or whatever needed to be moved. I got off at Tan Son Nhut airbase in Saigon. From Hotel 3 (the helipad) I made my way over to the Air Force officers club.

There I found about a dozen Air Force Spooky and Spectre pilots who had been on the charter flight with me from McGuire AFB in March 1967 to Bien Hoa. We took a long table in the O-club, each put about twenty dollars on the table, and asked Co, the waitress, to bring champagne until the money ran out. We proceeded to drink champagne most of the afternoon and engaged in obnoxious behavior, including shaking a bottle of champagne and spraying an Air Force brigadier general who happened to come by and refused our offer of champagne. He wisely did not confront such a large group of inebriated aviators.

Around 1700 hours I decided to go back to Cu Chi. I went out to Hotel 3 and caught the courier flight back to the base. When I arrived at the flight line, I was told to grab my gear and fly an urgent mission immediately. I protested that I was quite buzzed after drinking champagne all afternoon. If you think that excused me from the mission, you would be wrong. The answer was that I would be paired with a sober aircraft commander because there was no one else to fill the seat.

The mission was a seven-ship formation of UH-1D helicopters for an emergency extraction of an infantry company which was in contact with a VC or NVA regiment. The company was beyond the range of artillery support and night was near. If not extracted, the company was likely to be overrun and probably annihilated during the night.

There was tall grass in the LZ and unknown junk hidden in the grass. Therefore, we could not actually set the aircraft down in the LZ, but just come to a hover down in the grass and allow the infantry to clamber up on the helicopters.

We flew in a staggered trail formation to the LZ and landed tightly bunched up in the grass, while the infantry troops boarded. The troops remaining in the LZ awaiting the next lift were laying down a huge base of fire on each side of the LZ. In addition, we had UH-1C and Cobra gunships in racetrack patterns on either side of the LZ, firing 2.75 inch rockets and miniguns to suppress the enemy fire.

We made seven trips into that LZ to extract all of the infantry company. In the final two landings in the LZ, our aircraft were taking shrapnel from our friendly rocket fire. Our crew chiefs and gunners on each slick helicopter were laying down furious fire on the edges of the LZ on either side. I could see tracer fire crossing in front of the nose of my helicopter no more than six feet away. I don't know why the enemy didn't direct that fire to hit our aircraft. Maybe the story, perhaps apocryphal, that the VC/NVA were all taught to lead helicopters and other aircraft was true. Anyway, I am thankful that none of our flight was downed or seriously damaged although some took a few hits. My aircraft took no hits.

Daylight was almost gone when we departed the LZ after our last pickup. I was completely sober and had been since our first formation landed in the LZ. I don't recommend a combat helicopter extraction under heavy enemy as a cure for inebriation, but it worked for me.

I was awarded an Air Medal with V Device for valor for that mission. The aircraft commanders were nominated for Distinguished Flying Crosses for the mission.

ROBERT O. "BOB" BABCOCK

US Army — Infantry Officer
Dates of Military Service (Active Duty and Reserves Combined): 1965 to 1974
Unit Served With in Vietnam: Bravo Company, 1st Bn, 22nd Inf Regt, 4th Inf. Div.
Dates Served in Vietnam: Jul 1966 to Jul 1967
Highest Rank Held: Captain
Place of Birth: Heavener, OK — 1943

LITTLE THINGS MEAN A LOT

Major events and happenings, such as the stories in this book, are easy to recall and write about. But to get a complete picture of my year in Vietnam, there are a number of little things that influenced me that are not significant enough for a complete story but need to be mentioned in some form. Many Vietnam vets will identify with these "little things." For instance:

LETTERS FROM HOME

Probably the greatest single morale builder for any Soldier in any war are the letters he gets from his loved ones at home. My mother committed to write me every day and she steadfastly stuck to that promise. However, her letters were always read second because on most mail call days, I got a letter from my wife, and that always had top priority. Others wrote periodically and few mail calls found me without at least one letter from home.

A LETTER FROM MY GRANDDAD

Letter writing was not one of my Granddad's favorite past times, but I could count on a letter from him about every two or three months. His education had stopped at the fifth grade (born in 1890's) and his hand-writing was a barely legible scrawl, but the letters he wrote meant more to me than any I got—I could tell he was really proud of me.

PONCHO LINER

When we first got to Vietnam, we slept on a canvas sleeping bag liner with a rubberized poncho hooch over us to keep out the rain and wetness. After about six weeks in country, we were issued poncho liners—silky feeling, fast drying nylon blankets that felt so much better to snuggle up under than the rough sleeping bag liner. Unless you were sleeping on the ground and in the elements all the time, it probably would not have made the impression it did on me. But, it gave a feeling of civilization to the rugged existence we found ourselves living in.

THE PHASES OF THE MOON

As a kid, I never had paid much attention to the moon except to notice when it was full. However, the phases of the moon took on a whole new meaning in Vietnam. A full moon and the few days before and after it meant there would be light to brighten up the dark jungle at night. The Vietnam War was fought long before the perfection of the night vision goggles used by our current military.

When there was a new moon, the darkness was so intense you could hardly see your hand in front of your face under the heavy jungle canopy. And, it almost angered me to look up during the daylight hours and see the moon—that meant it would be down before the night was over and many of the long night hours would be lit only by the stars. Even today, I still appreciate the phases of the moon and think about how much a friend the moon is to an infantryman securing a defensive perimeter in the jungle, or on any hostile frontier.

A SUPPLEMENT TO C-RATIONS

C-Rations were a way of life to us—three meals a day. Unlike stateside duty where you knew you would be back to civilization within a few days, you learned to make the most of C-rations and learned how to mix and match the different meals to make them interesting and tasty.

However, it was always a great treat to have a box of warm sliced roast beef, fried chicken, or a box of fruit dropped down through the trees from a hovering Huey with our (almost) daily resupply of C-rations. We were like kids in an ice cream store when Staff Sergeant Steve Angulo and his cooks went above and beyond the call of duty to give us some variety to add to our routine C-ration meals.

COCA-COLA AND SCHLITZ

Troops in the rear always seemed to have Coca-Cola, Pepsi-Cola, Budweiser, and Schlitz, but there never seemed to be enough to make it out to the Infantrymen in the jungles of the central highlands. We learned to like RC Cola and Ballentine beer and the other off brand soft drinks and beer that made it to us. When we occasionally got a Coke or a Schlitz, even without ice, it was like heaven. (For young people reading this, Schlitz and Budweiser were the premier beers back in the 1960s).

WET TOILET PAPER, WARM BEER, AND WISE ASSES

I always said there were three things I can't stand—wet toilet paper, warm beer, and wise asses. Life as an Infantryman made me learn to tolerate the first two; I still do not like wise asses.

A METAL MIRROR

Prior to leaving for Vietnam, I bought a metal shaving mirror in the Fort Lewis PX. It would not break, was excellent for shaving, and gave me another little benefit that showed how you grasp at anything in tough times. I always carried the mirror in my jungle fatigue shirt pocket, right over my heart.

My brain told me that, just maybe, if I got shot, the shaving mirror would deflect a bullet and not let it hit my heart. I am sure a bullet would have zipped right through it but I convinced myself it was a little added protection, and I felt safer having it over my heart as I slogged through the jungle.

AN ELECTRIC RAZOR

Shaving five to seven days beard off in cold water out of a helmet with a dull razor blade was a painful experience. In a care package, my wife sent me a battery operated electric razor. It was so handy—you did not need water and I found myself shaving every two or three days. It was a nice luxury that lasted until my rucksack was dropped fifty feet through the jungle from a helicopter and broke it—so much for luxury in the jungle.

MY CAMERA

Prior to leaving Fort Lewis, I purchased a $12.95 Kodak Instamatic 104 camera with a cheap leather carrying case. I always kept it with me, carried in a third ammo pouch (two carried M16 ammo close at hand). Despite the times it got wet, it always took excellent pictures. As a result, I have a picture history of this very significant year of my life.

LETTERS AND KOOL-AID FROM MOTHER

As mentioned above, my mother wrote me every day, without fail. In an early letter home, I had mentioned the water tasted bad. From that day forward, she always included a package of presweetened Kool-Aid with my letter. After a while, I got used to the water and tired of the Kool-Aid but always found someone who was happy to have my mother's thoughtful addition to her daily letter. Mother, I really appreciated your letters and the Kool-Aid—little things mean a lot.

A VISIT FROM THE CHAPLAIN

"There are no atheists in foxholes," is a famous quote from someone. God has always been important to me. In Vietnam, He became even more so. As His messenger, Chaplain Sauer was always a welcome sight. I do not recall ever missing a service either when he came out into the field or when I was in the relative safety of the rear.

In addition to his sermons, his portable altar, his tape recorder that played church music, and hymn books carried in an ammo can, I enjoyed the time he spent talking to my men and me. When I became Executive Officer, he and I became good friends. After having not seen nor heard from him in over 25 years, I finally located him in November of 1994. We have talked on the phone, traded letters, and visited with each other at a 4th Infantry Division reunion in 1996 and again in 2002. I need to get in touch with him again.

DOUGHNUT DOLLIES

Red Cross volunteers, American women in powder blue uniform dresses, silly games to entertain the troops, sincere caring about making the lives of the GIs a little better, a bright spot when they came around with stale doughnuts is another memory that sticks with me. I never learned any of their names, never talked to them one on one except to arrange for a meeting with some of my troops, but I knew they cared and were there to help take away some of the realities of the War—little things mean a lot.

PLAYBOY

Even though it was usually very late, I eagerly anticipated the monthly issue of Playboy my wife sent me. (I am not sure she was as eager to buy it and send it as I was to receive it). I guess it gave me another view of what we were fighting for. When I became XO, we maintained a bulletin board of centerfold beauties (which was quickly turned around to reveal the normal Army bulletin board items on the other side when Chaplain Sauer came around.)

JUNGLE BOOTS AND JUNGLE FATIGUES

The most comfortable clothes I ever wore. After the first month of wearing leather jump boots and stateside fatigues, our newly issued jungle boots and fatigues were a Godsend. The boots dried out quickly, the fatigues were light weight, had pockets everywhere to store essentials, and also dried quickly after monsoon rains or wading through a jungle stream. I still have the two jungle fatigue shirts I wore in Vietnam—I still kick myself for leaving my jungle boots sitting by my cot the day I left for home.

THE ABILITY TO READ A MAP

After struggling with map reading as a ROTC cadet at Fort Riley, Kansas, it suddenly clicked with me how to read terrain features on a map while I was in Infantry school at Fort Benning, Georgia. My great map reading ability served me well throughout my Army career.

I always knew where I was so could always accurately call in artillery or air strikes, find an objective, call for resupply or medevac, and generally make myself and my people feel like I knew what I was doing. My CO, Captain Sandy Fiacco and I often debated about exactly where we were in the jungle and, more often than not, I was more accurate in my assessment than anyone.

DUST-OFF

Fortunately, we seldom had to use it, but it was a great comfort to know medical evacuation via helicopter was always on call. When the monsoon rains totally closed down flying, it significantly raised our concern factor, knowing we could not be evacuated if we got wounded. But most of the time, we could always count on a brave pilot and crew coming in to help us out when we needed it.

RUNNING INTO FRIENDS

The world is a small place and you can never totally get away from people you know. It was always a welcome sight to run into an old friend from my hometown, from college days, or from my earlier Army days. It made you feel not so alone in that far-off place.

EXCELLENT LEADERSHIP

I cannot say enough about Sandy Fiacco and Buck Ator, my two excellent company commanders. But also, I worked with other excellent leaders both above and below them—Colonel "Rawhide" Morley, Colonel Jud Miller, First Sergeant Bob MacDonald, Platoon Sergeant Frank Roath, Squad Leaders Jim Benge, Doug Muller, Al Burrell, Willie Cheatham, Bill Bukovec, and Aubrey Thomas (to name only a few of the excellent leaders in Bravo Company).

Leadership was often cited as one of the serious flaws of the American military in Vietnam—that was not the case in my experience. I worked with, around, and for excellent leaders during most of my Vietnam experience.

I did experience some leadership that, in my opinion, left quite a lot to be desired. But, there were always good leaders around to fill the holes left by the less than qualified leaders. If there is one thing that enabled so many of us in Bravo Company to come home alive, it was EXCEL-

LENT LEADERSHIP—and that is a BIG thing that means a lot, not just a little thing.

A WIFE TO DREAM ABOUT

Even though we have since divorced and gone our separate ways, the one constant dream that kept me going was knowing I had my wife at home waiting for me. Early in my tour, when I lost my billfold, I was more concerned about losing my pictures of her than about the money I had lost. I had carved her name on my rifle stock (which I often looked at as we slogged along on patrol). Dreaming about coming home to my wife let me focus on life in the world rather than always on what was happening in Vietnam.

DEROS

Date Estimated Return from Overseas—a date indelibly carved in the brain of every Soldier in Vietnam. Rather than an assignment for the duration of the war as our counterparts in World War II had, we could start pointing toward our DEROS date as a known time we would be out of Vietnam.

It has been written that because of the DEROS system, we did not have a sixteen-year war in Vietnam; we had a one year war sixteen times—and I cannot disagree with that. About the time people got good at what they were doing, they went home and new, fresh people came in to make the same mistakes the old timers had learned to avoid.

The system also kept pressure off our Congressmen to put an end to the war—if a person were expected to stay for the duration, public opinion might have been much different—and the outcome might also have been different.

But, as a twenty-three year old lieutenant, DEROS was one of those little things that meant a lot.

CHRIS NOEL

The sultry voice of Armed Forces radio. There was something about her sultry, sexy voice as she opened her show with "Hi, Love" and closed it with "Bye, Love." Whenever possible during the last half of my tour serving as XO, I was part of her nightly radio audience. And, I was more interested in the first and last minutes of the show than what was in between. She really knew how to make you think of home.

SHORT ROUND

Soldiers love dogs. Soon after we arrived in Vietnam, a small dog of obvious mixed breed was adopted by our company. I do not know who named him "Short Round," and I am sure they did not realize how we would learn to hate that reality of war. Wherever we went, Short Round went on our supply truck. He logged more than a few missions on airplanes and helicopters. He was a pet of the entire company and was loved by us all. It really hurt when he was hit and killed by a truck.

"COMBAT LOSS"

Keeping track of every piece of government issued equipment was almost impossible. With people scattered across the jungles, the fire base, the forward base camp, and Camp Enari, not to mention the people in hospitals with malaria or wounds, we always had equipment missing.

One little rule made it bearable for those of us responsible for the inventory—"combat loss."

Any time we were in direct combat, either a firefight or a mortar attack, we could claim equipment was lost in combat. It is amazing how much equipment we lost every time a shot was fired.

Without that safety valve, I would probably still be paying for equipment I was signed for. (It was embarrassing when a pistol showed up a few days after I had sworn it had taken a direct hit from a mortar round.)

THE CONSTELLATION ORION

In the southern sky is the constellation Orion. It is always there and is easy to pick out. On many a dark night in the jungle, I would look up and find "my" stars and feel a little better knowing a friend was there. I told my wife to locate Orion because I would be sending messages to her via those stars.

I can't say why human nature clings to familiar things, but the constellation Orion always had a warming and calming effect on me when I saw it. Today, I still look up in the southern sky and remember the comfort Orion gave me so long ago.

Little things mean a lot.

VANCE S. GAMMONS

U S Army- Infantry Officer/AviatorDates of military service: 1959 to 1982
Units Served with in Vietnam: 117th, 119th, 335th Assault Helicopter Co; 1st Bde, 25th Inf. Div.
Dates Served in Vietnam: Jan 1964 to Dec 1964; Jun 1966 to Oct 1967; Jun 1969 to May 1970
Highest Rank Held: Lt. Colonel
Place and Date of Birth: Cowan, TN – 1936

TUNNEL RAT

In Vietnam, especially in and around the area of Cu Chi, the Viet Cong had constructed many tunnels. Frequently our troops discovered them and in many cases, they tried to find out what was in them. The men who went in were volunteers and generally were the physically smaller troops. There are some people who think that they either were very brave or in some cases a bit thoughtless.

I was assigned to the 1st Brigade of the 25th Infantry Division in 1967 and what follows is the story of the division commander's encounter with a tunnel rat who had just emerged from a tunnel and was resting.

After some general conversation about what the man had encountered and some verbal pats on the back, the general—who was accompanied by the battalion and company commanders, asked the trooper if he could do anything to make his job easier. The trooper then told the general what it was like to have to fire the .45 caliber M1911 pistol in a very close space. He also went into great, and colorful, detail concerning the effects on his eyes, ears, and nose. He then spoke of the times that he and others had asked for a .32 caliber revolver, but kept being told that they were due to arrive at any time.

The general heard him out completely and following observation and questions concerning other things, proceeded to his helicopter. Before he climbed in, he took the time to advise the commanders that he would be very displeased if either they or anyone else in the chain of command were critical of the way the tunnel rat had appraised him of his pistol problem. That afternoon, when the division briefing was coming to a close, the general asked for the status of the request for .32 caliber pistols for the tunnel rats. A Lieutenant Colonel from the G-4 section (supply) advised that they were due at any time. The general advised him that he was to personally leave and not return until those pistols were in his possession, and he wanted to know when they were in the tunnel rats' possession. I believe they were issued the next day.

CLINTON E. DAY

US Army — Signal Corps Officer

Dates of Military Service (Active Duty and Reserves Combined): 1965 to 1971

Units Served with in Vietnam: MACV, 36th Combat Signal Battalion and 972nd Sig Bn (attached)

Dates in Vietnam: Feb 1966 to Feb 1967; Nov 1970 to May 1971

Highest Rank Held: Captain (O3)

Place of Birth: San Francisco, CA — 1942

AIR COURIER EXPERIENCES

When people ask about my 19 months in Vietnam as a U. S. Army Signal Officer, I invariably tell them about two airplane flights to deliver crypto supplies. Early in the war our crypto logistics were pretty crude. A little, eleven-man unit then was eventually expanded into a garrison of a hundred at MACV headquarters by war's end.

Most codes were classified Top Secret, and required either a Warrant or Commissioned Officer to act as courier. As a result, another and I Lieutenant shared most immediate resupply deliveries. Many times I'd get a call in the middle of the night to gather codes or equipment and head out to Tan Son Nhut to fly "up-country."

Our orders were high priority, and we could hop a ride on any aircraft controlled by the USA. I remember flying in a one-engine State Department plane, the 2nd Signal Group twin-engine, and all manner of U. S. Air Force aircraft. Usually it was either a C-123 or a C-130, depending on destination (and the length of the airfield).

Both episodes made strong impressions on me. The first was a return flight from Pleiku to Saigon after resupply to the ROK (Republic of Korea) Division. Things near the airfield were "hot" due to VC activity, and only the C-130 being used was able to climb out fast and high. They dropped the rear gate, put me on, and I "belted-in" near the door. After take-off, we rocketed straight up. First indication of something different with the cargo was maximum air-conditioning so strong the air was condensing from ceiling vents.

Leveling out, I noticed there was no cargo. I saw only one Air Force crew at the end of a large C-130 bay. Then, I looked down on the floor, and saw a sight I think of every Memorial Day (and have recounted to audiences). The load was wall-to-wall body bags, each containing a very young GI Killed In Action. I could have sworn there were seventy or more, and I was shaken by their numbers and the finality of life. I eventually walked to the base of the flight deck (about a three-foot ladder up), and asked to sit in the cockpit on a jump seat. To this day that flight back from Pleiku is my most vivid memory of Vietnam. So many boys ended their lives too soon, never to grow up, marry, have a career and family.

The second experience was another last minute hop from Tan Son Nhut. Arriving one morning, MAC (Military Airlift Command) brought a C-123 back from the end of the runway. It taxied up, lowered the tailgate, and I ran on. These aircraft are configured for paratroopers so we sat facing each other (versus forward). That particular week in a tent somewhere I had seen the cover of Sport Magazine with Brooks Robinson and Frank Robinson, both MLB All-Stars, on the cover.

Well, I sat-down, looked-up, and realized I was facing both of them. Next to them was Joe Torre, Harmon Killebrew, and not far away, Hank Aaron. Mel Allen was at the end of the row, and, if you know baseball, these were the All-Stars of 1966 on a USO flight up to Da Nang to visit and entertain troops. We got into some weather off the coast, and many of us (including me) became "white-knuckle" flyers clutching the straps. But, not the most famous player in the group. He was standing with a headset on, telling jokes to the crew and laughing like heck.

So, when you hear Stan Musial was "the Man," he really was! Pure relaxation in bumpy weather and ingratiating all the airmen. He went on to enter the Baseball Hall of Fame and received a Medal of Freedom from President Obama before the end of life at age 92. Mel Allen was the escort and Master of Ceremonies.

Now fast forward to a family Thanksgiving Dinner about fifteen years after my Vietnam service. My nephew, also named Clint, was being told about a one-sheet list of autographs of famous baseball players sent home to him the week he was born (1966). Apparently, he had been too young to remember, and his mother told him it was in his baby book. He left with it that night, and it now resides in a bank safety deposit box. While I would be inquiring as to its value with a sports memorabilia expert, my nephew has plans to bury it with other time-related items in a time capsule!

As proof of this historic flight, here is a copy of the autograph sheet written on the back of a Month-at-a-Glance calendar in my Army-issued briefcase with (typewritten note added later in the week):

JAMES DAVID ELY

US Army — Signal Corps Officer
Dates of Military Service (Active Duty and Reserves Combined): 1966 to 1972
Unit Served with in Vietnam: 69th Signal Battalion, Tan Son Nhut Air Base
Dates in Vietnam: Sep 1967 to Aug 1968
Highest Rank Held: Captain
Place of Birth and Year: Caledonia, NY — 1945

KEEP YOUR ASS DOWN AND DO WHAT YOU'RE TOLD

What goes through the mind of a 22-year old from a farm town in Western New York on a 21- hour flight to an unpopular war, halfway around the world, in a country called Vietnam? Needless to say, I had a myriad of thoughts playing in my head, caused largely by fear of the unknown. The Army had trained me to be a Microwave Radio Officer, but I never dreamed that much of General Abrams' communication would be my responsibility.

For 11 months, I operated an Army Signal Site at the end of the Tan Son Nhut runway called Site "Octopus." Upon my arrival in-country, I had 130 men working with me. Over the course of 11 months, our personnel strength was reduced to 30 with no change in our mission. Needless to say, we all worked very hard, often to the point of exhaustion.

Some memories of my year in Vietnam:

1. Experiencing the fear of the Tet Offensive where no one in Saigon was safe.
2. The fun of singing every Saturday night with an all-girl Philippines band.
3. The joy of playing Santa Claus to 98 orphans.
4. Building a new signal site from the ground up, while understaffed, yet keeping our radios up and running. This project saved the government thousands of dollars.
5. Being taught and role-modeled by my First Sergeant, Walter C. Camp (Top). He gave me this advice upon my arrival: "Lieutenant Ely, Sir—Sign everything I put in front of you and keep your ass down and I will get you through this year." I followed his instructions and he kept his word.

As I get older and reflect, I realize how "special" it was when my father, a Sergeant in WWII's Battle of the Bulge, met me at the Rochester, NY airport. I made the rank of Captain while in the reserves and was later awarded a Bronze Star for Meritorious Service which has become much more meaningful to me as I get older. I am 74.

As I reflect on my life, Vietnam played a very important role in making me the man I am today.

I am proud to have done my part and served in the Vietnam War! I love to hear the words, "Welcome Home!"

BENJAMIN H. "HAM" MCDONALD

US Army — Infantry Officer

Dates of Military Service: 1963 to 1968

Unit Served with in Vietnam: MACV Team 1, Hoi An, RVN — Advisor to 2/51st ARVN Battalion

Dates in Vietnam: Feb 1967 to Feb 1968

Highest Rank Held: Captain

Place of Birth and Year: Atlanta, GA

A LIFE CHANGING YEAR AS AN ARMY ADVISOR — ORDERS FOR VIETNAM

In July of 1966, I received orders for Vietnam. I was to be assigned to the US Army Military Command (MACV). At that time I was serving as Company Commander of C Company, 2nd Battalion, 13th Infantry, 3rd Brigade, 8th Infantry Division based in Coleman Barracks, located in Mannheim, Germany.

ORDERS FOR JOHN F. KENNEDY WARFARE CENTER

I returned to my home in Lithonia, Georgia, and dropped off my wife and daughter to stay with her parents while I attended the John F. Kennedy Warfare Center and School at Fort Bragg, North Carolina. This school's purpose was to train and educate the US Army Civil Affairs, Psychological Operations and Special Forces by providing training and education, developing doctrine, integrating force-development capability. The school had been expanded to train officers who were to be assigned as military advisors with Vietnamese army units and civilian agencies.

VIETNAMESE LANGUAGE TRAINING

After completing my training at Fort Bragg, I received orders to attend the Defense Language Institute which was located in Monterey, California. As fate would have it, the Monterey school was full and I was reassigned to a new training center at Biggs Field, Fort Bliss, Texas (The Armpit of the World). I returned to Lithonia and picked up my wife and daughter for our drive from Lithonia, GA to El Paso, Texas. More than 20,000 service personnel studied Vietnamese through the DLI's programs, many taking a special twelve-week military adviser "survival" course.

The Vietnamese language is a very difficult language to learn since it is a tonal language, and one word can have several meanings base on the inflection of how it is said. My language instructor was from the Saigon area in the southern section of Vietnam. I was assigned to Advisor Team 1 in Hoi An located in I Corps, in the northern part of Vietnam which has a different dialect. This would be like learning English in south Georgia and being assigned in Maine.

OFF TO VIETNAM

After language school, I drove with my family back to Lithonia, GA where I left my wife and daughter to live with her parents while I was in Vietnam. After a few days leave, I said goodbye to my family and friends and flew from Atlanta to the Army and Air Force Base at Fort Dix, NJ for transit to Vietnam.

The next day I boarded an Air Force C-141 for Vietnam. We had a refueling stop in Anchorage, Alaska. We were wearing our summer uniform and it was brutally cold. After the refueling stop we flew to Yakota Air Base, Japan, for another refueling stop. Again there was snow on the ground, and we froze to death while the plane was being refueled. Our next stop was Tan Son Nhut Airport in Saigon. I will always remember walking off that plane into 100 degrees heat and humidity like a wet blanket hitting me in the face.

Upon our arrival, we were transported to the Military Assistance Command Vietnam (MACV) headquarters. At this location we were given our unit assignments, issued our uniforms and weapons and equipment, and given a general orientation.

During my few days in Saigon, I was joined by my first cousin, LTC Wheeler Davidson. LTC Davidson was a MACV advisor to a Vietnamese District located in IV Corps, the Vietnamese Delta. We sat on the rooftop bar of the Rex Hotel drinking a Ba Moui Ba (Beer 33) beer and watched the war being fought around Saigon. Artillery, flares, planes bombing targets, tracer fire was the introduction to the war I will always remember.

After my MACV assignment, I was transported to Tan Son Nhut for an Air Force C-130 ride to Da Nang airfield some 500 miles north of Saigon. Once in Da Nang I was met by the MACV Area Coordinator and taken to the MACV Officer Temporary Quarters in the Palace Hotel. The next day I received a briefing at the I Corps MACV Headquarters which was co-located with the ARVN 1st Division HQs. Later that day I was flown by Huey helicopter to Hoi An, the 51st ARVN Regiment Headquarters.

ADVISOR TEAM UNIT ASSIGNMENT

Arriving at the 51st ARVN Regimental Headquarters, I was briefed by the Senior Regimental Advisor and his team and assigned to the advisor team for the 2nd Battalion, 51st Regiment. I met my three other unit advisors, CPT Norman Lemueix, Australian WO Lofty Moran, and Army SGT Will Clark. The Regimental Advisor briefed us on the planned search and destroy mission we were to be given for units of the Viet Cong (VC) and North Vietnamese Army (NVA) units operating in our TOA. Quang Nam Province was a hot bed of enemy activity, and numerous battles between the Marines and ARVN had recently taken place.

My Battalion, 2/51st ARVN, was assigned to sweep and clear the TAO between Hoi An and the Da Nang airfield. The area consisted mainly of rice growing fields with hamlets spread throughout. The VC and NVA would use the area as a food supply for their units locally and those based in the mountains a few miles inland. The 2/51st was one of the best ARVN fighting units in I Corps, consisting of some 600 men.

As such, it was tasked with some critical missions against the VC and NVA in the I Corps TAO.

A DAY TO REMEMBER

In February, after a hard day of searching several villages, we had limited contact with the exception of VC snipers who would take our men under fire and then disappear in to tunnels which were almost impossible to detect. I did have a near-death experience as we cleared one village. Following my Vietnamese counterpart down a trail leading into this village, I could tell the trail had seen limited use. I stepped on something.

I reached down and discovered the edge of a large piece of fabric covered in dirt. As I pulled back the covering, I found a pit about a foot deep with two strands of communication wire at the bottom. I alerted Sergeant Clark as to what I had found. He asked what I thought it was, and I told him it looked like an antipersonnel mine. Sergeant Clark came over and looked at the mine and stated that he had tracers in his carbine and thought he could set the device off. We backed up to what we felt was safe distance. Sergeant Clark then fired into the hole.

The explosion was unbelievable for what we thought was an antipersonnel mine. It turned out to be a 105 mm artillery round that blew a hole in the ground some four feet in diameter. Shrapnel flew by all of us and fortunately none of our advisor team was hit. I realized how close to death I had come that day and thanked the Lord for his saving grace.

DA NANG AIRFIELD ATTACKED

After clearing two more villages, we stopped for the night and established a defensive perimeter. Since the ARVN troops were providing our team security, that required the four advisors to normally rotate being on alert and remain awake until 2-3 AM. If our unit had not been hit by the enemy by that hour, we felt pretty safe. I had just laid down and hit the floor of a pagoda in a Vietnamese cemetery where I hoped to get a couple of hours 'sleep. I was awakened by huge explosions near the Da Nang airfield. My first impression was that the North Vietnamese Air Force had come south and was bombing the Da Nang Airfield. We could see the fires and continuous explosions for hours.

Our ARVN unit immediately went on alert, expecting to be hit at any minute. The next morning we learned from LTC Beasom, the 51st Regiment Senior Advisor, that the Da Nang airfield had been hit by 120 rounds of 140 mm rockets launched not far from our current location. There were fires started from secondary blasts by bombs and fuel. I later learned that the aircraft destroyed included F-4 Phantoms and C-130s. The Da Nang base suffered eight killed, 175 wounded, ten aircraft destroyed and 49 aircraft damaged.

The next morning the 2/51st Battalion Commander was given the mission to join with the US Marines to conduct a joint search and destroy mission to find the location of the rockets and destroy any enemy units remaining in that area. Since I spoke Vietnamese better than the other members of my team, I was always assigned on the flank between the Marines and my Vietnamese unit. In preparation of our mission, the 2/51st was assigned to sweep the South side of the Song Yen River and the Marines the North side.

The MACV Advisor Team mission was to make sure that a joint operation with US Forces was coordinated. The VC and NVA would often position themselves between the US and Vietnamese Units and fire a few sniper rounds in each unit's direction and we would end up with a major fire fight between friendly units. The Marines did not have a great deal of respect for the ARVN units, but I would put my 2/51st ARVN Battalion up against many US Marine units.

As the 2/51st maneuvered to sweep and clear the south bank of the Song Yen River, we came under fire by the Marine units sweeping the north bank of the same river. I could say the Marines attacked their side and we started receiving 81 mm mortar fire on our side of the river. My Vietnamese units started running by me away from the river shouting, "Marines shoot... Marines shoot." It took a few minutes to stop the incoming friendly fire and we continued the mission.

Later in the afternoon, we found the location where the VC/NVA had launched the rockets. The VC/NVA had crudely dug trenches in the river bank but those trenches were very effective. We also located several rockets that had not been fired. We failed to locate any VC/NVA units in the area and assumed they had fled back into the mountains to the West. The Nha Trang Valley in Quang Nam Province was a hot bed of both Viet Cong and NVA activity, and I participated in numerous operations in this area during the remainder of my assignment to the 2/51st ARVN Battalion.

From 1954 to 1973, more than 100,000 members of the U.S. Military were advisors in Vietnam. Of these, 66,399 were combat advisors. Eleven were awarded the Medal of Honor, 378 were killed, and 1,393 were wounded.

MILTON JONES

US Marine Corps — Telecommunications Cryptographic Technician
Dates of Military Service: 1964 to 1970
Unit Served with in Vietnam: HQ Co., Regimental Landing Team 26 (Forward), 3rd Marine Div.
Dates in Vietnam: Aug 1967 to Dec 1967
Highest Rank Held: Sergeant
Place of Birth and Year: Augusta, GA — 1946

ORDERS TO WESTPAC!

I joined the Marine Corps in the spring of 1964 during the Cold War, a time of relative peace. There was, however, this little-known far-away country called Vietnam where a civil war was heating up. Like most Americans at the time, I expected "John Wayne and the Green Berets" to quickly quell this little flare-up and all would be well.

After Boot Camp, Infantry Training, and a year-long series of Communications-Electronics technical schools, I was assigned to Marine Aircraft Group 31 at Marine Corps Air Station, Beaufort, SC. I'm an experienced professional in my field with a broad array of skills. Life is good! I'm on the East Coast, far away from California where many, many Marines were now being sent to Vietnam.

Then it happened. I got orders to WestPac. Oh, Damn! After readiness training at Camp Pendleton, my orders were to report to the Duty Officer, Fleet Marine Forces Pacific for assignment in the Western Pacific (WestPac). To all Marines, WestPac meant 13 months in Vietnam!

I had absolutely no desire to go to Vietnam. But hey! I'm a Marine. I go where the Big Green Machine says go. Off to Camp Pendleton I go for pre-deployment training. Interestingly, I almost always traveled alone from one assignment to another, due mainly to my occupational specialty being fairly scarce (telecommunications cryptographic repair technician).

In late November of 1966, I boarded ship with thousands of other Marines, none of whom I was actually traveling with. We sailed on board an old rust bucket, USNS General Leroy Eltinge, an old World War II Liberty ship that had been re-purposed, and now operated by the Merchant Marines. We were packed into this ship like sardines; the chow was not very good and not nearly enough. We were at sea for roughly a month and arrived at Okinawa on Christmas Eve.

On Okinawa, I'm at Camp Hansen with an organization called Provisional Service Battalion. Now the way I get there is, we're at sea with nothing to do and I'm thinking, "I got to find a way off this ship before we get to Vietnam." Then, I ran into John Ivey, my classmate since junior high school. Literally bumped into each other on deck, neither having known the other was in the Corps, although we'd already been in a couple of years. We were elated to see each other, but after our voyage I never saw him again until some years after we both got out. Thank God, we made it back. I repeatedly told John and others, "Man, I got to find a way off this ship before it gets to Vietnam." I'm just being candid. I love the Corps, but I really did not want to go to Vietnam.

Then I met this Captain, in a card game I think, and I heard him saying he knew somebody in an outfit on Okinawa and that's where he was going. "Hmmm," I thought, "It would be good to know this guy." Turns out he was the Brig Warden while on ship. This was just temporary duty he'd been assigned. We had about a dozen Marines who were over the hill or in some other trouble as we prepared to sail. The Shore Patrol picked them up and dumped them into the brig. I chatted with this Captain, asking him about the prisoners. He said, "Oh, they were just fighting or over the hill; just blowing off steam. They'll be held in the can until disembarkation." I asked, "What do they do?" He said, "Well, I take them out for PT and work details around the ship." I said, "You need some help?" I went to work for him as assistant brig warden.

I march these guys out to PT and work details, which is good 'cause we thought we'd starve on this ship. Periodically, I took them to clean the crews' mess (good duty). The ship's crew were eating like kings—ham, prime rib, roast duck, whatever—and we helped ourselves. As I talked more with the Captain, I learned about this Provisional Service Battalion (PSB). I may have ended up there anyway, but that's exactly where I went when we docked at Okinawa.

It turns out that PSB was part of the 9th Marine Amphibious Brigade (MAB). The 9th MAB was responsible for all Marine troops and assets in the Western Pacific outside the country of Vietnam. Meanwhile, all Marine troops and assets "in country" reported up to III MAF—Third Marine Amphibious Force—the Marine component of the joint Military Assistance Command—Vietnam (MAC-V). Part of 9th MAB's responsibility was to outfit the special landing forces... the reinforced battalions which, collectively, were Regimental Landing Team-26 (RLT-26) that I ultimately ended up with in Vietnam, but I didn't know this at the time.

RLT-26 (Rear) operated from Camp Schwab, Okinawa. PSB provided materiel, equipment, maintenance, and other support for those floating battalions. They would go ashore in Vietnam via amphibious landings or helicopter assault. They may be ashore a day, week, or more. Once ashore, they often would be chopped up, shot up, and eventually, what was left of them would be pulled out. Then they would sail to R&R in some exotic location—Manila, Bangkok, Hong Kong, and the like. By then many of their tours would be up.

Meantime, on Okinawa, replacement BLTs were being spun-up.

There were constantly two of these BLTs—Special Landing Force-A and Special Landing Force-B—with different Marine battalions rotating through the process of beefing up, gearing up, training up, and embarkation. We were continually preparing for the next set of battalions. As we continuously outfitted these BLTs, we also repaired, refurbished, or replaced equipment that might have been salvaged from earlier groups.

A lot of work, but great work! This was our "contribution to the cause." Okinawa was outstanding! Great for partying; great for work; great for extra assignments. I could have stayed there, but eventually got orders to RLT-26 Headquarters (Forward), "in country"—Vietnam—just where I didn't want to go. Oops!

I was doing a great job on Okinawa, and loving it! I just worked and worked. I found stuff to do beyond my regular assignment. I took a part-time job working in the Camp Hansen Enlisted Club—fondly referred to as the "Animal Pit," because 95% of all Marines going to and coming from Vietnam processed through the transit area adjacent to the Club. Regularly, the Animal Pit closed with a Western-style saloon brawl underway. I was in my zone and doing great stuff.

Finally, I was promoted to Corporal. I had not made any rank back stateside and was still a Lance Corporal—(E3). Amidst all this, I completed NCO School. An inter-service school including Army, Air Force, and Marines. I finished first in the class of 57. Considering all these other extra duty activities I'd taken on, I was awarded a meritorious mast and a meritorious promotion to Sergeant with only three months as Corporal ("... *his performance was more than exemplary and reflected most favorably on the Marine Corps.*") I was jubilant, but didn't realize, along with promotion to Sergeant came orders for Vietnam. Saddle up!

At Kadena Airbase, I boarded a Continental Airlines charter with flight attendants and the whole kit and caboodle. I'm one of very few people on-board with full gear on. Flak jacket on. Helmet under my seat. Everything except a weapon. People are looking at me like, "Who is this character?" I'm just sitting there, dumb and numb. We landed at Da Nang. Within hours after we land, there's a little incoming into the air field. Not much, but enough to cause people to go diving into bunkers. After that, I felt pretty good about all that stuff I had on.

I'm at Da Nang and *(as I talk with people at recent reunions)* I find this is typical. Like me, many Marines went to war individually, not as a unit. This was one of the weird things about the Vietnam experience. I report to the Marine Duty NCO. He looks at my orders and says, "you're up at Khe Sanh." I said, "Where is that?" "About as far North as you want to go," he said. I asked, "How do you get there?" "Probably get a rough rider." My internal alarm went off; ding, ding, ding!! I'm not too swift, but I know I don't want to be in a convoy going into "Indian Country."

Well, you're left to your own resourcefulness to get to your unit. You've got to find them and get to them. So, I'm kind of bumping around, asking people, "Where are you going, where are you headed?" "I'm headed back to the world. I'm getting out of here," they might say. Eventually, I asked someone, "Where are you going?" "I'm going to Phu Bai and I know some people up there. There's a chopper that goes to Khe Sanh from there. Get yourself to Phu Bai." Sure enough, I hop a chopper with him, and get to Phu Bai, just outside the old imperial capital of Hue.

At Phu Bai, lo and behold! Here is my next door neighbor from back home! I'd heard that he'd joined the service. I had no idea where he was. Apparently, as many as half a dozen guys from my neighborhood joined the Corps after me. Frank Richard Jordan, my home-town neighbor was already there in Vietnam. Also, there was John Kelly with whom I spent a lot of time together in San Diego in Tech School. John was a radio technician. I was a telephone, teletype, and crypto tech. Richard and John hosted me for several days. They were living like kings in Phu Bai and took good care of me. Made sure I knew where the good chow and bunkers were. Then I boarded a chopper into Dong Ha and eventually, worked my way over to Khe Sanh on another chopper.

On arrival at Khe Sanh things were fairly quiet. I knew about the First Battle of Khe Sanh, the Hill Fights. Hills 881 South and North, Hills 861, 861A, and 951. All were firefights that we heard about back on Okinawa; this area had been hot. I now realized these hills were all higher altitudes overlooking Khe Sanh Combat Base. It had been bad up here for several months, but much quieter now. Thankfully, it remained quieter throughout my tour. We'd get some incoming here and there, but not sustained. I just worked and worked. Thank God, our equipment and work area was in a bunker. We were located adjacent to the command center, the combat operations center.

I left Khe Sanh on Christmas Eve, 1967, headed back to Okinawa and, about 10 days later, back Stateside. Two weeks after I returned home, the Siege of Khe Sanh began. Suddenly, I had (and still have) conflicting emotions about my departure—joyful to be out of there, but ashamed that I left my Marine brothers when the going got rough.

I have very little memory of my time at Khe Sanh—not even the names or faces of those I served alongside or my immediate superiors. However, I cannot forget our commander, Col. David Lownds. A real hero. Excellent leader! "John Wayne" with a big red mustache. I had my 21st birthday at Khe Sanh, but don't remember it either.

Decades later, I received a copy of my service record book. In my NCO fitness reports by two different senior officers at Khe Sanh (neither of whom I remember), my overall "General Value to the Service" was rated as "Excellent." Also, they each recorded, "Considering the possible requirements of service in war, [he] would be glad to have this non-commissioned officer (Milton Jones) under [his] command."

I got orders to WestPac, but I really didn't want to go.

VANCE S. GAMMONS

U S Army- Infantry Officer/AviatorDates of military service: 1959 to 1982
Units Served with in Vietnam: 117th, 119th, 335th Assault Helicopter Co; 1st
Bde, 25th Inf. Div.
Dates Served in Vietnam: Jan 1964 to Dec 1964; Jun 1966 to Oct 1967; Jun
1969 to May 1970
Highest Rank Held: Lt. Colonel
Place and Date of Birth: Cowen, TN – 1936

SMOKE FILLED COCKPIT

I was an army aviator and this story took place in the III Corp area of
then South Vietnam in 1967. It is about flying a UH-1B helicopter gunship when the cockpit began to fill with smoke.

Our assault helicopter company was comprised of two slick platoons
of eight UH-1D's and one platoon of eight gunships, UH-!B's. Each
chopper had a crew of four—aircraft commander, pilot, crew chief, and
gunner. The crew chief and gunner had hand held machine guns when
we were in flight, while the gunships had their machine guns fixed but
movable. The gunships also had rockets or a 40 MM gun rig, plus two
fixed mounted machine guns on each side that were hydraulically operated by the pilot.

We had a flight with ten slicks and four guns to insert some 173d
Airborne Brigade troops into an operational area and were on our way
back to our base just NE of Bien Hoa, flying about 2000 feet above the
ground—just out of small arms range—when the cabin began to fill
with smoke.

Each of us wore a flight helmet so we had inner aircraft communications; however, in this case I could not contact either the crew chief
or gunner. I was the aircraft commander so I immediately advised the
company commander of our situation and though my gauges did not indicate a problem, that I was going to land immediately and keep another
gunship for high cover.

I flew in a slip—slightly sideways—to keep the smoke from filling the front of the cabin. That worked until we were near the ground; however, the place that I picked to land had six to eight feet thick grass and some trees. Our landing was going to have to be at zero ground speed so when our skids were in the grass, our cabin filled with thick smoke and we went down vertically to good solid ground contact.

I'm securing the controls and was shutting off the engine, and I looked to my left in time to see the backside of my pilot as he exited the aircraft before I could get my door open.

He had shut off all necessary switches in record time and cleared out. He had previously been burned in a crash and spent a few months in the hospital.

Our standard procedure when flying was for the crew chief or gunner to throw a white smoke grenade when we were receiving ground fire so that others could avoid or suppress the area. Since ours were holding the machine gun with both hands they usually hung the grenade—by the grenade pin ring—on the machine gun handle.

The gunners grenade pin had vibrated or been knocked loose, and the live smoke grenade had fallen into the rather large box of machine gun ammunition in front of the gunner. In his haste, and that of the crew chief, to get the grenade out of the box before the heat set the ammunition off, both had disconnected their communication cords.

When we found this out we immediately notified the company commander then checked the aircraft. In doing so noticed that we had missed a rather large tree stump by about two feet, then cranked up and proceeded to join our covering gunship and went home to the 'old corral'.

The aviation company was the 335th 'COWBOYS' and the gun platoon was the 'FALCONS'.

MAJ. GEN. DAVID R. BOCKEL, SR. (FORMER 2ND LIEUTENANT)

US Army—Materiel Readiness Officer

Dates of Military Service: 1966 to 2003 (37 Years)

Unit Served with in Vietnam: Co A, 25th Supply and Trans Bn, 25th Inf

Div — Dec 1966 to Jun 1967; Co B, 7th Combat Spt Bn, 199th Light Inf
Bde — Jun 1967 to Dec 1967

Dates in Vietnam: Jan 1967 to Dec 1967

Highest Rank Held: Major General

Place of Birth and Year: Dallas, Texas — 1944

DOIN' THE HOOCHEE CU CHI

I was assigned to the 25th Supply and Transportation Battalion in the 25th Infantry Division at the end of 1966 and reported in to the Division Replacement Battalion on January 4, 1967. The total extent of my military experience was four months of training to be a Quartermaster officer. The experiences in the 25th were nothing like I had prepared for. So, I will take this opportunity to tell you about some of my second lieutenant experiences after I initially arrived "in country."

"Hoochee Cu Chi" — It didn't take long to learn the terminology. I learned that the tent I slept in was known as a "hooch." Not too long after I arrived I made a little sign to go on the front of the tent. It was now the "Hoochee Cu Chi." The good news after arriving was that several of the lieutenants that I served with at Fort Lee were also in my new unit, Company A, 25th Supply and Transportation Battalion. We were supply officers, except for me. We had more lieutenants than were authorized so I suddenly became a "motor officer," even though we weren't authorized one and there were two very capable maintenance NCOs running the motor pool.

"Do You Want Me To Write Home For You?" — My company commander and I didn't get along very well. I'll get to that in a minute. The new lieutenants were given the responsibility of being the staff duty officer. This entailed finishing the duty day and going immediately into the tactical operations center (TOC) for an all-night duty which entailed monitoring communications, but also riding along the bunker line every hour on the hour and checking to make sure the soldiers in the bunkers were awake. If not, we had to wake them up. This was an hourly event which didn't end until the next duty day started around 0600. Then we were to go directly to our regular jobs until 1800.

The battalion S-2/3 saw that the lieutenants were wearing out. So, to start a new program, I was told that after completing my all-nighter, I could go to my hooch and sack out for a few hours. That worked great until around 0700 when my company commander yelled in my ear, "LIEUTENANT! What in the hell are you doing in the rack?" I told him that the S-2/3, a major, had told me that when we finished our tour as staff duty officer, we could sleep for a couple of hours since we had already been up more than 24 hours. In the spirit of understanding, my CO then said, "YOU DON'T WORK FOR MAJOR JOHNSON. YOU WORK FOR ME! NOW GET YOUR ASS UP. YOU'RE GOING TO THE FIELD TO INSPECT SUPPLY STATUS OF UNITS DOWN IN THE DELTA!"

In the evening after the duty day had ended, several of us would congregate and drink a beer or two. One grizzled old maintenance warrant officer from the truck company heard of my conversation with my company commander, and he cautioned me, "You're going to have to take the C&C chopper down to the Mekong Delta to visit those units. We've been having some trouble with those Chinooks you'll be flying in. They haven't determined if they are being shot down or if they are just shaking themselves apart. Let me know if you want me to write a letter home for you if something happens." That was when I learned that regardless of what I had been told, a warrant officer really doesn't have to acknowledge the rank of a lieutenant, particularly a brand new one. They are there to harass them.

So, the next day I reported to the helicopter pad, and there was the Chinook. I got in and positioned myself on the straps between the door gunners where I could look out their doors to see if we were being shot at and also at the instrument panels in the cockpit where I could see how high we are and how fast we were flying.

I was waiting for the horrible shaking gunfire and we did shake a bit, but I didn't see anything fall off or any bullets coming through the floor. Later on my adventure, I stood on a crossroads waiting for my next ride.

There was a bunker just off the LZ crossroads with a couple of soldiers inside. Suddenly I heard a shot and rushed to get in the bunker. The soldiers were laughing their asses off. I was the joke they were laughing at. Shortly after that the next C&C chopper arrived. It was a "slick" and right after it touched down, the door gunner waved for me to get in. I frankly couldn't see how to affix the seat belt when the Huey went airborne and immediately made a sharp turn away. Door open. No seat belt. Luckily, I didn't fall out! So much for my first adventures in helicopters.

"Y'all Get Hit Much?" — Back in the base camp at Cu Chi, I was getting used to the routine of being an excess lieutenant in a position that was designated for a sergeant first class. Then one day after about two months, we were notified that some of us would be going to a forward support base near Tay Ninh to support "Operation Junction City." I was ready to get out of Cu Chi. And Junction City was the biggest operation yet in the war. The 25th was the lead organization with two of its brigades, a brigade from the 4th Division (which later was swapped with the 4th Division up North for the real 3rd Brigade of the 25th), the 196th Light Infantry Brigade, the 173rd Airborne Brigade, a unit from the Australian Army, and a unit of South Vietnamese Rangers.

Believe it or not, I was happy to have a break away from the base camp in Cu Chi and my company commander. Speaking of whom, he actually came out to the FSB because he had to pay me IN PERSON! He had never been outside the base camp in Cu Chi. He was due in a couple of weeks to rotate out and go home, and he didn't relish the idea of going into a live fire area to pay that lieutenant he didn't much care for. He got off the helicopter dressed in full combat gear, a strange site on a Quartermaster officer. I met him as he got off the helicopter, and his first question was, "Do y'all get hit much?" Naturally I said, "Oh yes, sir. All the time." Except we had only been hit once and it wasn't anywhere close to my tent.

"Combat Infusion Program" — Back in 1965 and 1966 many of the divisions, separate brigades, and other organizations arrived with their full complement of soldiers on a specific date. That meant that after 12 months all the personnel would rotate home at the same time. Somebody had the foresight to see this coming. So they invented the "Combat Infusion Program." What this did was transfer a number of soldiers and officers from one organization with a specific DEROS ("Date Eligible to Return from Overseas") to another organization with a much later DEROS. The 25th was going to have a major turnover in the Spring of 1967. So, to even things out, some of us were transferred to other units.

Since I was excess to my unit, I was scheduled to transfer to the 199th Light Infantry Brigade, which had arrived in-country the previous September. I didn't know anything about the 199th, but I did know that I was designated to report to them in June on the same date as my scheduled R&R and meeting my family in Hawaii. My R&R was cancelled. PANIC! I was told that I could still have my R&R if I could find someone who would give me theirs! Enter SP4 Tanaguchi from the 25th Maintenance Battalion. I was advised that he was from Honolulu and might not need to take his R&R back home right away. What a nice soldier he was. He was more than happy to let me have his R&R, and he would put in for another later. So all was well, and I made it to Hawaii as scheduled.

I had yet to spend any time with the 199th so I was fearful of what I might be getting into. So I spent most of the six days worrying! If I had only known! My last six months spent in Long Binh and Saigon as a "Material Readiness Expediter" was a low-stress job, and I made a number of good friends.* And that was how I spent my last six months in-country.

I have many, many more stories from my year in-country in both organizations. And I'll quote a disc jockey from WQXI Radio in the late 1970's who was also a Vietnam Veteran during the same time as I, by the name of Gary McKee. He used to do his program every Veteran's Day devoted to all of us Vietnam Veterans. I'll never forget a quote he made one Veteran's Day: "You could not give me a million dollars for the experience of serving in Vietnam. And you couldn't give me ten million to go back." Well said.

"I DON'T BELIEVE IN COINCIDENCES"

One of the good friends I had when I served in the 199th was a young man by the name of Second Lieutenant Timothy Maude. Both of us had the same attitudes regarding our Vietnam experiences. Tim was an OCS graduate, and I'm not sure he was yet 20 years old. Tim and a couple of other friends came to see me at the 90th Replacement Battalion the night before I came home. We all signed the placemat in the little club at the 90th. It had a map of Vietnam on it and a printed message that said, "The 90th Replacement Battalion. Your first and last stop in Vietnam and always welcome in between." I still have that placemat. Years later, I believe it was around 1998, I learned that Tim was still serving and was stationed in Germany. An officer at FORSCOM gave me his email address and I sent him a message. His response was, "WOW!" Both of us were now Two Star Generals! We reconnected, and when Tim was re-assigned, it was to the Pentagon. He became the G-1 of the Army and a 3-Star, and I was commanding the 90th Regional Support Command in Arkansas (NO relation to the 90th Replacement Battalion in Vietnam). We saw each other several times over the next few years.

On September 10, 2001, I was at a general officers' conference at the Pentagon. I was waiting for my ride back to my hotel when Tim's sedan drove up and he got out and came over to me. We chatted for a few minutes, and then I didn't think about him until the next day. Those of us who were there in the Pentagon on September 11th and were able to evacuate the building safely went back to our hotel and congregated in the lounge to watch the news regarding the attacks in New York and at the Pentagon (we didn't know about Shanksville yet). I heard someone behind me mention Tim's name so I went over to this person and said that I had been with LTG Maude the previous day and asked about him. The person responded, "Well, he's dead. The plane flew in his window." Since that date I have stopped believing in coincidences. There are NO coincidences.

WILLIAM "BILL" LUSK

US Navy — Builder
Dates of Military Service (Active and Reserve): 1964 to 1970
Unit Served with in Vietnam: US Naval Mobile Construction Battalion Four
Dates in Vietnam: May 1966 to Aug 1966; Jan 1967 to Aug 1967
Highest Rank Held: Petty Officer Second Class (E-5)
Place of Birth and Year: Rochester, NY — 1943

THE PLACE, THE TIMING

I was a Third Class Petty Officer (Builder rate) attached to U.S. Naval Mobile Construction Battalion Four (Seabees). On my second tour, our base was at Camp Hoover in DaNang. The day after we arrived in-country, my squad deployed to a small outpost called Khe Sanh. It was located in the mountains close to the Laotian border, only a few miles from the Ho Chi Minh Trail and the DMZ. A company from the Third Marines held that position with a battery of 155 mm howitzers. Despite the ground-shaking, ear-piercing volleys, they provided a certain sense of safety and comfort for us. There was not much hostile activity then. Perhaps, it was a good timing.

Our mission was to build two bunkers for the USAF pilots with mahogany that they furnished, and maintain the short airstrip. Life was fairly routine, as well as it could be in a combat zone. We worked all day with only the most essential construction tools and limited equipment. Materials were at a premium, too. The closest lumber yard was 14,000 miles away. After a long working day, we would bathe at the spigot of a Water Buffalo with water from the nearby creek. Then, we would retire to our hooch, read, play cards, tell lies, or write letters. One didn't really go for walks outside the wire. In February of 1967, that area of the world was heavily forested. Battle scars from bombing were yet to appear and Agent Orange had not yet been applied to defoliate the jungle. It was as pristine and wild as wild would get. It was a pretty area. A contractor that I had met in Chu Lai the previous year had told me that he first arrived in French Indo China in the early fifties and had gone lion hunting frequently in this part of the country. It was remote. It contained flora and fauna that I had never before seen. But, it also contained its surprises for the unwary.

There were two "Recon" squads in that Marine company. At different times of the day and night they would go out on patrol into that overgrown sea of greenery. It was always good to see them all return in basically the same shape as when they left. Many times they would return with some trinket or "treasure" from their foray. It was one evening, after supper, and as dusk was dropping, that I saw the squad returning down the dusty path, in close single file formation. They all had their arms up as if toting something that looked like a log. It didn't take long for a crowd to gather and to identify their treasure. It was a fifteen foot long Python. Apparently it had recently eaten something, possibly a rabbit, but I was sure that it still had an appetite. The squad dropped their load between their hooch and ours. I slept with my eyes open all night. That incident made me realize there was more than the enemy lurking out there.

It was one year later that that base lay under siege for 77 days. One hundred fifty-five Marines lost their lives defending that outpost. Several hundred more were wounded. I think about how life all hinges on timing. I could have been there twelve months later. Whenever I see a snake, I think about that python and Khe Sanh. It serves as a reference point in my life when timing made a difference.

When asked by someone "What was it like over there?" I say, "It was hot; it was cold; it was dry; it was wet, and it had an odor that I will never forget." I was one of the lucky ones that came home with everything that I went over there with. It wasn't my time to go. I am blessed.

JOHN BLAIR

US Marine Corps — Infantryman

Dates of Military Service: 1966 to 1970

Unit Served with in Vietnam: India Company, 3rd Battalion, 4th Marines, 3rd Marine Division

Dates in Vietnam: Oct 1967 to Nov 1968

Highest Rank Held: Staff Sergeant

Place of Birth and Year: Sumter, SC — 1945

"MY FIRST KILL IN VIETNAM"

I landed in Da Nang, Vietnam, October 30, 1967. My Marine "MOS" was 0311, a grunt rifleman. On November 2, I was transported north to Don Ha and assigned to the Third Battalion, 4th Marines, 3rd Division. A few days later transported further north to Con Thien Marine Base, near the DMZ to join India Company, 3rd Platoon. The months of November and December of 1967, our company ran patrols and ambushes outside Con Thien. We had a few skirmishes but nothing major. This close to North Vietnam, you were in a free-fire zone, which meant no restrictions on when to fire your weapon, plus we only fought NVA uniformed soldiers.

December 28, 1967, India company was ordered to leave Con Thien and move north to Camp Carrol and Khe Sanh. The morning of December 30, 1967, I was assigned to walk point for the company. When you walk point, you are normally 100 yards ahead of the company, by yourself, going through jungle like woods and fields. The reason you are that far out, if the enemy engages, you might be killed, but a lot of lives would be saved.

Not everybody wants to walk point, but when it comes to your time, you had to do it. As point you go slow and quiet, looking for any sign of NVA soldiers. It is intense being out there alone. As I was walking point, I came upon a field of tall grass and a few trees. I stayed low for a few minutes and to my surprise I saw two NVA soldiers on the other side of the field get up and move slowly through the field with grass on their helmets. I slipped down the ridge and ran back to my company to our company commander, Captain John Pritchard (later killed on January 27th, Mike's Hill Battle). He said. "Blair, what are you doing back here."

I said, "Sir, do we have any friendlies in this area?" (keep in mind I am a young PFC)

He said, "What do you mean?"

I said, "I saw two soldiers ahead in a field. What do you want me to do?"

He said, "Kill the son of a bitches, get your ass back up there."

As I'm going back up there through the woods, I'm thinking: if there are two, there might now be 20, or might be a 100. My father, retired Army officer at Fort Benning, in Columbus, GA, loved to hunt. We would go deer hunting a lot. My hunting skills came into play, and I moved slowly to the edge of field. When I got to the edge of the field, I crouched behind a bush. To my surprise, the two NVA soldiers got up and were within 30 yards, moving slowly towards my position.

I was scared and my adrenaline was flowing. I raised my M16 to my shoulder, with a full magazine, as they turned to the right. I killed the two NVA soldiers. I reloaded, and as I was taught, you put a few more rounds in them to make sure they are dead. The saying going around was, if you kill your first NVA soldier, if you can take off his belt and wear it, you will make it through Vietnam. I weighed 140 pounds and it fit.

I immediately ran back to the woods and could hear India Company coming up fast to see why I was shooting. Captain Pritchard along with my platoon commander LT. Stewart and Staff Sgt Powell, searched the two NVA soldiers.

That night we camped near there and Capt. Pritchard had the two NVA dead soldiers laid next to the path so the company Marines would see them. The word spread through the company, PFC Blair, walking point, was the Marine who killed them.

It will always be with me, as will many other events during my tour. I left Vietnam in November 1968. I am so Blessed by God to be here. God's Angels watched over me through many battles in Vietnam.

I know this, I became a man quickly and understand what freedom means. God Bless our Veterans and especially our Vietnam Veterans.

Semper Fi

Silver Star Medal

2016 Inductee Georgia Military Hall of Fame for Valor

PHILIP H. ENSLOW JR.

United States Army, Signal Corps Officer

Dates of Military Service: 1951 to 1975

Units served in in Vietnam: Headquarters, 1st Signal Brigade; Headquarters, Regional Communications Group [Apr 1967 to Nov 1967]; 173rd Airborne Brigade (Separate) [Nov 1967 to Apr 1968]

Highest Ranks Held: Lt. Colonel

Place of Birth and Year: Richmond, VA — 1933

THE BATTLE OF DAK TO

One of My Sergeants Tells the Brigade Commanding General to "Shut-Up"

When I first arrived in Vietnam, I was assigned to the Operations Section of the 1st Signal Brigade Headquarters. I knew the Signal Brigade Commanding General, Brigadier General Robert D. Terry, from the days when I was a cadet at West Point, and he was the executive officer of the Department of Electricity. During the first full day in my job in the Operations Section, General Terry informed me that he was flying out the next day to inspect one of his Signal Battalions, and I was going with him. That was a quick introduction to signal operations in Vietnam. There I was with my fatigues still having stateside starch, and I was accompanying the Commanding General and telling a Battalion Commander how to do his job.

After just a few weeks at the Signal Brigade Headquarters, I was transferred to the Regional Communications Group. I stayed there for my first six months in Vietnam. At the RCG, I was responsible for activating the IWCS. The "IWCS," the "Integrated Wideband Communication System," was a very high capacity communication system covering South Vietnam, very similar to the commercial systems used by telephone companies in the continental US. The IWCS installations were easily identified by their very large antennas—about the size of a drive-in movie theater screen. My job took me all over Vietnam and into Thailand. I had just returned from a visit to Thailand (that could provide the material for another story or two) when I received a telephone call from the Signal Brigade Commander, Brigadier General Robert D. Terry. "Phil, you are jump qualified aren't you?" What experienced soldier would ever answer a loaded question like that? General Terry waited only a very short time for my reply. "It doesn't matter. You are now the Signal Officer of the 173rd Airborne Brigade." And that is where I went the next day.

When I reported to the 173rd, I learned why there was a sudden need for a new Brigade Signal Officer—the previous Brigade Signal Officer had been relieved, as had been his predecessor! Well, I was the Brigade Signal Officer now.

Very shortly after I joined the 173rd, the Brigade was moved to *Dak To*. Intelligence had reported that a very large force of North Vietnamese troops was going to enter South Vietnam from Cambodia passing through *Dak To*. The 173rd was "Westmoreland's Fire Brigade," and this looked like it was going to be a very big fire! The 173rd was to block the North Vietnamese advance. The intelligence prediction was quite accurate, and soon, the 173rd was engaged against a very large North Vietnamese force.

The battles that erupted on the hill masses south and northeast of Dak To became some of the hardest fought and bloodiest battles of the Vietnam War. [Battle of Dak To, Wikipedia, Quoted Sep. 15, 2019]

The intensity of the battles and the casualty rates were the top priority of everyone in the 173rd from the Commanding General, Brigadier General Leo H. Schweiter, down to the lowest Brigade trooper. We were continually answering questions from the "Brass" in *Saigon* about the high causality rates. The bravery of our troopers was attested to by the award of three Medals of Honor: Chaplain (Major) Charles J. Watters and two other troopers.

One afternoon, I had left the Brigade Operations Center and gone to my tent to do some planning and paperwork. Just as its name implies, the Operations Center monitored the actions of the units in the Brigade. To do this, the Ops Center monitored the important Brigade radio networks.

One of the other officers working in the ops center burst into my tent breathlessly and said, "Enslow—get over to the Ops Center now! You have to hear this!" I returned to the Operations Center and found everyone there crowded around one of the radios. They were listening to what was being transmitted on that network.

"Red Leg 6, this is Sparks 43. I told you to be quiet so that I could adjust this relay." "Red Leg 6" was the radio callsign of the Brigade Commander, Brigadier General Leo H. Schweiter and "Sparks 43" was the callsign of the Communications Sergeant who was operating a very important radio relay supporting radio communications over the long distances in which the Brigade was deployed. Obviously, the Communications Sergeant did not know that he was sending commands to the Brigade Commanding General.

There was silence on that radio net for a short period, but then the Brigade CG broke the silence and called one of his Battalion Commanders whose unit was in a very hot firefight! The Comm Sergeant still needed to adjust his equipment, and he required silence by all the radios on that network to complete those adjustments. "Red Leg 6, I am still adjusting this relay. Now, shut up!" Needless to say, gales of laughter broke out in the Operations Center, and a number of comments were directed to me about what was going to happen to this Brigade Signal Officer. "Better get your gear packed and ready to go."

Every evening, the CG held a review of the day's actions with presentations by the Brigade Staff Officers. After I gave my review, with no mention of the commands issued by the Radio Relay Sergeant, Gen. Schweiter said, "Major Enslow, may I see you in my tent after dinner." I could hear the muffled laughter by the other officers who had heard the radio exchanges over the air between the General and the Comm Sergeant—they were sure the next day would bring a new Brigade Signal Officer. I admit, I certainly did not look forward to my meeting with the general.

General Schweiter opened our conversation when I reported to his tent, "Major Enslow, would you please convey my sincere apologies to your Sergeant. The pressure was very high on all of us this afternoon. He was doing his job, and I did not make it any easier by not following his instructions." General Schweiter and I discussed several other communications and signal topics, but nothing further was said about the radio relay incident and the command to, "Shut up!"

The next day, the other officers working in the Brigade Operations Center were truly surprised that I was at breakfast and still in charge of the Brigade signal activities.

R&R SHOTS AND INOCULATIONS

All of the troops in Vietnam looked forward to R&R Leave. R&R (Rest & Recuperation Leave) was a chance to get away from the combat zone, and, if you were married, a chance for your wife to meet you in you in a much nicer location such as Hawaii or Hong Kong. My wife and I had plans to meet in Hawaii. Before you could go on R&R, you had to have all your shots and inoculations up-to-date. I had gone to the Brigade Aid Station to have my shot records checked. The last thing you wanted to have happen was to get to the airport to go on R&R and have them say, "You need some more shots."

Shortly after I got to the Brigade Aid Station, there was the familiar "WUMP—WUMP—WUMP" sound of helicopters approaching the Aid Station's landing pad. If you ever watched the TV series "*MASH*," based on the Korean Conflict, you undoubtedly heard the sound of incoming helicopters. Let me assure you that that is just what they sound like, and you can usually hear them long before you see them. (Remember "Radar" from the TV program?)

As the multiple helicopters landed at the Aid Station, everyone sprang into action. The obvious objective was to unload the helicopters as fast as possible and get the wounded into "triage," screening by a medic to determine how critical the troopers' injuries were so they could be properly treated as soon as possible.

I was trying to stay out of the way when the Captain in charge of the Aid Station grabbed the sleeve of my fatigues, and told me, in a rather commanding voice, "Get over there and get to work!" I quickly found myself holding one of the wounded soldiers in my arms and trying to figure out what I should do next. Even the Brigade Signal Officer could figure out I should follow the others, and I carried my trooper over to the triage station.

Later, when I was getting my shot record checked by one of the sergeants in the Aid Station, the Captain in charge saw me and came over to apologize for yelling at a Major. I told him that was precisely what he should have done. Personally, I was honored and humbled to be able to assist in this very critical activity.

MY HELICOPTER ADVENTURES

Even though I was not a pilot, I, like almost everyone else in Vietnam, spent a lot of time flying around in helicopters. All of us had some "adventures" during these flights. These are some of mine.

A CLOSE MISS

We were flying around the Area of Operations while I searched for good locations to install radio relay stations to support the troops on the ground. Things were going along quietly when the helicopter pilot and I were both surprised by a loud ping. Somehow, North Vietnamese troops had managed to hit our chopper. I was riding in the copilot seat and looking out the door hoping to detect the muzzle blast of who was shooting at us. All I could see of interest was an elephant working in one of the fields below us. I used my rifle to fire as much as possible in the general area where we thought the shot had come from. If nothing else, we might spoil the aim of any further shots. When we landed, both the pilot and I were thankful when we learned how close the enemy shot had been to the cockpit as well as close to some of the critical mechanical parts of the chopper. The enemy shot was only about eight inches from where I was sitting. It was even closer to the main drive shaft.

THE LOUDEST EXPLOSION OF THE VIETNAM WAR

One afternoon, we were returning to Brigade Headquarters in a helicopter. We were flying next to the *Dak To* airstrip, when suddenly our chopper was slammed by a very loud explosion, and we were moved sideways a rather large distance. Later, I had great praise for the pilot who managed to keep our helicopter flying safely. What happened?

Our food supply point was located right on the edge of the airstrip. Trying to get our ammunition and explosive supplies as safe as possible, they were located in the midst of the food. The ammunition and explosives were stored in a large metal container (a CONEX container). Unfortunately, a lucky mortar shot by the Vietcong had landed right on that metal container and set off all the high explosives. The explosion was so loud and powerful that it violently shoved the helicopter sideways. I was very impressed and thankful for the skill of the pilot who managed to keep control of the helicopter during that violent sideways movement. We were flying so that my right ear faced the explosion. Both of my ears "rang" for several weeks following that incident, and to this day, the hearing in my right ear is over a thousand times poorer than that of my left. The explosion blew an enormous hole in the middle of the food supply point. It was reported to be over 15 feet deep. I have read that this was the largest and loudest explosion during the entire Vietnam conflict. I continue to be reminded of this helicopter adventure every morning as I adjust my hearing aids. Not only did that explosion blow a big hole in the ground and destroy all of our high explosives, it also destroyed nearly all of our food that was stored at that supply point.

After several days of short rations, where the only thing we had to eat was American Cheese and Spam sandwiches, I realized why so many GIs from World War II did not like Spam. Our mess sergeant was trying his best to give us something to eat other than canned combat rations. But all he could "scrounge" together was a steady diet of Spam and American cheese sandwiches—three times a day. One day the Brigade Aviation Officer, sitting next to me in our Mess Tent, couldn't take it anymore and asked the cook if he could find something else, even some "beany-wee-ny." However, he was not very successful meeting that request, and was not yet the end of the Spam and American Cheese sandwiches.

BENJAMIN H. "HAM" MCDONALD

US Army—Infantry Officer
Dates of Military Service: 1963 to 1968
Unit Served with in Vietnam: MACV Team 1, Hoi An, RVN—Advisor to 2/51st ARVN Battalion
Dates in Vietnam: Feb 1967 to Feb 1968
Highest Rank Held: Captain
Place of Birth and Year: Atlanta, GA

MORE EXPERIENCES SERVING AS AN ARMY ADVISOR

Military Advisors in Vietnam normally had a one-year commitment in country. Six months of that time would be in the field with Vietnamese forces conducting combat operations and the remaining six months was normally assigned to a non-combat unit located in a more secure area.

I had completed my six months as Advisor to the 2nd Battalion 51st ARVN Regiment in July of 1967. Those six months had been quite challenging with numerous combat operations throughout the I Corps TOA (Tactical Operations Area). We had fought both Viet Cong and North Vietnamese units. These operations were in the rice patties, low land villages, and in jungle and mountains throughout I Corps.

ASSIGNED AS ASSISTANT G3 AIR ADVISOR I CORPS

I was called to Da Nang and I Corps Headquarters in early July 1967 and told that I was being reassigned to be the Assistant G3 Air Advisor for I Corps Headquarters. Responsibilities included all air support of Vietnamese forces operating in I Corps TOA. This included the TOA from the DMZ, west to the border of Laos and south some 250 km below Da Nang. I would also be responsible for scheduling close air support of Vietnamese units and clearing certain targets to ensure that no friendly forces were in the target area.

I Corps had a number of Army aviation units in direct support of the ARVN in I Corps. The 81st Aviation Bn was located at Marble Mountain just south of Da Nang. The 81st had both Huey slicks and Huey gunships along with OVE1 Birddog observation aircraft and other command and control aircraft. One of my primary responsibilities was to coordinate these resources for movement of Vietnamese forces in air assault operations.

Daily I had to brief the Senior Advisor of I Corps who in turn would brief the Vietnamese General commanding I Corps. We would show the days combat strikes and locations and estimated kills, any aircraft lost, any movement of ARVN units and planned operations upcoming and required air resources required. Located within the I Corps Headquarters was I Corps Direct Air Support Center (IDASC) and the Naval Gunfire coordination group. The Direct Air Support Center (DASC) was the principal United States aviation command and control system and the air control agency responsible for the direction of air operations directly supporting ground forces. The IDASC had Air Force and Marine pilots who would fly the daily Forward Air Control (FAC) missions directing the tactical air strikes in direct support of both Marine and ARVN forces.

I must admit after getting shot at on a daily basis and receiving incoming mortar fire too many times to recall, I looked forward to a job in a more secure area. I also looked forward to American food, a hot shower, sleeping in a bed and not on the ground. A cold beer or two at the Navy Officers Club was also a welcome treat.

BOREDOM SETS IN

After a few weeks in my new job, I got the briefings down pretty good and only lost the Hospital Ship Hope on one occasion. But having experienced daily combat for the last six months, I really wanted to see if could get back in the fight in some way. I got clearance from my boss, the G3 Air Advisor, to fly with the O-1 FAC pilots in their back seat when possible. This opportunity got me out from behind my desk and put me out where I might use my Vietnamese language skills to clear targets that might develop. Flying over the northern part of South Vietnam really showed me the beauty of that country and it was in that beauty the war was being fought.

CLOSE CALL OVER THE A SHAU VALLEY

In early August, I checked in with the Da Nang IDASC to see if they would be putting in any missions that day. I was told that several targets had been planned in the A Shau Valley where the NVA was building a road to connect to the Ho Chi Minh trail. I asked if I could come along and off we went to the Da Nang Airbase. With my helmet, weapon, flak jacket, and box of grenades, we were off. It was Air Force SOP that two O-1 FACS would fly together anytime we went into the area of the A Shau Valley. If one FAC was hit or shot down, the other FAC would remain on station until a rescue could be undertaken. That day we flew north from Da Nang up to Hue and on to Quang Tri and along the border between North and South Vietnam toward Khe Sanh.

As we approached the A Shau Valley, Captain Roberts, the FAC pilot I was flying with, saw what looked to be road building equipment as we passed over the valley. The O-1 is a very light aircraft and only goes about 80 knots and is quite vulnerable to ground fire. Therefore the pilot and back seat observer normally sit on their flak jackets for protection since ground fire is the greatest danger. Once the Air Force F-4 Aircraft from Da Nang arrived on target, FAC Roberts rolled in to mark the target with 2.75 mm rockets. Once the rocket was fired to mark our target, we pulled up to turn away from ground fire. With a loud bang a round came up through the floorboard of the aircraft between my legs, hit and ignited a red smoke grenade that was on the back of the pilot's seat, and then exited the plexiglass roof of the plane. The plane immediately filled up with red smoke. You couldn't see anything and you were choking to death. It seemed like hours, but minutes later the smoke grenade burned through the floor of the aircraft and fell out the bottom. Once the plane cleared of smoke, Captain Roberts checked his controls and thanks be to God all were working. Our plane took the second plane cover mission and our sister O-1 completed putting in the remainder of the fighters on the targets. Once the mission was complete, we headed straight back to Da Nang. I must admit the pucker factor was really bad and I did soil my fatigues that day. After a hot shower and a few prayers of thanks to God for His saving grace, I went to the Navy Club and drinks were on me.

INFILTRATION ON THE SONG DIEM BINH RIVER

The RVN 51st Regiment Commander requested assistance in helping to stop the VC/NVA from using the Song Diem Binh River to move weapons and supplies to the VC/NVA units operating in the area west of Hoi An. Local intelligence related that night boat traffic on the river was evidence of the movement of men and supplies into our area.

A planning session was held at I Corps to see what strategy could be developed to stop this movement. Commander of the 81st Aviation BN and his S3 were brought into the planning. One of the Huey units came up with an idea to rig up a Huey light ship. The maintenance chief was able to rig up C-123 landing lights on a light bar which we could mount under one of the slicks. The following night we decided to try out our Huey light ship. I was able to ride along on the Huey light ship to see if the concept would work.

We lifted off the airfield at Marble Mountain at around midnight.

We had the Huey light ship and two Huey gunships. Flying south along Route 1 to Hoi An we stayed low with no running lights. Once we got to the Song Diem Binh River we gained altitude and turned on the light ship. There were over 30-40 boats moving down the river toward Hoi An. Since there were no friendly boats or friendly troops in the area, we were given permission to engage the boat traffic. On returning to Marble Mountain, our after action report listed that 30 boats were sunk with little enemy fire on our helicopters. We considered the night to be a real success.

We were so successful the first night we decided to try our operation again. The next night we lifted off from Marble Mountain with our Huey light ship and two Huey gun ships and headed for the same river. Surprise, surprise, what a fireworks display did we draw. As we turned on our light ship all we could see were green .51-caliber tracer rounds coming at us from both sides of the river. I have never experienced anything like that and our light went out as fast as it went on. Our whole team had a very high pucker factor for the evening. We quickly left the area and returned to Marble Mountain.

What would be our next course of action? We confirmed that the VC/NVA were using the river for transport, but our weapons were limited and we needed more firepower to combat the automatic anti-aircraft fire. The Air Force and Marine Air Wing were brought into the planning. The Air Force FAC who had been in the area the night before suggested napalm and 500 pound bombs might do the trick. Coordination between the Air Force, Marines, ARVN, and the 81st Aviation Battalion was put into action. The operation was planned for later in the week.

Two days later, all coordination was complete. We lifted off from Marble Mountain with our Huey light ship and four Huey gunships. Our FAC was on station and the Air Force and Marine Air Wing had birds in the air. Again we flew south to Hoi An, but this time we went south of Hoi An before turning west toward the Na Trang Valley and the River. When we got to altitude and started our run down the river toward Hoi An, we turned on the light. And again it was like the 4th of July Fireworks except the fireworks were aimed at our light ship. The

FAC called us to say we could cut off the light, and he had the problem under control for now on. We orbited above Hoi An to observe the action. Air Force F4's, and Marine aircraft proceeded to burn and sink both sides of the river with napalm and 500 pound bombs. Numerous boats were sunk, and the operation was considered to be a success in helping to curb the movement of arms and troops into our operations area.

So my safe and secure non-combat assignment back in I Corps Headquarters was more enjoyable since I helped to bring the power of US air resources to support our Vietnamese troops in helping them to fight against the aggression of the North Vietnamese.

I was proud to be an Army Infantry Officer Captain who was awarded the Air Medal for the above actions and other combat operations which I participated in.

ROBERT L. HOPKINS

US Army — Infantry Officer
Dates of Military Service: Aug 1965 to Jun 1968
Unit Served with in Vietnam: Bravo Company, 1st Bn, 39th Inf Regt, 9th Inf Div
Dates in Vietnam: Dec 1966 to Dec 1967
Highest Rank Held: First Lieutenant
Place of Birth and Year: Baltimore, MD — 1943

AMBUSH

Ap Binh Son is a small village located southeast of Bien Hoa in III Corp area of South Vietnam. The village is surrounded by a rubber plantation managed by the Michelin Company, we know today as a manufacturer of high-quality tires. The trees provide sap used as the raw material for the manufacture of tires. The livelihood of the village was earned by tending the trees that involved skinning the bark in a pattern allowing the sap to flow into a cup beneath a spout driven into the tree. The trees are neatly planted in orderly rows.

My unit, Bravo Company, 2nd Battalion/39th Infantry, 9th Infantry

Division was assigned to provide security for the village. I was platoon leader for the first platoon. The unit occupied the overseer's house perched on a knoll overlooking the village. We spent about a month at Ap Binh Son, patrolling the area during the day and setting ambush patrols at night. Everything was quiet. One of our more entertaining activities was to swim in the creek that flowed through the village. It was fun until a rumor was spread the creek was used as a sewer for the village.

September 5, 1967 dawned as a normal day for the first platoon. We were standing down from a long patrol the previous day and relaxing in the overseer's mansion. Mid-morning there was some gunfire, but we thought little of it since it was distant, not lasting long. Later we learned a long-range recon squad had not checked-in for a regular sitrep (situation report). There was general concern from the battalion S-3 and if there was no contact with the lost squad, a search patrol would be initiated. The radio remained silent and eventually a search mission was initiated with my platoon leading the effort.

For the first time, the first platoon was assigned a K-9 to accompany us. We saddled up and moved to the last location the squad reported. This location was at the far edge of the plantation, close to an area that had been defoliated by Agent Orange. Using the dog as a guide, we located the point where the squad left the rubber plantation and entered the underbrush. The dog had no problem following the trail and alerted several times as we followed on.

After about an hour following the trail, we received a radio message. A worker from the plantation notified our interpreter at the manor house that there was an incident in a far corner of the plantation. Another patrol was sent out to investigate and the worst was discovered, the squad had been ambushed; there were no survivors. A deuce-and-a-half was sent to the location with support troops to receive the bodies. My platoon was still on the trail of the squad. When we learned of the massacre, we hurried along the trail left by the squad, arriving at the site where we observed eight dead American soldiers. They had walked into a well-staged "u" shaped ambush and had little opportunity to defend themselves.

To fully understand what had happened, I walked the ambush site, noted the position of each soldier, and checked out if they had been able to return fire. Not all had an opportunity to defend themselves. Most had been killed on the dirt road the squad had been following, killed instantly when the ambush was sprung. Several were able to seek cover off the road behind a rubber tree. These individuals were able to defend themselves for a brief moment since the ammunition in their magazines had been expended. But the odds were overwhelming. Because of the element of surprise in an area that was generally quiet, our soldiers were not alert and did not stand a chance in defending themselves. Scattered around the ambush site where each enemy soldier hid was a small pile of spent brass from an AK-47. At the top of the "u" was a pile of cartridges from a 30-caliber machine gun. The enemy left quickly since they failed to police the brass or take the weapons from the fallen soldiers.

The medical personnel who had been dispatched were fearful of the bodies. They assumed they had been booby-trapped and stayed away. My platoon saw the squad as brothers, checked each for booby-traps by pulling each body with a rope. Nothing. My troops reverently placed each in a body bag.

Rumors followed after the ambush the bodies had been desecrated and dumped into a drainage ditch. This was not the case. The Viet Cong or North Vietnamese must have thought there would be an immediate action by US forces in retaliation and left the area. The ambush site was not far from the 9thDivision base camp at Bearcat.

The US Army lost eight of America's finest soldiers in a Michelin rubber plantation. Could the ambush have been avoided? Probably not. The area of operations was relatively quiet and the squad let their guard down. The enemy unit that staged the ambush was probably passing through and took advantage of an opportunity.

My thoughts go out to the families of those who perished. They never had the opportunity to live a full life. No girl-friend, no marriage, no children. The casualties of war are devastating. The casualties of Ap Binh Son, average age 21.5 years:

Sgt Bravie Soto, 26
SP4 Edwin P. Prentice, 24
SP4 William T. McDaniel, 19
SP4 Kenneth J. Krause, 23
SP4 Arnold Benson, Jr., 20
SP4 Elmer D. Byrd, 20
SP4 Willie L. Jones, Jr., 20
PFC William R. Brennan, 20

This short poem is from the British Remembrance Day Service for World War I, but I find it very appropriate for the Lost Squad of Ap Binh Son.

They shall grow not old, as we that are left grow old; age shall not weary them, nor the years condemn.
At the going down of the sun and in the morning, we will remember them.
We will remember them.

The names of these eight Soldiers can be found on the bottom two lines of Panel 25E and the top five lines of Panel 26E on the Vietnam Memorial Wall in Washington, DC.

RUSSELL F. "RUSTY" REDDING, JR.

US Navy — Seal Officer
Dates of Military Service (Active and Reserve): 1965 to 1976
Unit Served with in Vietnam: Seal Team One
Dates in Vietnam: Jun 1967 to Dec 1967; Jun 1968 to Oct 1968
Highest Rank Held: Lieutenant (O-3)
Place and Date of Birth: LaGrange, GA — 1942

NAVY SEAL OPERATIONS, VIETNAM

Echo Platoon and our gear boarded a four-engine twenty-five-year-old propeller powered U. S. Navy transport at the Navy's North Island Naval Air Station in Coronado, Ca., in June, 1967. We were the only passengers on board and were to make a 36-hour flying time trip to Saigon, the capital city of South Vietnam, to begin our 6-month tour of duty during the Vietnam War. I was an Ensign (0-1), the bottom rung of naval officer ranks, and the assistant platoon commander.

My platoon commander was a 37-year-old former enlisted man, Richard Brereton, who held the rank of Lieutenant (0-3). He was an aggressive operator tempered with prudent judgment. I held him in high regard and was pleased to be in his platoon. We had 12 enlisted men in our platoon, most of whom had made a previous tour in the Da Nang area. I'm sure that Echo was the last platoon to deploy to Vietnam from Seal Team One with so many senior enlisted men. There was a demand for the SEAL Team capabilities in Vietnam after six platoons had proven themselves earlier in combat. The increased demand for operational SEAL platoons forced our command to provide younger and more inexperienced personnel with each ensuing six-month platoon deployment.

We were going to a place where our team had taken some serious casualties. The fact that I was going to war as fast as four propellers could grind on air at 10,000 ft. was becoming a reality. No more instructors or training exercises, this was serious business.

The memories of the past two years began to occupy my mind dulled by the boredom of the flight. My path to the SEAL TEAM was not a straight one but a series of twists, turns, events and choices. In fact, the public was not aware that the SEAL TEAM existed in 1965. Nor was I. The irony of the path was that the last step to the team was not a choice but an order. An order that I gladly and proudly accepted.

While at OCS, 400 of my classmates were given the opportunity to volunteer for the Underwater Demolition Teams known as UDTs. We were shown a UDT recruiting film depicting all the capabilities trainees would have to master during the six- month training course. Nothing in this 1965 presentation was said about the SEAL TEAM. The extra pay, out-of- doors aspect, demolition work, parachuting, scuba diving, working with submarines, and living on a beach all appealed to me. After the recruiter asked for volunteers, I decided I'd better get up to the front of the line in a hurry before the quota was filled. I hustled to the front and began to hear hoots and jeers from my class mates behind me. Looking around to my rear, I saw that I was the line and the only volunteer.

After looking down at the vast Pacific Ocean from ten thousand feet, making several refueling stops and many box lunches all with one boiled egg, the bright lights of Saigon were a welcome sight.

Seal Team One's Commanding officer, Franklin Anderson, had told Echo that he would try to get us home by Christmas. Being in a new and dangerous environment the first few days of our deployment raised concern about our safety. Only three weeks earlier, I attended funerals back in Coronado for my friends, Dan Mann and Ralph Boston, teammates killed in the same area of operation where my platoon would spend the next six months. All of us were saddened by their deaths and, of course, we felt the stress of being in a new and dangerous situation.

Our home for the next six months was a small U. S. Navy base named Nha Be. The base and the village just outside the security gate, also called Nha Be, were located on the deep-water shipping channel which runs fifty miles from the coast of the South China Sea up to Saigon. The 500 square mile mangrove swamp surrounding the shipping channel was not conducive for regular army operations due to the deep mud when the tide was out, the many streams when the tide was in, and the extremely thick foliage. This swamp was called the Rung Sat by the Vietnamese meaning "killer swamp." It had been a sanctuary for pirates and law breakers for centuries.

In 1966, the senior officers in Seal Team One decided that this would be a good place to cut our combat teeth in Vietnam. Echo was one of three operational platoons in Nha Be. One platoon was routinely rotated out on an operation while one other platoon rested and the third platoon planned the next operation. At that time a platoon was 12 enlisted and two officers. We were transported to our ambush/patrol sites by our own heavily armored landing craft called a "Mike" boat, light Navy patrol boats, or helicopters, (when we could borrow one from the army). At high tide all but a small portion of the swamp was under water. The first and primary reason for the U S Navy's presence at the edge of this swamp was to keep the 50 mile long deep water shipping channel open for large ocean going ships between the South China Sea and Saigon. We were there to support this effort and make life as miserable as possible for the communist (Viet Cong aka "VC") who used this swamp to hide food and weapons.

A routine day would start with the platoon officers meeting with the base intelligence staff. If the staff could advise where recent activity by the Viet Cong was suspected, the feasibility of the next operation would be studied. All our operations were in what was called a "free fire zone." This meant anything that moved could be shot from the air or ground (after clearing with the operations center). It was important to clear our operation locations ahead of time with our tactical operations center in Nha Be for our own protection. If seen from the air by U.S. forces, we did not want to be mistaken for the enemy.

The planning for setting up our ambushes and the preparation of weapons took most of a day. We might depart that night or the next morning as early as 4 am in order to arrive at our insertion site before daylight. Darkness was our friend as the VC might hear our boat approach but they could not determine our precise position. The only exception to this advantage of the cover of darkness and the element of surprise was when the enemy obtained prior knowledge through a spy/double agent. If we were betrayed by a spy in our midst, the VC could be waiting to ambush us at our precise insertion location. In fact, this did happen to us and on one occasion, we were in big trouble.

Most of our ops involved inserting into contested enemy areas and setting up ambush positions. Patrolling with six or more men was considerably more dangerous that sitting on a static ambush. Sloshing through the mud and breaking through the underbrush could get noisy and alert the VC of our presence. In two instances, a point man (the man in the front of the group with the most experience) for two other platoons was ambushed and killed while patrolling.

All our operations were in the Nha Be area except for one that stands out in my mind. That operation was planned by a top- secret intelligence staff in Saigon, the Studies and Observations Group aka "SOG." Protocol for all special operations missions required that U. S. planning staffs coordinate with their Vietnamese Military counterparts. This coordination involved numerous Vietnamese officers with knowledge of future operations conducted by U.S. forces. Because this protocol created a higher chance of a security risk, top- secret missions planned by SOG were coordinated with only one high ranking Vietnamese officer. Unfortunately, it was later learned that this one high ranking Vietnamese officer was a double agent! He was responsible for the deaths of numerous Americans, and was, ultimately, killed by a U S Special Forces officer.

In our briefing for this operation, we were told that, on the shore of a large bay north of Cam Ran Bay, there was a suspected exchange location of Viet Cong material and messages. We were tasked with capturing a VC suspect during such an exchange at an old water well about 50 yards from shore. The platoon went by Swift Boat to within half a mile from shore. Then a team mate and I were selected to swim to the beach as Swimmer Scouts to patrol about 50 yards to our hiding place. We swam with World War II 45 cal. automatic "grease guns" because they could take in water and sand and still function. Once on shore, as we patrolled single file at a slow careful pace, we felt for booby trap trip wires, and strained our ears listening to make sure we were alone. Every twenty yards we would stop and listen. We agreed to share point man duty. I was always happy after my twenty yards to turn the lead over to my swim buddy. After it seemed safe to bring the other seven men to the shore, we used an infra-red light to signal them. The moon was bright, and we could clearly see the rest of our squad waiting off shore. I wondered who picked a night with an almost full moon for this operation?

Nevertheless, we got settled in our hiding place about midnight and put all eyes on the old well. About 1 A.M. we began to relax a little bit because of the comfort of the cover of darkness. Fighting sleep became the next challenge until 2 pm the next day when we heard a small U.S. spotter airplane. My first thought was that he should have received word at his pre -op briefing that "friendlies" were in the area and all U.S. forces were to stay away. Before I had time to get angry about his presence, automatic weapons fire came from the direction of the beach we had just crossed, directed at the spotter plane. This meant the enemy was only one hundred yards away. Now the adrenalin pump was on at full force.

Soon we heard a loud boom which we knew was a rocket from the spotter plane shooting at the VC who were shooting at him. Now we had two problems: one was that the VC might be looking for us and the other was that the pilot might call an artillery or air strike on our location. Either type of strike against the VC was a danger for us because of our proximity to the VC. Thanks to good planning, three members of our platoon were about a quarter mile away high on a hillside on an "over-watch" position. (They had inserted by helicopter before our squad arrived). We were too close to the VC to talk on the radio, and we depended upon the "over-watch" team to use their radio. Within a few minutes the spotter plane left the area, and we heard no more gun fire. After a long fifteen minutes it appeared that there would be no artillery barrage from our own U.S. Forces on our position. That was a big relief and an exhale moment.

Now our concern shifted to the nearby VC. Had they moved on or were they looking for us? We had a good position for surprise at the well. However, we were at the base of a steep hill with vegetation for cover but no protection if we were attacked from above and behind. The rest of the afternoon was quiet. Our platoon commander signaled that we would stay put for the night as the sun was going down.

Darkness is the friend of those who are in territory controlled by the enemy, however, at midnight of night two, came the sound of wooden boxes bumping together and voices from the shore. We were sure there would be some activity now at the well and once again, my heart began to race. In twenty minutes, the noise faded away, and there was no activity at the well. The rest of night number two was peaceful.

At dark thirty on night number three, our last night for the operation, we patrolled to the shore and got into our positions to watch for boats coming into the beach to offload supplies. Now we had two of our operational friends working for us, darkness and escape by water. This was comforting. As I settled into my position as rear security, and looked landward, I saw the outline of a small pagoda not too far away and thought, "What a peaceful scene this is: white sand, clear calm water, cool night air, mountains that came right up to the bay, no illuminating flares, no aircraft noise and no sounds of gunfire." It made me think, "This is wonderfully tranquil in your sleep deprived state of mind, but let's not forget you have one more night to stay alert."

The tranquility ended about 11pm when a sampan approached the beach with several passengers. As rear security, I was looking inland and away from the beach, but I could hear our CO issue the command in Vietnamese to "come here" several times. When it was obvious that the sampan was evading, we fired upon it. Two of the passengers surrendered. Our pick-up boat arrived soon, and we went to the Swift Boat with our two captives. They were turned over to the South Vietnamese for interrogation.

BOB LANZOTTI

US Army Aviator
Total Time in Military: 1960 to 1985
Unit Served with in Vietnam: 1st Avn Bde, 1st Cav Div
Dates in Vietnam: 1967 to 1968; 1969 to 1970
Highest Rank Held: Lt. Colonel
Place of Birth: Taylorville, IL—1937

WAR STORY (PG RATED)

There's an old cliché that says, "Nothing screws up a war story more than an eye witness!" But, this war story involves me, and I'm pretty certain that none of you readers were there to witness this action. So, I suppose I could do a little embellishing. But I won't, I promise.

During my first tour to Vietnam, I was serving in the 11th Combat Aviation Battalion, 1st Aviation Brigade, as an assistant S-3. In that capacity, I found myself as a permanent pilot of Smokey, a Huey configured with an oil pump that directs oil into a ring attached to the engine's exhaust stack. When the oil is directed into the super-heated exhaust, it creates a smoke screen that would obscure the takeoff and landing of a Boeing 747. Smokey's job was to deliver smoke screens to obscure the aerial helicopter assaults of three Infantry Divisions we supported, the 1st, 9th, and 25th Infantry Divisions.

The old adage that "Flying helicopters can be hours and hours of boredom, interrupted by moments of stark terror" is as true for me as for anyone who has ever flown in a combat environment. I've certainly had my hours and hours of boredom, but a moment or two of terror as well. One of the latter moments occurred while we were supporting the 25th Infantry Division on a beautiful fall day during 1967. I was flying with a close buddy, Steve Stoudt, and we were laying a smoke screen along a number of unoccupied villages located between Cu Chi (due north and about 25 miles from Saigon) and Tay Ninh, about 50 miles northwest of Cu Chi and near the Cambodian border.

The plan was to obscure the landing of our helicopters to the north of the village, allow the infantry to walk through the smoke, clear the village, then be picked up (extracted) by the Huey lift ships south of the village. The pickup on the south of the village was accomplished without Smokey as the village was deemed clear of enemy.

It was about the second village that all hell broke loose. During the pickup operations, the troops were taken under enemy fire. The fire was coming from the village that they had supposedly just cleared! Two Hueys were hit when they touched down. One was able to take off, but the other one was disabled. Steve and I could see the rounds hitting in the rice paddies where the troops were pinned down.

I was flying and both Steve and I recognized immediately what we needed to do... obscure those guys hunkered down with smoke so they could be extracted. I remember Steve shouting over the intercom, "Hey Bobbie, it's Show Time!" That was a true axiom if there ever was one! I dropped down to the deck and flew as fast as that Huey could go toward the pinned down troops. The bad guys turned their attention our way and we became the new target. I was leaning way forward in my pilot's seat and probably looked like a jockey on his last furlong of the Kentucky Derby and going for the Roses. I could see the rounds hitting the wet ground in front of us and hear the rounds finding their mark on our helicopter. Actually, my leaning forward had to do more with me trying to get all of me into my chest protector (flak vest) that covered just my upper torso.

We put down a beautiful smoke screen that was undoubtedly even more beautiful to those hunkered down in the rice paddies, as well as those Huey pilots orbiting aircraft above. We had a few bullet holes in our Huey, but the crew came away unscathed and no vital parts were hit on our ride, the Huey.

After we laid that smoke screen I finally relaxed and sat upright in my pilot's seat. Steve observed that and said, "Hey Bob, back there you looked like you were trying to hide behind that cyclic stick, and you were gripping that cyclic handle so tight that if it had been an olive and the red trim button on top had been pimento, the pimento would have been ejected right through the roof!"

I don't know what the rules of engagement were back then, but unfortunately for the bad guys, their position was within artillery range of the 25th Infantry Division at Cu Chi and its several artillery batteries. We watched the village go up in smoke as artillery pounded just shortly after we delivered our smoke screen. Payback is hell. And that, readers, is a happy ending to a war story.

WILLIAM (BILL) H. BROWN

US Army – Signal Corps Officer

Dates of Military Service: 1966 to 1974

Unit Served with in Vietnam: 2nd Plt, B Co, 121st Sig Bn, 1st Inf Div; 1st Div

Combat Photographer

Unit, C Co, 121st Sig Bn, 1st Inf Div.

Dates Served in Vietnam: Aug 1967 to Aug 1968

Highest Rank Held: Captain

Place of Birth: Savannah, GA – 1944

CHRISTMAS IN VIETNAM

Do you remember that wonderfully sentimental Christmas song made popular by Bing Crosby, "I'll Be Home For Christmas"? As the words to that song run through my mind, I'm reminded of another Christmas over fifty years ago. . .

It was Christmas of 1967. I was in Vietnam. I was surrounded by the members of my platoon, some fellow officers, and a few doughnut dollies. I had an old guitar that my platoon had given me. It was just a guitar body with no strings, tuning knobs, or hardware when they gave it to me. But I had written home to Mom and Dad and asked for a set of strings and the hardware to attach them to the guitar.

We sat around the Photo Lab in Lai Khe, had a few drinks, and sang some Christmas songs as well as other songs that reminded us of home. Lai Khe was 1st Division Headquarters at that time. We had moved 1st Division HQ from Di-An just before Christmas. I was a platoon leader for a communication platoon while we were in Di-An, and my platoon and I spent a lot of time in the field. And now I was a platoon leader of a platoon of combat photographers and lab technicians. At least we were in the relative safety of the base camp this Christmas Eve, and not out in the field somewhere in Vietnam.

Of course, little did we know what lay ahead a month and a few days from this Christmas Eve. There would be fireworks of an unwelcome kind celebrating a different holiday. 122 mm rockets, mortar rounds,

AK-47 and other small arms fire as the Viet Cong began what would be known as the Tet offensive, the most intense fighting of the entire war.

All we knew was that it was Christmas Eve, and we were a long way from home and family and friends. And while our hearts ached, we were comforted by the Christmas music, the memories it stirred, and the comradery we shared in our little group so far from home.

Back in the U.S. in a small South Georgia town was a young blond-headed girl I had met just before going on active duty in 1966. After graduation from college, I had about six weeks before I was to report for active duty. My home church in Albany, Georgia, asked me to drive a group of teenagers to a church camp in St. Petersburg, Florida. I was to stay at the church camp for the entire week, and I was referred to as a counselor, although I was really just a glorified driver. This blonde-headed teenage girl was one of the campers. I found out that she lived in Waynesboro. My first active-duty assignment was at Fort Gordon, and as it happens, Waynesboro is right along the way from Albany to Fort Gordon.

So, I stopped by to see her, and met her parents. I'm sure they were glad to see me leave. After all, I was a college graduate, and she was still in high school. A year later, I'm in Vietnam. I didn't have any Christmas cards, but I was a pretty good artist. I drew a picture of a soldier, head and shoulders, in jungle fatigues and steel pot. I took a photograph of the drawing, developed it in the photo lab, printed it on good paper, cut it out, and sent her a home-made Christmas card. That's the first she knew that I was in Vietnam. After that we exchanged a few letters until I came home.

Soon after I came home from Vietnam in August of 1968, I went to see her. She was a little older now, and the age difference didn't seem as important as it had two years or more earlier. She was a college girl, away from home, and spreading her wings. That was the beginning of an on-and-off courtship that lasted another five or so years until we were married. I've shared every Christmas with her for the past forty-six years, and I have no regrets.

Now, fifty years after that Christmas in Vietnam, I'm here at this

Vietnam Veterans Christmas party sharing good music, good food, and good times with other Vietnam veterans and their loved ones. We have so much to be thankful for. We survived the Vietnam War, we're home, and it's Christmas time.

So thank you for your service, welcome home, and may God bless you, your family, and friends, and the United States of America.

Merry Christmas,

Bill

RICK WHITE

US Army — Infantry Officer

Dates of Military Service: 1966 to 1997

Unit Served with in Vietnam:1st Tour: 3rd Platoon, C Company & Recon Platoon, 2nd Bn, 35th Inf Regt, 4th Inf Div; 2nd Tour: B Company, 503rd Abn Inf Regt, 173rd Abn Bde

Dates in Vietnam: Dec 1967 to Dec 1968; Jan 1970 to Dec 1970

Highest Rank Held: Colonel

Place of Birth and Year: Atlanta, GA — 1947

A 21-YEAR OLD LIEUTENANT REPORTS TO HIS UNIT IN VIETNAM — *IN CONTACT!*

Just after Christmas in 1967, I, Second Lieutenant (2LT) Rick White from Norcross, Georgia, arrived in Vietnam and was assigned to the 2nd Battalion, 35th Infantry Regiment (2-35), 4th Infantry Division at Camp Enari in the Central Highlands. Within a few short days and after the required in-country training and equipment issue, I, along with another 2LT, boarded a "Huey" helicopter which was full of ammo, C-Rations, and water and flew for about 40 minutes to where the 2-35 was located in the field.

As we were about to make the approach into the Battalion Headquarters Landing Zone (LZ), I noticed a burning helicopter on the ground that had just been shot down. About that same time, the Door

Gunner grabbed me by the shirt collar and yelled into my ear that this place was under enemy mortar attack, that many of our guys had been KIA or WIA, and that as soon as this aircraft landed, the other 2LT and I were to get off fast along with the resupply. The Huey was on the ground for mere seconds and was gone.

After getting our bearings in the thick swirling dust of the LZ, we were directed to and then raced toward the Battalion Headquarters Bunker. There, we soon met cigar-chewing LTC (later General) Bill Livesey. He asked me where I was from in the States and, for some reason, what were my best sports. I responded with, *"Sir, I'm from Norcross, Georgia and I played football, ran the two-mile, mile, half-mile in track and also cross country."* He responded that he was from Clarkston, Georgia, was a baseball player and now his unofficial call-sign for me was, *"The Norcross Flash"*!

He then immediately assigned me to C Company and sent me on my way across the perimeter to find the Company Commander, 1LT Homer Kraut. As soon as I found 1LT Kraut, he said that I was now the Platoon Leader of 3rd Platoon, that the previous Platoon Leader had been killed a few days before, handed me a map, and said that Hueys were in-bound in just a few minutes for a Company Combat Air Assault, and *"Get to your platoon."* Upon locating 3rd Platoon, which was now in position to board the birds, I quickly found the Platoon Sergeant who said that he had been with the platoon for about a week. He then pointed me to the landing spot for the bird that I was to board. Within just few minutes, about fifteen Hueys landed, we quickly loaded and were off.

As we were flying, the gravity of the situation that I now found myself in flooded my thoughts, and I began to take note of what I knew, which was little, and what I did not know, which was great.

My Known Facts: My name; that I was in Vietnam; for about 10 minutes now, I was the 3rd Platoon Leader, C Co, 2-35, 4th ID; I knew the names of the Bn and Co Commanders and the Platoon Sergeant; and that I had a map.

My Unknown Facts: How many men were in my platoon; their names; their positions within the platoon; where we were going; and the enemy situation.

My Major Concern: If we made contact and my men were KIA or WIA, I did not know their names.

God was merciful and must have felt pity for this very young Platoon Leader because the LZ where we landed was "cold," there was no enemy action, at least on that day. Then, for the next 47 days we patrolled the hills and valleys of the Central Highlands during the day and lay in ambush at night. During that time, we never stopped to wash, change our rotting jungle fatigues, eat a hot meal, or drink any jungle stream or rice paddy water that did not have the pungent taste of iodine. However, 2LT White did learn his men, how to be an Infantry Platoon Leader in combat, and many, many other vital lessons well taught by 1LT Kraut and those wonderful and dedicated young Soldiers all around me.

Those first two months of hard combat experiences, out of my 24 months in Vietnam, proved their worth for on 27 February, during the 1968 Tet Offensive, C Company found itself in an all-out, three-day battle against a well-entrenched, well-equipped and well-trained North Vietnamese Army Regiment. Again, God was good, C Company prevailed, and this once young 2LT, who had to grow-up quickly, was richly blessed with a 31-year Active Duty Army career of leading Soldiers, America's finest.

God Bless the USA and our Selfless, "Defenders of Freedom."

Colonel (Retired) Richard H. White is Director & Chairman of the Board of the Georgia Military Veterans' Hall of Fame — http://gmvhof.org — each November they honor Georgia military people with induction into the Hall of Fame.

JAMES B. "JIM" STAPLETON, JR

US Army — Infantry Officer

Dates of military service: 1964 to 1994

Units served with: 1/22 Inf, 2nd Brigade, 4th ID; G-3 4th ID; Aide to CG, First Field Force Vietnam; HQ; Advisor, Vietnam Airborne Division (Team 162)

Dates in Vietnam: Aug 66 to Dec 68 (28 months)

Highest Rank Held: Colonel

Place of Birth: Fort McPherson, GA — 1942

A 28-MONTH TOUR OF DUTY

It all started when my father, Army surgeon James "Buck" Stapleton, was assigned to West Point to command the Hospital. Impressionable and young, I soon struck up a friendship with several neighbors who attended the Military Academy ahead of me and with me. I used to ride to high school with them, and they left a lingering impression. They thrived at West Point. I wanted to be part of that; I just knew that I wanted to go to West Point.

When the time was right, I entered West Point and graduated as an Infantry officer in 1964. Once I got to West Point, I realized that I wanted to make a career of the Army. 'The Vietnam War, how can we get there?' everyone would ask.

While I knew the risks — and lost friends to the war while still at West Point — I accepted the risks and pushed forward. Fort Lewis, Washington, was my first assignment, with the 4th Infantry Division to prepare for deployment to Vietnam.

We spent nine months training in "smizzle" — smog and drizzle — at Olympic National Park to get used to environmental conditions before deployment. The entire Second Infantry Brigade I trained with was sent to Vietnam. We left Tacoma and spent 16 days on the Navy troop ship USNS General Nelson M. Walker before we reached Vietnam. We were close knit and well trained. Everyone knew everyone. Confident and competent.

When I arrived in the summer of 1966, I quickly became familiar my new home in Vietnam, Pleiku and the Central Highlands, with a six-week mission on the coast south of Tuy Hoa.

Unlike most servicemen who spent a year tour in Vietnam, I, not married, extended and spent 28 months—without taking a break, spending my 25th birthday in the Highlands, and my 26th in Saigon. I think my tour in Vietnam framed my career in the Army as an officer, and it framed me as a leader.

After the first nine months as a rifle company Executive Officer and S-3 Air on the battalion staff, I was promoted to captain and assigned as Company Commander of Charlie Company, 1st Battalion, 22nd Infantry Regiment of the 4th Infantry Division. I was happy that I was able to stay in the same battalion I had been in for over two years by then. The assignment was sobering, because my predecessor was Killed in Action on his first day in the field as commander. It motivated me to be all I could be for my Soldiers and our combat operations.

Seven months of Rifle Company ground combat operations covered about 2,000 kilometers—or six miles a day. In over 30 combat engagements with the enemy, not one Soldier was lost. I am proud of the way we were always alert and focused on force protection skills. I focused on one thing, mission first, people always.

As we traveled through jungle, woods, mountains, and highlands, we would establish a routine where we would execute combat operations until about 2 p.m. before clearing an area, setting up a perimeter, putting up shelters, and establishing Listening Posts. I knew the North Vietnamese had their soldiers on reconnaissance and had us under surveillance. They were watching to see if we showed any weaknesses. That's why we did things like not putting the same people on night watch who had been cutting overhead cover as we set up in the late afternoon. We also dug in every night in case we were mortared. We always plotted on call artillery "def-cons" and Air Force support.

As the weeks went by with no casualties, my platoon leaders, NCOs, and I continuously focused on battlefield proficiency. We had to sustain our combat skills as new personnel rotated into and older ones ended their tour and left the company. My leaders and I understood we had to sustain combat skills. As a leader, you must have the mentality to sustain readiness.

After completion of my company command time, I was selected to be an aide-de-camp to Lieutenant General William "Ray" Peers. Peers later led the investigation of the My Lai Massacre. I recall being with Peers on high ground at Pleiku Airbase when we saw rockets flying in the air on January 31, 1968 — the Tet Offensive."

Tet 1968 was a coordinated series of North Vietnamese and Viet Cong attacks on more than 100 cities and outposts in South Vietnam. While American and South Vietnamese forces quickly subdued the attacks, 246 Americans were Killed in Action.

News coverage of the massive TET offensive shocked the American public and started the erosion of support for the war effort.

In the final six months of my tour, I joined the Vietnamese Airborne Division as a Battalion Advisor, defending the U.S. decision to go to Vietnam. I still believe we went over for the right reasons: to keep communism out of that country and to keep it from coming here.

SOMEONE HAD TO DO IT — MY 28-MONTH TOUR HAD SOME BENEFITS:

Assignment to the Vietnamese Airborne Division as the 1st Battalion Advisor immediately following Tet 1968 had some benefits. Immediately following the Tet 1968 attacks on Saigon, my Battalion had the mission to secure Saigon with a curfew each night. So, in daylight hours I kept fit swimming, golfing, and playing handball. We had a weekly swim meet with Bobbie, the AFN Weather Girl — she gave the weather each evening. Yes, golfing at Saigon Country Club was dangerous, but... if not me, who, and if not now, when....

Never Volunteer is the lesson learned.

RUSSELL F. "RUSTY" REDDING, JR.

US Navy—Seal Officer

Dates of Military Service (Active and Reserve): 1965 to 1976

Unit Served with in Vietnam: Seal Team One

Dates in Vietnam: Jun 1967 to Dec 1967; Jun 1968 to Oct 1968

Highest Rank Held: Lieutenant (O-3)

Place and Date of Birth: LaGrange, GA—1942

NAVY SEAL OPERATIONS, VIETNAM—PART 2

Rusty Redding continues with his experiences as a Navy Seal work-ing in the Rung Sat zone, located on the deep-water shipping channel which runs fifty miles from the coast of the South China Sea up to Saigon (now known as Ho Chi Minh City).

Four months into our deployment, our platoon was split into our two squads, one squad in each patrol boat. We were in an area where we usually made contact with the enemy. My squad was going to insert first while the platoon leader's boat would be standing 50 yards away to provide covering fire.

En route to the ambush site we were ahead of schedule several miles away from our insertion location. We waited for the cover of darkness before inserting. While waiting, I remember sitting on the side of the patrol boat during this time thinking, "We've been lucky so far in this deployment and I hope our luck holds."

The most dangerous time of our operations was the transfer from patrol boats to insertion sites on land or from helicopters to the ground. At the time of insertion either vehicle is moving slowly and is an easy target. Once we were on land we felt secure knowing that we had the element of surprise and superior firepower. As we approached the bank for this night's insertion, my medical corpsman (medic) was standing on the bow ready to jump onto the river bank. A good long jump to the bank meant you might not get totally wet to start the long night. Dry-ness was happiness.

The rest of us were moving forward on the boat to follow the corpsman off the bow. Just as the bow touched the thick foliage on the bank, the sky lit up with tracers coming directly at our boat and the jolting sound of automatic weapons fire coming from our precise insertion point. Our boat could not have been any closer to the VC on the bank. Green tracers were making a zipping noise inches above our heads. I thought, "This is where you are going to die" because our boat is going to be hit with a shoulder fired rocket any second.

Somehow and thankfully, the VC were shooting high in the dark and no one was hit. Our corpsman immediately fell overboard, which probably saved him, as he was on the bow ready to be the first man off the boat. We were getting fire from a 300-yard- long VC ambush. We were clearly in the kill zone at the extreme right edge of the ambush and were experiencing first hand being the "ambushees" rather than the "ambushers."

The heavy tracer fire spread along the bank three hundred yards to our left indicated that there was a large body of VC waiting for us. The bow (front) gunner on the patrol boat immediately returned fire from fifteen feet away with twin 50 cal. machine guns directly to our front. The two machine guns were going through the mangrove bushes like a chain saw. The VC were smart to keep their heads' down. This is probably why no one on either boat was wounded.

Another factor for our good fortune was that it appeared to me that our boat inserted us at the extreme right part of the VC ambush. Due to the thick mangrove bushes, the VC could only see straight ahead. The boat driver backed out to the right, away from the kill zone and the VC field of vision. As we were backing out, I expected we would be hit with a rocket any second as we were still very close to the bank. Rockets were not very accurate but at this close range, how could they miss?

Within seconds, with both boats returning covering fire, our boat went back to get our corpsman. We were certain he was dead or badly wounded and overjoyed that he experienced such a close encounter unscathed. When Navy helicopter gunships that were on call arrived, a pilot, and good friend Cdr. Peter Shay USNR Ret., told me that the tracer fire from the VC ambush was so thick and widespread that it looked like you could walk on it.

The VC ambush was probably the largest enemy force any of our seal platoons had faced in our 18 months of operating in this swamp. The odds of a VC ambush at the precise site on our insertion in a 500 square mile area were too high to be a coincidence. There had to have been security breech in our Nha Be operations center, but we never heard of a culprit. Thirty years later, Cdr. Shay visited Nha Be and learned that at least half the village of Nha Be were covert VC. (In order not to arouse old ill feelings for Americans on his visit, he told the locals that he flew transport helicopters, not gunships.)

This was not the only time we were on the receiving end of enemy fire. Our base was hit with mortar fire five times during our tour. One round penetrated the outside overhanging shed roof five feet from a teammate's bunk. The round exploded below on the first floor at the edge of the building's concrete slab. The explosion demolished all the liquor in the officers' club bar. The VC had the exact distance from our base to their mortar positions. They would "walk" about eight rounds across the base and stop because they knew our Navy gunship helicopters would launch and see their muzzle flashes from their mortars.

Fortunately for us, the VC preferred to live rather than lob a few more mortar rounds at us and risk being attacked by our gunships. There was an Army artillery base several miles away. When they would fire at night over our base into the swamp, it sounded as if we were under mortar attack. After being awakened and listening for a few rounds, I could discern between the artillery fire and incoming mortar rounds. If the blast were mortar rounds, we would all pile into a bunker just outside our barracks. One night the artillery fire went on for such a long time I got tired of listening and deciding what kind of fire I was hearing. In order to get some sleep, I took an air mattress below deck on a Navy ship tied up at our base. I thought if the rounds are mortars, they will explode on the top of the steel deck, and I will not have to scramble to the bunker. Peace of mind and a peaceful night's sleep ensued.

One sleepy Sunday afternoon, we set up for an ambush in the western edge of the swamp closer to civilian population. Closer to civilians usually meant closer to VC activity. About 3 pm a sampan, with two paddlers and a passenger sitting in the middle, was seen hugging the shore line about 50 yds away in the free fire zone. When our eight automatic weapons opened up at once breaking a long period of silence, the noise was jolting.

It was hard to know if the passengers were hit or rolled out of the boat into the protection of the water. Fortunately, the sampan remained upright. We put a swimmer, Gary Shadduck, in the water to retrieve the drifting sampan. (I wrote him up for a Bronze Star for this op and Gary did receive the award.) There were numerous documents indicating the passenger was a North Vietnamese Major and probably a paymaster due to the list of names and amount of money found in the sampan.

This was at a time in the war when the North Vietnamese said they did not have any troops in South Vietnam. We collected enough money to pay for our laundry for the remainder of our tour. The base intel officer was glad to have the documents but irritated with us for tampering with the transistor radio we found in the document bag. He said there might have been a setting on the dial where VC messages were received. Sorry sir, we blew that one.

Late in our tour, we used an army "Slick" (a troop transport helicopter) to insert into the western edge of the swamp. This was near the area where we had been ambushed earlier. The terrain was a little higher than the eastern part of the swamp that was closer to the coast. We usually made contact with the Viet Cong in this part of the swamp. That day proved to be the case.

The positive aspect of a helicopter insertion over a boat insertion was that the Viet Cong were accustomed to hearing helicopter traffic over the swamp zone. On the negative side, the helicopter is a very big slow-moving target as it makes a landing. It is a time of extreme concern for all hands on board. If we had been making a boat insertion into an area this deep in an area the VC controlled, we would have been tracked by the VC to our ambush site. We wanted to make contact on our terms.

Within minutes of our insertion, we heard automatic weapons being fired about a quarter mile away. We knew that there were no friendly forces in the area and that the gunfire sounded like VC troops were being trained on a fully automatic weapon in a Viet Cong base camp. After about 30 minutes on our ambush site, we saw two young men emerge on our left from a small stream onto our larger ambush stream. They were holding a fish net in waist deep water, moving very slowly towards us on the opposite creek side of our position. Their actions indicated they were looking, as their heads kept turning from side to side, rather than fishing.

Our helicopter insertion, which is a very loud procedure, must have produced concern in the enemy base camp and these men had been sent out as scouts. They obviously had a good idea of where the helicopter insertion took place as they passed directly in front of us on the opposite side of the stream. They made easy targets but we wanted them as prisoners, hoping they would cross to our side. After going a few hundred yards to our right, they crossed to our side and turned in our direction. When they were 10 yards in front of our position, our Vietnamese interpreter told them to stop and surrender. We wanted a prisoner and gave him every chance to stay alive.

The fish net they were holding went flying straight up. They were looking at men with green faces with guns pointed at them. (We later learned that the VC referred to us as "devils with green faces"). Finding what they were looking for, they both attempted to escape. The younger of the two ran up a small stream bed and escaped. The other older man was crawling up the ten foot muddy bank twenty feet away. The tide was out, making for his slow progress in the thick mud. We were watching the last moments of his life in slow motion.

Our interpreter told him three times to stop, giving him every possible chance to stay alive. He could not escape unless he made it to the tree line. That was not going to happen. About the time he reached the safety of the mangrove foliage at the top of the bank, our platoon leader gave the order to fire, our machine gunner ended his Viet Cong activity.

We had now compromised our position, and worn out our welcome. It was at most a quarter of a mile to the VC base camp. (Much too close for comfort) Certainly it was time to use the return part of our round trip helicopter ticket back to our base.

TET AND MAJOR BATTLES

(1968-1969)

January 21, 1968 — Siege of Khe Sanh began. The isolated Marine outpost was under siege into April 1968. 205 Americans died related to the battle.

January 30, 1968 — Tet Offensive began by the North Vietnamese and Viet Cong. Over 100 military and civilian installations across South Vietnam were struck in coordinated and surprise attacks on the first day of the Lunar New Year. This caused news anchor Walter Cronkite to announce that the US could not win in Vietnam, even though each of the attacks of the Tet offensive resulted in military victories for the South Vietnamese and Americans. The real impact was on American popular opinion at home and the will of American political leaders.

March 31, 1968 — President Johnson announces he will not seek re-election. His announcement was, "I shall not seek and I will not accept the nomination of my party for another term as your President."

April 4, 1968 — Martin Luther King, Jr. was assassinated in Memphis, TN. Riots erupted in the streets of every major US city. Many Americans thought the country was being torn apart by divisions over race, equality, and the war in Vietnam.

June 5, 1968 — Senator Robert Kennedy, presidential candidate, was assassinated in Los Angeles, CA. Brother of slain President John F. Kennedy, Robert was a source of hope that he could start healing our country.

August 26, 1968 — Riots broke out at the Democratic National Convention in Chicago, IL. The National Guard was mobilized the help the Chicago police in quelling the riots and restoring order.

November 5, 1968—Richard Nixon elected President of the United States, running on a platform of withdrawal from Vietnam. His plan was called "Vietnamization" with a plan to equip South Vietnamese military and withdraw American troops.

December 31, 1968—American deaths in Vietnam reached 30,000. By all accounts, 1968 was the most tumultuous year in American history. The number of Americans killed in Vietnam had doubled in a single year and the damage to American popular support was something that would not be recovered for the rest of the war.

April 15, 1969—Woodstock… more than 400,000 people gathered in Bethel, NY for the Woodstock Music Festival. It is considered a definitive moment for the counterculture generation.

April 30, 1969—American troop strength in Vietnam peaks at 543,282 in-country.

May 10, 1969—Battle of Hamburger Hill in Vietnam's A Shau Valley. With over 70 killed in action, it sparked public debate of America's strategy in the war. Reacting to public and political pressure, GEN Creighton Abrams altered US strategy from "maximum pressure" to "protective reaction" in an effort to lower casualties.

July 20, 1969—Neil Armstrong becomes first person to set foot on the moon.

September 3, 1969—North Vietnam leader Ho Chi Minh dies at age 79.

November 12, 1969—News of My Lai Massacre in South Vietnam reaches the US. The massacre was the mass killing of Vietnam citizens by US Army soldiers. Despite the atrocities, a US helicopter pilot tried to stop the killings and rescued civilians.

Source: www.vvmf.org/VietnamWar/Timeline

STEVE PRESSER

US Army — Medic

Dates of Military Service: 1967 to 1970

Unit Served with in Vietnam: D Company, 2nd Bn, 16th Inf Regt, 1st Inf Div

Dates Served in Vietnam: Mar 1968 to Mar 1969

Highest Rank Held: SP5

Place of Birth and Year: Blytheville, AR — 1944

COMBAT MEDIC WITH FIRST INFANTRY DIVISION

I arrived at Bien Hoa Airport in March 1968, three days after it was burned during the Tet Offensive. All the buildings at the airport had been burned and some were still smoldering. I will never forget that stench, thinking I had landed in Hell for sure, which turned out to be a preview of what the next year was going to be like. I was a naive young man full of adventure and thought I was prepared for the challenges ahead of me. Little did I realize that I was quite ill-prepared for what was to come.

I was assigned to the 1st Infantry Division, 2nd Brigade, D Company, 2/16 Infantry Regiment. I went to my base camp in Di An as a combat medic. I was only there for a couple of days before I was sent out to the field as the platoon medic. Upon arrival, we immediately had a mortar attack and there were some casualties, so I immediately went to work patching up guys and putting them on the medevac helicopters. We were always so glad when the medevac team arrived, knowing that the guys would get the medical help we could not provide in the field. This was my first introduction to the war in Vietnam, but it was only a preview of what was to come. I quickly became friends and formed close relationships. I felt like all the soldiers wanted the medic to like them so I would come to their aid in case they got wounded!

There were many small battles which usually lasted only a few minutes, but there were three battles that I will never forget. One of them came when I was pulled out of base camp and taken to join our troops who had had a significant battle the day before. I had no idea why they sent for me until I got to my unit and I was told that the medic had been killed the day before. We immediately came under fire and I was never so scared in my life. The medic who was killed was my best friend and he was gone. The impact of hearing about his death and then the ensuing firefight was more than I could digest. The near panic and fear just numbed me. Fortunately, we had no casualties that day since I was not in good shape to treat anyone.

The next battle came when my company had surrounded a village for an ambush, because there was word that the regular NVA was wanting to overrun the village for food. We were in position when we spotted them coming directly towards my platoon. We threw up flares and could clearly see them no more than 50 yards away. A fierce gun battle ensued and we threw all the fireworks we could at them. I even fired my M16 which was unusual since medics were not required to engage. We eventually ran low on ammo and it looked like we were going to be overpowered and all would die.

Just as hope was fading, in came the APCs right behind us and began firing with their 50 caliber machine guns, so the NVA retreated and we were all hysterical with relief. I have never seen such bravery from an American soldier as I did that night. No soldier panicked and nobody ran away as we coordinated with each other with conserving as much ammo as we could. The problem for me was that not one of us was recognized for holding up under these severe circumstances. Every guy in that platoon deserved a medal!! So Be It. The most important thing was that we had no casualties, but the NVA left many dead on the field as they fled. Fortunately, the village was fortified with mounds of dirt, which probably saved many lives.

The third significant battle came when my company received gunfire off and on for several hours and we had 21 casualties. I was the senior medic so I was treating wounded soldiers under fire and also had the responsibility of organizing them in order of the seriousness of their wounds for rapid aerial evacuation. I was experienced enough at this point to know what to do and was awarded the Bronze Star with V device plus received a Purple Heart because of shrapnel in my leg.

We were attached to the 11th Armored Cav for about a month. During that time we marched many miles per day and trudged over many creeks requiring us to hold on to an overhead rope or hold on to each other to cross. We were not allowed to take our boots off at any time. As a result, many soldiers became infected with jungle rot. It was so bad that I told the Commanding Officer that a soldier had to be sent back to the rear. He balked at the idea so I told him I would report the illness to the doctor in the rear. He relented so a medevac copter was sent out to pick the soldier up.

Then the fireworks really began, because the Sergeant Major told me he was going to have me court marshalled for insubordination to a Commanding Officer. I have never been cussed out like that before as he circled around me and threatened to hit me. All I was thinking was that I wanted him to hit me so I could be sent back to the rear. Nothing was done, but the doctor in the rear told me that the Sergeant Major told him that I was doing a great job and to keep me in the field!!

Medics were supposed to be rotated out of the field after six months, but the Commanding Officer of my Battalion at that time wanted me to be the Battalion Medic in charge of the forward Aid Station for the remainder of my tour. I was now exposed to a lot of tactical information I would not have otherwise known since I was now part of the Command Group. Some this information just turned my blood cold. I finished my tour of duty as a SP5 while filling the shoes of an E6. In order for me to be promoted to an E6 I would have had to reenlist, which I declined to do!

THOMAS A. "TOM" ROSS

US Army — Special Forces

Dates of Military Service (Active and Reserve): 1966 to 1992

Unit Served with in Vietnam: Detachment A-502, 5th Special Forces Group

Dates in Vietnam: Jan 1968 to Dec 1968

Highest Rank Held: Major

Place of Birth and Year: Huntington, WV — 1945

EVER DONE ANYTHING REALLY STUPID?

The sun was just rising over the Citadel's east wall and pouring golden rods of sunlight across A-502's inner compound. The only sounds were the low hum of the camp's 100kw generator and the occasional crow of roosters in the nearby village as they announced the arrival of morning.

The Citadel was a very old Vietnamese fort that looked as though it could be part of a Hollywood set for an old Cary Grant French Foreign Legion movie. In fact, the French had occupied the fortress before the Americans arrived in Vietnam. Now, it was home to Vietnamese units and their American advisory team, U.S. Army Special Forces Detachment A-502. I was the team's Operations and Intelligence Officer.

It has been said that war is hell. Actually, it can be much worse than that. The night before this glorious morning, the war had tested me to a level I had never experienced or expected. At about 1:00 in the morning, our radio-man woke me to tell me that my best friend in Vietnam, Lieutenant Bill Phalen, was in contact. That meant that he and the unit he was advising were engaged in battle with an enemy unit.

Our radio man woke me because I was the Duty Officer that night and was responsible for providing the support any of our units might need. As it turned out, Bill's platoon size unit of about 30 men had encountered an enemy unit of about 100-125 men and they were seriously outnumbered.

To save my friend, I contacted three different artillery batteries for support and had all three fire high-explosive rounds without any marking rounds. The situation was that desperate. Normally, marking rounds are fired to be sure you aren't going to hit friendly positions with killing high explosives. When all that ordinance hit the ground, we felt the earth shake at the A-502 compound, several kilometers away from Bill's position. When I felt the concussion in the radio-room, I feared that I had killed my friend in my effort to save him.

Later in the night or early morning, a FAC (Forward Air Controller) reported to me that he thought some of the artillery I had fired to save my friend Phalen had landed in one of the nearby villages. I knew the village and many of the children living there. If the worse had happened and I had killed women and children, I didn't know how I could ever explain that to my mother. So, when I say that war can be worse than hell, I know exactly what I'm talking about.

As it turned out, I hadn't killed my friend—I had saved him and, shortly after sunrise, the FAC reported to me that I hadn't hit the village with artillery. What the FAC had seen and mistaken for an artillery strike was a flare parachute that had landed in the village and started a fire.

However, in the time that it took for me to learn that I hadn't actually killed a friend or killed women and children, I experienced the crushing belief that both had actually happened. There aren't words in the English language to fully express how that belief feels—it is a feeling you have to experience to understand. And, I don't wish that experience on anyone.

Knowing that my friend was safe, when I was given the added good and incredibly welcome news that artillery hadn't hit the village, my spirit soared. I was sure I could fly. But, since I really couldn't fly, I went to the next best alternative—a teammate's motorcycle.

Before I went to Vietnam, I often rented a dirt bike and rode through the piney woods of Northwest Florida. It was always relaxing and great fun. This day, I needed to be reminded of home and the values of home. So, attired in combat boots, cutoff jungle fatigues as shorts, and an olive drab T-shirt topped off by my green beret, I headed out to ride. And, for security, I had my .45 caliber pistol strapped around my waist. I was a real military fashion plate.

Before I knew it, the beauty of the day had beckoned me beyond the streets and safety of the local village. The wind and sun felt good on my face as I raced along the trails and rice paddy dikes. Feeling as though I had left the war back in the A-502 compound, I enjoyed the beauty and pastoral serenity of the Asian countryside. Time and location seemed unimportant, but it soon became evident that I had ridden much too fast and had definitely gone much too far from camp.

As I came to a sharp turn in the trail, I downshifted and slowed just enough to navigate the curve and once again began to accelerate. Upon rounding the turn, I was surprised by what at first appeared to be a group of Vietnamese farmers. They were all squatted in a circle and a couple of them had sticks in their hands. I waved as I approached from the distance, hoping not to scare them and to let them know I saw them on the side of the trail. As I rode closer, they stood up. Clearly, something was strange about the situation. The men were all similarly dressed in greenish faded khaki clothing that I had seen before—on the bodies of dead NVA soldiers.

The hair on the back of my neck immediately stood upright, telling me all was not right. Just then, I noticed what appeared to be the muzzles of weapons lined up on a log near the men who were now standing at the edge of the trail. Because the weapons were on the far side of the log, and because I was moving by so quickly, I only had a glance. But they looked remarkably like AK-47s—the enemy's weapon of choice. Exactly what type of weapons they were seemed of little consequence.

It is at times like this that you realize how magnificent our brains are and how quickly they can process information—quicker than any computer I've ever used. I immediately knew I had ridden into an extremely dangerous situation and could either stop and turn around or ride past the danger. In a split second, I considered those options. If I tried to stop, turn around, and go back the way I came, the men would have more than enough time to reach their weapons and probably kill me before I got out of sight. If I continued towards them, they would clearly see my beret and recognize my shiny American lieutenant's insignia, jump for their weapons and maybe kill me before I got out of sight. I don't know how or why, but a third possibility occurred to me.

Because of the quickness of our encounter and my speed, neither the group nor I could do much except to acknowledge each other's presence. Hoping to surprise and distract them, I showed them a huge smile and waved again as I rode past within an arm's length of the small assembly. None of them returned my smile, but two gave me a half wave, perhaps a reflex response, while another one kicked sand across whatever they had scratched in the dirt. It seemed obvious he didn't want me to see whatever it was that they had drawn.

After passing the men, I accelerated even more and waited to hear the crack of gunfire and feel the sting of a bullet. Luckily, I was able to disappear around the next turn in the trail. Without ever looking back, I left the area as quickly as I could.

As I rode a paddy dike headed back to A-502, there was little doubt in my mind that the men were NVA from the Dong Bo Mountains on a daylight recon patrol. I had been—very lucky.

I had momentarily allowed myself to become disconnected from the war and had gone where I shouldn't have been, an error in judgment that could have cost me my life. While we knew that NVA and VC routinely infiltrated the city of Nha Trang, our local village of Dien Khanh, and the other outlying villages, such action typically occurred at night. I didn't expect to see the enemy on such a bright sunny day. And, from their reaction, I feel quite certain that they didn't expect to see me in what amounted to their backyard.

Weeks passed before I said anything about the incident on the trail to anyone. I didn't want Major Lee, the team's commanding officer, or my teammates to find out because of my concern that they might think I was just dog-stupid. Although, a dog would surely have had the good sense to avoid an area where he knew wolves prowled. Ironically enough, after sharing a few bottles of beer with him one Saturday afternoon, Major Lee would be the only other person in Vietnam to whom I ever told the story. His response, "Damn it, Ross! Don't do that again! I don't want to have to write a letter to your family."

While he was appropriately firm, Lee wasn't really angry. In fact, he seemed amused by the tale—maybe because of the beer. Whatever the case, he never mentioned the incident again and neither did I. While I continued to take motorcycle rides, I never rode that far away again.

There's a biblical saying that God takes care of children and fools—I am a very thankful believer.

NORMAN E. ZOLLER

US Army—Field Artillery Officer (later Judge Advocate General Officer)
Dates of Military Service (Active and Reserve): 1962 to 1993
Units Served with in Vietnam: Detachment B-130, US Special Forces; 3rd Bde, 82nd Airborne Div.
Dates in Vietnam: Sep 1964 to Mar 1965; Mar 1968 to Feb 1969
Highest Rank Held: Lt. Colonel
Place and Date of Birth: Cincinnati, OH—1940

DUTY WITH THE 3RD BRIGADE OF THE 82ND AIRBORNE DIVISION

Attacking 13 cities simultaneously in central Vietnam, one of which was Hue on the Perfume River, the Tet Offensive initiated by the Viet Cong formally began on January 30, 1968 and continued until September 23, 1968. Many historians consider the Tet Offensive, which has been compared to the Battle of the Bulge in 1944, the turning point of the war. During the battle of Hue alone, it was estimated that 216 Americans and 452 South Vietnamese soldiers were killed. On the Viet Cong side, an estimated 5,000 were killed, most of them by artillery and air strikes. By mid-February into the Tet Offensive, enemy casualties had risen to about 39,000, including more than 33,000 killed. Allied casualties amounted to 3,470 dead, one-third Americans and 12,062 wounded, almost half of whom were Americans.

It was against this backdrop that General William Westmoreland requested additional combat forces to be deployed immediately from the United States. He also specified that a portion of these units be airborne troops, reflecting his confidence in the proven ability and reputation of these hard-fighting soldiers. As it turned out, the only stateside unit from which such soldiers could be drawn was the 82nd Airborne Division about which General Westmoreland already knew.

And so after an alert and within 24 hours, an advance party of the 3rd Brigade of the 82nd Airborne Division deployed from Fort Bragg to Chu Lai on February 12, 1968, with remaining elements of the Brigade of about 5,000 soldiers arriving at Phu Bai located 10 miles south of Hue in late February and early March. It was a war-time and dangerous environment at and near Hue with offensive and defensive combat operations expanding apace. And more was to come in the months ahead until the principal enemy attacks were quelled by mid-September. By that time, there were more than 500,000 American service members in-country.

I joined the 3rd Brigade at Phu Bai in early March 1968 as a replacement officer having been diverted from my initial orders to the 25th Infantry Division at Cu Chi. Although I became the Brigade Inspector General in December, following the Brigade's relocation in September to Phu Loi on the Long Binh security perimeter, I served first as the Assistant Personnel Officer (S-1) from March until November 1968. The Personnel Office was responsible for all manner of human resources support including personnel policies; individual staffing, assignments, and replacements; promotions; finance and pay services; chaplain activities; legal services; records; information management; enemy prisoner of war transportation and control; and casualty reporting.

In considering just one of these responsibilities, the importance and sensitivity of casualty reporting cannot be over-emphasized. In this area one must scrupulously guard against mistakes because to do so would cause untold angst and grief were an error made in reporting (even inadvertently) incorrect or false information to a soldier's next of kin or family member.

Tragically, the 3rd Brigade suffered more than 100 soldiers killed during its first three months in-country. Overall, the United States suffered 58,220 fatal casualties in Vietnam.

When a soldier suffers serious injury or is killed, communications with family members are extraordinarily critical and sensitive. Among other measures, a survivor assistance officer is appointed who helps guide the survivors through an extensive roster of assistance and benefits for the soldier's spouse and his or her dependents. Consider, for example, extracts of a letter from one of the Brigade commanders to the widow of one of its fallen solders:

"Dear Mrs. _____:

It is with a very heavy heart that I sit here tonight to tell you how I feel about the loss of your husband…I felt a special closeness to him because …

There isn't much I can say. I was in the area of the battle. One of his men was wounded and _____ ran to get the man out of danger. He gave his life to save one of his men. His classmates, many of whom are with the brigade, wept openly, and I lost another piece of me.

… I think we both know that he would never pause, even for an instant, to leap to help one of his men. May his courage and selflessness give you strength in this hour of sorrow. Those of us who knew him will never forget him and will continue to be inspired by what he taught us. God bless you."

For anyone who has served as a military commander, the duty of having to send words of condolence and comfort to a soldier's next of kin is one of the most difficult and heart-wrenching responsibilities. Not incidentally, soldiers who serve in combat support roles, likewise, can be affected by incoming artillery and enemy soldiers attempting to infiltrate and gain access to purportedly protected compounds and are not immune to death and serious injury.

I have been in such circumstances, and I would only observe that in peace and in war, the role of a commander can be a lonely place, but also a heart-warming and personally enriching experience. I consider myself very fortunate and honored to have been in our Nation's military service.

JAMES F. CRAWFORD

USMC — Communications Officer

Dates of Military Service (Active and Reserve): 1966 to 1999

Unit Served with in Vietnam: 7th Comm Bn, 9th Engineer Bn

Date in Vietnam: Jul 1968 to Aug 1969

Highest Rank Held: Colonel

Pace of Birth and Date: Washington, DC — 1941

DETERMINED TO SERVE

I was born in Washington D.C, and raised in Northern Virginia. I graduated Augusta Military Academy and Wheeling Jesuit University. In my senior year of college, 1965, I was summoned before the draft board. Having had knee surgery due to a football injury, I was declared 4-F: physically unfit to serve in the military. Upon graduation from university, I secured a sales position in the steel industry. While in sales training I elected to have a second surgery on the same knee. I had a goal in mind. I began extensive physical training and soon was in good shape. I had decided if I failed to serve my country in a time of war, as so many young men had, I would forever be ashamed of myself. Soon afterward, I visited a Marine recruiter and applied to Officer Candidate School. I was told two surgeries on the same knee made my acceptance unlikely. Several weeks' later in Quantico, VA, while undergoing a pre-induction physical, I was told I was Not Physically Qualified to serve in the Marine Corps. The sales training program proved its worth as I convinced the orthopedic surgeon he was wrong and I could withstand the rigors of OCS.

Things went well in OCS until the last week when I crushed a disc in my back and I was admitted to Bethesda Naval Hospital for a laminectomy. Upon returning from medical leave, the neurosurgeon informed me he had arranged a disability rating and medical discharge for me. I convinced him I should at least be given the chance to proceed with my training. He relented and I reported to The Basic School (TBS). After a few months of limited duty and rehabilitation, I graduated TBS, reported to Communications School and two months later upon graduating received orders I had requested for RVN.

I reported to 7th Communications Battalion, 1st Marine Division, on Hill 10 south of Da Nang in I Corps. We provided our own security as we worked outside the wire daily. Additionally I had radio personnel attached to units of 1st Marine Regiment. I visited them frequently by chopper or jeep. Our rear was adjacent to a battery of Army 175 mm guns so we received rockets nightly. I remained there for five months.

Third Marine Division operated along the DMZ. During Tet of 1969, they lost a great many company grade officers. The CG 3rd Mar Div requested temporary replacements from 1st Mar Div so I was one of many ordered north for five months. Our rear was LZ Stud, AKA Vandergrift Combat Base. I flew each day into the Ashau Valley to visit my Marines in LZs and fire support bases in support of 9th Marines. In Vandergrift we lived in bunkers.

My relief arrived at Vandergrift in the spring of 1969. As I was about to depart, he informed me 7th Comm Bn was on China Beach waiting embarkation to Okinawa. I reported in to battalion and while debriefing my Commander, I told him I was not well-suited for garrison duty and requested he find me a billet to remain in Vietnam. Within a week, I reported to 9th Engineer Battalion based in Chu Lai where I served the remainder of my thirteen-month tour. I was the Communications Officer and Assistant S-3. The Battalion was tasked with clearing Highway #1 of mines each morning and hard-surfacing the fifty-five miles north to Da Nang. Each morning I caught a chopper, flew north to assess damages to Highway #1 and bridges caused the previous night. For the remainder of the day, I ensured communications were operational throughout our AOR and with subordinate units.

I was fortunate to have moved extensively within I Corps. I was privileged to have served with some of the Corps' finest enlisted and officers. I lost forty-four classmates in Vietnam. I participated in seven major operations, to include Dewey Canyon.

"Warriors never forget their fallen brothers. They carry a sorrow, which cannot be put into words, to their graves."

DAN BENNETT

US Air Force — Flight Engineer

Dates of military service (Active and Reserve): 1967 to 1975

Units served with: 4th Special Operations Squadron, Da Nang AB and 18th SOS, Nha Trang AB

Dates in Vietnam: 1967 to 1968; 1972 to 1973

Highest Rank Held: Tech Sergeant (E-6)

Place of Birth: Atlanta, GA — 1947

WE OWN THE NIGHT!

I served two separate tours of duty in Vietnam. My first tour of duty (67-68) was just in time for the big TET Offensive. During my second tour I experienced the Easter Offensive of 1972. Lucky me!

My duty was as Flight Engineer on the AC-47 "Spooky" gunship. The aircraft was a converted C-47 (the same airplanes used to carry paratroopers into Normandy during WW-II). The AC-47 was equipped with three 7.62 mini-guns: aimed and fired by the pilot. As flight engineer, I was responsible for all aircraft systems including mechanical, electrical, fuel, hydraulic, and pneumatic systems, and I monitored aircraft performance during flight and supported the pilot in whatever capacity needed. I also calculated the weight and balance of the aircraft prior to each flight. On the ground, I supervised aircraft servicing and all maintenance that was needed. The C-47 was very basic and simple to operate. Our missions were mixed day and night sorties providing close ground support for troops and dropping flares to illuminate the night. The mini-guns would produce 4,000 rounds per-minute per gun.

My second tour I was Flight Engineer on the AC-119 Gunship. As such I was responsible for all aircraft systems during flight. This included the electrical, hydraulic, pneumatic, fuel transfer, engine monitoring as well as calculating the aircraft's weight and balance prior to each sortie. The AC-119, call sign "Shadow, "had a flight crew of three (pilot, copilot, flight engineer), and a gun crew of eight in the back.

The C-119 (C for Cargo) was built in the late 1940s and used during the Korean War. The AC-119 was an upgrade from the AC-47 and equipped with more firepower and night vision cameras. We had two 20 MM Gatling guns, each capable of 6,000 rounds per minute and four 7.62 miniguns, each capable of firing 4,000 rounds per minute. The underside of the aircraft was painted black making it completely invisible at night "until" our guns opened up. That made it easy for the enemy gunners to see where the tracers were coming from. That is when the NVA would open up with their anti-aircraft artillery (AAA) and the dreaded .51 caliber {Russian made} machine gun. Taking hits became routine and expected on every mission. Our main target was the Ho Chi Minh trail by seeking and destroying trucks, armored vehicles, and troops. The Ho Chi Minh trail was well fortified by many enemy gun emplacements.

Our gunship crews had very little free time to go downtown or to party, because 99% of our sorties began after 8 PM, lasting five to six hours. The AC-119 was equipped with night vision infrared camera's allowing us to be the "eyes of the night" and Charlie couldn't hide under the trees as they once did. The large infrared lenses allowed us to see the heat signatures of troops and trucks on the Ho Chi Minh trail.

By the time of the Easter Offensive 1972, the North Vietnamese gunners had drastically improved their shooting skills by adding radar to their guns. The exploding shells (flack) caused our aircraft to shake violently each time a shell detonated in close proximity. This was nerve racking because there was nothing we could do about it. All too often, holes would appear in the aircraft floor and walls. That would work on you mentally and physically. It was routine for our flight crews to go without adequate crew rest, irregular meals (if any), and each sortie was fatiguing due to constant and sudden shifting of gravity as the aircraft ducked and dodged enemy fire. This caused your body weight to go from 'zero' pounds one second to 'three times your body weight' the next. Receiving AAA fire was horrible.

On a brighter note: I had a night off and was in my hooch writing a letter home. It was dark except for a small lamp on my desk. Something caught my eye and when I looked, I saw nothing. A few minutes later, something caught my eye again and this time I saw it. It was a rather large snake poking out its head through the slits in the sand bags. Our hooch had sandbags on the lower portion. I didn't know what kind of snake it was, but I panicked, pulled my revolver and emptied all six rounds. Of course, I missed the snake and it pulled back into the sandbags. However, I did hit my locker several times and the sandbags had holes pouring sand out on the floor. No snake in sight!

The firing of my weapon caused our compound to sound the alarm. I had to stand before our squadron commander the next morning and explain why I was shooting my weapon in the middle of the night. After I explained what had taken place, my commander laughed and suggested a little target practice might be in order.

The next evening two guys killed a seven foot Cobra coming from my hooch going to a neighboring hooch. I shudder to think of how many times that snake came into my room while I was asleep. It gives me shivers just thinking about it!

MY WELCOME HOME

Late December, 1972, I departed Vietnam on a military charter flight, destination Hawaii. We were advised to change into civilian clothes when we hit stateside to prevent harassment. "Harassment, why, would people want to harass us?"

In Hawaii we boarded a Delta 747, next stop Los Angeles, California. Some of the guys changed clothes in Hawaii while others (like me) did not. I was going all the way to Charleston, South Carolina, and I didn't want to have to wear the same clothes for that length of time. I decided I would change in LAX.

We arrived at LAX about 4:30PM and as I was departing the plane I noticed a group of long haired, young men wearing bright orange robes. They were chanting and banging on tambourines. I saw they had cornered several Marines in uniform, causing a scene. Being in uniform myself, I didn't want any of it, so I ducked around to the far side of the jetway, making my way to the men's room to change.

I thought I had made a clean get-away when suddenly someone

grabbed me by both shoulders and pulled me back rather abruptly. Without thinking or looking, I threw my elbow back as hard as I could. I had just spent the previous 13 months in a combat zone. My elbow hit the guy in the throat and down he went, gagging and choking. It was a "Hare Krishna" that grabbed me. No serious harm done except I knocked him on his ass. Within seconds all the other "Hare Krishna's" had me surrounded, yelling and giving me pure hell. The airport police arrived, and I explained what exactly had taken place. The policeman handcuffed me, and I was taken away. I could not believe this was happening to me.

I was taken to one of the precincts and locked away in a holding tank for three nights. It was one of the most foul-smelling and nasty places I had ever seen. I was in with drunks, drug users, and criminals. That was New Year's Eve 1973. I wasn't allowed phone access, no showers, no change of clothes, and unable to brush my teeth for three days!

That was the only time in my life that I have had any trouble with the police. After being there three days, I was released and told, "You can go home now!" My family was waiting for my flight to arrive in Charleston, and when I wasn't there, they had no idea of what to think!

Looking back, perhaps I should have departed the plane skipping along, singing and tossing flowers to the people in the jetway. I know the Christian thing to do is to Forgive and Forget, but I think Lee Harvey Oswald received better treatment than I did. I just didn't have it in me to remain passive when suddenly grabbed from behind and pulled back.

JAMES J. "JIM" HOOGERWERF

USAF C-130 — Pilot

Dates of Military Service: 1966 to 1973

Based off-shore in the Philippines at Clark Air Base: 773rd TAS, 463rd TAW

Unit Served With in Vietnam: TDY to 834th Air Division, Det 1 TSN, Det 2 CRB

Dates Served in Vietnam: May 1968 to May 1970 (not inclusive)

Highest Rank Held: Captain

Place of Birth: Detroit, MI — 1943

"CAN'T BELIEVE I AM ACTUALLY HERE"

Clark Air Base is located approximately fifty miles to the north-north-east of metropolitan Manila along the western edge of a broad verdant plain on the island of Luzon in the Republic of the Philippines. Long a site of American power in the Pacific, Clark Field (as it was still referred to locally, harking back to Army Air Corps days) was an important support facility as the war in Southeast Asia escalated. With the war at its peak, peaceful harmony had long since given way to wartime bustle. It was into this setting that I arrived early on the morning of Friday, May 24, 1968.

A Second Lieutenant commissioned through AFROTC, I was a twenty-four-year-old college graduate, a rated USAF pilot, and a combat qualified C-130 copilot. To the United States government, I represented a $1.5 million investment. Whatever the uncertain future held in store for me, would be in exchange for that American taxpayer money. A volunteer, I was ready to meet my obligation.

Getting off the plane was not an unpleasant experience. The warm tropical setting was idyllic. Still, it was somewhat disquieting to realize the war was drawing me closer. I had harbored the thought during my training that it might be over before I got there. About that, I had mixed feelings of relief and regret — relief I would be safe, but regret I wouldn't be a part of it. Whatever emotions I felt were put aside as I made my way through the arrival formalities at the MAC terminal and was met by Maj. Joe E. Johnson, the 773rd Tactical Airlift Squadron (TAS) administrative officer. He drove me to operations where I tried to appear as sharp as possible in my travel wrinkled 1505s.

I came to attention before the desk of Lt. Col. Merle B. Nichols, the squadron commander, saluted, and announced: "Second Lieutenant Hoogerwerf reporting for duty, sir!" Still being a gold-bar Lieutenant, I was something of a novelty for the senior officers. The squadron was in transition as many people rotated out and new ones arrived. I was one of the first group of young replacement pilots coming in to take over from the older career officers. I was told to report back on Monday, and on Thursday I would be going to Vietnam.

In my first letter home, I wrote: "Now I am in a house off base where I shall live from now on. There are other bachelors living here and a room was vacant. It is a really nice place if you can imagine bamboo furniture, ceiling fans, and exotic wood carvings. Well, I think I will like it here. We have a maid that does all the house work."

In-processing on Monday involved making the rounds to make sure everyone knew I was there and had my records. For a newcomer, unfamiliar with the base, this was a tedious, but necessary chore. One stop was at the photographer's to have my picture taken for identification purposes should I become missing behind enemy lines, or killed in a crash (disquieting thoughts). These were taken in my flight suit, but all patches had to be removed. Back at the squadron, I was put on the schedule for a flight in the morning. My training in the US had been in a C-130 "E" model. Here the 463rd Tactical Airlift Wing (TAW) flew the earlier "B" model and I needed a familiarization flight.

Lt. Col. William J. Bush was the assigned instructor. I met him and the other crew members at operations. Col. Bush briefed the flight. I would be in the right seat and we would fly to Naval Air Station Sangley Point at Cavite City on a peninsula in Manila bay and shoot landings (it was off Sangley that Admiral Dewey defeated the Spanish fleet in 1898). The runway length was well within the capability of the C-130 although, situated along the shore, there was water immediately off either end. A raised sea wall deflected the surf, but created a potential hazard for air operations. On touch-and-goes, Col. Bush would set the power and re-set takeoff trim and flaps.

A crew bus pulled up for the short drive to the hardstand where our assigned aircraft was parked. Immediately evident, as I entered the cockpit, was the seat armor. "It wasn't heavy enough to stop the big stuff," Col. Bush explained, "but it will protect you from small arms fire." Armor plates were also placed on the floor and in front of the lower cockpit windows.

At Sangley, I made five landings. All went well enough except for one of the touch-and-goes. I touched down long on a full flap landing. Col. Bush, intent on my flying, set the power, but the plane wasn't accelerating like it should. We were quickly running out of runway without gaining enough airspeed to fly, but going too fast to abort...something was wrong. The flight engineer saw what it was and simply said, "flaps." Col. Bush calmly selected them from full down (high drag) to the take-off position. We quickly picked up speed and I pulled back on the controls to raise the nose just before the sea wall flashed by underneath.

Three things about the incident impressed me. First, it was the flight engineer who saw the problem and spoke up. Every member of the crew was a valuable resource. Secondly, Col. Bush, without hesitation, calm-ly and deliberately reset the flaps.* Third, once the flaps were set, the aircraft responded immediately. The C-130's reputation as a forgiving airplane was well deserved. After the flight, it was confirmed I would leave Thursday for sixteen days temporary duty at Tan Son Nhut (TSN) to fly airlift missions within South Vietnam. I went home to get ready.

In a letter to my parents from TSN, I wrote: "Can't believe I am actually here."

About my initial experience, I later wrote: "The Philippines is a very beautiful and peaceful country after Vietnam. It is like being on vaca-tion." I did not realize then how the "vacations" were little more than reprieves from which I could never totally unwind. It was an unusual way to fight the war.

* Later in my civilian flying career with Delta Air Lines, the concept

of Crew Resource Management (CRM) grew out of instances of human factors as the cause of aircraft accidents and incidents.

RON SHERMAN

US Army — Signal Officer

Dates of Military Service: 1966 to 1969

Units Served with in Vietnam: 1st Military Intelligence Bn, HQ Company, 45th MI Detachment/ARS

Dates Served in Vietnam: Nov 1967 to Oct 1968

Highest Rank Held: 1st Lieutenant

Place of Birth and Date: Cleveland, OH — 1942

A PHOTOGRAPHER'S VIETNAM ADVENTURE

My path to my military service is an interesting story. When I got out of college in 1964, the University of Florida offered me a position of Staff Photographer with a possibility for an academic position. Theoretically, that would keep me out of the service, but that never happened. So when I saw that I couldn't stay out of the military that way, I went to graduate school. Then my draft board said, "At the end of the school year whether you graduate or not, we're getting you." In 1966 I applied to the Peace Corps and told them I was going to be drafted. They were taking their time so I looked around at the various services. I felt like the Army had the best program for me.

I had been doing photography since I was 14 years old, and went to college for photography, so I figured the Signal Corps would be the way to go. I enlisted for the OCS program in the Army Signal Corps, figuring two years and ten months was better than being drafted and carrying a rifle. I thought it was a better way of using my ability than the other services. About a month into basic training, I got a letter from the Peace Corps saying, "We're ready for you." I wrote them back and said, "Well, I'm not ready for you." They responded, "Well, when you get out of the

Army, come see us." I responded to them, "It was the Army or the Peace Corps, and you lost."

Vietnam, at the time, really wasn't on my radar in terms of going over there and documenting the war. Given the choice that I had, I just felt like entering the Signal OCS program was the best path to serve. The biggest shock to me was when I finished the OCS program, and my duty assignment came down, it was 8500 Photo Officer. I was really happy. I felt like, gee, somebody goofed. I had an assignment that matched my experience.

My first assignment after OCS was to the 45th MI Detachment, 1st Military Intelligence Battalion, Aerial Reconnaissance Support, which was forming up at Ft. Bragg, North Carolina for duty in Vietnam. The 45th MI detachment was a support detachment and their mission was to interpret aerial photography made from Air Force missions. My job as the photo officer was to produce, with the help of a sergeant and five enlisted men, prints from the long strip of Air Force negatives. Since this was a new detachment going to Vietnam and the darkroom and inter- pretation facility were in a tractor trailer, we traveled by troop ship. We left California in November 1967 with the first three weeks in Vietnam as a cruise, with a stop in Okinawa.

When I first arrived in Saigon, I didn't have specific duties, so I let the staff know that I was an experienced photographer. I started doing some aerial photography on my own and doing "volunteer" projects. One particular project which I'm most proud of was decreasing the time di- vision commanders needed to view after-action photographs. At that time Air Force photography took anywhere from three to five days to get from the Air Force plane, to us, to the interpreters and then to the division commanders. Those officers were having a problem with that time frame, because things could change very quickly. I asked myself, "Is there a way we can speed up the process for the Commanders?" Digital photography was just a dream in 1968, but Pentax Corporation had just come out with a camera that had a motor winder and automatic expo- sure control so I experimented with the camera.

The result was what I called the "Hand-Held Photo Program." All

Divisions had aircraft, photo lab, pilots and observers/photographers. The camera could be turned into a point and shoot machine. It used 35mm film, came with a telephoto lens and with a set shutter speed, the aperture would open and close depending on the light conditions and the photographer would tape the focus on infinity to insure sharp focus. Over the target area, the pilot would fly, the photographer would shoot, return to base and in a couple of hours the commander would have his photos.

I was then asked to develop an SOP (Standard Operating Procedure), get it printed and then travel to the divisions to explain the program.

Saigon was fairly safe to travel throughout the city. We would go to the restaurants and bars, shop in the open air the market and Cholon, a thriving black market trading in American Army issued supplies.

Then came the TET offensive in January 1968. That changed everything. The Officer Billets were located on the main road between downtown Saigon and Tan Son Nhut airport. We had a secure compound and were celebrating the Chinese New Year on the roof and watching the fireworks when we received word that the city was under attack. Down came the American Flag and we went to defensive positions. The only bullets we took were from the armored cav vehicles going past our compound to defend the airport. Our shouts to them went unheard because of the noise of their vehicles.

In early May 1968, I was reassigned to Hue Phu Bai where the 45th MI Detachment was finally settled, and took charge of the photo reproduction operation located at the airport. The lab that we brought from the US was set up and my main job was to make sure we didn't run low on photo paper and chemicals. When the supply chain broke down I would have to go out and scrounge what I needed. The nice thing about the lab was that it was air conditioned. Besides making prints from aerial films, we also had capabilities to process 35mm black and white film and make 8x10 prints. At 80 degrees instead of 70, which was ideal for a regular photo lab, we had to use a special technique for high-temperature film processing to prevent the emulsion on the film from sliding off.

The most exciting project for me was a classified "Top Secret" assignment that sent me to Ubon Thailand. An Air Force pilot returning to Ubon's American airbase from a mission in Vietnam spotted a shining reflection in the tree line just before entering Laos's airspace. He mentioned this observation in his after action report, which made its way to the intelligence unit at MACV in Saigon. The detachment commander asked me if I would volunteer to go the Thailand to get photos of this reflection. Without thinking, I said, "Yes."

The target was a shining spot in the tree line in the vicinity of the A Shau Valley and seen at sunset. The pilot who made the report didn't have the exact coordinates, but did have the flight path of his flight back.

My pilot had two missions, my mission authorized by J-2 MACV in Saigon to photograph the shining object, and their mission to support the jets to mark the targets with their rockets. The first day we were up, and I see all this flak going off above us, I said, like a dummy, "What's that?" The pilot said, "Well, we're waiting for the jets to come in. They wanted to try to interfere with that."

I said, "Do you guys ever get hit?" He says, "Oh, occasionally." That was my exposure to aerial combat. The second day we spotted a man on the ground pointing a manual rifle at us. The chances of him hitting us were statistically impossible. I was using a mirrored telephoto lens which gave me a clear image of the man and his rifle. I wish I could have kept a copy of that photo but everything I made that day was classified and all images were sent to Saigon.

On the third day as we're about to fly back to Ubon, we saw the light in the tree line but the sun was already down, so getting a usable photo was not possible. On landing I call Saigon, told them we got the exact coordinates and got permission to stay another day. The next afternoon, the pilot's only mission was to return to location so I could get usable images. What the North Vietnamese had done was run a phone line through the trees, from North Vietnam down the Ho Chi Minh trail into South Vietnam.

My understanding was military intelligence wanted to tap the line. I returned to my lab, developed the film, made some prints and sent everything to Saigon. I put the story away, back in my head, and said, "Okay. Nice adventure. Too bad I can't tell anybody."

In February 1969, seven months after my Thailand adventure, I read in Newsweek Magazine, The Periscope section, "HANOI STRINGS A PHONE LINE. The North Vietnamese, in order to avoid U.S. eavesdropping on their radio messages, have strung a telephone-telegraph land line down the Ho Chi Minh Trail through Laos and down into South Vietnam. U.S. photo reconnaissance pilots have picked up the shining reflection of glass insulators flashing through heavy foliage."

I called my detachment commander at the 45th and asked him if I could tell my story since the Newsweek article was published. He told me it was ok, because the CIA used stories like that to poke the North Vietnamese and, obviously, it was no longer a secret.

I had six months to serve when I left Vietnam in October 1968. My next duty assignment was in Cleveland, Ohio as a Communications Officer at an Air Defense Unit in the suburb of Warrensville. I had asked for duty as close to Cleveland as possible because my father had passed away while I was in Hue Phu Bai. I will always be grateful to the American Red Cross for getting me from Phu Bai to Cleveland in twenty-four hours. It was like having an express pass, no wait list and no delay, first class all the way.

To summarize my assignment in Cleveland should have been a no brainer. Having no experience as a Communications Officer, I appointed the newly arrived second lieutenant as the Crypto Officer, so that was one less duty I had to perform. It was just a very quiet six months. But it was also the worst time in my life being in the Army. The reason was there were only a few officers in this unit, and when you were the Officer of the Day, OD, you had a number of responsibilities.

The hardest duty was to notify families of their missing or killed family member. In 1968 I was 26 years old and there was no training or preparation for this duty. The first time I rang the doorbell, a pregnant lady came to the door and all I could see was my older sister. A few weeks later, an older couple came to the door and all I could see was my parents. The third notification was worse. The lady lost her composure when she saw me and I had to hold her so she would not fall. The fourth time the family was not home and on a snowy, cold winter night I had to wait hours and was totally unprepared for what I had to do. I could hardly stand or speak when I approached their door. What saved me that night was the family must have seen the pain in my face and my body language and was very kind to me, even though I could not comprehend their pain.

In the six weeks of notifications, no other officer had to make a notification. After the fourth one, I went to the Commanding Officer, and I said, "I've got a problem, and I hope you can help me out." After explaining the situation he said, "You're off notification duty."

The Army did not train me for that situation. I had no idea what I was getting into. I didn't know how to handle it. It stayed with me for years. I still have nightmares about it. I see my Mom, I see my sister coming to that door. I did get help. In Atlanta there was a Vietnam Support Group and I did go there for a while, and they were very, very helpful.

For my family there is no secret about the most positive benefit of my Vietnam experience. Shortly before I shipped out I made a trip to Milwaukee to visit my sister. She invited several of her friends, including a really cute girl named Myra, to sort of a going away party. As they were leaving that night, I invited them to support the troops by sending me chocolate chip cookies. About a month before I returned home the cookies arrived. In the box was a note from Myra saying she'd heard my tour was almost over and wanted to fulfill my request. Back in the States, it occurred to me that Milwaukee was on an almost straight line between San Francisco and my next duty assignment, Cleveland. Myra and I celebrated our fiftieth wedding anniversary on August 16, 2019, surrounded by our three children and seven grandchildren and many friends.

TONY HILLIARD

US Marine Corps — Combat Engineer Officer

Dates of Military Service: 1965 to 1992

Units Served with in Vietnam: 7th Engineer Bn, 1st Marine Div, III Marine
Amphibious Force

Dates Served in Vietnam: Jan 1968 to Jan 1969

Highest Rank Held: Lt. Colonel

Place of Birth and Date: Philadelphia, PA — 1944

MARINE ENGINEER STORIES

I graduated from Marine Corps Engineer Schools at Camp Lejeune, NC in early January 1967. I spent a year there learning from Sergeant Adams, my Platoon Sergeant.

I found out that I was going to RVN in the fall of 1967. I became very excited about the prospect of getting into the fight and immediately called my wife of less than a year to tell her. After a long pregnant pause, she started asking lots of questions that I could not answer.

When I got my orders, they required me to report to Travis AFB, outside San Francisco for further transportation to WESPAC. After a stop in Okinawa, we landed in Da Nang, Vietnam in early January 1968.

After a quick sorting process and a quick "welcome" briefing, I was taken to the 1st Marine Division Headquarters at Hill 327. The next morning I was directed to report to the 1st Marine Division Engineer Officer, a Lieutenant Colonel who would decide what unit I would be assigned to. He told me, I would be going to the 7th Engineer Bn. at Camp Love, located on the outskirts of DaNang.

I reported to the Company Commander, Captain Tony Lopez. My primary responsibility was to "learn" how to do my job. I was to shadow 1st Platoon on their daily mine sweeps of Route 5.

After my training week, I was sent down to An Hoa on the administrative helicopter run. When I arrived, a Gunny from the company who was overseeing the platoon met me and took me over to the platoon area. I was now officially "Delta 2 Actual."

The An Hoa combat base was located about 25 miles southwest of DaNang. Its reason for existence was to provide protection for the development of the An Hoa industrial complex being constructed just outside the wire of the base. Our job was to sweep the roads daily to clear them of mines and explosives, and then develop those roads for the anticipated industrial traffic.

Our detachment consisted of a "light "platoon of combat engineers (about 12 Marines) some dump trucks and drivers, some heavy equipment operators, a rock crusher, "et c'est moi" as the officer in charge. The senior enlisted man in our detachment was Sgt. Warren Chapman. The troops dubbed him "Sgt. Rock" because he was a good guy.

A typical day would begin shortly after sunrise. The engineer platoon was usually heading out of the An Hoa compound around 7AM, walking North on Route 5 toward Phu Lac 6, a little less than 10 kilometers from the base. Our mission was to sweep North on the road until we met the Delta Co. sweep team coming South from Hill 37 or until we got to Phu Lac 6. A sweep could last from five to eight hours, depending on what was going on.

The road was in various stages of development. The section close to An Hoa was in good condition with a pretty good unpaved surface. It was close to the base, so it was under friendly surveillance most of the time. As the road proceeded North, the condition was less well prepared. From about ½ mile north of An Hoa, almost all the way to Phu Lac 6, the road was in poor condition. Depending on where we were along the sweep route, the condition of the road worked for us or against us.

The state of the art for mine detection equipment was not very good. Mine detection was sometimes more of an art than a science. The continuous tone in the headset of the detector often made it difficult for the operator to pick up any variances in the tone, which would indicate something below the surface. Mine detector operators had to be rotated frequently. We found that being able to visually detect changes in the road surface ahead of the team proved very successful. Visual detection became very difficult in those sections of the road where rock and gravel had been recently put down.

A mine could be placed by removing the rocks and digging a hole in the dirt. An explosive device would be planted deep in the road bed and covered with the rock and gravel. In these cases, it was very difficult to pick up a visual clue. These deep devices also created a problem for the detectors, because in most cases, the VC would form a block of explosive from dud ordnance and wrap them in paper or rags and insert a fuse. Usually, the only metal in the explosive device was the fuse, so there was only a small metal signature to pick up. We were dealing with a cottage industry in mine and booby trap manufacturing.

The formation for the sweep consisted of our platoon on the road, led by two Marines with mine detectors, followed by a couple of engineers with probes. The rest of the platoon was part of the security section. My radio man, the corpsman, and I were close behind the detectors in the formation. The area outside of An Hoa would best be described as hostile. When I first got there, our security augmentation consisted of a company of ARVN Rangers. They deployed a platoon on each flank at some distance from the road to preclude ambushes. The remaining elements of the company followed the sweep platoon on the road.

As we cleared stretches of the road, the dump trucks would begin to lay crushed rock on the roadway to prepare the road bed. The rock crusher back at the An Hoa base would provide the crushed material and we had a tractor that would rumble out to work the material and prep the road. The rock crusher was a beast. It had been seriously damaged by an explosion during the trip to An Hoa and was operating like a Rube Goldberg invention, requiring lots of TLC.

We had a variety of encounters with the VC. On several occasions the VC would attach notes or printed propaganda material on sticks and place them along the shoulders of the road. We sent the handwritten notes back to the 7th Engineer Battalion intelligence section. We never received a report about the content of the notes. The pamphlets were discarded or kept for souvenirs.

We sometimes found "things" along the road. On one occasion, we saw a US military jungle boot sitting on a slight rise about 30 feet from the road. The first inclination was to go up and check it out. I was absolutely adamant that no one leave the road, ever, unless there was an absolute necessity to do so. I believed that walking on the road was dangerous, but walking off the road increased the risk exponentially. We walked past that boot for many days until it became an invisible part of the landscape. On another occasion, as we rounded a bend, the lead detector man discovered a hand grenade lying in the road. ALL STOP! We looked at it from some distance to see if the area around it looked like it had been disturbed. It looked OK. L/Cpl Bellini started forward to check it out, but I stopped him and said I would go. Leader of Marines and all that stuff! We all had to take risks sometime.

I walked out ahead of the sweep team very slowly, my sphincter tightening with every step. When I got to the grenade, I stooped down and looked at it very hard, looking for any telltale sign of booby trapping. I moved it with my bayonet to see if there were any wires or if a fuse was under it, and there was none. I felt it was safe and picked it up and gave it to Bellini who was carrying the demolitions that day. I told him to take it back down the road and blow it in place along the shoulder. He wanted to know why we shouldn't just pull the pin. I explained to him that I was sure someone was watching us to see how we handled the situation, and also that the grenade could have had the fuse removed so it would go off when the pin was pulled. All in all it ended up being a good experience, hopefully we disappointed the SOB who may have been watching us, and we had a teaching moment about how to minimize risk when we found "things" on the road.

The good guys did not always win. One beautiful, sunny day we were well into the sweep. We had a tank with us behind the sweep team for added security. As we came around a bend, we found a large crater on the left side of the road, about three feet deep and the diameter extended almost to the center of the road. There was great excitement and talk about how the "dumb ass" VC had blown themselves up trying to get to us.

The testosterone was flowing. We went over the crater with the detectors to make sure there were no booby traps or additional explosive devices in the hole and radioed the heavy equipment guys back at An Hoa that they would have to come up and fill in the hole and level out the roadway. When we resumed the sweep, we started slowly and were very cautious in the event another device might be ahead. As we moved out and the column stretched, the tank swung right, over the far shoulder of the road, and BALAAM! The VC had planted a large mine off the road parallel with the crater. It had been an elaborate booby trap and we had fallen right in to it.

The blast blew the forward road wheels and the sprocket off the tank, and the front fender was bent all the way back over into a shepherds crook. The crew was protected by the hull. The tank driver had his bell rung because he was right next to the blast. The tank commander, who was a Native American gunny, stared down at me with a look of disgust on his face. We had failed. The sweep was stopped for several hours as we waited for a tank retriever to come and tow the tank back to base. We got back late and it was a long time before we had a tank with us again.

Beside chow, mail was the only other thing we looked forward to. Our mail delivery system was based on our ability to hook up with the sweep team coming South from Hill 37. They would hand off our mail to us when we made contact. On many occasions, when enemy activity prevented the two teams from meeting, we went without mail. During one period we had been without mail for almost two weeks because the sweep teams were unable to finish their respective sweeps for a variety of reasons.

As we were approaching the point where we were to meet the South bound team, we came under fire from a village off to the left. It was not an extensive firefight and no one was hurt. It was more of a hit and run ambush, but it stopped everything in both directions. We could see the other team in the distance, but it was late in the day and we would not have time to finish both sweeps and connect. The Company CO directed us to terminate both of the sweeps and to return to our respective bases since it was late and he wanted to make sure everyone got back "home" OK. I called the CO and the other sweep team and asked if we could each send one of our men to meet somewhere between the two teams and transfer the mail. The request was denied because it was believed to be too dangerous. The up side of the situation was that we got a large bag of mail via the helicopter admin run the next day. There was joy in An Hoa.

Most of our direct encounters with the VC were minor and involved a flurry of firing from a distance and return fire by us. No one was wounded by gunfire, and the sweep continued when we were sure the ambush was over. The 2nd platoon was relieved by Delta 1 in late May and we returned to the good life on Hill 37.

The mini TET offensive was still underway but being in the Dai Loc area was preferable to being down at An Hoa. The roads were in better condition and area was more "pacified." One morning in late May, we woke to find a NVA flag flying at the top of a tall tree in a wooded area about a mile from the Hill. Big pucker! We held off on the sweep until the 3/7 Marines could check out the area, and then headed up the road. Another long, tension filled day.

DENNIS SHOUP

U.S. Navy — F8 Crusader Pilot

Dates of Military Service: 1964 to 1970

Unit Served with in Vietnam: USS Ticonderoga (CVA-14)

Dates in Vietnam: 1967 and 1968

Highest Rank Held: LT (03)

Place of Birth and Year: Detroit, MI — 1943

OUR HUMBLE HERO

Dennis Shoup died on July 29, 2019. He is missed. His story was submitted by his wife, Linda Shoup.

My husband, LT Dennis E. Shoup, served aboard the aircraft carrier USS Ticonderoga (CVA-14) as F8 Crusader fighter pilot and squadron Landing Signal Officer (LSO) during both the 1967 and 1968 deployments to Yankee Station during the Vietnam War. His fighter squadron was VF - 191 with the nickname "Satan's Kittens."

He had over 200 combat carrier landings and flew many high risk missions, often at night or in very bad weather—especially over North Vietnam and the heavily defended areas where many pilots were killed by advanced Soviet surface-to-air missiles. There were also many missions on the Ho Chi Minh Trail or to provide close-in ground support. He participated in the USS Ticonderoga (CVA-14) one day record of 175 sorties during May 1968 and the large Task Force sent to Korea during January 1968 in response to the very tense Pueblo incident with nuclear bombs reportedly under consideration.

Dennis was the wingman for VF—191 Executive Officer CDR Richard Mullen when Mullen was shot down on June 6, 1967. Dennis remained in the area as long as fuel permitted. CDR Mullen was captured and spent six years as a prisoner of war—he came home in the prisoner release in 1973.

Dennis was awarded the Navy Commendation Medal with Combat V for his successful air strike on the Thanh Hoa railroad complex in North Vietnam on February 4, 1967. He was also awarded a Gold Star in lieu of a second Navy Commendation Medal for destruction of the Haiphong Cement Factory on April 25, 1967. He later received the Air Medal (Bronze Star in lieu of First Award) for successful action against enemy forces to enable the rescue of pilots after demolition of the Vinh Airfield on May 7, 1968.

He made over 320 total carrier landings and later served as Flight Instructor at Naval Air Station Pensacola to train navy pilots how to make their first carrier landings.

Our family considers him to be a genuine hero who served with distinction and humility.

SHEPARD BRYAN BENEDICT

US Army — Personnel Specialist

Dates of Military Service: 1967 to 1969

Unit Served with in Vietnam: 25th Admin, 25th Inf. Div.

Dates in Vietnam: Jul 1968 to Jul 1969

Highest Rank Held: Specialist 5 (E5)

Place of Birth and Year: Atlanta, GA — 1945

"MY GOD, YOU ARE FILTHY!"

I enlisted September 1967 in the College Option Program of the US Army, which entailed Basic Training, Infantry AIT, Officer Candidate School (OCS), followed by a two-year service commitment. You could pick any OCS from the 15 OCS's, and I picked Ordnance OCS. I was sent to Fort Dix NJ for Basic and Infantry AIT. I graduated in mid-December 1967 as an E2 11B10 infantryman. However, during AIT (advanced individual training), the Army decided to close many of the 15 OCS's, and I was reassigned to Engineer OCS at Ft. Belvoir, VA. I reported for duty in early January 1968.

After about six weeks of OCS, I became ill with an upper respiratory infection which I had initially contracted in AIT. Eventually, I decided to drop out of the program and was placed on 'medical hold' in the OCS holding Company while I recovered. My commitment to the Army then reverted to two years from when I entered the Army.

I worked in the Replacement Company orderly room HQ and got a 70A10 Clerk Typist secondary MOS on my records. During this time Martin Luther King was assassinated, Ft. Belvoir went on high alert due to riots in Washington DC. In late May, I got orders for Vietnam, and I enjoyed the month of June 1968 on leave.

I reported to the Oakland, CA, Army personnel center about July 1, 1968, and was reunited with Paul Matson from Ann Arbor, MI, whom I met and became friends with at OCS. He too dropped out of OCS. Several days later we flew from Travis AF Base north of San Francisco to Tan Son Nhut airport outside of Saigon and were bussed to the nearby

Bien Hoa Replacement Detachment. We stayed there several days, and Matson and I both got orders to the 25th Infantry Division.

We flew on a Caribou to Cu Chi Base Camp about 25 miles northwest of Bien Hoa. This was a large base camp with an airfield and the HQ of the 25th Infantry Division. The strategic importance of Cu Chi and the 25th's area of responsibility was that the NVA and Viet Cong came south in Cambodia on the Ho Chi Minh trail, and infiltrated through our area above ground and through tunnels in order to attack Saigon and surrounding areas.

We were housed in the 25th Replacement Detachment with the other 75 or so enlisted men who came that day and reported to Finance the first evening. The next day I snuck away to find the Personnel Section of the 25th, which was in a Quonset hut next to division HQ. I introduced myself to SP5 Bob Greenlee, who incredibly had my personnel file on his desk. He led me to CPT Byron A. Malogrides, the AG of the 25th. After a short interview, he offered me a job in the 25th Admin Company, which I accepted. That day Matson interviewed with a dog platoon company that was soliciting new men, and he was accepted.

The Cu Chi area was geographically flat, with the exception of Nui Ba Dinh, the Black Virgin Mountain, at 3,000' of elevation, adjacent to Tay Ninh City. We were inland from the South China Sea to the east; to the west was Cambodia. Vietnam six months of the year was the dry season with no rain, and everything was covered in dust. The other six months was the rainy season, and it rained most every day, with mud everywhere. The VC and NVA were more active in the dry months, mostly in the winter.

The 25th Infantry Division had approximately 15,000 enlisted men. My initial assignment for three months was to create DEROS (Date of Estimated Return from Overseas) orders for the E-3 thru E-6 troops nearing the end of their one-year assignment and returning to the USA. Next I was put in the job of assigning all the new incoming E-3 thru E-6's to the units in the 25th, working with my friend Stan Zak. These new men were primarily infantrymen, but also mortar-men, mechanics, engineers, cooks, clerks, medics, artillery men, radio men, and supply specialists. The 25th Division consisted of the nine Infantry battalions, as well as the 25th Admin, 25th Supply and Transport, 65th Engineers, 25th Medical, the ¾ Cav armored unit, 25th Aviation, an MP unit, an Artillery unit, and more, all of which were housed on Cu Chi base camp. Six infantry battalions were on the ground, and three were on mechanized armored personnel carriers (APCs). For the most part, they were out in the field or operating from either of our three Brigade HQs.

When Zak and I took over, our predecessors had kept up with unit strength and daily casualties via the 'morning reports' of every unit, gathered by various means. These reports were always several days old, or more. Zak and I devised a better system. Every unit had a Personnel Sergeant (PNCO), who was responsible for that function in the unit. We began using the somewhat primitive phones we had to stay in daily contact with PNCOs in order to know their daily need for new troops which included casualties in the field via KIAs and Medevacs, or vacancies from men leaving as their year was up. The day before we initiated this practice, we had every PNCO in every unit that we were responsible for to give us an accurate body count of all men in their unit, with their rank and their MOS, i.e. their Military Occupational Specialty, which was a number/letter system of their skill set.

This was our base line, and daily we updated the troop strength of each unit, adding the troops we sent to the unit that day, minus the troops we knew were lost or rotated out. We got to know the PNCOs, and they communicated with us regularly. In fact, they sucked up to us via gifts of beer and steaks. They would let us know if they needed a senior level cook or a good personnel clerk, etc., and we would review the files for that person, or men with other skills that would be useful in a field unit, such as a college degree. The field units needed qualified admin people too. Capt. Malogrides was proud of our innovation and brought 'full bird' Colonel Fair over to listen to us describe how we were handling the personnel function.

Likewise, we sucked up to guys at USARV in Bien Hoa in order to keep us at the 'front of the line' when we, as a division, needed men. I flew on a Huey (sat by the door gunner as the pilot demonstrated his flying abilities with acrobatics over rice paddies) down to USARV to deliver some reports and visit the guys upstream from us, who assigned the hundreds of men arriving every day from the States to the 1st Division, American Division, 9th Division, 4th Division, the 25th Division, and others.

We received anywhere from 10 to 100 men every day. They arrived up from Bien Hoa mid to late afternoon, so often we worked late into the evening to make unit assignments, have orders cut, and off they went the next day. We picked replacements for our own office with care to keep our team harmonious. However, when the 25th Division was in heavy combat, every new troop went to the field. A medic could go to 25th Medical or to an infantry unit where he was the same as an 11B10, which is a basic infantryman. Of the nine infantry units, some we replaced whole companies on a monthly basis and some were rarely in action. Lots of luck was involved, being in the right place or the wrong place.

From time-to-time, we were asked to draft 'outstanding' enlisted men to be waiters in the General's mess. This was a reward for a decorated enlisted man, and we went to the nine infantry PNCOs and asked for candidates. One kid from South Carolina, who had been awarded a Silver Star, ended up as a waiter. He had a stunned look on his face when he passed through our office. I was able to get my friend, Paul Matson, out of the field from the dog platoon. He had had enough and had told me so earlier.

We normally worked 6 ½ days with one afternoon off. Cu Chi Base Camp had a PX, a library, a combo steam/massage facility, some or all of which I would visit on my half-day day off. Normally, after work we drank beer and listened to Armed Forces Network TV and radio. Every Sunday was the beer run to somewhere, and the truck came back with cheap cases of primarily Budweiser. Bob Hope's show came to Cu Chi just before Christmas 1968.

The "Tunnels of Cu Chi" were real and around much of the base camp. We regularly had rocket and mortar attacks as the VC or NVA came by on their way to blow up Saigon. The 25th Admin Battalion was responsible for about six bunkers and one watch tower on the west side of base camp, in the direction of Cambodia. I had guard duty every sixth night or so. During Tet 1969, we were extremely busy with the flood of new personnel that replaced the troops who arrived during and immediately after the more severe Tet 1968. Zak and I were able to get relief from guard duty at that time.

There were three men assigned to each bunker, with two up top awake at all times and one asleep. I usually stayed awake all night. There was a Korean base to the west outside the bunker line that always seemed to have some action. Frequently, illumination flares would appear over the bunker line, and we would call on the crank phone to the guard duty HQ and ask, "What the fuck, over?"

Many times I watched Cobra helicopters at work on some target outside the bunker line, night and day, with their mini-gun cannon on the nose spewing out a stream of red tracers and then dive bombing the site with their side mounted rockets blasting away. I also saw AC-47 Spooky in action firing a 50-caliber (I think) stream out of the side door. Some nights on the bunker line we were instructed to open fire with our M16s for Harassment and Interdiction (H&I) fire, though we never knew what we were shooting at or what we might have hit.

At night from several places on Cu Chi, we could watch the mountain Nui Ba Dinh get blown up. We held the top with 11C10's mortar men, the VC held the middle, and our Tay Ninh Brigade men were at the base of the mountain. It was compelling to watch, but very dangerous and deadly to be there. Often during the day we would experience outgoing artillery going over the base camp to some target on the far side of the base camp, or a B-52 strike which shook the earth and was sometimes visible in the distance (i.e., where the bombs landed, but not the B-52s), or a Phantom strafing a close by target (very loud and powerful to watch).

There were daily Eagle Flights in and out, normally 10 Hueys lined up in a row. Huey helicopters were coming and going all day long, including many Light Observation Helicopters (LOHs). I had an in-country R&R at the Vung Tau beach resort east of Saigon on the South China

Sea. We were flown there in Caribous. In March 1969, I had R&R to Hawaii to meet my fiancée and my father and new stepmother. Our jet stopped in Guam on the way to Hawaii, and I bought five bottles of Johnny Walker Red for my dad (a WW II US Navy vet) and me to enjoy.

Every day there was a mail run down Highway 1 on a 'Deuce and a Half' truck to Ton Son Nut airport to pick up the mail for the 25th Div. I rode shotgun once, carrying my M16 and wearing ammo bandoliers, down to Tan Son Nut. I then made my way into downtown Saigon to see my Atlanta friend, Zach Thwaite, who was in a small specialty unit that was responsible for big construction equipment throughout South Vietnam.

He lived in a hotel taken over by the Army. I walked up the stairs to his room on the 5th floor, knocked on his door, and when he opened it he said to me, "My god, you are filthy!" I got cleaned up in his bathroom, and of course I had not seen a bathroom in a while. That night we visited the notorious Tu Do Street, where many bars were. Almost everybody we passed knew Zach, and though he was very comfortable, I was not.

I was wounded during a rocket and mortar attack one night. I tore the toenails and a wad of skin off one foot low crawling across the plywood floor, and over our mini-refrigerator filled with beer, in order to get down into the bunker next to our tent. The next morning I limped over to the 25th Medical Bn., climbed on to a gurney, and a medic plucked the toe nails out and bandaged my foot. He tossed the paper work for a Purple Heart on to my stomach, which I never filled out.

Our 25th Admin assignment during a ground attack was to form a perimeter around base HQ. Cu Chi Base Camp had a major ground attack on February 26, 1969. At approximately 3 AM, we were hit on three sides by the VC/NVA, who proceeded to blow up nine Chinook helicopters. They were reduced to total scrap. I was in the culvert along the road at the front of Division HQ with rockets and mortars falling very close by and shrapnel bouncing off of the culvert and me, thick smoke everywhere, helicopters flying overhead, and VC in the base camp. The 3/13th Artillery bunker line just north of our 25th Admin bunker line had many killed. We were on red alert for several days while the enemy was expelled and order restored. At about the same time frame, the 25th Aviation mess hall, which adjoined our mess hall, was booby trapped with mess trays that triggered explosions when they were pulled out. Several men were killed.

Sometime in early 1969, new hooches were built with 20 or so capacity, concrete slab floors, and tin roofs. I liked the eight-men tents better. Also in the effort to provide the troops with more amenities, an in-ground pool was built in front of the HQ shack. I never liked it and thought it was stupid and a waste.

Toward the end of my year I became fixated on improving my physical condition through exercise. I also quit smoking which is ridiculous as why not smoke until you leave that place. I slept in the bunker for the last two or three months and had many 'short-timer' calendars to mark off my remaining days. The bunker was full of very large cockroaches, but my paranoia that my luck would run out via a rocket or mortar attack was greater than my dislike of the cockroaches. I DEROS'd out in late June 1969 from Ton San Hut to Oakland where I was discharged from the Army. I flew home in the rear section of a Delta jet with Gracie Slick and the Jefferson Airplane in first class. They got off in Texas.

I was awarded two Bronze Star Medals for Meritorious Service. One was for the Tet Offensive 1969, when we worked continuously for about 40 days, and one for my contributions for the whole year.

I was stunned upon returning to the States. Things had changed, like more diversity in TV ads and the like. Not a big deal, but it emphasized that while in Vietnam you were completely cut off from "the world." There was no instantaneous communication with home. We were just out of touch the whole year.

I have great pride in my almost two years in the US Army, and my year in Vietnam.

R. ZACHRY THWAITE

US Army — Interpreter/translator

Dates of Military Service: 1967 to 1970

Unit Served With in Vietnam: MACV Advisor Team 9

Dates Served in Vietnam: Sep 1968 to Mar 1970

Highest Rank Held: SP5

Place of Birth: Waycross, GA — 1946

MACV DUTY IN SAIGON

My name is R. Zachry Thwaite and I was an SP 5 in the US Army, an enlistee. I did my Basic Training at Fort Benning in July and August of 1967 and my AIT at Fort Jackson. After AIT my Top Sergeant had my orders changed from OCS and Jump School at Fort Benning to assisting him in the clerk's office of my old AIT unit. He told me I was crazy to have signed up for all that stuff. I was young and gung ho and was ready to go. I remained in that capacity for about ten months or so until I shipped out to Vietnam on September 23, 1968.

I arrived in Bien Hoa and was processed in and moved to Saigon. My orders called for me to be assigned to MACV, Advisor Team 9, but I had been repeatedly warned that any orders were subject to change upon arrival in VN. Mine were not. My quarters were to be the St. George Hotel in Cholon, in the Chinese sector of Saigon.

My office was on Le Van Duyet Street in downtown Saigon. It was one hell of a change from the good old US of A. On the trip from the air base, I remember the different smells and people squatting on the side of the road, the fumes of Lambretas and motor cycles, the noise in town, the absolute chaos at the traffic lights and roundabouts, all of it so different. And the fish market smelled to high heaven!

But I was one of the lucky ones. I had my own jeep and driver, Thinh, and he and I would be together over the next 523 days. A tour of duty was 12 months but when the "early out" program was instituted, I extended my tour of duty another 158 days to fall one day under five months remaining on my return to the US. Under the program it was deemed not worth retraining a soldier for five months remaining, so we would simply be sent home five months early.

I was assigned a room on the 7th floor (but no elevator for security purposes) of the St. George Hotel, which happened to be situated behind the Capitol Hotel where the MPs were housed, along with two Korean units, the Tigers and the White Horses. Interesting thing about the Korean units, they rotated in and out of the field on a monthly rotation. They were the crack Korean units and used to complain that "Charlie" would go miles out of the way to avoid confrontation with these guys. They wanted to engage! I remember marveling that they were all about the same size, about 5 ft 10" and 180 lbs, and they kept all their equipment in immaculate condition. As time went on I noticed I could tell which jeep belonged to whom, with the spit shined jeep belonging to the Koreans, the clean jeep to the US, and the poorly maintained one to the ARVN. So much for that.

My office was in an old classic French Building in a courtyard on Le Van Duyet street. I quickly learned when I was without Thinh and taking public transportation to my office to say "Toi di doi bo chi huy ba tiep van" and that got me there. I had no idea what I was saying but it was one of the first things my pool of five interpreter translators taught me.

Advisor Team 9 was charged with advising our Vietnamese counterparts and the interpreter translator pool handled documents from Vietnamese to English and vice versa. We were advising the 3rd Area Logistical Command, or simply 3ALC. For logistics purposes the US divided Vietnam into five (some said six) corps and Saigon was in III Corps. We had commissioned officers as counterparts to our Vietnamese allies up and down the line for Ordinance, Transportation, Engineering, etc. We had about 15 officers and usually about 10 non-coms. In my office were myself and our Top Sargent and the translator pool. Top's name was Anderson and he was a laid back, easy going gent. All day long, back and forth, documents English to VN and VN to English. All hand written. Over time, as fate would have it, I developed a relationship with one of our translators, but nobody on the team seemed bothered by it so all was good. She had an interesting story.

Kim Vui was her name. Her family was fleeing Hanoi when they were captured by the NVA while heading south toward Saigon. Over the course of their confinement, the NVA killed her father, her brother, and her fiance. The women were allowed to continue. From that point forward, Vui had decided to do anything in her power to assist the US forces and that is how she wound up with Advisor Team 9.

Her family was very talented. Vui was fluent in seven languages, including North Vietnamese dialect, South Vietnamese dialect, Chinese, French, English, Korean, and Japanese. Her sister was the lead singer in one of the most popular singing groups in Saigon. They had a fantastic knack for replicating on stage all the popular US groups such as the Supremes, Martha and the Vandelas, etc. Together, her family went a long way toward making my life in Vietnam more meaningful.

Advisor Team 9 was an interesting group. We had a Bird Colonel in command, two LTCs, three or so Majors, two Captains, and one Lieutenant. Plus the non-coms such as myself. It was a very casual environment with no saluting. We also had the Saigon EOD team. They were an interesting bunch. SFC Miller had been in Vietnam for eight years. No telling how many bombs and booby traps he had disarmed. One incident comes to mind when we had a sniper situation at the old French horse track in Saigon.

There were allegedly three snipers in the upper reaches of the stands. The US Army answered with their own sharpshooters and soon eliminated the threat. SFC Miller approached one of the downed snipers to retrieve his weapon as a trophy and while bent over to pick it up he was hit in the back of the head by yet a fourth sniper that no one had seen. Miller had only been grazed and was soon back on the job.

We also had two mobile units that spent most of their time in the field and we saw very little of them in the office. They had M60s mounted on both jeeps and were in need of some heavier fire-power and miraculously, through a little horse trading here and there, we were able to provide them with the 50 Cal they had wanted so badly.

And then there were the two prisons under our advisement, Phu Quoc Island and Con Son Island. There were days when I would report to work only to find one or more Charlies blind folded and hands tied sitting on the floor outside the office, awaiting questioning. They were bad about spitting.

III Corps was a rather large expanse. We had on issue from the 2d Field Forces one Huey Slicky chopper at our disposal on a daily basis. It was required that a US soldier be on every flight, so many times on days when our Vietnamese counterparts were using the chopper with no advisor accompanying them, I would go. We had many sites to visit, but the ones I remember the best were Thay Ninh City and Vung Tau.

Thay Ninh City was the site of a large ammo depot that we had noted on many occasions was improperly configured. Sure enough, it received a direct hit from a mortar round and the ensuing explosions almost leveled the city. Just outside of the city was Black Virgin Mountain, known locally as Nuy Ba Dinh. It was controlled on the top and the bottom by US troops but was said to house in its interior one of the biggest Viet Cong hospitals in Vietnam. The VC and NVA wounded were transported to and from the mountain via an extensive tunnel system that went for miles.

Now, about Vung Tau. It was on the beach and we always told the pilots when Vung Tau was the destination to bring their bathing suits.

My Vietnam tour was the first time in my life that I spent virtually the entire time without a close friend nearby. My good friend Bryan Benedict was with the 25th Infantry in Cu Chi and though we spoke on the phone once or twice it wasn't until toward the end of our tours that he was able to make it to Saigon to visit. He rode shotgun on the daily mail truck (deuce and a half) down Highway 1. He came to my hotel dusty as hell, with ammo bandolier and in field dress. He was shocked to find I had a shower and even more so to find I had hot water. He said he hadn't seen either of those since he had been in Vietnam. I took him out on the town and I could tell he was a little uncomfortable in that environment of chaos. I took him to a few bars and we had a great visit. Thu Do street in those days was filled with tea girls ripping off the soldiers on in-country R & R drinking colored water called Saigon Tea for umpteen dollars a glass. We locals knew better. No thanks!

Though we were a non-combat unit, we still had our moments. On one flight to Thay Ninh, my chopper was filled full of holes from enemy groundfire along the way. We knew not where. We flew either at contour level or 5,000 feet or higher to avoid as much as possible any small arms ground fire. In Saigon, we had an incident one day when a VC cowboy, motor cycle hotshot, rolled a grenade in the path of my first Top Sergeant who suffered considerable shrapnel and was partially blinded. E8 Anderson was replaced by E8 Steinberg just about the time LTC Andrews was replaced by a West Point full bird named Broderick and the whole atmosphere of our cozy Team 9 tightened up considerably. I was happy to be getting "short" by then and welcomed my departure soon thereafter to the US.

I had trained my replacement and gave him as many assurances as possible that he was moving into a good job, all things considered. I likened Saigon to the bad side of Harlem, not the greatest place to be but when it came to being in Vietnam, not too bad. About three months after my return to the US I received a letter from Vui informing me that my replacement had been killed in a random VC attack in Saigon.

My stay there was an experience I will never forget. I learned an awful lot at a young age as a result, but I wouldn't want to do it over again. And oh, by the way, it will always be Saigon to me, not Ho Chi Minh City.

BILL MILLER

US Army — Engineer Officer
Dates of Military Service: 1962 to 1992
Unit Served with in Vietnam: Alpha and Charlie Companies, 93rd Engineer
Battalion
Dates Served in Vietnam: Jan 1968 to Jan 1969
Highest Rank Held: Colonel
Place of Birth and Date: Denver, CO — 1939

A COMBAT ENGINEER IN VIETNAM

My story depicts how a typical combat engineer company of 155+ men supported the war effort. I arrived in-country on January 1, 1968. I was a 29-year old Captain in the Army Corps of Engineers, MOS 12 Bravo, Combat Engineer. After three days at the reception center in Long Binh, I received orders to the 93rd Engineer Battalion and was flown by helicopter to Camp Castle in the Delta, about 50 kilometers south of Saigon near the village of Long Than. The 93rd provided direct support to the 9th Infantry Division (Mechanized). The Battalion had been engaged for nearly six months constructing a fixed wing C-130 capable airfield at Long Than North. My initial assignment was to Command Alpha Company, the Battalion's Heavy Equipment and Maintenance Company.

The Tet Offensive began soon after, on 30 January 1968, and all hell broke loose. The Battalion had completed work on the airfield and was in final preparation to convoy further south to a new 9th Division base at Dong Tam. As Tet moved closer to Saigon, we received an urgent mission to go north to clear and push the jungle back around the perimeter of the huge Long Binh complex. My company took up defensive positions outside the wire at Long Binh to provide support to the land clearing teams. As the fighting in Saigon and Long Binh intensified, we began to take heavy sniper and mortar fire. Combat Engineers have a secondary mission to fight as infantry and so we held our own and continued land clearing operations for two weeks — fortunately the NVA was defeated in Saigon and failed to move further south into the Delta.

After Tet, the Battalion relocated 100 kilometers further south to support the 9th Infantry Division and Naval Mobile Riverine Force (MRF) at Dong Tam. Charlie Company was the lead company to convoy south in mid-February. I assumed command of Charlie Company when the company commander was wounded and evacuated. Charlie Company had three primary missions when we closed at Dong Tam in early March 1968: (1) land clearing; (2) mine clearing; and (3) base support.

Land clearing was a vital mission to enable mechanized movement of the Division through dense jungles of the Delta. This mission was carried out by a reinforced combat engineer platoon augmented with heavy equipment and maintenance teams. Much of our work focused on the extensive Michelin Rubber Tree Plantations in Binh Duong Province, clearing travel lanes and building artillery fire bases. Heavy D-8 Cat dozers, Rome Plows, Letourneau Tree Crushers, chain saws, and demolitions were our tools. Equipment operators welded 1/2-inch steel side plates to their cabs to protect them from RPG and small arms fire. Occasionally we barged into small Viet Cong camps in the deep jungle, finding still warm cooking pots and fires. The VC vacated quickly when they heard our rumble, often leaving behind maps and notebooks that were valuable to Division G-2. We never ate any of their food.

Mine clearing was a daily task along the Lines of Communication (LOCs) essential to maintaining freedom of movement and resupply in the Delta. The Viet Cong would put in antitank and antipersonnel mines nightly along the major roads and, to their credit, they were skilled in hiding them. Every morning at first light, two convoys moved out, each consisting of a mine sweeping team, a demolitions team, four dump trucks loaded with crushed stone, and, if we were lucky, an MP detachment for force protection. Roads were assigned priorities by Division G-2 each night based on patrol reports and logistical needs.

If mines were detected by the dismounted teams they were carefully flagged, the demo team would blow them in place, and then crushed rock was dumped into the crater. Booby traps (grenades and improvised explosives with trip wires) were a constant worry; when found they were disabled by "hooking" the wires from a safe distance or shooting up the explosives. By 1000 hours we were required to report to the Division G-3 those roads that could be reopened to supply convoys. Some days after mine sweeping we would be diverted to repair a bridge abutment or span that had been damaged overnight.

The base support mission was carried out by our platoon of skilled electricians, carpenters, and equipment operators. The nightly mortar attacks on Dong Tam necessitated revetments and overhead cover to protect troops and vital facilities such as the rubber inflatable MUST Hospital. Shrapnel deflated the air supported frames causing havoc in the wards and operating rooms. It took a few weeks of work by our engineers to completely surround the hospital with protective revetments, and to construct heavy timber and sandbag overhead cover. This solved the problem! The magnificent Army nurses and doctors loved our men and thanked them with movie nights and beer. Much of our base support work was to build protective bunkers for critical facilities such as the division and brigade headquarters, generator stations, and communications centers. We built troop bunkers, a chapel, and numerous troop comfort facilities (showers, latrines, and mess halls).

JOHN VAIL

US Army — Finance Corps Officer

Dates of Military Service: 1963 to 1969

Unit Served with in Vietnam: 126th Finance Section, Pleiku, Vietnam

Dates Served in Vietnam: Jan 1968 to Jan 1969

Highest Rank Held: Captain

Place and Date of Birth: Jersey Shore, PA — 1941

THEY CALLED ME PAYCHECK

The day was sunny, a cool day in January, 1968, as we arrived at Dulles

Airport for my departure to Vietnam. Nervous, concerned and definitely scared, not only for myself but for my small family I was leaving behind; a wife and an 18 month old daughter, who simply said, "Good-bye, daddy. See you later!" It was really happening and I wouldn't see them again for probably a year.

Our departure was uneventful, quiet and solemn since many of my fellow passengers were also going in the same direction as I. We tucked ourselves in with our thoughts and prayers for the long flight to Long Binh, South Vietnam and hopes that a return flight would occur.

Approaching the airstrip in Long Binh, we came under enemy fire and was forced to divert our landing to Bien Hoa where we landed safely. The doors of the cabin opened and I was hit with the smell of burned ammunition, "There had been some 'action' here in Bien Hoa." What a welcome! But it got worse within a few days.

Processing, adjusting to the time change and my new "life" in Vietnam took several days, but once everything was in order, I, along with fellow military personnel, started out in the evening to the landing strip when almost immediately the entire area came under attack. Two bunkers were nearby and as a group we scrambled to get in. As it turned out, most of us ended up in the first bunker! We were packed in like sardines. The attack lasted through the night. It was hot, stuffy, and hard to breath. We also wondered if there would be anything left of Bien Hoa when we were cleared to come out. Our departure from Bien Hoa was definitely delayed.

Two nights later, we all returned to the airbase and left for Qui Nhon, and again came under fire while in the air and diverted to Phu Cat with a plane load of soldiers I was now responsible for their "safe delivery." Upon arrival at Phu Cat, I asked the Top Sergeant if we could be issued weapons. His response was surprising. "No," he said. "This base is protected by the Republic of Korea military." FYI, that base was never hit the entire year I was in Vietnam due to their outstanding protection and the Viet Cong fear of them.

Finally, we flew into Qui Nhon safely the next day and found out that I was going to be the Commanding Officer of the 126th Finance

Section. What? A trained Airborne, Ranger Officer? This can't be happening! But it did! This group was made up of 85 of the best enlisted men, several officers, 10 Vietnamese women, and four Vietnamese men. This was to be my "home and family" for the next year!

My job was to pay all the troops in the Central Highlands of Vietnam. Needless to say, I wanted all my personnel to be friendly to every soldier who came to us for assistance. Our barracks and Finance office were at Camp Schmidt, fondly referred to as Camp S—t!

Other assignments given to me were, Responsible for the Perimeter of Defense, that required very little sleep on a daily basis. My personnel were required to take their rotation on the perimeter, putting themselves in serious danger. However, during the entire year at Camp Schmidt, our perimeter was never breached and no personnel were lost.

Responsible for the security of the "Donut Dollies," (Red Cross women) who lived in a separate hooch in Camp. From time to time the Executive Officer of Camp Schmidt and I would go over in the evening and play bridge with these Red Cross women. The XO wrote a note to his wife and mentioned that he and I played bridge with the women. In his next letter from his wife, she let him know that she was "upset" that he would participate in such an activity, but we continued to enjoy this innocent diversion.

As the year went on, I was asked to head-up the building of an American Express Banking Facility in our Camp. Once completed, a ribbon cutting ceremony was planned and an American Express executive, Dale Doll, came to Camp Schmidt for the event. Dale and I visited while he was there. He gave me his business card and asked that I come and see him when I was stateside in New York. During that visit, he asked me to interview with various other folks. I did and was offered a position with them beginning after my separation from the Army. Upon acceptance of the offer, I worked for American Express for 37 years and consulted with them for an additional three years.

I was also the "official entertainment" officer for Camp Schmidt. About once a month, a Pilipino band would come and play for the enlisted personnel from 6-8 PM, and then the small Officers Mess from 8-10

PM. Prior to the entertainment, I would seek out a military bus, drive to the 71st Evacuation Hospital, go door to door in the nurses' quarters and announce that a band was playing at Camp Schmidt. Loading the nurses into the bus, I then headed back to the point of entertainment.

At the end of a fun filled evening of dancing, I would take the nurses back to the 71st. Many times, some of the male officers were included on the bus. Once back at the Nurses' Quarters, I would tell the men that I would be leaving for the Motor Pool in one hour; most returned. Others, I had to find.

During my year in Vietnam, the MPCs being used were changed twice; that meant no sleep for three days! Once a month, pay officers would arrive in "choppers" from the field to take the spending money back to their units' personnel. In some cases, the choppers would be shot-down and the officers, pilots, and pay would end-up being lost. For five years after my separation from the Army, I was held responsible for the $20,000 that was lost in those events.

As noted above, I had Vietnamese men and women who paid the local civilians working for the military. Not accustomed to our western culture's use of bathrooms, they would walk to a near-by field and leave tissue paper where they had been. OK, my job was to build them a latrine. Once completed, they were advised to use the latrine and not the field any longer. However, again, culturally to them, a latrine was just private but didn't solve the problem. My First Sergeant came to me explaining the current problem and said, "Sir, you will need to demonstrate to them what the cut-out areas are used for and where the tissue paper was to be put. Gathering the local employees together, Captain Vail proceeded to demonstrate the use of a latrine after which, problem solved.

New Year's Eve, 1969, a group of us celebrated the New Year and being "short-timers." After a rather rowdy party, I proceeded to the motor pool, borrowed a jeep, loaded-up my buddies and headed for the airbase. Light hearted, a bit intoxicated, and with less than two weeks left in country, I decided to race down the landing strip blowing the horn. Not able to hear anything above the jeep's motor, the singing and yelling and screaming of my companions, it was only by heavenly interference that we weren't hit by the incoming aircraft bringing our replacements. Sitting for what seemed like an eternity, we all realized what could have happened. I drove back to the motor pool and never spoke of our close call again.

Leaving Vietnam was bitter sweet; so happy to be going home but sad to be leaving so many of our troops behind. As the wheels of our "silver bird" lifted off the tarmac, a huge cheer went up from us all. We had done our jobs to the best of our abilities and in some small fashion helped the mission succeed.

Next stop, home! My little family had changed. Our 2 ½ year old was now a busy toddler, with a little brother born while I was away and a wife who had kept the family on solid ground. But how would they receive me after not seeing them for so long? In their minds, no time had passed as they welcomed me with open arms and an innocent, "Hello, Daddy!" from my toddler!

JUDY VAIL
Wife of John Vail

THE LONG AND LONELY YEAR

Christmas 1967 was festive, happy and fun, especially for our first born, 18-month-old daughter, Kimberley. That time of year is always busy and exciting, catching up with family and friends and experiencing this special holiday through the eyes of our daughter. Fortunately, she was not able to sense the anticipated major change coming to our little family when January 1968 would roll around. Mine and John's discussions were centered around where Kim and I would spend the next 12 months during his deployment to Vietnam. It was decided that since John's brother, Ron and sister-in-law, Peggy lived in the DC area, had offered to find housing for us, and lend support and help during John's deployment, that DC would be our home base.

Moving in went without a hitch; setting up our bed, Kim's crib, and the little furniture we had accumulated during our first years of marriage was pretty easy. It kept us busy and focused on other things rather than the "white elephant" in the room.

The day we had tried to forget came and we headed to Dulles for John's departure. He requested that none of us go into the terminal with him, and so it was to be; off he went with a simple, "Good-bye, Daddy," from our little girl. Kim and I traveled back to Ron's and Peggy's house and enjoyed a lovely dinner before heading back to "our little home."

The next few weeks went well; all things being considered. The washer and dryer were connected, food in the frig, and a warm home during January and February and most importantly, Kim was able to watch her favorite cartoon, the Flintstones, every evening at 5:00.

The TV was a source of entertainment for me as well, but so often the sights and sounds of the fighting in Vietnam, soldiers returning stateside to a barrage of insults and thrown objects, and especially, the ceremonial return of our casualties was never entertainment for those of us on the home-front. In fact, it was down-right frightening. I found myself a bit paranoid when returning home some days fearing to see a "green sedan" with military men waiting in front of my address.

As for real communication from Vietnam, mail was three to four weeks behind and certainly no WIFI, internet, or cellphones were available at that time, so I had to rely on prayer and the adage, "No news is good news." Word did finally arrive that John had arrived safely and was assigned to command a Finance Section at Camp Schmidt in Pleiku, Vietnam, located in the Central Highlands. It sounded fairly "safe" to me, although safe was a relative term in Vietnam. It certainly was not without anxiety and stress for either of us.

Toward the end of February, I began to feel ill and after a week or so of nausea 24 hours/7 days, I realized I was pregnant. The first three months of my pregnancies were always wrought with three months of morning sickness, day and night. My little daughter and I would muddle through those trying days and evenings when all I really wanted to do was go to bed and cover my head! My letter writing was not the most consistent to begin with, but John understood when I wrote about the pregnancy. The doctors at Fort Belvoir were especially kind to the pregnant spouses of deployed military. During one of my visits, the doctor asked if I was getting enough exercise. I looked him straight in the eye and said, "I have a two-year-old!" He and the nurse laughed and said, "We understand!"

While sitting in the waiting room during one doctor's visit, I was introduced to another military wife whose husband had also been deployed. She told me about enjoying the friendship and company of other wives of deployed servicemen. We exchanged phone numbers and within a few days, John's Commanding Officer's wife, who was also in the DC area, contacted me and invited me to join their "Waiting Wives" group. They met once a month, shared dinner, and then played Bridge. It was a wonderful break in the normal monotony of my daily activities, a great way to share with other women going through the same anxious months, and definitely some "grown-up" conversation.

At the time, Alexandria, VA was home to many military families, "Waiting Wives" and military retirees. Two neighbors, living in our same building, sought me out and became very good friends and confidants during that year. One woman, Hue, was Vietnamese with a sweet 4-year-old daughter who loved playing with Kim. Hue's husband was a Marine on his second or third tour of duty in Vietnam. She taught Vietnamese to the military who needed the ability to converse with the local Vietnamese and, sometimes, while out on patrol with their units.

We had some fun day outings that I would never have ventured out alone with Kim had it not been for Hue. But adventures were definitely what we both needed at the time; a visit to Chinatown in DC, where we bought the best barbequed pork from a hanging pig in a lower level shop where no one spoke English. Fortunately, Hue spoke many languages so we did just fine. The barbeque was heavenly! Unfortunately, I ate too much and suffered for it that night. I know the little girls loved all of our outings, and those days spent away from the confines of our townhome were a much-needed distraction for me.

Our other neighbor was an Air Force family with four children; a teenage girl who was my go-to babysitter for occasional evenings when

I needed her. Lucille, their mom, helped me out during the day many times, especially when the OB appointments came due. She enjoyed taking care of Kim since her youngest was 10 by then and she missed "those years." Her husband, Jim, was active duty in DC so he was around when challenging problems came up for me, such as my car stopping in the middle of Alexandria for apparently no reason and that same car being stolen. The conclusion from the Police is that the thieves ran into the same problem as I was having. The car stalled and they couldn't get it started so they bailed. No damage to the car but a very upsetting experience for me. How would I explain it to John!

John and I had originally planned on meeting in Hawaii in mid-October. That would have left only two plus months until he was scheduled to return home. Since I was now pregnant and my due date was October 6, that trip had to be moved up to late July. Fortunately, I never "showed" as a very pregnant woman and traveled to Hawaii in late July, early August.

When I was welcomed at the Military's so-called "hotel" in Honolulu, I still had 12 hours to wait before John's plane was scheduled to arrive. Immediately, I knew this Military "hotel" was not going to be acceptable accommodations for us. I called a hotel on the beach and immediately moved over to what would be our Rest and Relaxation week together.

5:00 AM the next morning, I was lined up with 40 other wives waiting for our first glimpse of our husbands. I burst into tears as he deplaned and came into my arms. What a relief to see him in person! The stress and anxiety of his time in Vietnam for both of us was on the back-burner, at least for the week we were together. But saying good-bye this time was more difficult since we both knew I was going to go through childbirth alone and at that point his arrival home was still going to be two plus months away. I was on my own!

My mother came with my niece to stay with Kim while I was away. One day, John's mother and Peggy decided to go for a visit and see Kim and my mother. Whatever went on during that visit, even 50 years later, still is unsettling and unclear. Apparently, John's mother came into town from Pennsylvania to stay with Ron and Peg. When they visited my home, there was some sort of verbal altercation or insults thrown around and everyone was unhappy and upset and ended up holding a grudge for many years. I was thrown in the middle and kept hearing snippets of what transpired from both sides, but no real clarity of exactly what had gone on. After that, I never heard from Ron or Peg while I was in Alexandria. It certainly would have been comforting and supportive to me had they put aside their anger and followed through with their original commitment to John, that they would be there for me. A very unpleasant, unnecessary and unhappy situation for all. I certainly could have enjoyed their companionship and support those first few months after our second baby's birth.

John's brother and sister-in-law had a difficult time getting past that day, even though neither John nor I had any part of it. This made John's relationship with them a bit strained for many years, also. John still gets upset remembering that his brother and sister-in-law were going to "take care of his 'family' no matter what," while he was in Vietnam, and that certainly didn't happen. My take-away from this incident is that there are long-term peripheral effects of "war" that are sometimes never understood nor resolved!

That summer was especially hot in DC; although, after living there later in our marriage, I know DC is always hot in the summer. But that particular summer was extremely hot, not only for me but for many families living in DC. The assassination of Martin Luther King earlier that year had left many in the black communities saddened and unsettled. This, compounded with the lack of jobs for these folks ignited the pent-up anger and frustration in the form of bonfires and vandalism. From my bedroom window I could see into the boiling cauldron of DC. In the hot darkness of a summer's night, the city of our Nation's Capital was glowing with fires. I was quite frightened for Kim and me watching this event from my vantage point in Alexandria, even though we were quite a safe distance from where it was actually happening.

As days went by, this event gave me a much better understanding and appreciation of how families in poverty, drug infested communities, as well as war torn countries, must have to cope with these daily occurrences. How do these people keep their sanity and their families safe?

When my due date was within a week, my mother came back to be with me and to stay with Kim when I went into labor. October 5th arrived and it was going to be an evening with my "Waiting Wives" friends. Not feeling very well and wanting to just rest, I was going to stay home. But my mother, knowing full well what was planned, encouraged me to go. I arrived at Cindy's home where we were to have dinner and play bridge and was greeted with a baby shower; not only for me but for the other two wives in the group who were expecting within two months of my date. What a wonderful time we had!

That night would also be our second child's arrival into the world. At 12:30 AM, I called the hospital to let them know my symptoms and they told me to come in. Within five minutes of my call to Lucille and Jim they were at my front door. Off to the hospital we went. "It's a full moon tonight," the hospital staff informed me as I was quickly wheeled off to the labor/delivery floor. A truer statement was never spoken! The hospital's labor and delivery wards were packed. Women in labor all over the place; labor rooms full and the overflow, including me, on gurneys in the hallways being attended to by very efficient nurses.

Fortunately, I was one of the first to deliver. At 2:30 AM on October 6, 1968 our second baby was born... a son! After delivery, and due to the overcrowded conditions, they whisked him away to the nursery. I spent the rest of the night on a gurney in the hall of the ward listening to a number of women in labor. Ugh! No rest! The next morning, I was finally wheeled into a beautiful, sunny, single bed room where I was finally able to meet, hold, and cuddle our son, John Douglas. John and I had decided that Douglas, meaning strength of character, would be this baby's name, if it were a boy. My mother was first to suggest that our son should bear his dad's name, also. And so, it was decided that John Douglas came to be his name.

After being served a wonderful breakfast, a cheery Red Cross volunteer walked into the room and told me that John would be notified about the birth of his son. I later found out that a "field phone" was the means by which John was informed about his son's arrival and that his son would bear his name. Imagine, no cell phones, internet, or Skype to share our good news together!

Two days later, I arrived home with baby Doug and was greeted by his excited two-year-old sister. The fun was just beginning! Mom and my niece stayed three days longer, making sure all was well and then left the three of us on our own. The first few days were a blur; many feedings (Doug was a big baby and seemed hungry all the time), diaper changing, diaper washing, and playtime with Kim and baby Doug. My wonderful neighbors, Hue, Lucille, and Jim were a great support those first few months. I was able to take Kim to a Mothers' Morning Out twice a week for two hours and do grocery shopping while not having to pack up baby too. As much as some people pass off Post-Partum Blues as non-existent, they do exist. My joy was tempered with sadness and loneliness for quite a while. But my neighbors kept up with me and made sure I had food and companionship. What a Godsend these friends were to me during that year!

My letter writing was never very good and I know John was very disappointed that I didn't write more often. Honestly, after Doug was born, the events of that nine months without John caused me to be quite angry toward him. I didn't like him very much. He was doing his job as an officer in the US Army and kept his emotions in check. John thought this would be his career. I was thinking much differently!

There were days of anger, days of resentment, and days of unhappiness that I couldn't get past. So, writing a "cheery letter" was definitely off my "to do list." I felt it better not to write at all rather than fill a letter with unpleasant feelings toward him. He finally was able to get through to me on a phone; each person would talk and then say, "over," so the other person could talk. Of course, you didn't talk too much about personal things but just was happy to hear each other's voice and to know he would be home soon. That gave me hope for our future.

Kim and I enjoyed the upcoming Thanksgiving and Christmas days with our good neighbors and counting down the days until "daddy would be home." On Inauguration Day, 1969 at 6:00 AM in the morning, a knock came on the front door. I ran down the stairs and there standing in the light of the porch was my wonderful, whole husband, John. Kim heard the commotion and started down the stairs. I asked her, "Kim, who is this?" Without hesitation came, "That's my daddy!" The year of separation for her was nothing more than "yesterday." In order to keep her daddy in her mind, I had taped a picture of John on the refrigerator. Every day, we would look at it and send a "kiss" to daddy wherever he might be, and to remind her that he would be home "soon."

The whole, long, lonely, sad, frustrating, challenging, frightening year was probably one of the best experiences of my life. I really became a mature wife, mother, and woman. Taking on responsibilities, solving problems, learning patience and acceptance of things you can't change, being in-charge and knowing that my husband would be there to support our family's life ahead was such a comfort. I never envisioned that a year of separation could be so rewarding. It made us stronger. Yes, we had some challenges upon John's return, but nothing we couldn't overcome.

John's first decision was that he was going to resign his commission and work in the civilian world. It was his decision; not mine! I lost track of Hue after that year. Her husband came home safe at the beginning of January and immediately they were transferred out of DC. Keeping in touch with the Fox family was easier; they, too, were transferred but we were able to keep in touch for many years through Christmas cards. Once Jim retired, they moved back to Iowa and I never heard from them again. The take-away, military families always take care of military families. For John and me, The Long and Lonely Year was over and we could face any challenge that was to come in our future together.

TONY HILLIARD

US Marine Corps — Combat Engineer Officer

Dates of Military Service: 1965 to 1992

Units Served with in Vietnam: 7th Engineer Bn, 1st Marine Div, III Marine
Amphibious Force

Dates Served in Vietnam: Jan 1968 to Jan 1969

Highest Rank Held: Lt. Colonel

Place of Birth and Date: Philadelphia, PA — 1944

SHORT STORIES FROM AN ENGINEER ROAD CLEARING UNIT

When I was a Marine road clearing platoon leader near Da Nang, the company had an area that was used to dig fill material (dirt) for construction purposes located some distance from the Hill. That area was swept by another platoon as part of their daily road sweep. We returned from our sweep one day and were immediately directed to get into dump trucks and get out to the fill area. A Marine had been injured when he sat on a booby trap out at the site. We got there and deployed, and began to sweep the area for mines and booby traps. It was a tedious, time consuming process.

I was scheduled to go on R&R in Hawaii to meet my wife in a few days and began daydreaming and wandering around the un-swept area. A snap over my head brought me back to reality. It was a sniper shot that missed, but was close enough to hear it go by. I also realized where I was. I had wandered away from the platoon and was out in an unsecured area by myself. I had violated the cardinal rule that I had repeatedly impressed on the platoon. NEVER LOSE YOUR FOCUS! I very carefully retraced my steps back to the secured area and breathed a sigh of relief. I had been very lucky!

In June I was assigned as the Company XO and told I would be heading to Alpha Company back at Camp Love for duty as the XO there. This was a typical rotation policy for officers who had been in the field for about half of the tour. I was in the Company office with Captain Lopez one day while the Battalion Commander was making a routine visit to the outlying units.

As we were talking, a blood curdling scream come over the radio in the "Skipper's" office. He had a radio there to monitor all units on the company net. The scream was followed by some incoherent babble. At first, Captain Lopez told the person to get off the net, thinking it was some kind of prank. Then the transmission became clearer and it was the radio operator from the 1st platoon out on the An Hoa road. He was crying and said, "Sgt. Hollingsworth stepped on a mine and it blew him in two pieces."

The company radio section quickly started the process of ordering an emergency medevac to get him out. It took a few minutes to get the operator calmed enough to get the coordinates for the pickup. The net went quiet as everyone was doing what they were supposed to do. A short time later after the medevac left the scene, the platoon leader, 1st Lt. Art Morrison, made a report to the CO. Sgt. H had suffered severe trauma and passed away at the site. All Art could do was talk with him and try to keep him calm while the corpsman did the best he could to ease the pain and work on his wounds. A deep sadness fell over the company.

The casualty occurred when the leader of the USMC infantry platoon that was providing security for Delta 1 walked in from the flank and came up on the road at some distance in front of the sweep team. Sgt. Hollingsworth walked out in front of the sweep team calling to the Lt., telling him he was in a dangerous area. As he walked forward, he stepped on a large explosive device. His death came as a shock to everyone in the company. Marines like him were not supposed to be lost like that. I left Delta Company at the end of June.

THOMAS A. "TOM" ROSS

US Army — Special Forces
Dates of Military Service (Active and Reserve): 1966 to 1992
Unit Served with in Vietnam: Detachment A-502, 5th Special Forces Group
Dates in Vietnam: Jan 1968 to Dec 1968
Highest Rank Held: Major
Place of Birth and Year: Huntington, WV — 1945

RESCUE IN THE VALLEY OF THE TIGERS

1968 was one of the darkest years in our country's history. The war in Vietnam raged, Bobby Kennedy and Martin Luther King were assassinated, and rioting over the Vietnam War filled streets and parks outside the Democratic National Convention in Chicago. So, when one morning in August of 1968 dawned, there was no reason for me to expect that it would become one of the brightest, most inspirational days of my life.

Early in 1966, I felt compelled to join other young men and women who were serving in Vietnam and joined the military. After months of training, I applied to and was accepted by the U.S. Army Special Forces, the elite unit also known as the Green Berets. Immediately upon completing nearly a year of intense training, I was sent to Vietnam and assigned as the Operations and Intelligence Officer of Special Forces Detachment A-502. It was while assigned to this unit that I would have the privilege of witnessing countless deeds of selflessness and courage by others serving our country.

Having arrived in Vietnam during the infamous Tet Offensive of January-February 1968, by August of that year I had become a seasoned veteran. As a matter of routine, I often received calls and reports from our remote outposts. However, the call I received early one August morning was not routine. The outpost reported that three enemy soldiers had surrendered and they requested that I come to the outpost as soon as possible.

Upon arriving at the outpost, I discovered that the three men weren't enemy soldiers at all. Rather, they were Montagnard mountain villagers—and they had a story to tell.

Recognizing the men as Montagnards, I asked my radio man, who was Montagnard and also served as my interpreter, to ask the men why they had come to us. What followed was a tale of servitude and abuse. It seemed that the VC and North Vietnamese had made slaves of not only these three men, but also the entire village. For the past eight years, the villagers had been forced to carry ammunition and supplies as well as grow food to be used by various enemy units passing through the area. Their story of captivity included tales of brutal beatings and other more violent atrocities perpetrated on village inhabitants.

Mang Quang, the primary spokesman and apparent leader for the three, was a village elder who claimed to have been kept restrained in the village. He explained through the interpreter that after years of abusive captivity, he had grown weary of the VC treatment and began to defy their orders. Rather than killing him when he became defiant, they told him that he would serve their needs better as a living example for the rest of the village. They responded by beating him and keeping him tied to a post at night for the better part of two years. He had rope burns and the other two men as witnesses to support his story.

Mang Quang said he recently convinced the VC of his contrition and asked to be released from his in-camp duties and permitted to work in the fields. He said he decided to make the plea after his two friends told him about the information contained in the Chieu Hoi passes they found. Chieu Hoi passes were leaflets dropped from aircraft that promised food and good treatment to enemy soldiers who turned themselves into the allies. The Montagnards viewed the leaflets as a way to seek help for their village.

Early in the morning after the VC released him and allowed him to return to the fields, Mang Quang and his two friends quietly disappeared into the dense jungle. They hoped to find the allies and the things promised on the pieces of paper they carried with them.

Despite the fact that the three had been successful in their attempt to reach our outpost, Mang Quang was noticeably upset as his story continued to unfold. The reason for his distress was quick to follow and easily understood. He explained that, when released to work the fields, his captors gave him an extremely stern warning about checking in on a routine basis. They told him that if they hadn't seen him in three days, they would kill his family.

Mang Quang said the VC would think he slept in the fields one or two nights since that was a common practice and because he had just been released. But, if he didn't return by the third night, he felt sure they would know he had run away and would very likely carry out their threat to kill his family. One of the other Montagnards quickly added that the VC routinely threatened the villagers and they had killed other families before. He said that was the reason they came to us. They wanted to "turn themselves in" because they needed help for their village.

My interpreter, translating the villagers' emotional plea, turned to me and repeated it in English, "He says they are not able to defend themselves or their families from the VC. They need help because they don't want their wives or children to be hurt anymore." His lips were quivering as he spoke the last words of the villager's plea for help. The interpreter appeared visibly shaken by the story he had just translated. He wasn't the only one. I and the other American advisors around me were moved as well. I imagined my own family in such a horrible situation.

After more questioning of the villagers, it became apparent that their village was located in mountainous jungle somewhere in enemy territory to the far west of our camp and far outside of our area of operation. It was clear to me and the other advisors that an attempt to rescue the village would be both extremely difficult and extremely dangerous for many reasons. Standing before the pleading Montagnards, I pondered all that would have to be done and all that could happen if we were to attempt to help the villagers. And, due to the threat to Mang Quang's family, we would have to be at the village the next day—not much time to prepare an operation of the magnitude required.

At the time of my encounter with Mang Quang and the other two villagers, I was twenty-two years old. I suddenly felt the tremendous weight of responsibility for what might happen to the villagers if we did not go for them—along with the equally tremendous responsibility for what might happen to members of the rescue team if we did go after them and met disaster. After considering all of the possibilities and looking into the eyes of teammates who were present, I decided that if we were wearing the uniforms of American soldiers, there was only one possibility—we would go after the villagers.

Upon verbalizing my decision, every American present immediately volunteered to be a part of the rescue team. I can't explain how extremely proud that made me feel about the men with whom I was serving. Despite the many risks and challenges they knew we would face; they didn't hesitate to step forward. Their commitment made me absolutely certain that I had made the correct decision.

I told my interpreter to tell Mang Quang that we would honor his request. Upon hearing the news, he took my hand with both of his—and he began to cry.

If we were to attempt the rescue, planning had to begin immediately to give us any hope of saving Mang Quang's family. As an American advisor, I did not command Vietnamese troops—I simply advised them. To form a rescue team, I would need the approval and help of Major Ngoc Nguyen, the Vietnamese camp commander. So, I quickly headed out to find him.

Over the many months I had spent at A-502, Major Ngoc and I had developed a very close and warm working relationship. So, when I told him about the Montagnards and what I wanted to do for them, his only question was, "Is this mission important to you?" When I assured him that it was and that it would give meaning to my service in his country, he told me that I could have whatever I needed to complete the mission. Then, he surprised me by saying that he would come with me to command the Vietnamese troops, rather than simply assigning one of his Vietnamese officers as my counterpart. When I told him that he may want to reconsider because I had no idea what resistance we might face, he said, "If you're going…I'm going!"

With the needed troops committed, the next requirement was a way to get them out to the village. There was only one place to go for the transportation and that was the 281st Assault Helicopter Company. The 281st was the Special Operations helicopter unit attached to Special Forces to fly the secret and very dangerous Delta Project Missions. They could provide both the ride we needed along with gunship support. However, even though they were one of my assigned support units, I couldn't order the unit to fly the mission because it was outside of my assigned area of operation. But—an order wouldn't be necessary.

After leaving Major Ngoc, I went straight to my radio room and placed a call to the 281st AHC. After explaining the mission to one voice on the radio, I was then asked to explain it to another voice on the radio. By the callsign the second voice used, I knew immediately that I was speaking to the 281st's commanding officer. I held nothing back as I told him about the dangers involved in attempting this rescue. He seemed unintimidated by the risks involved because his response was, "Tell me where and at what time you need us…281st will be there." And they were there!

The next morning at sunrise, our rescue team lifted out of the A-502 compound and headed west into enemy territory. As morning mist rose off the jungle beneath us, we swooped down, secured our landing zone, and quickly offloaded our ground assault unit.

As the ground rescue team fanned out to locate and gather villagers, an Air Force plane with speakers attached flew low over the trees to alert the villagers to the opportunity for escape. The message was one pre-recorded by Mang Quang.

We were fortunate that our rescue attempt caught the enemy by complete surprise and we only faced light resistance as many simply fled into the jungle. We captured seventeen others.

When the sun set on the rescue mission, our team had extracted 165 Montagnard villagers from enemy control, Mang Quang's family among them. They were medically treated as needed, fed, and temporarily housed at A-502's compound. Ultimately, they were relocated to the coast of the South China Sea where they joined a Montagnard fishing village.

Before I left Vietnam, I went to visit Mang Quang at his new home.

The coastal villagers were teaching the mountain villagers to fish while the mountain dwellers were teaching them to grow corn and crops. They were living peacefully and everyone seemed perfectly happy with the arrangement.

This account of the very challenging Rescue in the Valley of the Tigers has obviously been abbreviated. However, what I hope you find by this brief retelling is that a group of selfless young Americans risked their lives for those of the primitive inhabitants of a remote mountain village. Every American member of the rescue team volunteered, including the pilots, copilots, and crew members of the 281st Assault Helicopter Company. The rescue could not have occurred without every one of them. This is not the typical story you hear about Vietnam. It is a good story about service in Vietnam and there are many others to be told.

The story of the rescue aired on the CBS Evening News with Walter Cronkite on August 13, 1968. David Culhane, the CBS reporter, concluded the story by saying, "This is a rare occurrence in this war, an act designed to give life and freedom in a place and time noted mainly for death and destruction."

JAY PRYOR

U.S. Naval Reserve — Communications Officer
Dates of Military Service: 1967 to 1970
Unit Served with in Vietnam: USS Hissem (DER-400) & Commander Destroyer Squadron 29
Dates in Vietnam: Three deployments in period of 1967 to early 1970
Highest Rank Held: Lieutenant (Junior Grade)
Place of Birth and Year: Albany, GA — 1944

MARKET TIME — A LESSON IN LEADERSHIP

The Ho Chi Minh Trail is an iconic feature of the War in Vietnam. The Trail was the main supply route for troops, weapons, and supplies supporting the North Vietnamese forces in the South of the country. The Trail was not a single road nor a single trail but a network of roads, some of them going through the neighboring countries of Laos and Cambodia. In the early years of the war, the sea was the primary means used by the North Vietnamese to get supplies to the South and the trail was used primarily for the infiltration of manpower. But when Operation Market Time was put into place, the use of the sea as a supply route virtually ended.

I was a brand-new Naval officer in early 1968 and following training at Communications Officer School in San Diego, I headed to the Western Pacific to find the ship that would be my home for most of the next two years, the USS Hissem (DER-400). The ship's mission when I reported onboard was Taiwan Patrol, a program started by President Truman to "show the flag" and prevent conflict between the Republic of China and the People's Republic of China. When that assignment ended, the Hissem headed south for an operation having the name of Market Time, in the waters that were in the combat zone of Vietnam.

The purpose of Market Time was to stop Hanoi from using the sea to resupply troops in the South. It turned out that the Hissem was ideally suited for Market Time operations. The ship, 306 feet in length and having a displacement of 1,700 tons, was powered by twin diesel engines and could remain on station for a long period of time without needing fuel or other supplies. The ship could more closely be likened to a marathoner than to a sprinter. When our speed got in the neighborhood of half the speed of a gas turbine destroyer, the shaking could be felt throughout the ship.

I had only been on the Hissem for a short time, but I already was assigned a number of collateral duties, things I was responsible for in addition to being responsible for all the ship's external communications. One of those collateral duties was as the Market Time Boarding Officer.

Every minute of every day, the sailors in the Hissem's Combat Information Center (CIC) were keeping a close eye on the ship's radar, looking for any ships, large or small, that were going from the North to the South. When one of those vessels was found, we would stop them, board them, and search them. I was one of those responsible for searching those boats.

When the Watch Officer in CIC gave word to the Officer of the Deck (OOD) on the Hissem's bridge, the OOD would notify both the ship's captain and the South Vietnamese Naval officer we had onboard as an interpreter. The Hissem would maneuver close enough to the other ship so that our interpreter could tell them via bullhorn that we wanted them to stop so we could board and search.

I am glad to say that my boarding experience was like most: uneventful. The Hissem boarding party would get into the Hissem motor whaleboat and be lowered to the water. I was armed with a .45 pistol and one of the deck crew was armed with a rifle. We also had the interpreter with us. When we boarded the boat to be searched, the Hissem motor whaleboat would maneuver out of the way because on the Hissem was a sailor with a machine gun trained on the boat being searched. Most of the boats we stopped were fishing. Two things that stand out in my memory are the fact that they would use wood shavings in the hold of the boat to preserve the fish; and the one time I encountered a boat crew member who was infected with leprosy.

On the first of March 1968, the North Vietnamese tested the Market Time force and sent four gun-running trawlers to the south. Two of the four trawlers were destroyed in gun battles with U.S. Navy ships and a third trawler was scuttled to avoid capture. The fourth trawler reversed course and escaped to the north.

A post-War study concluded that at the beginning of 1966 approximately 75 percent of the supplies going from the North to the South went along the South Vietnamese coast. But by 1967, following the implementation of Market Time, this number had dropped to approximately 10 percent. One of the results was the increased reliance on the Ho Chi Minh Trail.

SONNY DELLINGER

US Army — Medical Service Corps

Dates of Military Service: 1967 to 1969

Unit Served with in Vietnam: 1/503rd Infantry, 173rd Airborne Brigade

Dates in Vietnam: Aug 1968 to Aug 1969

Highest Rank Held: 1st Lieutenant

Place of Birth: Annapolis, MD — 1945

AID STATION — LZ UPLIFT

My father was a retired Navy Chief and met my mother at the Baltimore Naval Hospital where she was serving in the WAVES during WWII. After my father retired from 22 years in the Navy, he went to work for the Public Health Service, today's CDC, and our family moved to Chamblee, GA.

I attended Cross Keys High School and then went on to attend Furman University. At Furman, I joined the ROTC program and was commissioned a 2nd Lieutenant in the Army Medical Service Corps in June of 1967.

As soon as I went on active duty in August of 1967, I was sent to the Medical Field Service School (MFSS) at Fort Sam Houston, TX. "Ft Sam" was the training center for all the medical-related personnel in the Army. All doctors, dentists, ophthalmologists, nurses, MSC officers etc., attended a "basic" course which covers some medical procedures but mostly tries to teach these new officers (specifically doctors) what it means to be and to act like an Army officer.

After the basic class at MFSS, several MSC officers in my class were chosen to attend an additional four-week course for "Battalion Surgeons' Assistants." This course gave us additional training in emergency medical procedures and taught us the basic medical operation at the Battalion Aid Station level. This was also an alert to us that we were on the short list to get orders to the Republic of Vietnam sometime in the next few months.

My next assignment was to teach "Introduction to First Aid" and

"Chemical Biological and Radiological Warfare" to basic trainees in the Sand Hill and Harmony Church areas of Fort Benning, GA. I remembered from my days in MFSS at Fort Sam that I always seemed to pay more attention to an instructor that came up with a joke or amusing story, so being a nervous 2nd Lieutenant in front of a couple of hundred trainees, I started each class with the story of the General who called the motor pool checking on the types of vehicles there.

The General asked the private who answered the phone to tell him what vehicles were there.

The private said "We've got three Jeeps, two Deuce and a Halfs and a Staff Car that belongs to that fat-ass General."

The General is a little taken aback and asks, "Soldier, do you know who this is?"

The private says, "No Sir."

"Well, this is the General."

The private asks, "Well General, do you know who this is?"

The General says, "No."

The private replies, "Good-bye, FAT-ASS" and hangs up as fast as he can.

This always seemed to get the class started off on the right foot.

While at Fort Benning, I decided to go ahead and volunteer to be sent to Vietnam as well as Airborne Training as sort of a "package deal." I felt sure that I would be getting orders to Vietnam soon anyway and I wanted to be assigned to an Airborne unit. I had seen the training received by the enlisted men and Special Forces Medics at Fort Sam and I felt that an Airborne Unit would have medics that were very well-trained. I was accepted for both "Jump School" and Vietnam and in August of 1969 I was on a plane headed to Bien Hoa AFB near Saigon.

As we were landing at Bien Hoa, it seemed surreal that I was in the middle of a war zone but the bomb craters around the runway assured me that I was there. At Long Binh, I was assigned to the 173rd Airborne Brigade and became the Medical Platoon Leader of the 1st Battalion/503rd Infantry located at LZ Uplift in the northern part of II Corps.

LZ Uplift was the headquarters for the 1st Battalion and its support units of Armor and Artillery. The Battalion itself was made up of Infantry companies, a mortar platoon, a long range patrol platoon and the medical platoon which was part of Headquarters Company. The Battalion Aid Station and medical platoon had a Battalion Surgeon (doctor), Medical Platoon Leader (me), Platoon Sergeant and between 20-25 enlisted medics. As the medics rotated in and out during my twelve months in-country, they were assigned to Infantry companies. They usually spent about six months in the field with an Infantry company before coming back to work at the Aid Station for the balance of their tour.

Early in 1968, the Tet Offensive had involved all US and allied troops but by the time I got there in August of 1968, things had slowed considerably. The largest engagement for the 1st Battalion at the time was in the Suoi Ca Valley not far from our LZ. Casualties were sent to our Aid Station via MEDEVAC ("Dustoff") where they were stabilized and transferred to the Evacuation Hospital located in Qui Nhon which was about an hour away by helicopter.

I had been at the Aid Station for only a short time when we received two casualties, a pilot and co-pilot from a "Bird Dog." The "Bird Dog" was a fixed wing plane similar to a Piper Cub that flew slowly over an area and "spotted" for the artillery. In this instance, the plane had received small arms fire that damaged the engine and forced them to crash land not far from our LZ. The co-pilot was wounded pretty seriously and the pilot said he was hit but he insisted he was okay. The doctor and our team went to work to stabilize the co-pilot and within five minutes the pilot, who had been standing nearby, passed out and before we could do anything, he died. He had on a wedding ring. Reality set in. I was wondering to myself how I could make it through the next ten or so months but finally decided that I just had a job to do. I could not let my feelings for the casualties get to me.

For the balance of my tour at LZ Uplift, I went about each day/week assisting the doctor with daily sick call, attending the daily Battalion Commander's briefing, conducting administrative duties concerning the enlisted members of the Medical Platoon, organizing Medical Civilian

Aid Programs (MEDCAP), and conducting malaria prevention procedures. Our Brigade Surgeon had initiated a program of unscheduled urine testing to make sure that the soldiers in the field were taking their malaria prevention pills. It was the job of each Battalion's MSC officer to go to the field unannounced, collect urine samples, and take them back to their respective Aid Stations for testing.

It was on one of these "urine collection" missions that one of our young paratroopers asked me, "Sir, what are you going to tell your kids that you did in the war?" I gave it a little thought and replied, "I guess I'll just have to tell them I was a Piss Tester." In any event, this program was quite effective and we were able to lower the number of malaria cases in the 1st Battalion by over 80%.

Finally, I have to commend all the Medical personnel involved in the Vietnam War. The medics from our platoon received a number of commendations including a number of Bronze Stars and one Silver Star. We had a Medevac "Dustoff" crew that rotated through LZ Uplift usually every two or so weeks. The Dustoff pilots and crew were absolutely fearless as I saw them go out in all types of weather, at all hours of the day and night to retrieve casualties.

Sherman said it, "War is hell." But if I had to go back, it would be with the 173rd Airborne.

JOHN FRASER

US Army — Chaplain

Dates served: 1953 — 1983 Texas National Guard (Enlisted) 1953-1956; Army
Ready Reserve (Enlisted) 1956-1960; Army Active Duty (Commissioned) 1963-1983.
Units Served with in Vietnam: 2/14 Inf and 1/27 Inf, 25th Inf Div; 8 RRFS.
Dates Served in Vietnam: Apr 1966 to Apr 1967; Nov 1969 to Nov 1970
Highest Rank Held: Lt. Colonel
Place of Birth and Year: Greenville, TX — 1936

BLUEBELL REPORT

Between tours in Vietnam I was the Post Chaplain at a small installation
near Warrenton, VA, about an hour drive west of the Washington Belt-
way. This small one-chaplain post was in the beautiful North Virginia
horse country. Vint Hill Farms Station consisted of about one thousand
troops plus their families. This was an Army Security Agency post where
even the cooks and bottle washers had top secret clearances. It was a
wonderful place to be assigned.

It didn't take long to establish a family routine. Donna, my wife, had
quickly made new friends, my two daughters were in the post kinder-
garten where they spent weekday mornings with new friends. It seemed
that life couldn't get much better than this. Then after about a month, my
chapel telephone rang with some unwelcome information.

Post Adjutant: "Chaplain, we just got a Bluebell Report." Me: "What
in the world is a Bluebell Report?" Post Adjutant: "It's a report of a KIA
(Killed In Action) and you have to tell the family." Me: "How do we go
about that?" Post Adjutant: "Call the motor pool, order a car and driver,
pick up the information at my office, and go tell the family. The address
is in the Bluebell." I don't remember which little town we went to, but it
was a little northwest of Shenandoah National Park. On the way, I studied the information that was provided so I wouldn't have to just do a cold
reading to the family. The report showed that the soldier had been in my
Battalion in Vietnam, and I was only three months from Vietnam. I was
thinking, "I'll bet I knew him and just can't place the name." Finally we

arrived at the address I had been given. I swallowed a great big lump in my throat, took a deep breath, and walked up to the door. A very pleasant-appearing, late-middle age man opened the door. "Sir, I'm Chaplain Fraser from an Army post not far from here. Is your wife at home? I need to talk with both of you."I recognized the picture of their son that was on the coffee table. After I told them that their son had been killed, I told them that I had been his Chaplain and that I remembered him. To make a long story short, they asked me to come back to hold his funeral. I agreed to do that.

The day of the funeral I was astonished to find another young soldier that I knew. He was a Buck Sergeant who was a SP4 when I was his Chaplain in that same Battalion. He had grown up in the same county as the other soldier and was given the job of escorting his friend's body back home. After the funeral, we talked for a while about the recent activities and people from my old Battalion. He told me that my replacement, Chaplain (Captain) John Durham had been shot, but had survived and been evacuated to a hospital in Japan.

About two months later I received another Bluebell. This time it was that same sergeant that had escorted the body. His parents asked me to come back for the funeral. I did.

When I was with the Infantry in Vietnam, I thought the worst thing I would ever have to do was lose soldiers in combat. I was wrong. The worst thing was telling parents, wives, and children that their soldier had been killed. I suppose I delivered a Bluebell every couple of months before I went back for my second tour in Vietnam. Most of the time the families graciously thanked me, told me they were sorry that I had to deliver the news, and asked me to come back for the funeral. I always did.

GLENN PEYTON CARR

US Army—Aviator

Dates of Military Service: 1958 to 1986

Unit Served with in Vietnam: 213th Assault Support Helicopter Company (Chinook);B Troop 7th Squadron 17th Cavalry Regiment (Air)

Dates Served in Vietnam: XO & CO 213th May 1967 to May 1968; CO B Troop 7/17, XO 52nd Combat Aviation Battalion 1971

Highest Rank Held: Lt. Colonel

Place of Birth: Shawnee, OK—1934

VIETNAM WAS NOT ALL DRUDGERY AND KILLING

We had a few moments of fun and enjoyment

I had recently read several articles in the Stars and Stripes about Martha Raye that touted her love for the grunt and her selfless efforts to apply her surgical nursing skills where ever she could in Vietnam. In addition to her Hollywood career, she was a Lieutenant Colonel, surgical nurse in the US Army Reserve. I had the pleasure of encountering her on more enjoyable conditions.

I was the XO of the 213th Chinook Company at Phu Loi when we were alerted to host Martha and her USO Hello Dolly show early January 1968. I assembled my construction crew consisting of two Spec 5's to sort out what kind of facility we were to build and plan our procurement of needed assets. Those two Spec 5s both had what I considered to be PHDs in procurement. (Read that they were the best thieves in the US Army).

It was decided to gang four flatbed trailers together for the main stage, use two M109 shop vans for the right and left dressing rooms, and stair type maintenance platforms for ground level to stage access. Ms. Raye's lead man said that was great, but the truck beds were too rough for the ladies in high heels. My lead man said, "No problem, we will cover the truck beds with ¾" inch plywood."

Well, that was easier said than done as ¾" inch plywood was the most sought-after construction commodity in Vietnam. That didn't seem to bother my PHDs except one load was legitimate and had to be signed for by a field grade officer. That field grade officer was yours truly to the tune of about $600 at stateside prices, and it was returned personally to the owner. I was told not to ask any questions about the other two loads.

Soon we decided we had conjured up enough support that we could go full blast with all the luxuries one would expect in a Hollywood production, although cloaked in a combat environment.

The stage was spacious with two wings on each side turned slightly inward to facilitate the actors being close by, observing and waiting for their que. On each side of the stage was a hallway leading from the wings back to the dressing areas. We went over all the flooring areas insuring they were free from any trip hazards. Believe me they were as good and flaw free as any modern stateside home.

The dressing areas were two M109 shop vans mounted on 5-ton trucks full of machine tools, lathe, band saw, small milling machine, and welding equipment. This was all removed and replaced with field tables and chairs, as far away from Chippendale as you can get. We built a six-foot-long dressing table against the wall with a sink in the middle and two mirrors on the wall. A small shelf was added and well stocked with washcloths and towels.

The inside of the vans was the easiest task. Now we are going to plumb the sinks with hot and cold running water. A rack was built to span the roofs of the two shop vans and raised about four feet. The Air Force came forth with two unserviceable aircraft wing tanks, one for cold and one for hot water.

The hot water was achieved by cutting a large hole in the top of one wing tank and inserting an emersion heater. The emersion heater was a simple device used in mess halls consisting of a fifteen-inch hollow donut with a chimney and air vent rising about three feet. Over the air vent was a one-gallon gasoline tank with a needle valve. You would throw a lighted piece of paper down the air vent and turn on the needle valve for a slow drip. That slow drip would keep a nice fire burning in the donut and heat a wing tank of water in about fifteen minutes. Now we had H&C running water.

The next item of necessity was a latrine. PA&E (Pacific Architects & Engineers, the guys who built our mess halls and other facilities, brought us an outhouse very similar to the portable toilets in use today. We situated it right behind and centered on the shop vans. PA&E begged to do something more than just provide the latrine so we asked them to wire the stage and dressing areas with lights. Were they ever happy! That stage was as bright as any Broadway theater.

We took up a collection to pay for one of the waitresses at the officers' dinning facility to serve soft drinks and light snacks during the show and later at the reception in the 213th "Black Cat" Chinook Company area. She was present for the walk-thru by the command group consisting of my company CO, the battalion CO, 1st Infantry Division Artillery CO, who was the Camp Commander, and Martha's show boss. They were all very pleased with what they saw and especially impressed with the waitress dressed in her finest formal Vietnamese attire to serve light snacks.

The show boss was speechless with what he saw saying, "Martha is going to go crazy over this facility. I can't think of a thing you missed. Superb job, Major!" Needless to say, that made my day as I turned and introduced my team lead by my PHDs. Out of the corner of my eye I noticed my company CO grinning like a possum eating guts.

The evening of the show, my team was standing by early in case the cast needed anything. Martha come over and gave me a great big hug, exclaiming she had never had a stage facility like this built for her show. She was ecstatic! The show was extremely well received by some five hundred soldiers. I believe the cast put their all into the show to return the favor. And Charlie behaved himself—no incoming rockets!!

The reception went well with lots of food and booze. Martha was so friendly and forth-coming she told us to call her Maggie or Old Big Mouth. Obviously, we choose Maggie, but that set the tone for a most relaxed evening. Room assignments were made and a walk-thru on where the bunker was located was conducted. The cast was to occupy the officers' quarters, and we were to grab our fart sacks (sleeping bags) and sleep in the Chinooks.

As the party was breaking up, I noticed Maggie was looking at our lighted fountain in the middle of the yard between the two rows of officer billets. I went over to explain how we made it. Sidewalks paralleled each billet with a crosswalk in the middle with a six-foot circle in the middle.

This was what I found when arriving last May, except the center was a mud hole. So, I dug that out, pool shaped, and cemented it in. We placed a 12-inch culvert pipe vertically in the center and put a Vietnamese cone hat on top. I then placed a bilge pump from an M113 personnel carrier inside the culvert pipe with the water discharge pipe going up thru the center of the hat and two Chinook landing light bulbs under water on each side facing up and center. The transformer from a shop van was placed in the XO's quarters (mine) to power the pump and lights. Add water and *voila*, you have a lighted fountain in a combat zone.

Maggie stood in silence for a few minutes shaking her head then said, "Seeing that stage complex you and your team built and this fountain, I shall never ever doubt the ingenuity of the American GI."

I had the number one aircraft the next morning which means the earliest take off with the longest list of missions, so I didn't get to bid the cast farewell as they left our company area and Phu Loi.

Our association didn't end there. About two weeks later, I again had the number one aircraft and when I was hauling my last sling load about 4:30 pm, I got a call from Black Cat operations. "Black 5, I got an add on for you." Immediately all my crew started hollering all sorts of profanity because we were tired. I responded, "After that party last night and sleeping in the aircraft, you're gonna leave us out here for another couple of hours." Ops responded, "5, I think you will like this one." I replied, "OK, try me out." Ops said, "Maggie and cast are stranded at the Air Base in Can Tho (I think that's the correct name). The Air Force fell flat on their ass and can't fly her out to Saigon, and they are leaving country the next day. She asked the AF to call the Black Cats and specifically asked for you." By that time, I had dropped my last load and was at altitude. I responded, "We will gladly take that mission. I'm going into Lie Ka now to refuel and then heading south."

Arriving at the Air Force Base, the tower gave us a priority landing.

At the parking spot, ground control told us to shut down and all crew get into the ramp wagon as we had been instructed to meet Maggie at the Officers' Club. Recognize the fact that we had been flying for a little over eight hours and were dirty as pigs. I expect the tower could smell us coming on short final approach. What will our reception be at the O-Club?

Hesitatingly, we opened the door which placed us directly in the bar. Maggie quickly spied us and at the top of her voice, which was mighty, shouted, "Hurray, the ARMY Black Cats are here." Emphasis on Army for obvious reasons. Maggie said rather loudly, "Glenn, you will never guess what I did last night." Having no idea, I said, "Tell me." She replied, "We had a rocket attack last night, and I spent most of the night in the Medical Facility emergency room attending to the wounded." She was known to have done that several times while in Vietnam.

We had a few Cokes and small chatter with a few AF pilots who were interested in the capabilities of our Chinook. As we departed the club, Maggie grabbed me and put her arm around me as did the rest of her female cast with the rest of the Chinook crew. An obvious *screw you* to the Air Force.

Equipment all loaded and tied down and passengers belted in, I cranked up the Chinook. Seeing many of the folks from the club out to watch us take off, I asked the tower if they would approve a vertical takeoff from the taxiway. They approved, and I tickled the engines to get the fuel controls headed up and then pulled the thrust up under my arm pit, starting an ascent of about 3000 feet per minute and doing petal turns round and round, then in the other direction. Leveling off at 3000 feet, I bent the nose down and headed north to Saigon. I wish I could have been near the club bunch to see how many Air Force jaws hit the pavement.

Landing at Hotel 3 (the main helipad at Saigon), We all helped the cast unload their gear, which was massive. Maggie invited us to walk out the gate to the Cholon PX for a big milk shake. We couldn't turn that down. After a cool shake and chat, we bid them a fond farewell and headed to Phu Loi. That was the last time I saw Maggie. I am told she is the only non-Special Forces soldier permitted to be buried in the Special Forces cemetery at Fort Bragg. NC. A most fitting tribute**.

In two weeks, I would again go into Hotel 3, then carrying a sling load of ammunition and had two Cobra gunships escorting me firing into the perimeter fences and barriers. The PX was on fire. The TET-68 offensive had started three days ago, and all of South Vietnam was in flames.

That made the time we enjoyed with Maggie and her USO entertainers much more memorable.

Editor's Note: Martha Raye (born Margy Reed) was born August 27, 1916, in Butte, Montana, and died at the age of 78 on October 19, 1994, in Los Angeles, California. She was buried with full military honors at the Fort Bragg Main Post Cemetery at Fort, Bragg, North Carolina. She is the only civilian buried at this location who receives military honors each Veterans Day.

GEORGE CONSUEGRA

US Army — Field Artillery Officer

Dates of Military Service (Active Duty and Reserve): 1965 to 1972

Unit Served with in Vietnam: American Division,198th Light Infantry Brigade, A Battery, 3rd Bn., 18th Arty.

Dates in Vietnam: Oct 1967 to Sep 1968

Highest Rank Held: 1LT

Place of Birth: Havana, Cuba — 1944

FIRE MISSION!

I had the "best assignments" — Field Artillery Officer — served as a Forward Observer with an Infantry Company, Battery Executive Officer, and Battery Commander — even though I wasn't very good in math, I spoke Spanish and French fluently.

I served in the American Division, I Corps. , 198th Light Infantry

Brigade. Yours truly went through Infantry training at The Citadel, excellent artillery training at Ft. Sill, and "troop training" at Ft. Hood, Texas.

We sailed from Oakland, California on 10/6/1967 and arrived at Chu Lai on 10/24/1967. There were 500,000 protestors in Washington, DC when we left Oakland. I couldn't understand the protesting.

I was assigned as a Forward Observer (FO) — E Company, 1/46th Infantry. We were based at Hill 69 between Chu Lai and Tam Ky (west). Captain Monty Wolfe, 33 years old, was the Company Commander, they called him the "old man." He was always in the first helicopter to land into the landing zone and was on the last helicopter to fly out when we completed our "search and destroy" missions.

My FO Team was always with him. I was always scared, especially when the helicopter M60s softened up the LZ. Several times my RTO (Radio Telephone Operator) froze on the helicopter and I had to kick him out. We encountered lots of sniper fire and several fire fights. Everything happened very quickly. I was well-trained, brought in the first "smoke round" and then "Fired for Effect" — no "short rounds." I learned very quickly how to adjust "Close-In" artillery fire with no binoculars. The "grunts" loved my artillery team. Captain Wolfe and his Platoon Leaders continually asked us — "Artillery — where are we?" I became very handy with a map. It was my job.

I remember one of our "tunnel rats" in E Company. He borrowed Captain Wolfe's 45 to go down a tunnel. He broke the gun handle. He was scared to tell Capt. Wolfe about it.

Also, yours truly remembers stuffing C-rations in socks before going on "Search and Destroy" missions. I liked the Cs except for the ham and eggs and lima beans. My favorite was hot chocolate and fruit cocktail. I've only been hungry several times in my life — my freshman year at The Citadel and while on Search and Destroy missions when we were unable to be re-supplied due to enemy fire.

Several times, we brought in the Medevac helicopters at night to evacuate our wounded. Those pilots and their flight crews were very brave and dependable.

I was an FO for about four months before being assigned to an artillery Battery. I learned a lot while serving with the Infantry. I used this experience immediately when I returned to the Battery. I knew how best to help the Infantry when they called my Battery for Artillery support.

I subsequently became Battery Executive Officer and Battery Commander of A Battery 3/18 Field Artillery. We were located at Hill 54 North. The Battery consisted of two 155s, two 8-inch, two 175s, and two 105 guns. We had tubes pointing in every direction since our fire missions were requested from different locations. We shot fire missions during the day and H&I missions at night. We kept track of the "body count."

The food was much better at the Artillery Battery. I worked with the Executive Officer, Lt. Knight, for a few days before he left for the U.S. The Troops nicknamed him "Sky King." He slept in the Fire Direction Center Bunker because he was terrified of rocket and mortar attacks.

Unfortunately, I developed "jungle rot" on both of my legs shortly after arriving at the Battery area. The Doctor gave me a lotion to apply on both legs, but I had to cut off my fatigues at knee level. So, I walked around the Battery in cut-off fatigues and was asked many times by visiting Officers for the Doctor's written orders. I carried those orders with me at all times.

We were busy 24 hours per day and I alternated duties with the Fire Direction Officer. The days were long, very hot, little sleep, etc.

In May, 1968, Major Gavin and the Vietnamese District Chief thanked us for helping them kill 60 Viet Cong who almost overran a village. We fired 64 rounds on that mission. They presented us with a VC flag which we hung in our mess hall.

On July 3, 1968, we received a direct rocket or mortar hit on our #3 powder bunker. It happened around 10 or 11PM. I ran down from the Observation Tower and drove one of our 8" guns out of the pit to a safe location. We had eight guys wounded that night. They were all evacuated to the hospital in Chu Lai. They were awarded Purple Hearts. I wrote up their citations and visited them as much as I could at the hospital. I was awarded a Bronze Star with a "V" Device. That puzzled me since I was only doing my job.

On July 29, 1968, Stanley Resor, Secretary of the Army, and General Charles Gettys, Commander of the American Division, inspected our Battery. It was a very hot day. General Gettys was obese and sweated profusely while inspecting us. I thought he was going to pass out. We passed the inspection with "flying colors." He had a huge water bottle in his private helicopter.

While at the Battery, I passed my vision exam in Chu Lai to qualify me for helicopter training — the vision standards had been reduced since so many of our helicopters were being shot down. I was awaiting orders to extend my service and attend flight school. The orders didn't come in time so I decided to try "civilian life." This decision probably saved my life! I was very lucky — didn't get killed or wounded, except for the "jungle rot."

JAMES J. "JIM" HOOGERWERF

USAF — C-130 Pilot
Dates of Military Service: 1966 to 1973
Based off-shore in the Philippines at Clark Air Base: 773rd TAS, 463rd TAW
Unit Served With in Vietnam: TDY to 834th Air Division, Det 1 TSN, Det 2 CRB
Dates Served in Vietnam: May 1968 to May 1970 (not inclusive)
Highest Rank Held: Captain
Place of Birth: Detroit, MI — 1943

WHEN THE RUNWAY ENDS...BEFORE THE AIRCRAFT STOPS... OOPS!

Sixty C-130s were lost in the Vietnam War; about half from operational causes. On this day we almost made the latter listing. Why we did not is a tribute to the pilot skill of Lt. Col. Luther H. Waechter, my aircarft commander, and a generous dose of lady luck's charm.

On December 6, 1968 I was the copilot on a Lockheed C-130B mission in support of airlift requirements to Dong Xoai, Vietnam. My crew was TDY from the 773rd Tactical Airlift Squadron (TAS), 463rd Tactical Airlift Wing (TAW) based at Clark Air Base in the Philippines. This was my sixth rotation (16 days each) TDY in-country since arriving in May at Clark, my first duty assignment out of flight school. My aircraft commander, Lt. Col. Luther H. Waechter, was a trim, spit-and-polished officer and a good pilot. His previous assignment had been with air/sea rescue picking up downed airman off the coast of North Vietnam in UH16 Albatross amphibians. Maj. Joe E. Johnson was the navigator. A flight engineer and a loadmaster, both sergeants, rounded out the crew.

We were "Terry 608." "Terry" was the permanent call sign for C-130 "B" model aircraft operating in Vietnam. "608" was our "frag," short for fragmentary order, assigning us our mission for that day. It delineated several stops to be flown. On this day, before we got to Dong Xoai, we had already flown from Tan San Nhut (TSN) to Cam Rahn Bay (CRB), Phan Thiet (PNT), and Bien Hoa (BNH). At BNH we were to pick up a load of 155mm artillery shells, destined for the big guns at Dong Xoai, approximately 40 miles to the north. This was not an unusual load or destination for the C-130. We carried everything the troops needed to wage war, and the troops too, to airfields all over South Vietnam. The shells were secured on five cargo pallets for easy loading and quick off-loading at the rear of the C-130.

That morning I checked for NOTAMs (notices to airmen), but there was nothing special to note. We referenced the Tactical Aerodrome Directory for specific airport details. Dong Xoai was a 3250' assault landing strip, reddish in color from its laterite surface. Hazards were noted as:

Opposite ends of rwy not vis fr touchdown. App to Rwy 07 over road and ruf field, 250' fr thld. App to Rwy 25 over road and 4' concertina fence 300' fr thld. 1'-2' ditches aja to both shoulders. Roads run entire length of rwy, aja to both drainage ditches. 6' concertina fence 40' fr S edge. 8' CONEX 40' fr N edge, midfield. Not recommended for ngt ops.

At BNH we shut down at the ammo dump for our load. Ground time was a short thirty minutes; soon we were airborne again. Time to Dong Xoai was a quick thirty-five minutes⊠the C-130 was very efficient at "hauling trash" (i.e., cargo). Overhead, Col. Waechter banked the aircraft to survey the field. The runway was a widened straight section of a road that continued off each end. As we circled, it was obvious a section had been peneprimed, a black material to seal the surface, distinguishing it from the laterite. We noticed also, the standard runway markers were missing. The panels are placed 100' and 500' from the runway threshold. The desired touchdown on an assault landing was between 100' to 300.' Anything less than 100' was unacceptable; a touchdown more than 500' mandated a go-around. We did this routinely and were good at it. It was a skill of necessity since a balked landing would unnecessarily expose us to potential enemy ground fire.

Col. Waechter and Maj. Johnson discussed the situation and determined to land on the peneprime surface, reasonably assuming it was the useful part of the runway. I concurred. "Hacking the mission" (i.e., completing the mission under all conditions) was a point of pride of airlifters in Vietnam. As Col. Waechter flared it was obvious the laterite was serviceable so Col. Waechter touched down short of the peneprime. Per standard procedure he applied maximum reverse thrust (60% of full forward thrust was available in reverse), and full brakes. The airplane began decelerating immediately, but not enough to stop on the runway remaining. There is a truism in aviation that nothing is more worthless than runway behind you and we were about to prove the point.

Col. Waechter, I remember his words to this day, exclaimed: "We're not going to make it!" Recognizing the impending consequences but short of the end of the runway, he turned the airplane sharply to the right...still in reverse and braking...off the runway. The wings wobbled dramatically, but not enough for the prop tips to touch the ground. As the C-130's fat tires dug into the soft berm, we slowed considerably more. Brilliantly, before coming to a complete stop, with still enough momentum available, Col. Waechter maneuvered the aircraft (more wing wobbling) on to the hard surface of the roadway and to a full stop... miraculously just in time. Ten feet more and the number one prop would have hit a water buffalo (not the animal, but an Army trailer carrying a 400 gallon fresh water tank). The consequence of that, with all that ammo cooking off from the intense fire that would have surely followed, is not pleasant to contemplate.

Col. Waechter set the brakes, jumped out of his seat, and, along with the navigator and flight engineer, abandoned the cockpit to survey the situation in the back. I was left in the cockpit to monitor the brakes and engines and otherwise look after things. Uncomfortably I thought, what more could go wrong? Were the brakes overheated, what if hydraulic lines ruptured spewing fluid on the hot brakes, would the engines overheat? In front of me outside, our situation had attracted the attention of local Army personnel and some Vietnamese. It was a spectacle no doubt, something to break up their day: this big airplane on a road, like a beached whale out of its element.

As I sat there hoping nothing would happen to require my immediate attention, a Vietnamese started walking around the airplane on the left heading right at a spinning propeller! There was nothing I could do to warn him. I was certain he was going to walk directly into the prop, but amazingly he passed under it oblivious to the danger of the spinning blades. Fortunately, he was of smaller stature than even most Vietnamese.

Behind me the crew and ramp personnel safely off loaded all the pallets. Soon Col. Waechter and the others returned to the cockpit. What we had to do now was get the C-130 back on the runway. With all that weight removed, we might be able to back up the road. It was our only choice. The Army had moved the water buffalo and I warned the observers in front of us to get out of the way of the prop blast coming their way.

The procedure in backing under difficult situations, was to let the plane roll forward slightly to get off the flat spots on the tires, and then immediately go to full reverse thrust, holding it to see if the airplane moved back. Col. Waechter did this several times, rocking the airplane back-and-forth. After several tries, finely, imperceptibly at first, the C-130 slowly began to roll backward up the road and onto the runway.

We taxied to the ramp area to do an inspection and check with "Hilda" our operations control. It was determined to fly to TSN gear down. This whole scenario only took forty minutes (I kept a daily flight log).

Later we learned the only damage was from the front tires being deflected so far off the rims it trapped debris in the wheels; fortunately they had not gone flat. Probably, since the runway was improperly marked, the work not reported, and the aircraft not severely damaged, nothing was ever said that I heard of about our overrun that day.

Between May 1968 and May 1970 I flew over 1100 sorties and accumulated more than 900 combat hours flying in South Vietnam. I upgraded to aircraft commander in-theater and completed my tour in command of my own airplane and crew in the war zone. Fortunately we never ran off a runway again, but I don't think I could have matched the skill Col. Waechter displayed at Dong Xoai that day.

THOMAS A. "TOM" ROSS

US Army — Special Forces
Dates of Military Service (Active and Reserve): 1966 to 1992
Unit Served with in Vietnam: Detachment A-502, 5th Special Forces Group
Dates in Vietnam: Jan 1968 to Dec 1968
Highest Rank Held: Major
Place of Birth and Year: Huntington, WV — 1945

DIVINE INTERVENTION

There are so many opportunities to be killed in war that I felt extremely fortunate and blessed to return home after serving in Vietnam. As I consider some of my closest brushes with death, there is one in particular that stands out above all the others. The reason it is so memorable is because other men had to risk their lives to save mine.

During almost all of 1968, I served as the Operations and Intelligence Officer of Special Forces Detachment A-502 in South Vietnam. One late afternoon in August of that year, fate would find me standing in the middle of a cornfield we had used as an LZ (Landing Zone) during a rescue mission. The cornfield was situated on the side of a jungle covered mountain far out in enemy territory. I had just sent the major portion of troops used in the rescue mission back to our basecamp. Still with me on the LZ was one other American Special Forces advisor, approximately ten Vietnamese soldiers, and an unarmed three-man CBS television crew. The CBS crew had been assigned to cover the rescue mission—but didn't know when to leave.

Our small remaining team was awaiting pickup by two final extraction helicopters when I received word by radio that they had been grounded by bad weather while refueling. Well, that wasn't a good thing because we were in enemy territory and knew that the enemy was very active in the area. Earlier in the day, one of our patrols had sighted an enemy unit of approximately thirty men and another patrol had seen others. So, at the moment I was notified that there would be no immediate extraction of my team, we were already outnumbered by at least three to one. Such is the starting point of this story.

The CBS Camera crew was still with me when they should have been long gone. After spending some time in the jungle around the lower side of our perimeter organizing a defense with our few remaining troops, I walked back up toward the center of the LZ to wait for the helicopters that were due to arrive at any time. Back on level ground and able to see across the LZ, I was surprised by what I saw.

There, near the center of the cornfield stood David Culhane, a well-known CBS reporter at the time, and his two-man crew. When I left to go down over the hill, they were packing their gear and I was certain they had been taken out on the last lift. After getting close enough to be heard, I asked, "Why are you guys still here?"

"I told you we wanted to cover this operation as close to the end as possible," Culhane said.

"Yeah, well, this isn't close to the end. This is the *very* end. This isn't safe for you. It isn't safe for any of us. If there had been two more choppers, we would all be gone," I said.

"But the choppers are due back about now, aren't they?" he asked.

"Yes, they are…but, do you realize we're down to about twelve men?" responding with a question of my own.

It was clear from the change in his facial expression that he understood the reality of our situation as I continued, "And ten of those men are only very loosely deployed around our entire perimeter to provide security until we are extracted. There are a lot of holes in our fence."

Culhane said nothing, but exchanged glances with his crew. Now, he seemed more anxious. Frankly, I wasn't very excited about our situation myself. And—things were about to become worse.

I picked up the radio handset and attempted to reach the A-502 basecamp radio-room to notify them that the CBS TV crew was still with me. But, nothing—only static. Off and on, I tried several times to reach A-502, our only link to support, but was met with the same results. The weather had become so heavy that it had restricted our ability to communicate by radio. We had become a very small unit, now with absolutely no support and a very long way from home. On top of that, we were in an area that was a very familiar landscape to the enemy. Very clearly—we were now in need of rescue.

Despite our dilemma, I did what I could to reassure Culhane and his crew. Trying to lighten the moment, I told them we might have to give them some quick jungle training and said, "You may get more of a story than you came for."

There were a few weak smiles and the cameraman said, "No, the story we have is great. Just get us outta here with it … in one piece."

I laughed and told them we would be fine until the helicopters made it back. However, in reality, I was very concerned about our situation. The reason for my concern was that an intelligence report given to me earlier in the day indicated that a regular NVA (North Vietnamese Army) unit of 125-150 men was due to arrive at our location—at any time. That would certainly make our situation about as bad as it could get. Only I, the other American advisor, and the Vietnamese officer had that information and I told them not to share it with the TV crew. That news would surely scare them unnecessarily—it scared me.

As has been said by many others, there are times in life when minutes tick by like hours. This was one of those times. It was late in the day and it was now raining, heavily at times. As I looked out from the mountainside into the sky before me, I couldn't image a helicopter being able to fly through such weather, let alone find us. And, it wouldn't be long before night settled in over us. It was time to consider other options and there was really only one.

I pulled out my map and with the other American advisor and the Vietnamese officer at my side, I told them and the CBS crew that if the helicopters didn't arrive before dark, we would go down the mountain and hide in the jungle until morning. Then at sunrise, we would begin walking back to our basecamp. I told them that if the weather had broken by that time, we would call for a helicopter extraction. Otherwise, we had about a thirty-kilometer hike through enemy territory ahead of us. Again, Culhane and his crew exchanged understandably grimacing glances. None of us were excited about that prospect. I assured everyone that the 281st Assault Helicopter Company, the unit supporting the rescue mission, was one of the most daring in Vietnam. I told them, "If they can fly...they will fly!"

As reported to me later, while we waited out in the middle of nowhere, Lieutenant John Weir, the 281st lead pilot in charge of air operations for the rescue mission, sat in his grounded helicopter at the Nha Trang Air Base. As Southeast Asian rain poured down and lightening flashed around him, there was little he could do for us. John's callsign was Bandit Leader. Unfortunately, like us, Bandit Leader could only hope that the weather would break.

Back out on the LZ, after waiting for what seemed like hours and when it seemed more likely that darkness would arrive before the helicopters, I called in the troops from around our perimeter and told everyone that we were going to move down into the jungle. We had picked up our gear and were preparing to move down over the hill when the radio on my radio-man's back seemed to boom with Bandit Leader's voice, "This is Bandit Leader, Bandit Leader...looking for Bunkhouse Zero Two, Zero Two. Over."

My callsign was Bunkhouse Zero Two. So, while his transmission was still a little scratchy, it was clear and loud enough that I knew that he was up there somewhere in the clouds — and he was close.

Again, as reported to me later, Bandit Leader (John Weir) grew weary of waiting for the weather to clear. He told the 281st unit commander that he wasn't going to leave our team out in the jungle overnight and he asked for volunteer pilots and crews willing to follow him to our location.

Obviously proud of their skill and daring, the 281st had a reputation for disregarding any regulation that prevented them from achieving their mission. So, in a driving rainstorm, Bandit Leader lifted off with another troop-carrying helicopter and gunship and headed into the sky.

At this point, I will take a line or two to say that those who flew helicopters in Vietnam were among the most skilled, selfless, and courageous men I have ever met.

Bandit Leader led his rescue team to a river that he knew would lead them out to the area near our location. When he arrived in the area where he thought we were, Bandit Leader found it still filled with clouds and couldn't see the ground. He directed the trailing helicopters to follow at a greater distance and began to descend through the clouds. Weir's crewchief, Jay Hays, told me that John told the other two helicopters that he would report his speed and altitude constantly as he descended. He told them that if his transmissions stopped, it would be because he had hit a mountain and he directed them to attempt to find another way down.

On the LZ, when we received Bandit Leader's first message, there were spontaneous cheers of "Okay!" and "All right!" from those of us standing at the edge of the LZ.

I responded immediately, "Roger, Bandit Leader. This is Zero Two. Where are you? Over."

"This is Bandit. I'm not exactly sure. We're still in the rain and clouds. We've been using the river as a reference and feel like we are near and to the north of you. Can you see us? Over."

"He's out there somewhere," I told everyone. "Look for him."

There was a sea of clouds over the valley. I scanned the swirling mass from west to east, focusing at various distances hoping to see him. Nothing, I couldn't see anything but clouds.

"Does anybody see him?" I asked.

There was an assortment of negative responses. Then, off in the northeast at some distance, I thought I saw something moving between the clouds. Looking further ahead for what I hoped was Bandit, I waited to see if he would fly across my line of sight.

What happened next was extraordinary. Hollywood would have a difficult time depicting it as dramatically as it occurred. While I'm reasonably certain it was caused by the reflection of light from the setting sun penetrating the clouds, I have never disregarded the possibility of divine intervention. But, between two clouds directly in the center of my line of sight there was a bright, swirling, starburst flash of light that was clearly visible—for only a split second. The flash was as bright as any lighthouse beacon I've ever seen. Even though it was very likely the reflection of sunlight on the wet Plexiglas nose assembly on Bandit's helicopter, the occurrence was remarkable!

"I've got him!" I shouted to the others.

Then, quickly, I called Bandit Leader.

"I've got you, Bandit! Break southwest. We're at your ten o'clock position. Over."

"Roger. Breaking! I'm coming out of the clouds now," he answered.

Taking the only smoke grenade I had with me, a red one, from my web gear, I pulled the pin and threw it up the hill behind us.

"Bandit, we've got smoke out. You should have no problem spotting us when you break through the clouds. Over."

"Roger, we're coming out now."

There was a short pause.

"Roger, I've gotcha. I assume all that red smoke belongs to you, Zero Two. Over."

"Roger, it does. We didn't want you to miss us. Over."

"Not a chance. Get ready. We're comin' down and don't want to be there any longer than we have to. Over."

"Roger, no problem. We're ready. Come get us!"

At some distance behind Bandit Leader, the other two helicopters burst out of the clouds. The gunship quickly caught up and took the lead. He made a low pass over us to confirm our identity. He then pulled up and out of the way sharply to allow the other helicopters to quickly drop in for pickup. I did a count as we loaded, watched the other helicopter rotate and leave, signaled Bandit Leader we were clear to go, then jumped aboard and we were gone.

As we flew through the clouds on our way back to A-502's basecamp, I am not at all embarrassed to say that I whispered a prayer of thanks. And, about two weeks later, I would discover that prayer was not without very good cause. During a night ambush, one of A-502's units captured an NVA soldier who was a part of their infamous 18B Regiment, known to operate in our area. During interrogation of the soldier, he revealed that he was part of the larger unit we feared might arrive when we were stranded. He said that they had watched from several kilometers away as the earlier helicopters left the area. And, the part that made the hair on the back of my neck stand up — they were on their way to the LZ when they saw Bandit Leader and the other two helicopters pick us up. He said they were at the base of the hill!

Had John Weir (Bandit Leader), Jay Hays (his Crew-Chief), and the other brave men who flew with them not come after us, there would very likely be no one to tell this story. While I would like to think that the twelve of us on the LZ could have fought off the 150-200 men we would have faced — that kind of thing only happens in the movies. But — we would have done our best.

As I have reflected on this incident throughout the years, it has never ceased to amaze me that other men risked their lives for mine and those of the men with me. I believe that is the most selfless thing one person can do for another. And, you know, that kind of thing happened every day in Vietnam. It is far past time that this type of story is told.

HON. T. JACKSON BEDFORD, JR.

US Navy—Aviator (Naval Flight Officer, Electronic Warfare)

Dates of Military Service (Active and Reserves): 1966 to 1976

Unit Served with in Vietnam: VAQ (Electronic Warfare), Detachment 11, USS INTREPID CVS-11

Dates Served in Vietnam: Jun 1966 to Dec 1966

Highest Rank Held: Lieutenant, USN (O-3)

"THE WAR IS OVER!" (OR SO I THOUGHT)

In October 1968 I was flying off the USS INTREPID (CVS 11) on Yankee Station in the Gulf of Tonkin. The ship was stationed about 50 miles off the coast of North Vietnam. I was designated an NFO-E which was a Naval Flight Officer-Electronic Warfare. I flew right seat in what we affectionately called the "Fat Spad" which was an A1 Skyraider, a single engine prop, configured to hold a pilot, NFO, and two aircrewmen in the rear side by side. I was responsible for the mission which was to jam the fire control radars in North Vietnam when our attack squadrons flying A4's were in the air and conducting attacks on targets south of Hanoi and Haiphong. The two aircrew in the rear cockpit operated the intercept and jamming equipment at my direction.

As the loiter time for jets is substantially less than that for props, two hours vs six hours, we usually covered two strike cycles, which means we would launch first and get on station, usually a half mile off the coast. We then waited for the first strike to launch and at the appropriate time to coincide with their going "feet dry," crossing over the beach, we would start jamming on the fire control radar frequencies used by the North Vietnamese anti-aircraft guns. We would also try to monitor for missile radar control activity and call appropriate warnings. After the first strike was completed and the last plane was "feet wet" we then had to wait for the second strike to launch. Usually during this time which could be as long as an hour, being for the most part young and stupid, we would go sightseeing up and down the coast of North Vietnam to kill time. The beaches were beautiful as was the coast generally. We got shot at on a number of occasions, but I will save those stories for another day.

On the day in question, October 31, 1968 off the coast of Vietnam, we were on a break between strikes and boring holes in the sky. As I was basically co-pilot and navigator, I had access to our ADF radio which was a navigational device used to tune in AM radio stations for purposes of direction finding. In theory we could lock on to the strongest part of the radio station signal and fly to the station or use several signals for vectors to establish a position. It was the first "modern" navigational aid after they quit using bonfires! I could listen to the radio on the ADF and this day I was tuned in to the Armed Forces network (Hanoi Hannah was tiresome). I can't remember what the programming was but at some point an announcer's voice broke in with the announcement, "Ladies and gentlemen, we interrupt this program to bring you an important message from the President of the United States." Then this voice in a Texas twang came on saying, " My fellow Americans, as of 8:00 AM, Washington, DC time on Friday, November 1, 1968, I am ordering a cessation in our bombing of the Republic of North Vietnam…" (or words to that effect—and it was November 2, 1968 Vietnam time).

Of course in retrospect I had no clue as to the meaning of the bombing halt except it sounded to me like the war was coming to an end. I was so excited that I wanted to share the news with my pilot and crew so I depressed what I thought was the intercom button (there were two buttons at my feet, one for intercom and one for external radio communication which was always set on Channel 16, the "guard channel" which everyone in the fleet and air could hear as they all monitored "guard"). I started yelling excitedly that, "The war is over, the war is over!" Then I realized in my excitement I had depressed the radio button and not the intercom. To my everlasting chagrin and embarrassment, I had just broadcast to the entire fleet and all aircraft in the Gulf of Tonkin that the "War is over!" After realizing my faux pax, in a more subdued but still excited state I told my air crew what I had heard.

After completion of the second strike, we recovered about two hours later. When we got back aboard ship there was this general chatter about someone yelling over guard that, "The war is over!" I was asked if I had heard it. I quickly said I had and wondered aloud where it might have come from and what it was about. I never confessed!

Within a few hours the Captain of the INTREPID made a ship-wide announcement that a bombing halt had been called and that we would be departing Yankee Station for Dixie Station to await further orders.

I was a Lieutenant (junior grade) USN at the time. I was with a five plane Detachment of my squadron, VAQ 33, home ported at Quonset Point, Rhode Island. I was the Aircrew Officer and Detachment Legal Officer. I flew roughly 105 Combat Support Missions. While on the INTREPID we transited the globe, going east around Africa and then around Cape Horn on the trip home.

My plane was known as an EA1F. It was configured with jamming and intercept equipment plus a navigation radar which I operated as needed. The plane had roughly 2800 horsepower and swung a four-bladed prop, each blade of which was approximately 14 feet across. According to some history books, I flew the last single engine prop combat mission of the US Navy. When we returned to the States we stood down our Squadron and gave our planes to museums and the South Vietnamese. One of the planes I flew in now hangs in Hangar One at the museum in Pensacola.

JEFF 'BIC' BICKERTON

US Army — U-1A 'Otter' — Crew Chief

Dates of Military Service: 1966 to 1969

Unit Served with in Vietnam: 18th Aviation Company, 1st Aviation Brigade

Dates Served in Vietnam: Feb 1968 to Apr 1969

Highest Rank Held: SP5

Place of Birth: Queens, NY — 1946

MARBLE MOUNTAIN (VN): A SORRY MESS

This happened in 1968, more than midway through my 'tour'. I was 21 and a SP5 Crew Chief of an Army U-1A DeHavilland 'Otter' Aircraft (a S.T.O.L. (Short TakeOff and Landing) bush-type utility aircraft). We supported the Special Forces' A-Camps along the Cambodian/Laos border in I, II, & III Corp. I flew with the plane every day. I was partial co-pilot, maintenance man, fueler, loadmaster, steward, document checker and document keeper, and waist gunner.

Late in the afternoon, we flew into an airfield called **Marble Mountain Marine Airbase** in Da Nang near the coast in the northern part of I Corp. It was a new stop for me and our destination for the night.

While the pilots went off to the O-Club and before I tied down the plane for the night, I taxied over to the fuel truck to fill-up. The fuel Specialist was a 'druggie' (you could tell) named 'Duke', and I asked him if he could get me some pot. I wanted to have a small supply to try it on for size, as it fit my 'always on the go', different airport every night life as a Crew Chief. I was not a smoker, never had been. This was not a wise decision point in my life.

He told me to meet him at the NCO club at 2030. I did. After a watered down with a cold beer, he told me to follow him. We made our way out to a bunker just inside the wire. It was an elevated revetment with sandbags on top in a 'C' shape forming a two-foot wall as a firing position. There were a few Grunts and GIs sitting up there. We joined them and their conversations of 'home' and girls. The skies were beautiful, there was an ocean breeze, and the conversation was good as a joint was passed around. I was not a big pot fan, but this was nice.

After thirty minutes or so, Duke tapped me on the shoulder and said this guy walking up could get me what I wanted. This guy had a head band, beach-boys type look and was wearing round sunglasses in the dark. Once seated, he told everyone to pass 'this' joint and tell him what they thought. Little did I know it was the last joint I would inhale in Vietnam. Once it reached me, I took a drag and held it down. I felt it fill my lungs and go directly to my brain. My head whirled.

I rolled off the bunker and dropped down to the sand at ground level. Duke jumped down and asked if I was alright. I was staring blankly at him and said, "Oh my God, what was that stuff," when the first rocket went over our heads and hit a hooch some 50 yards away. The explosion ripped off one end of it.

We hit the dirt but everyone else came off the bunker and over us. We followed. Boots, butts, and sand was all I saw in front of me against the light from the hooch. We were running away from the perimeter as the second rocket came in. I don't know where it hit, I was running. Everyone was scattering, and I fell and rolled up against a building with an overhang of some sort. I didn't know the base, and I didn't know where the interior bunkers were. I was new and hadn't taken the time to look. I laid there in the sand with my head against the ground for what seemed to be an hour, listening to the gun fire and the rockets hitting on and near the base. I had no weapon, except a pocket knife.

The adrenalin started to clear my mind but that last joint had some kind of psycho, mind-altering additive, and it blew me away. I was a 'mess'. . . a 'sorry mess'. I laid there. The only thing I did was think. And what I thought was, "What in God's name are you doing?"

I told myself that I had one 'personal mission' while I was here in this place on the other side of the world and that was to make it HOME. Here I am, **lost and helpless** (I have never liked being either of those since), in a mind altered state, eating-dirt, and scared to death.

That night's 18 rockets and sand, self-talk, and my primal goal-setting was powerful and has stuck these past 50 years. It will always be with me. The lesson I learned: When in a combat zone or other danger-ous situations, stand 'frosty', be alert, don't lose focus, be aware, and be armed. Make it home and stay alive.

I made it back to the 'world' and left the Army in 1969 in time for the Moon Landing viewing and married my girlfriend. I became a com-mercial Pilot and worked in the air cargo industry the rest of my career. I have lived a good life. Never smoked again. 'Bic'

LARRY TAYLOR
Air America Pilot (civilian)
Laos 1967 to 1968

Memory fades, but happy to talk about it. My log says the aircraft was H-44 and I was flying with Billy Pearson. The Flight Mechanic, I'll have to look up; I remember it was an American (wore glasses?), but not one of the guys who usually shows up at the reunions. It was a special mission on the Sihanouk Trail, grid was XB8324. My log says it was near the Se Kong River (I don't remember a river near the LZ). We were the lead bird in a big gaggle (several AAM 34s and a few USAF Pony Express H-3s; escorted by USAF A-1s) going in to pick up a team from PS-22 that had been on the Trail several days. The team (Nung mercenaries based on the Bolovens Plateau, as I recall) was not a road-watch team; much bigger than that; they had been put in to close the Trail.

I speculate that the bad guys had been following the team without making contact, knowing that we'd be coming back to get them. We were in the flare just before touching down in the LZ when they opened up on us from what seemed like all sides. (If they had waited just another minute or two, they could have gotten several of us sitting in the zone). We waved off, knowing we had been hit, but no indicators of mechanical problems, other than a vibration. We were lucky. The A-1s immediately rolled in on the enemy and blew the hell out of them. We did *not* leave H-44 in the field; we flew it to the strip at Attopeu and surveyed the damage. A bunch of hits in the blades, cabin, and we were leaking fuel from some holes in the tanks. My recollection is that we sat there at Attopeu and ultimately took it to Pakse, after determining that it was flyable, and the self-sealing front tank would retain enough fuel to get us to Pakse. After Pakse, I don›t remember, and there›s no entry in my log about how we got back to Udorn.Rest in Peace, Billy Pearson; he and I had a lot of fun telling this story at a reunion or two. I think it was those bright pastel socks he always wore that was the enemy's aiming point.

On the 23d, I show I deadheaded with Charlie Weitz in XW-PFH Udorn to Savannakhet, and then flew with Mike Jarina in H-53 on a late-day special mission in the Mu Gia Pass area. We landed back at Savannakhet at night, according to my log.

WILLIAM J. BILLIONS

USMC — Rifleman, Interpreter
Dates of Military Service: 1966 to 1972
Unit Served with in Vietnam: HQ, 5th Marine Regiment, 1st Marine Division
Dates in Vietnam: Jul 1968 to Aug 1969
Highest Rank Held: Sergeant
Place of Birth: Memphis, TN — 1947

AN HOA, 23 FEB 1969

On the morning of 23 February 1969, our Combat Base at An Hoa came under intense mortar and ground attack from a combined Viet Cong and North Vietnamese Army force. The explosion of enemy mortar rounds caused a chain detonation of 31,000 artillery rounds in the ammunition storage area. There were numerous casualties, but the enemy retreated after several hours of combat. It took several days to extinguish the fires and to replenish the ammunition.

We thought being in the rear at An Hoa was going to be a safe place for some rest, but on this day in 1969, it may have been better to be in the Arizona Territory, which was considered to be a very dangerous area.

Semper Fidelis!

CARL H. "SKIP" BELL, III

U.S. Army—Armored Cavalry Officer, Aviator

Dates of Military Service (Active and Reserve): 1967 to 1998

Units Served With In Vietnam: A Troop, HHT, B Troop, 1st Squadron, 4th Cav, 1st Inf Div (First Tour); C Troop, 3rd Squadron, 17th Air Cav; 18th Corps Aviation Co; G3, HQ 1st Aviation BDE (Second Tour)

Dates in Vietnam: Feb 1969 to Feb 1970; Feb 1972 to Feb 1973

Highest Rank Held: Colonel

Place of Birth and Year: Decatur, GA—1945

MY VIETNAM EXPERIENCES

For me, my time in Vietnam serves as a way to keep things in perspective. Whenever I'm having a bad day at work or problems outside of work, I can look back at some days in Vietnam (especially those from my first tour of duty there) and truthfully say to myself, "you've seen worse."

Vietnam was a kaleidoscope of good and bad experiences, made all the more intense because those of us in combat units genuinely lived life on the edge: when we got up in the morning, we literally never knew if we would live through the day. On the positive side, I have never had more responsibility (I spent most of my first tour as an armored cavalry platoon leader and troop commander), made closer friends, or learned more about what people can do under adverse circumstances. On the negative side, I saw people killed or horribly wounded, saw what I considered to be stupid or self-serving decisions made by senior leadership (both in-country and back home in the U.S.A.), and I found myself treated as a pariah by many of my fellow citizens upon my return from Vietnam.

One of the positive things that we can take from the September 11 tragedies in New York, Washington, and Pennsylvania, is that the police, firemen, emergency aid workers, and airline passengers and crew who did what needed to be done under truly horrendous conditions are getting the recognition as the heroes they are by their fellow citizens. There were many true heroes in the Vietnam War whose courage and sacrifices were not recognized. I think it speaks well for America that we are still producing people with the courage to stand up and do what they have to do in spite of adversity the way the folks did on September 11 (and continue to do today). I was afraid that after the cynicism and politicizing of so much of our public lives, that the quality of heroism might be being removed from our national psyche; recent events prove that is not the case.

Back to my Vietnam experience, I did two tours of duty there. One tour was as a "camper" (on the ground with an armored cavalry unit which spent most of its time in the field), and the second tour was as a helicopter pilot. The lifestyles and the definition of "combat" for those two groups of people were quite different.

"Combat" for the soldiers on the ground included not only firefights and direct encounters with enemy soldiers, but also the rigors of the day-to-day life that a field soldier in that environment lived. Actual combat with the enemy was relatively rare (but quite intense when it happened). I've heard combat described as, "weeks of boredom and monotony interspersed by moments of stark terror." For ground soldiers, I believe that is an accurate description. Most of the time, one's life as a field soldier was endless filth, heat, mosquitoes, ants, leaches, encounters with the occasional reptile or scorpion, lack of sleep, dirty clothes, and the constant fear of hitting a mine or booby trap or of being ambushed. We almost looked forward to direct action against the enemy because it was a way to vent our frustrations over the life we led and actually to do what we were trained to do.

"Combat" for helicopter aircrews was different. Generally speaking, helicopters were shot at directly a lot more than the soldiers on the ground. I guess that's because helicopters ranged over a wider area in a shorter period of time, and because many of their missions involved support of ground units (infantry/armor/artillery) who were in actual contact with the enemy. By definition, the helicopter was much more exposed to enemy fire than the folks on the ground (that could seek cover and concealment). We might support several different ground units in one day, each of which might be in enemy contact.

Also, acts of heroism by aircrews were generally much more deliberate than those of the ground troops. A helicopter crew could always choose to leave the scene or abort the mission (though few did); the folks on the ground were there and in the middle of whatever was going on; heroism there was often more necessity than choice. The life we lived as helicopter crews when we weren't flying was good, compared to the life of the field soldier. We were generally based at airfields. This usually meant that we slept in beds or cots (except during rocket, mortar, or ground attacks), ate hot food, had access to showers and clean clothes, could get cold beer if we wanted, got mail on a fairly regular basis, and could listen to music in our "hooches" (the buildings or tents in which we were housed) at night.

I don't mean to infer that aircrews didn't do their share of fighting and dying—they certainly did (the "huey" helicopter has come to be the symbol of that war). They often worked all night back at the airfield to repair damaged/shot up aircraft so they would be flyable for the next day's operations. The life of an aircrew member was not easy; it was simply a different type of "combat."

As I said at the beginning of this article, for me my experience in Vietnam was multi-faceted. I wouldn't trade anything for the friends I made and the positive experiences I had, but I hope my children never have to experience war as it was fought there. (I was going to say that I hope my children never have to experience war at all, but I'm afraid that the events of September 11 may have changed that). I appreciate the opportunity to put these thoughts on paper, and to have had the opportunity to serve in the Armed Forces of this great Nation.

RAYMOND L. COLLINS

US Marine Corps — Logistics Officer

Dates of Military Service: 1960 to 1980

Unit Served with in Vietnam: 1st Tour: FLSC-A, FLC, IIIMAF; 2nd Tour: HQ IIIMAF

Dates in Vietnam: Aug 1968 to Nov 1969; Jan 1971 to Apr 1971

Highest Rank Held: Lt Colonel

Place of Birth and Year: Kinston, NC — 1938

MY TIME IN 'NAM

In June 1964, while serving as a First Lieutenant Company Commander in the Second Marine Division at Camp Lejeune, NC, I received Permanent Change of Station (PCS) orders to sea duty aboard the USS Albany CG-10, a guided-missile cruiser, as the Marine Detachment Executive Officer. Although sea duty, similar to embassy duty, was a prestigious assignment for a young Marine officer, like most young hard-charging Marines, I was highly disappointed that I was not going to war in Vietnam. And to make matters worse, after two years of sea duty I was sent to Parris Island, SC to train new Marine recruits. I was literally afraid the war would be over before I could get there and that would be a career-killer.

I eventually got my 'wish' and served two tours in Vietnam; first, in 1968-69 with Force Logistics Support Group-Alpha (FLSG-A), Force Logistic Command (FLC), III Marine Amphibious Force (MAF) located in I Corps. Alpha Group provided direct logistic support to the Fire Support Bases (FSB) located throughout the First Marine Division area of responsibility (AOR). As the Group S-4, I was responsible for all facilities, equipment, supplies, and material; in essence everything necessary for the Group to perform its mission, including an array of Logistic Support Units (LSU) located at the various Division FSBs.

The majority of my time was spent working with my staff verifying requirements, determining realistic solutions, approving priorities, arranging transportation, and coordinating timely delivery/resupply to our units with priority going to the LSUs at the FSBs. I traveled frequently via ground vehicle and helicopter to the FSBs and LSUs throughout the AOR, often encountering sporadic enemy fire.

On one occasion, our helicopter was hit by MG fire and forced to crash-land. Fortunately, it happened near friendly lines and there were no casualties. My unit initially operated near Phu Bai west of Hue and north of DaNang; later in 1969 we relocated to Red Beach in DaNang to assist in the backload of Marine units aboard ships returning to Okinawa, Japan and performed the Vietnamization program (turning equipment and supplies belonging to departing units over to the ARVN forces).

I departed Vietnam as a young Major at the end of 1969 and was reassigned to the G4 Staff at the US Pacific Headquarters located at Camp H.M. Smith, Hawaii. Initially I led a team that was implementing the logistic portion of a new DOD Forces Status Readiness Reporting System (FORSTAT). In 1971 I was ordered to return to Vietnam for Temporary Additional Duty (TAD) to assist the III MAF HQ staff in implementing FORSTAT. Our team traveled via helicopter to nearly every Marine Battalion and higher headquarters providing instruction and assisting in the implementation of FORSTAT. As expected, our team was met with mixed reaction: appreciative for the assistance but resistant and very negative to another SEC DEF McNamara administrative requirement. Needless to say, President LBJ and SEC DEF McNamara were not very popular amongst Marines in Vietnam (or elsewhere).

LTJG JIM DICKSON

U.S. Navy—Operations Intelligence Division Officer (Combat Information
Center)—Officer of the Deck Underway

Dates of Military Service (Active Duty and Reserves Combined): 1967 to 1977

Unit Served with in Vietnam: USS Ticonderoga (CVA-14)

Dates in Vietnam: Jan 1968 to Aug 1968; Jan 1969 to Jul 1969

Highest Rank Held: Lieutenant—Inactive Reserves

Place of Birth and Year: Lancaster, PA—1944

YANKEE STATION—VIETNAM WAR

I served aboard the aircraft carrier USS Ticonderoga (CVA—14) for
both the 1968 and 1969 deployments to Yankee Station during the Vietnam War as Operations Intelligence Division Officer for about 60 men
who worked in Combat Information Center (CIC)—the operations
center on most navy ships—and Officer of the Deck Underway.

There were normally three aircraft carriers on Yankee Station that
conducted air operations for twelve hours each day—one from noon
to midnight, another from midnight to noon, and another one during
daylight hours from 0600 to 1800 which provided 24-hour coverage and
additional effort during daylight hours when sorties were most effective.

It seemed the USS Kitty Hawk (CVA—63) was on line with us on
Yankee Station more than any other aircraft carrier—can still remember many times saying "Pawtuckett this is Panther" using our call signs
to contact her by radio. I also remember hearing countless times from
speakers in Air Operations located adjacent to CIC the words "On glide
path, roger, roger ball"—meaning the pilot landing at night was on the
correct course for a safe landing confirmed by the proper line up shown
on his instruments during discussion with the Landing Signal Officer
(LSO).

The twelve-hour flying cycles were done with 90-minute inter-
vals—45 minutes into the wind at the speed necessary to generate 30
knots of relative wind needed to launch and recover planes—and 45
minutes downwind at the same speed in attempt to maintain the same

operating location. Since the launch and recovery often lasted more than 45 minutes, we sometimes moved considerably off station—especially when going 30 knots when there was no actual wind—and there were several times when we came within 15 miles from Hainan Island—part of China.

The 1968 deployment involved sorties to South Vietnam and many high-risk missions to North Vietnam with targets heavily defended by advanced Soviet surface-to-air missiles. We lost several pilots—the first was an A-4 Skyhawk pilot who was able to reach the ocean but unfortunately got caught up in the parachute cords and drowned before rescue. President Johnson decided to end the bombing of North Vietnam on November 1, 1968.

The targets during the 1969 deployment were in South Vietnam—sometimes close-in ground support—or along the Ho Chi Minh Trail. One day I will never forget is the A-7 Corsair takeoff from the port catapult—lost oil pressure and quickly crashed into the ocean. The order of right full rudder to avoid the plane was immediately belayed by the experienced Captain Richard E. Fowler and changed to right 10 degrees rudder to prevent too much centrifugal force and tilt of the deck—something very dangerous during flight operations. When going at a speed of 30 knots and turning with even 10 degrees rudder produces a very strong centrifugal force and tilt that can be felt everywhere on the ship—and the diameter of the turn would be about one mile under those circumstances. I will also always remember the night in Combat Information Center when Radarman Chief A. E. Morton was able to coordinate and vector the rescue of a downed pilot about 100 miles away.

There were F-8 Crusader barricade landings during both the 1968 and 1969 deployments—both high-risk and near-fatalities. When the landing gear malfunctions and does not come down properly, the pilot can volunteer to land into the barricade net or eject. During the 1968 deployment, a damaged F-8 Crusader tore and stretched the barricade net until finally coming to rest only six feet from the super structure. During the 1969 deployment, an impaired F-8 Crusader landed into the barricade net and ended up dangling over the flight deck above the ocean before the crew could quickly stabilize and pull the plane back to safety.

There were trips to Korea in January 1968 as part of a large task force in response to the capture of the USS Pueblo and also in April 1969 as part of Task Force 71 ordered by President Nixon in response to the shooting down of an EC-121 Navy reconnaissance plane. The April 1969 naval action was the largest naval armada since World War II, with over 40 ships including four aircraft carriers and the battleship USS New Jersey (BB-62). Tensions were very high during both excursions to Korea with reported contingency plans for nuclear bombs and mines.

Something very unique happened around May 1968. We entered Subic Bay in the Philippines after coming off Yankee Station but were not able to dock since there was only one pier large enough for aircraft carriers. It was still occupied by the USS Enterprise (CVN-65) because of a threatening typhoon. As a result, we were forced to anchor in the middle of the very large bay. I vividly remember informing Captain Norman K. McInnis the wind was zero one zero relative at 30 knots—almost dead ahead and enough to launch the planes. After discussion with the meteorologist and Commander Air Group (CAG), he made the bold decision to launch all the planes located on the flight deck and send them to land at nearby Cubi Point Naval Air Station. The last plane to take off was the large A3D-1 Douglas Skywarrior tanker. We then pulled up anchor and went out to sea to ride out the typhoon that later fortunately moved away. To this day, I have never heard of this being done by another aircraft carrier.

My qualification letter as Officer of the Deck Underway was signed by Captain Richard E. Fowler who later became Rear Admiral Upper Half. He was a Torpedo Bomber on the USS Essex (CV-9) in World War II and under very heavy fire made a successful strike on the Japanese battleship Musashi that later sank during the Battle of Leyte Gulf in October 1944—the largest naval battle in history. He was awarded the Navy Cross by Vice Admiral John McCain.

I consider the original qualification letter as Officer of the Deck Underway signed by him and framed in my office to be a prized possession.

CARL H. "SKIP" BELL, III

U.S. Army—Armored Cavalry Officer, Aviator
Dates of Military Service (Active and Reserve): 1967 to 1998
Units Served With In Vietnam: A Troop, HHT, B Troop, 1st Squadron, 4th Cav, 1st Inf Div (First Tour); C Troop, 3rd Squadron, 17th Air Cavalry; 18th Corps Aviation Co; G3, HQ 1st Aviation BDE (Second Tour)
Dates in Vietnam: Feb 1969 to Feb 1970; Feb 1972 to Feb 1973
Highest Rank Held: Colonel
Place of Birth and Year: Decatur, GA—1945

A LONG DAY, JUNE 1969

It was mid-June, 1969. I was the Acting Troop Commander of A Troop, ¼ Cav while the Troop Commander, CPT Bill Newell was on R&R in Hawaii. I had been a platoon leader in A Troop since February and had to leave my platoon in early June when I made Captain. My next assignment in the Squadron was uncertain, so when Bill Newell got ready to go on R&R, his XO, LT Robert ("Road Runner") Rash asked me if I'd volunteer to take the troop while Bill was gone.

I told Bob I'd be glad to go back out to A Troop; I knew the troop, knew how it operated, got along well with the 1SG, and knew the other platoon leaders and most of the NCOs. Besides, since my permanent assignment within the Squadron was still unknown, I was ripe for every crap detail that an officer could pull at the basecamp of the Big Red One. LTC Haponski, the Squadron Commander, gave his blessing, and off I went, back to A Troop.

On the day I'm writing about, we had been conducting three separate platoon-size reconnaissance in force missions (RIFs) north of Lai Khe and on both sides of Hwy 13. At about 1530 hours (the time we would normally start looking for a place to set up our night defensive position

(NDP)), I received a call from the Squadron TOC telling me to assemble the troop at the south gate of Lai Khe by 1700 hours, and to be sure that we refueled all the vehicles at that time. Then we were to go into the Trapezoid to an infantry battalion NDP and reinforce them.

They were deep in bandit country and expecting to get hit that night. I was also advised not to use the roads after we crossed the bridge at Ben Cat (a single-span Bailey bridge which could only take one tracked vehicle at a time) because they were suspected of being heavily mined (in fact, the whole area was full of mines and booby traps). We had recently been briefed on a new mine which was thought to be in that area, called a tilt-rod mine. It was planted in the center of the road or trail and was designed to go off under the belly of an armored vehicle (where the armor is the thinnest).

The directions I was given to the infantry battalion NDP were to cross the Ben Cat bridge, move northwest across the northern portion of the Iron Triangle, and cross into the abandoned rubber plantation (which was rapidly becoming jungle again) at the base of the Trapezoid. We were to go north into the abandoned rubber plantation for several hundred meters until we found a road running roughly east west. That road would take us to our objective; nobody was quite sure where the objective was, but it was at the end of that road. I thought, "another hairbrained scheme from Division; it'll take us at least an hour to get all the tracks across the Ben Cat bridge, another 30-45 minutes to get across the Triangle and into the Trapezoid rubber, and who knows how long to bust that jungle to get to the infantry NDP.

Thus, we'll parallel the road they said led to the NDP, but stay one or two rows of rubber off of it, in case it's mined. It will most likely be dark a long time before we get there, and busting jungle in the dark in bandit country is a great way to get ambushed. What a bunch of idiots!" When we left, I had no frequency or call signs for the infantry battalion we were going to reinforce, and when I asked about that, I was told that they would be given to me en route.

We moved out around 1730, after topping off all the tracks. We got to the Ben Cat Bridge and crossed it one vehicle at a time, taking up defensive positions on the far side until all the vehicles were across (this *was* the Iron Triangle after all, and most anything was possible there.). The crossing of the Triangle was uneventful. We had to be careful to dodge the bomb craters that were difficult to see in the Rome plow cut that had grown back up so that it was even with the top of an ACAV. As usual, we led with a tank. In this case, I believe the third platoon was in the lead of our column. The TC was PSG Farler. Order of march was lead platoon, troop HQ, and the other two platoons, plus the VTR.

As we entered the rubber (plantation), the shadows were getting long. We moved due north until we found the east-west road, crossed that road, and turned left (west) and paralleled the road one row of rubber off. The going was incredibly slow; the jungle was thick, and Farler's tank was having a tough go of it. As it got darker, I got more nervous. I knew that flares were useless in the rubber (I learned that during an all-night foray through the Michelin plantation a month or so earlier) and was anxious to rendezvous with the infantry battalion (whose call sign and frequency I still did not have, nor did I have firm coordinates on their location). "This is very, very flaky," I thought. The idea of linking up with these dug-in grunts in the dark, moving my vehicles into their perimeter, and establishing positions within their NDP without rolling over a fighting position or an infantryman we couldn't see seemed like a mighty difficult undertaking.

I called 3rd platoon again to urge them on. 3rd Platoon (PSG Farler), recommended that we "thunder run" the road, because at the rate we were busting jungle, the trip to the NDP could take all night. I thought about it for about 30 seconds and said, "OK, let's do it." Bad mistake.

Farler busted his way to the road, turned onto it, and started moving out slowly to allow the remaining vehicles in the column to make it to the road. When I got the report that the last vehicle was on the road, I told 2nd Platoon to "kick it." At that time, I could still see PSG Farler's tank up ahead, but just barely. I happened to be looking at the tank when I saw a blinding yellow/red flash of light, heard a huge BOOM! and the whole rear end of the tank lifted off the ground! I also saw PSG Farler shoot straight up out of the cupola, almost level with the back of his thighs, and then fall back down into the cupola. I thought, "tilt-rod!" as the tracks in the column herringboned and opened fire. This was done without command by me—it was "normal" battle drill when we were in bandit country. When it became evident that we were getting no return fire, I started yelling, "Cease Fire!" As most of you know, that was easier said than done—once cavalry troopers started shooting, it was difficult to get them to stop. I jumped off my vehicle and went to Farler's tank. It was not burning, but it had severe damage to the engine compartment and a huge hole in the bottom. Evidently, the vehicle was going fast enough that the mine (which should have gone off under the crew compartment, went off under the engine). Farler was on the ground then, as were his crew. They were shook up pretty badly, but (except for some scratches and bruises and ringing ears) not hurt.

I went back to my track and reported what had happened to the Squadron TOC. I then ordered the platoon leaders to move their vehicles into the jungle on each side of the road so the VTR could move up. I advised them to have their men be vigilant; the bad guys knew for sure we were here now, and they might try to move against us. As I was formulating a plan to move another platoon into the lead and bust jungle for the rest of the night, I got a call from Squadron. "Abort the mission and return to Lai Khe." "Well god-DAMN," I thought.

We got the VTR up to PSG Farler's tank and were hooking it up for towing when the incoming small-arms fire began. It was sporadic, probably from no more than a squad-size enemy unit. They could get pretty close in the jungle, but the firing seemed to be coming from at least 50 meters out (which in that jungle was a good ways). We returned fire in heavy volume. The VTR crew got the tank hooked up in record time. We suppressed the enemy fire and turned the column around and moved off the road due south toward the Iron Triangle. So far on this mission I had combat-lost one tank and had no confirmed enemy killed to show for it. But the crew of that tank was pretty much OK, thank goodness.

Once we broke out of the rubber and crossed the road that formed the northern boundary of the Iron Triangle, I asked my FO (I believe his name was LT Borsos) to see if he could get some illumination for us. It was quite cloudy, and there was very little ambient light. I was concerned that we might lose a vehicle in a bomb crater (in fact, about a month later, B Troop had an ACAV go into a bomb crater in the Triangle; the vehicle flipped over and the TC was crushed).

The FO got some illum going, and about that time sporadic sniper fire broke out from the rubber. We got beyond the range of that quickly, but then were faced with another problem. The illum rounds were going off right over us, and the shell canisters were dropping around us. I asked the FO to move the illum back toward the north—we would still get the benefit from it, but it wouldn't be dropping the canisters around us. He said he would but couldn't seem to adjust the fire away from directly over our heads.

As we continued through the Rome plow cut, I got a frantic message on the radio from the troop medic, SP5 Gene Garrison. He said that someone had thrown a grenade into his track. This was entirely possible, given the height of the Rome plow growth we were moving through, but I heard no explosion. I asked if he was sure and was anybody hurt? He said nobody was hurt and that he'd get back to me. We continued to move across the northern part of the Triangle toward the Ben Cat Bridge. When he called me back he told me that one of the illum canisters had gone through the open troop hatch on his ACAV, through a 5-gallon water can, a case of C-Rations, the sandbags on the floor of the track, and stuck in the bottom of the track itself. That meant I had combat-lost a second vehicle that night (when the armor is penetrated, the vehicle is considered a combat-loss). Of course, since the vehicle was still drivable, we would continue to use it while we waited for its replacement.

Shortly after that, it started to rain, and the rain came in torrents ,and we quickly got soaked to the skin. We continued toward the Ben Cat Bridge, and when we got to the bridge, we crossed in the rain—one vehicle at a time. We turned the guns of the vehicles waiting to cross to the rear in case the bad guys decided to mess with us.

Getting the dead tank across was quite an operation, and a tribute to the professionalism of the VTR crew. The VTR towed the tank to the near end of the bridge, stopping just before getting on the bridge, and un-coupled the tank. The VTR then crossed the bridge, turned around, and the crew deployed its winch cable back across the bridge and hooked it onto Farler's dead tank. It continued to rain the whole time this operation took place. The VTR crew skillfully winched the dead tank across the bridge, unhooked the wench cable, turned back around, and hooked up the tank for towing again. The remainder of the unit crossed without incident, and we moved out for Lai Khe. Of course, we had to get concertina wire removed at several places, and then had to convince the REMF MP's at the south gate of Lai Khe to let us in. It continued to rain. We were all soaked, and pretty pissed off.

I reported in to the TOC via radio that we had closed Lai Khe and was told to park the vehicles in the troop motor pool. I asked if I should personally come to the TOC and was told there was nobody there but the duty officer to whom I could report. By this time, it was well after midnight. I was soaked, as was everyone else in the unit. I went back to the Troop HQ/Rear Operations Center with a couple of the officers (don't remember who just now), thinking, "This has been one hell of a day." There was some cold beer in somebody's refrigerator or ice chest, and we broke out a couple of those.

Suddenly, the sirens started wailing, and we heard explosions—rocket attack! I crawled under a table with a couple of other folks, and we waited out the attack. One of the troopers came running up from the motor pool and said that one of our tracks had taken a direct hit from a 122mm rocket. Don't remember which track or which platoon it was from. I asked if everyone was OK, and was told, "Yes, everyone is fine; nobody was on the vehicle." I said a silent prayer of thanks and walked down to inspect the damage. It was impossible to ascertain the extent of it with my flashlight, so I decided to postpone further inspection of my third combat-lost vehicle of the day until daylight (which, by that time, was not that far off). I found a cot, crawled into it, pulled a poncho liner over my head, and went to sleep. My last conscious thought was, "this has been a long day; thank goodness we didn't lose anyone."

Several days later we went back into the Trapezoid, but that's another story.

AL ARRINGTON

US Marine Corps — Combat Engineer Officer
Dates of Military Service: 1966 to 1971
Unit Served with in Vietnam: Alpha Company, 11th Engineer Battalion, Third
Marine Division
Dates in Vietnam: May 1969 to Nov 1969
Highest Rank Held: Captain
Place of Birth and Year: Dalton, GA — 1943
Semper cum superbia

VIETNAM

In May, 1969, I flew to San Francisco and stayed one night at the Marines' Memorial Hotel, where I had a drink at the bar with a Marine who had been in the Battle of Belleau Wood in June 1918. The next day, I flew from San Francisco to Hawaii, refueled, and thence to Okinawa. We stowed our civilian clothes and Service Uniforms in a warehouse in Okinawa and the following morning about 2:00 A.M., flew to Da Nang.

DA NANG

We arrived early in the morning. Looking out of the Continental Airlines airplane window, there were fires everywhere. I thought to myself, what have I gotten myself into? Getting off the airplane, it was reported that the temperature that day was expected to be 106 degrees. I changed my U.S. Currency to Military Payment Certificates (MPCs) and then took a C-130 the next day to Third Marine Division (3rd Mar Div) headquarters in Quang Tri.

QUANG TRI

At The Basic School we had been informed that all second lieutenants arriving in the 3rd MarDiv would spend their first three months as infantry platoon commanders (Military Occupational Specialty (MOS)

0302) before being assigned to a unit in their primary MOS. MOS 0302 Marines are the backbone of the Marine Corps. For some reason, I was assigned to the 11th Engineer Battalion at the outset (1302) and sent to 11th Engineer Battalion headquarters in Dong Ha, via jeep.

DONG HA

The 11th Engineer Battalion Headquarters in Dong Ha assigned me as Platoon Commander of first platoon which was stationed at Vandegrift Combat Base (f/k/a LZ Stud), to which I arrived via jeep in a convoy.

VANDEGRIFT COMBAT BASE

After five days of travel, I had a platoon, and it was a good one. Our first assignment was to build a 400' x 600' helicopter landing zone of 2' x 12' M8A1 metal planks for CH 46 and CH 53 helicopters to deposit supplies for UH1-E's to deliver to troops in the field.

Then, my platoon was tasked with a series of projects that included:

1. Rebuild wooden platforms at the "Rockpile" for their 175 MM self-propelled guns aimed at the DMZ (demilitarized zone), 17 miles north. We put six layers of 3"x12" lumber on top of concrete decks—the guns would squirm around on the wood to position themselves for firing. When one layer was worn down, its remnants were removed, and the next layer was used.
2. 10' x 24' Battalion Aid Stations (BASs). The Vandegrift Base Commander's Operations Officer bet me all the beer my platoon could drink if I could build five BASs in five days. My platoon was all for it: one guy did site preparation, levelling the ground (on the sides of hills) with a bulldozer, a team supervised the cutting of the frame at the lumber yard, another brought the lumber to the sites, and another assembled it at the sites. We completed the five BASs in five days. I went to collect the beer. The guy I made the bet with had gone on R&R—I never saw him again. But, I had to get the beer—I got a

scoop loader and drove it all over the base acquiring the beer wherever I could find it.

3. Bridge repair. A steel bridge had been damaged by the NVA (or VC). We went to the site, set up a security perimeter, and the welder began work. We had to return the following day to finish. By then, the villagers had set up a store made of C-Ration boxes. Then they made their way to each of the Marines on the perimeter to sell them whiskey. I followed them and poured out the whiskey. This bothered the Vietnamese a bit, but what really bothered them was when I broke a bottle. In route to the bridge site, we passed a dead cow in a ditch. An hour later, when we were at the job site, six Vietnamese and an entourage came along with the dead cow, built a fire by the river and placed the cow on it to cook. Then they stripped naked and took a bath in the river.

4. Rats. At night, the rats would come out looking for food. We set traps and baited them with peanut butter. The rats would eat the bait but were so large they would drag the traps into the woods, so we tied one end of a cord onto the traps and tied the other end to the building. During the night, we would hear the snap of the traps onto the rats. In the morning, each of us would count the number of rats we caught and keep score. Then we would toss the rats into the woods and a mongoose would come along and eat them during the day.

5. Road Repair. During the rainy season, a side road washed out in a village between Vandegrift and Dong Ha (in the Cam Lo area). Our repair work was in an open area and provided great entertainment for the villagers. After we finished, the village chief and his son invited us to his house for a lunch of chicken and rice.

6. Mine sweeps. I did one mine sweep—from Dong Ha to the DMZ (about 15 miles). When I was almost in sight of the DMZ, the Battalion called off the mine sweep for the day.

7. M79 40MM grenade launcher. One of my friends was riding in a jeep with an M-79 when he hit a bump and he squeezed the trigger.

The projectile hit a passing civilian in the chest. Fortunately, it had not gone far enough, or spun enough, to arm.

8. Currency Conversion—periodically, and unannounced, we would re-issue the Military Payment Certificates with a new issue, leaving the old issue worthless. I was the pay officer for one of these. We had to go to every single Marine in the battalion and collect the existing paper in nickels, dimes, quarters, five dollars, etc. Then we had to go to the paymaster and swap these into the same number of pieces of paper and go back to each Marine and give him the exact number of pieces he had given us initially.

BACK AT DONG HA

I spent a fair amount of time travelling back and forth between Battalion Headquarters in Dong Ha and Vandegrift.

1. In the Fall of 1969, it was announced that the 3rd MarDiv would be leaving Vietnam and going to Okinawa. Immediately thereafter we began preparations. One of our tasks was to dismantle multiple "Butler Buildings" to ship them to Okinawa. My platoon dismantled about six of them. Assuming we would be re-assembling them in Okinawa, I contacted the Seabees and got a bill of materials and assembly plans from them. None of the buildings I dismantled were whole—all had been hit by rockets at some time. One of the buildings was a PX supply facility. On the morning we were to disassemble it, we arrived to see a corporal sitting on top of a mountain of beer in the partial building. He bemoaned the fact that he had no one to help him to dispose of its contents. We agreed to clean up the whole mess if he would sell us the Schlitz and give us the Carling Black Label. Conveniently, a case of beer exactly fits in a sandbag, which we had plenty of. We stored all of the beer and sold off the Carling Black Label until we had recovered the cost of the Schlitz. We then had a free "beer bust" for the whole company.

2. Our next task was to transport all the battalion's equipment in "Mike

Boats" from Dong Ha via the Cua Viet River to Cua Viet, on the South China Sea. At Cua Viet, we loaded the equipment and Marines aboard the U.S.S. *Point Defiance* and, after a stopover in the Philippines, arrived in Okinawa.

AFTER VIETNAM – OKINAWA

On Thanksgiving Day, 1969, U.S.S. *Point Defiance*, LSD-31 anchored off-shore of Camp Hansen, Okinawa (Kin Village). We began transferring personnel and their gear from the Point Defiance "well decks" to LCM-8 "Mike Boats" to the shore. Personnel were taken to barracks at Camp Hansen and gear was taken to a holding area at Camp Hansen. Personnel then claimed their personal belongings and these were checked for contraband. About ten percent of the personal belongings were not claimed and, upon inspection, were found to contain contraband such as drugs and unauthorized weapons.

After the personnel were ashore and eating Thanksgiving dinner at the Mess Hall, we began transferring the heavy equipment such as bulldozers, dump trucks, cranes, graders, scrapers, and the like. At the outset, the Mike Boat was not fastened to the shore and as a dump truck was off-loading to shore, it pushed the Mike Boat away from the shore and the dump truck (and the officers' sea bags with our gear) went into the ocean. The officers formed a daisy-chain into the water and retrieved most of our sea bags before they sank. We called in a tank retriever to get the dump truck. Two scuba divers took a cable down to the bottom to the dump truck and secured it around the front axle and then began retrieving the vehicle, slowly at first, and then faster, grunting and groaning at first and then more relaxed. Then, as it surfaced, all it had was the front axle. The rest of the dump truck did not come up.

In Okinawa, I had three simultaneous operations underway:

1. At Camp Hansen, my heavy equipment platoon leveled about four acres and covered it with crushed coral. On the four acres we spread

out the pieces of the Butler Buildings we dissembled in Vietnam, but I left Okinawa before we began assembling them.

2. The coral was from a quarry operation we developed a few miles south of Camp Hansen.

3. Ammunition storage facility. We developed a firebreak around a large ammunition storage facility. At HQ, in reviewing on a map the progress of land clearing around the ammo storage facility and displeased with the progress, I visited the area and saw little trails off the main circumference clearing. What was occurring was the villagers were persuading the Marines to make these trails so they could easily harvest their pineapple crops.

In the evenings, after we secured work for the day, I would practice operating the bulldozers, cranes, scoop loaders, scrapers, and the like. One morning, I was summoned to the Battalion Commander's office. As I stood at attention in front of his desk, he looked at me sternly and said the Sergeant of the Guard had reported that I was operating heavy equipment after hours in a secured area. My career flashed before my eyes, and I said, "Sir, I believed it appropriate for me to become familiar with the equipment my men operate." Then he said, "I used to do that as a Lieutenant also — Good job." I left his office a couple inches taller than when I came in.

CHERRY POINT, NORTH CAROLINA

In June 1970, after Vietnam and Okinawa, I went to Marine Corps Air Station, Cherry Point, North Carolina, as Assistant Wing Engineer Officer in G-4. It was there that I observed that if there was a hot, barren, lonely place to build a military base, the Marine Corps would find it and build a base there.

EPILOGUE

In May 2019, my Basic School class, TBS 3-69, celebrated the fiftieth anniversary of our graduation. The attendees stayed in Fredericksburg, Virginia, and had a pre-arranged a structured visit to MCB Quantico, the Marine Corps Museum, Arlington Cemetery, The Pentagon, and the 8th & I Sunset Parade.

BOB LANZOTTI

US Army Aviator
Total Time in Military: 1960 to 1985
Unit Served with in Vietnam: 1st Avn Bde, 1st Cav Div
Dates in Vietnam: 1967 to 1968; 1969 to 1970
Highest Rank Held: Lt. Colonel
Place of Birth: Taylorville, IL 1937

STAFF SERGEANT COMMANDING SEVEN-STARS

This story involves a drafted Army enlisted man who performed magnificently during his three years of active duty, then doubled down when he re-entered the civilian world.

At the outset of my second tour to Vietnam, I was fortunate enough to be given a command of a CH-47 Chinook company in the First Air Cavalry Division. The command came with a company clerk by the name of Jerry Courington. Jerry was an accounting graduate of Abilene Christian University. He got his draft notice just weeks after graduation. Ironically, the same day he got the draft notice he also learned that he had passed the CPA exam. Jerry was not at all enthused about going into the Army, particularly since we were at war in Vietnam, but he thought he might serve well as a Chaplain's assistant.

But Jerry was very good at taking tests. And lo and behold, the Army in its infinite wisdom had different plans for Jerry. Apparently, the Army thought Jerry would make a terrific helicopter crew chief. So, after basic training, off he went to Ft. Eustis to become a 15U CH-47 flight engineer. Next came "shake and bake" schooling where he emerged as a Specialist E-5. Then off to Vietnam. Jerry never became a full-time flight engineer. Although we did allow him to get his flight time so he could earn his flight pay. You see, Jerry was smart and he could type, thus he found himself as the company clerk. He was already serving in that position when I took command of the company.

It didn't take me long to learn just how valuable Jerry was. He was indeed my Radar O'Reilly. And, I suppose, I was his Colonel Blake of MASH fame. Jerry served as my company clerk for about four months, but halfway through his 12 month tour we lost our supply sergeant to rotation back home.

A supply sergeant position is a pretty critical position in a Chinook company with 16 aircraft and some 325 men. My First Sergeant suggested to me, "Who would make a better supply guy than an accountant?" Great recommendation I agreed, so we promoted Jerry to Staff Sergeant E-6, and he became our Supply Sergeant. And he proved that we had made a wise decision... he did a splendid job. I left the Company in 1970 and didn't hear from Jerry again for 36 years.

He read my book, *Flying Through the Years*, sent me a letter, and we began communicating again. I learned a lot from our newfound reunion. Like, unbeknownst to me he wrote all my platoon leaders OERs for their men. And, all along, I thought I had a bunch of good writers in my unit. He also claimed that he could sign my name better than I could, and that meant that some of my pilots probably found their way to the wild side of Saigon with a pass supposedly signed by Major Lanzotti. Then he related his experiences as a supply sergeant. He certainly learned quickly how to wheel and deal, and I'm sure Donald Trump would agree that, "This guy ain't no apprentice!"

I also learned that after he was discharged from the Army in 1971 he went back to college on the GI Bill and earned an MBA. He then joined an international exporting company and ole smarty pants began working his way up the food chain. After several years, Jerry was promoted to CEO of a new start-up subordinate office in Houston, TX. He was told before he moved to Houston that he would have a few veterans working for him and one was a Navy Seal.

He met the moving vans and several of his new employees at the new company site and while unpacking noticed that one of his men was wearing a Top Gun T-Shirt. Well, that wasn't Seal paraphernalia, but Navy nonetheless. So, Jerry approached the guy and asked him, "What did you do before you joined the company?" The guy replied, "I was in the Navy." Thinking he may have found his man, Jerry asked, "What was your job?" And the guy responded, "I was commander of the Pacific Fleet!" A four-star Admiral!

A week later another Vet reported to Jerry and he happened to be General Schwarzkopf's Chief of Staff during Desert Storm, but later retired as a Lt. General. Now I ask you, what do you think the odds would be to find an E-6 Staff Sergeant today supervising seven stars?

I want to extract just one paragraph of one of the letters Jerry sent to me. "Although I did not do everything right while I was there, I believe it was the time I really developed the values that carried me through the rest of my life. I hope I was a good example and did not do anything to put a bad light on our unit. Not a day goes by that I don't think about my time over there. I value the time I had in the Army and especially with Charlie Company, 228th."

You know, there's one thing you cannot deny about the military... it certainly accelerates maturity. Personally, I think we may have made a big mistake when we said, "bye bye" to the draft in 1973. Let's just say that I think today's millennials may have missed an opportunity.

RABBI JEFF FEINSTEIN

US Army — Senior Jewish Chaplain's Assistant
Dates of Military Service: 1967 to 1969
Unit Served with in Vietnam: 1st Logistics Command
Dates in Vietnam: Jun 1968 to Jun 1969
Highest Rank Held: Staff Sergeant (E-6)
Place of Birth and Year: Stockton, CA — 1947

A RABBI'S STORY

I was born in Stockton, CA in June of 1947. After high school, I attended the University of California, Davis, but played around for the entire year and had to leave on an academic probation. I came home and attended a local junior college, redeemed my GPA, then went to the University of the Pacific. I got sidetracked with a young lady who dumped me and it turned my life upside down. The next morning I was knocking on the Recruiter's door. It was 1967, I enlisted in the Army and was promised I wouldn't have to go to Vietnam.

Basic: The Army send me to Fort Lewis, WA for basic training in December and January which is someplace you don't want to be in December and January. At the time, I was still okay with the Army, but still didn't have my head screwed on properly. I almost washed out of Boot Camp because I became ill and they were going to recycle me because I missed rifle qualification. But I asked the Captain, "If I qualify, will you put me through?" He didn't think I had a chance, but I proved him wrong. I ended up with a trophy for the highest score in our battalion.

Bad abbreviation: "After basic we were standing out in the rain while the Drill Sergeant called out our MOS designations. Then he gets to me and roars, "What is a 71 Charlie? And what in the hell is a Chap's Ass.?" It was Chaplain's Assistant, but someone had abbreviated the wording to Chap's Ass. I sort of guessed where I'd end up because my hometown rabbi had written a letter to get me into chaplain's duty.

Two forts: I was sent to Fort Dix, New Jersey for Clerical School but that didn't last too long since I could already type 80 words a minute. I became the Commanding Officer's secretary and typist and went to Fort Hamilton in Brooklyn for Chaplain's School. I was pretty much destined to be a Jewish Chaplain's Assistant. I quickly learned there was a right way, a wrong way, and the Army way, and the first two didn't count. One Jewish guy out of every other class went to Vietnam and I found out the quota in Nam had been filled by the guy two classes ahead of me. I was bound for Germany, which was fine with me. At the time, that's in 1968, there were only five Jewish Chaplains in Vietnam.

Broken promises: In the military area, Jewish soldiers at the time were heavily concentrated in the Medical Corps and the Staff Judge Advocate area. There weren't too many Jewish ground-pounders. Well, the guy bound for Vietnam slipped and hurt his back. No problem, "Hey Feinstein, you take Vietnam and he'll take Germany." Right. So, I'm off to Southeast Asia.

No air-conditioning: It's June 3, 1968, and the 707 airliner sets down at Bien Hoa, Vietnam where I step from a 72-degree airplane into 102 degree heat. I just about died. Albeit, I had an interesting first night in-country. I spent the night at BienHoa awaiting orders in a 'holding barracks' which is nothing more than a concrete slab with a canvas cover and triple-deck bunks. We got hit that night. Welcome to Vietnam. One guy in the holding barracks had reenlisted so he could get an early-out. He didn't run to the bunker when we had incoming, all he did was grab the side of his mattress, flipped over on the floor, and covered himself with the mattress. Thing was, he'd forgotten he was perched on the top of a three-tiered bunk. He was the only injury we had that night.

Rabbi Ostrovsky: I shipped out early the next day to the 1st Logistical Command in Long Binh and served as a chaplain's assistant for the 2nd highest ranking rabbi in the US Army, Lt. Col. Jack Ostrovsky. We went everywhere together, from the DMZ all the way down to the Mekong Delta because there were only five Jewish chaplain teams to cover the entire country. I remember flying in a chopper over an LZ (landing zone) that was under heavy fire. Rabbi Ostrovsky said, "I wonder if any of our guys are down there?" I replied, "If you go down there, it's just you and God because I'm not going!" That happened a lot.

Toting weapons: The few times we did go into a hot area we were told to leave immediately. We were not allowed in. We had one rabbi who carried a .45 automatic and an M16, which was against the Geneva Convention. He was also a paratrooper and well-trained. And, yeah, I carried, too. I was not ordained at the time, I was just a senior Jewish chaplain's administrator in Vietnam.

I was put in charge of all the chaplain's assistants, around eight or nine of them, plus I served as my Chaplain's bodyguard. Still, we lost one rabbi in Nam, I think his name was Singer. (**Rabbi Morton Singer** was one of 14 soldiers killed in the crash of an Air Force transport plane in December of 1968. Rabbi Singer was en route to conduct Chanukan services for Jewish servicemen; he was 32 years old).

Experiences: A couple of things. The worst job, actually there were two of them... the first was counseling prisoners at the Long Binh jail. It wasn't often that I counseled Jewish prisoners, but a few were in there. I remember counseling one young man who had blown away his Commanding Officer. I asked him bluntly, "Did you do it?" He replied, "Hell, yes, I did it. He was going to get us all killed, but nobody will ever prove it because there were too many bullets flying around." And he was right; they never did prove it. That gave me a different perspective about what was going on around me. I guess I'd been shielded from what was happening in Vietnam.

The real eye-opener was when I had to go to the mortuary in Da Nang and go through looking for Jewish toe-tags. That was difficult. In such a large area there was no sound, just tomb-like silence, and you're the only breathing person in the area. It was an ethereal experience. There were 300 to 400 boys inside that noiseless room. If I found a Jewish soldier we would notify the next-of-kin. Normally that was a job for a commanding officer, but sometimes they passed the responsibility to chaplains. The senior chaplains made the call... the government notice arrived later. I'm not sure I could have made those calls, I mean, I was still pretty young and I don't know if I had the wherewithal to accomplish that.

The good; the bad; the ugly: You know they say when you're in a war, you tend just to recall the good times. Our brains are amazing things, they shield us from stuff like that. One highlight in Vietnam was when we conducted the first, and perhaps the only, Passover Seder, which is the ritual meal for the holiday. I was in charge of that Seder and every commanding officer in Vietnam was told if they had Jewish personnel they were to arrange for them to attend the Passover Seder. It was held in Da Nang and over 500 were in attendance. I contacted the Jewish Welfare Board in New York City and arranged for them to fly the food into Da Nang, since the food had to be kosher. Wine and food, everything... it went off perfectly.

More poignantly: We arrived in Da Nang a day early so Rabbi Ostrovsky decided to take a swim at the beach. Signs all over the beach warned that the undertow was really bad. So, I'm standing there watching Rabbi Ostrovsky and suddenly he's being swept out to open ocean and couldn't do anything about it. There were three lifeguards on the beach; I used to be one, too, and I beat all three of them to the rabbi and pulled him back in. Then he told me, "I don't want this written up in Stars and Stripes because I don't want my wife to find out about this incident." Then he suggested, "So, how about a Bronze star and a promotion?" I said, "That works for me!" That's how I got my first of two Bronze Stars, that one for valor; don't know what kind of valor, but that's all they could put it under.

And the ugly: One incident is tough to talk about, but I will share it with you. Everyone in the detachment at Long Binh, from E-6 on down, had to pull guard duty. As an E-6, I had to. We were in four-man bunkers, dug about 4½ feet into the ground. Each corner had a six-by-six post that had a plank with sandbags on top of it, so you were pretty well shielded. Two or three strands of barbwire were in front of you, plus a Claymore mine. It was the responsibility of one person in each bunker to check the Claymore before sundown, but the guys next to us didn't do it. The VC had turned the Claymore around.

We got attacked and two things happened simultaneously. The VC ran a ground assault at us; the guys to our right fired their claymore that had been turned around and it blew a gap right through the barbwire then at the same time a rocket struck next to our bunker and buried into the ground before exploding. It blew out the corner posts of our bunker and the top came down and sealed us in a coffin. The guys in bunkers to our left and right were all killed that night, but the VC couldn't get to us and we could not get to them. We were sealed in like a crypt. When they got us out the next day, there were shrapnel holes through the sand bags at one end of the bunker and through the sand bags at the other end of the bunker, and not one of us got a scratch. I have no idea how that was possible. I guess God was looking out for us. From that point in my life, I knew I had to do more than I was doing, the incident gave my life a new purpose. You know, I volunteered to pull guard duty on Christian holidays, like Easter and Christmas, so other Soldiers could have time off. Ironically, nobody ever volunteered for me.

One funny story: Religious services in the military have to be conducted in English. It's okay to conduct a service in Hebrew or Latin, but they have to be translated into English. The holiest day of the Jewish year is Yom Kippur, the Day of Atonement. In the chapel, I had to go in and remove the cross and put up the Star of David. We had a packed house for Yom Kippur, starting at dawn and finishing at dusk. Yom Kippur is a day of fasting and intensive prayer, you can't eat or drink. Rabbi Ostrovsky would read in Hebrew and I would translate into English, then vice versa, all day long. The last thing done on Yom Kipper is the blowing of the Ram's horn, the Shofar, one blast only. Well, it's about 120 degrees in the chapel, we've had nothing to eat or drink all day long, my mouth was like sand. The rabbi gave me the command to "give the long blast." I put the Shofar to my mouth and I think the only thing that came out was dust. Not one sound. So there I am with 500 eyes on me, all thirsty and hungry, and they can't leave until I blow the horn. I tried again. Absolutely nothing. I sort of looked skyward and said, "Please," tried again, and got a very low 'squeek', but that was enough and we all enjoyed the feast.

Strange reunion: I'd been back from Nam about 15 years and was working for a rather large paper company. They flew me into New Jersey and I was picked up by limousine about a block long. I'm sitting in back and can't even see the driver, but after I did my business 'thing' and had to return to the airport, I said to myself, "I'm not sitting back there again," so I sat up front with the driver. We talked, the driver was about my age; I asked him if he had been in the service. He said, 'Yeah.' Vietnam? 'Yeah.' What years? 1968 and 69.' Me, too. Where were you? 'Long Binh.' What'd you do? 'Chopper pilot.' Who'd you fly for? 'Chaplains.' This guy had been my chopper pilot that I hadn't seen for 15 years. We detoured from the airport and found a bar for a couple of cold beers together.

After Nam: My diabetes had onset while I was in Nam. I went from 225 lbs. down to 160 lbs. in just six weeks. I ended up in Letterman Army Hospital for almost two months then received a medical discharge. I finished up college at the University of the Pacific with a Master's degree in Communication and went into the public business sector for several years. I raised two wonderful sons, but I was never happy. I realized the sadness came from chasing dollars. I wanted to be a rabbi, knew that when I was in Nam, but life got in the way. With the Master's degree, I entered the Rabbinic Seminary in Manhattan, and I've been doing this for about 16 years.

Rabbi Feinstein is the founding rabbi of Kehillat HaShem and serves a small but dedicated congregation at the base of the Kennesaw Mountain Battlefield in Marietta. He can be contacted through his web site at www.rabbiatlanta.com.

Final thoughts: You look at America's history and to me there are four things that changed this country. First, the Declaration of Independence. Second, the Civil War, which changed forever the way we looked at things. Third, the completion of a trans-continental railroad system that opened up the entire country. And fourth, the Vietnam War. For the first time, we realized we can challenge our government, and we have never been the same since. When I went to Vietnam, I was a little bit to the left of Jesus Christ, then when I came home, a bit to the right of Attila the Hun. You can look at something from a distance but you get a completely different view if you're there. I was petrified to go to Vietnam, but it made me who I am today. It changed all of us, some of us for the good; some of us for the bad. But would I do it again? Absolutely.

Shalom, Rabbi Feinstein, Shalom.

CARL H. "SKIP" BELL, III

U.S. Army—Armored Cavalry Officer, Aviator

Dates of Military Service (Active and Reserve): 1967 to 1998

Units Served With In Vietnam: A Troop, HHT, B Troop, 1st Squadron, 4th Cav,

1st Inf Div (First Tour); C Troop, 3rd Squadron, 17th Air Cav; 18th Corps Aviation

Company; G3, HQ 1st Aviation BDE (Second Tour)

Dates in Vietnam: Feb 1969 to Feb 1970; Feb 1972 to Feb 1973

Highest Rank Held: Colonel

Place of Birth and Year: Decatur, GA—1945

INCIDENT IN THE TRAPEZOID

On June 24, 1969, A Troop received orders to move into the Trapezoid under OPCON to Division HQ, as part of a task force consisting of A Troop, a platoon from C Troop, a Tank Platoon from B/2-34 Armor, an Infantry Company and an Engineer Land Clearing Company. I was the Acting Troop Commander while CPT Bill Newell was on R&R; he was expected back the following day. The mission was to destroy and then Rome plow the jungle covering an NVA base area in which A Troop had spent a week about six weeks earlier.

On that previous operation we discovered a huge enemy base camp laid out like a wagon wheel. It was full of bunkers designed for use by supply and service units. Only the outer ring of bunkers contained fighting positions. We had gone in on the heels of an Arclight (B-52 strike) and found a few bodies and casualties during the time we spent there. We also found lit cigarettes on the ground, NVA ponchos, packs, food on cooking fires, etc., all indicators that the enemy was close. We removed a lot of supplies, captured a hospital (complete with a wounded NVA soldier who had received some fairly sophisticated medical care), and generally wreaked havoc with the place. The bunkers were very well built, and we were unable to destroy many of them. We even tried neutral steering a 60-ton VTR on top of one and it held up.

We moved out early on the morning of June 25, moving through the area we had been in a couple of days earlier when our mission to reinforce the infantry battalion NDP had been aborted, and we had com-bat-lost three vehicles. This time we stayed off roads and stayed in the Iron Triangle longer, entering the Trapezoid near the northwestern end of the Triangle.

Our objective was a large open area southwest of the bunker complex. We were going to use it as a base area from which we could move the Rome plows into the bunker complex. We entered the large clearing at approximately 1400 hours and immediately began to set up our perime-ter. I placed the attached tank platoon from 12 o'clock to 3 o'clock (with 12 being north—they were in the northeast quadrant of the perimeter which was the portion facing the direction of the bunker complex lo-cation and the area I thought we'd be hit from if the enemy attacked us). The other three A Troop cavalry platoons completed the circle, and we had the infantry company dig in their foxholes between the tracks around the entire perimeter. The Engineer Company and the troop CP were in the center of the perimeter. I asked the engineer company com-mander to dig a trench that we could back a couple of HQ tracks into in order to make a covered TOC. They started on the ditch but didn't get it dug out enough to put the tracks into it. That ditch, however, was to come in handy later on that night.

The OIC for the operation was a major (I believe his name was Smith) from Division HQ. As I recall, he pretty much stayed out of the way and let us set up the perimeter as we saw fit. By dark, we had one strand of concertina wire up and all of the RPG screens erected in front of the tracks. The infantry guys had their fighting positions well under-way. Sometime during that afternoon, my former platoon sergeant, SSG Godwin of 1st Platoon, informed me that a couple of the troopers had seen some people watching us from the edge of the wood line. When the watchers saw that they were noticed, they disappeared.

As it started getting dark, the FO (artillery forward observer) regis-tered some on-call fires (again, mainly to the northeast of our position, and we settled in for the night. We did not put out any listening posts because the distance between the perimeter and the wood line would have made it impossible for us to pull the LPs back in if we were hit.

I normally slept in the A66 track (to be close to the radios), but with the major from Division there, the track got a little crowded so I decided to sleep on the ground beside the vehicle. I bedded down in a cluster of folks that included SSG Roderick "Pappy" Guy, HQ Platoon Sergeant, and SGT Gene Garrison, the Troop senior (medic) aid man.

At about 0210, I was awakened to a rapid TUNG, TUNG, TUNG, TUNG, TUNG, TUNG sound coming out of the jungle to the east of our position. Somehow I recognized the sound as mortars being fired. I yelled, "incoming!" and started to get up to get to the track. One of the rounds impacted close by, knocking me down and wounding both SSG Guy and SGT Garrison. The mortar rounds kept coming in, as did RPGs which were being "lobbed" in, as well as being fired directly at the vehicles. By this time, the platoons were firing out 360 degrees around the perimeter. I got to the radio and called for a sitrep from the platoons.

Several men were wounded, including one severely wounded man from the attached tank platoon. His name was SP5 Hipshman, and he was sleeping behind his tank when a mortar round landed next to his head. I directed that the wounded be brought into the ditch. SGT Garrison stuck his head in the track and said that he was hurting pretty badly. I told him to go get in the ditch, and that I would call for a medevac. Instead, SGT Garrison risked his life by going to the perimeter and dragging other wounded soldiers into the aid station we were establishing in the ditch. Sergeant Garrison was subsequently awarded the Bronze Star Medal with V Device for his courageous actions that night, as well as the Purple Heart.

By this time, the enemy had opened up on us with small arms and at least one heavy machinegun (from the sound of the bullets passing overhead, it was a .51 caliber, and there may have been more than one). We began to put up hand flares, and I told the FO to fire his preplanned targets and to order illumination rounds ASAP. Although he was unable to contact the DS (direct support) artillery unit, he tried repeatedly. Of course, I was screaming at him to get me some damn illum and to put some counterbattery fire on those mortars. We eventually discovered that one of the mortar rounds had blown off his radio antenna. Once we got another antenna on, friendly artillery rounds started falling, and we quit using our hand flares (though we used nearly all of them by the time friendly artillery arrived). The hand flare situation was made more critical when, earlier in the attack, an RPG had struck a box of hand flares that was hanging off the side of the turret of one of the tanks. The explosion was spectacular, and I thought the tank had taken a direct hit from the RPG and that it had penetrated the armor and hit some of the ammunition inside. I was about to move one of the HQ tracks to the perimeter to cover the gap but when the smoke cleared, I saw the tank still sitting there firing. Fortunately, everyone on it was OK.

Meanwhile, the dustoff (i.e., a medevac helicopter) came on station, as had a couple of gunships from the 1st Cav Division. They were working on knocking out the .51 caliber machine gun (or one of them, anyway), which they did. The dustoff said he was willing to come in, but I waved him off because the small arms, machinegun, mortar, and RPG fire was still too intense. That was the night I found out what being an officer really meant. I asked the dustoff to stay on station until things quieted down a bit, and he said he would. I then ran over to the ditch to check on the wounded. I went from man to man telling them that we had a dustoff on station and to hang on—that they were going to be OK. The medics were working on the more seriously wounded. When I got to Specialist Hipshman and told him he was going to be OK, the medic working on him (I believe it was SGT Garrison), said, "It's OK, sir; he just died."

That was the first man I had lost killed in action; I'd had several wounded before, but nobody killed. I can't describe the feeling. Fortunately, I was really busy at that time, and really didn't have time to contemplate it until later on. I still find myself wondering if I'd let that dustoff come in when he first got on station if Hipshman would have made it. Of course, I could have just as easily ended up with a crashed helicopter and four more injured, if not killed. I'm sure that's not much comfort to Hipshman's family. Fifty years later it's not that much comfort to me, either.

After the 1st Cav Cobra knocked out the .51 caliber, things quieted down, though we continued to take sporadic small arms and RPGs until daylight. About an hour before dawn, SSG Guy and I walked (jogged, actually) the entire perimeter talking to the troops and telling them to hang in there and conferring with the platoon leaders and the infantry company commander. Recall that SSG Guy was wounded in the first burst of mortar fire, but he showed his courage and did his job, just like the rest of the troopers, tankers, and infantrymen did that night. Like SGT Garrison, SSG Guy was also awarded the Bronze Star Medal with V Device and the Purple Heart. Both soldiers were awarded their medals by the Division Commander at Lai Khe a couple of days after the action while they were recovering from their wounds.

At daylight the enemy broke contact. We had redistributed ammunition, and we called for more, as well as bringing in the dustoffs (it ended up taking a couple of those ships) to evacuate the wounded. Bill Newell came out on one of the resupply choppers. He thanked me for taking the unit in his absence and told me he had heard I had done a good job. It was a bittersweet thing for me. I was glad to see him back and for my personal safety I was glad to be getting out of the field, but I really hated to leave that unit out there.

I thought at the time that this could have been my last night in the field. I was going back to a staff job (turned out to be the S4) and figured to spend the remainder of my tour doing that (though I really didn't want to—I had come to Vietnam to lead troops, not to push paper). But it was the Army, and we do what we are told to do.

That was the start of a difficult couple of months for A Troop. The NDP (I believe it was called Mons II) became a sea of mud (tanks were literally sinking up to the sponson boxes (typically installed for added bulletproof protection or storage), and they made frequent contact with and hit lots of mines. They did succeed, however, in completing the Rome plow operation and denying that base area to the enemy (at least while they were there). "Hardcore Alpha" once again proved itself Prepared and Loyal.

RICHARD H. MOUSHEGIAN

US Army — Quartermaster Officer

Dates of Military Service (Active and Reserve): 1968 to 1993

Unit Served with in Vietnam: HHC, Long Binh Post, USARV

Dates in Vietnam: Oct 1968 to Oct 1969

Highest Rank Held: Lt. Colonel

Place of Birth and Year: Camp Wheeler (deactivated after WWII; near Macon, GA) — 1943

TWO DIFFERENT PATHS

As a Quartermaster Captain in the Army while in Vietnam (1968-69), I was the Logistics Officer (supply and maintenance) for a 40-unit club system in Long Binh (near Saigon) for American officers, Senior NCOs, and enlisted troops. The clubs provided a friendly environment for the troops after their duty hours to purchase and consume adult beverages and food items such as pizzas, hamburgers, packaged snack items, and cigarettes. The majority of the clubs were located in the sprawling Army post of Long Binh with a few clubs at the Air Force base at Ton Son Nhut. Did I have problems? Yes.

(1) As a result of shenanigans of a former purchasing officer, I inherited extreme over stockages of American cigarettes and beef jerky when I assumed the logistics assignment. Due to limited space, the huge supplies of cigarettes were stored in two CONEXs (Container Express which are large, steel shipping containers that fit on flatbed trucks and ships to transport military supplies around the world). Since the large supplies of cigarettes could not be sold fast enough, the tobacco leaves were rotting in the extreme heat and high humidity of Vietnam. I had to schedule an Army *veterinarian* to inspect the spoilage so that I could delete the rotting items from my inventory. (Yes, the Army uses them to inspect and certify food spoilage when they are not currently treating security dogs in today's Army and mules and horses during the Army of WWII.) (No, I had NO interest in the cigarettes for ANY reason, and I have never smoked any vegetable substance.)

The other item that I inherited was a 5 YEAR supply of beef jerky also for sale in the club system. The item stored well, but during every inventory, the stock levels never seemed to significantly drop. (I do not eat it, to this day.)

(2) My boss insisted that I continue to order 2,000 cases of frozen pizzas (12 per case) every month for the entire system "for the troops," *regardless of the usage factor*. At one point, the shipments were delayed until I received a call to pick up 6,000 cases of FROZEN pizzas at the dock in Saigon! Where do you store 6,000 cases of an item when your refrigeration-capacity is only 2,000 cases? I called all the major units at Long Binh, including the *General's Mess* (dining facility), to temporarily store as many of the frozen items as they could reasonably store. My only "bargaining chip" was to tell them not to throw a party with my pizzas, but they could help themselves to one or two boxes. You can bet on two things: (1) the General's Mess was the *first* storage site that I pulled the items, when possible. (I wanted to minimize the exposure-time that the pizzas were in their freezers so that the generals would not start discussing the situation with each other and take some action that would make the condition even worse.) (2) I was happy to recover *anything* from all the satellite sites. In the end, we recovered about 95% of the frozen pizzas!

(3) My boss eventually retired from the Vietnam assignment and started working for the same pizza company that so enthusiastically shipped us frozen pizzas. Unfortunately, he decided to make a trip to Long Binh as a *civilian agent of the vendor*. The first I heard of his arrival is when CID (Criminal Investigation Division) asked me if I had seen this individual. I had been out of the office doing my job and had not seen him. CID's interest in him caused him to race to the airport for the next flight out of Vietnam! (I understand the CID missed him by 30 minutes.)

Hence, my additional duty was to clean up the mess of others who had made shady deals in Vietnam.

In March 1969, I was fortunate enough to have my younger brother, 1LT Stephen Moushegian, join me in Vietnam since he had recently graduated from the Army's helicopter school. Since I arrived first, I requested that he be assigned nearby. Then, I mentioned to him that both of us did not need to be in a combat zone at the same time. His response was that when he signed up for helicopter school, he already knew where his first assignment would be: Vietnam. So, he was staying. I countered that I got there first, so I was also staying. (The banter was not surprising since our father was a retired Army, infantry officer who was a West Point graduate of 1939 and had fought in Germany in WWII. We were *expected* to do our duty, not find a way to shirk our responsibilities to our country.)

I was successful in getting my brother assigned to the 25th Aviation Company next to Long Binh Post in which his mission was to transport high-ranking officers and civilians (or VIPs- Very Important Persons) around Vietnam within 1-2 hours of Long Binh Post. Sundays were my off-days because we employed foreign nationals in our front office in the club system, and they did not work on Sundays. On those days, I would fly with him (with the knowledge of flight operations) by relieving the door gunner (enlisted) on the Huey helicopter (UH-1) until he transitioned into the Loaches (OH-6A). That arrangement continued for about three months. Since he was a Type A personality (that is, he HAD to be in-charge), the transition from Huey helicopters (two pilots; he was not immediately in-charge) to Loach (one pilot) was perfect for him, he was back in-charge!

Since he was at the top of his helicopter flight class (he already had a private pilot license before entering the Army), he was an extremely good pilot and yearned to make a difference in the war, not just act as a taxi driver. One evening 1LT Moushegian met a visiting company commander of a combat-flight unit in the officers' club and knowingly requested a transfer on the spot to his combat unit. The visiting Major was able to transfer the lieutenant *within two weeks* to the 11th Armored Cavalry Regiment in the aero-scout platoon as a screening force from the air. His flight operations were based in Quan Loi (an old French rubber plantation) somewhat close to the Cambodian border.

He absolutely loved the assignment since he felt that he was making a positive contribution to the war effort, and he flew with honor and distinction by searching for the enemy by flying *below tree-tops* and "reading the vegetation" for any troop movements (direction, size force, and how long ago), directing ground-combat units from the air while being exposed, and neutralizing enemy machine gun and artillery emplacements with the on-board mini-gun, when necessary. He loved the 11th ACR and continued in the assignment until he flew into an ambush on Sept 3, 1969. The enemy force was a North Vietnamese DIVISION, unknown to be in the area. 1LT Moushegian flew into a curtain of small-arms fire and barely escaped with his life and that of his PFC (Private First Class) aerial observer. (The 19 year-old PFC was receiving OJT, on the job training, on how to fly the helicopter since the Loach was equipped with dual flight controls!)

Just the week before the ambush, 1LT Moushegian gave the young man practice on how to hover as an "insurance policy" for their survival in the event that the single, authorized pilot was incapacitated. As the situation unfolded, the pilot suffered wounds to the ear in which a bullet passed between the flight helmet and his head, split his ear open, and exited the rear of the flight helmet. Additionally, he suffered severe wounds to his right forearm, left shoulder, and left eye with a metal fragment coming from the exploding radios that hemorrhaged his left eye. The result of the ambush was that the pilot was *incapable of controlling the aircraft* with his arms, his RIGHT eye kept closing as a sympathetic reaction to the damage to the left eye, he was losing blood from all of his wounds, and he *was in danger of losing consciousness at any moment.*

Fortunately, the young PFC kept an even temperament and *accepted responsibility and control* of the aircraft based on OJT! While 1LT Moushegian was still conscious and could intermittently glance at the instruments, he realized that despite the heavy damage to the aircraft and radios, the engine, transmission, and fuel tanks were relatively intact, and the aircraft was still holding together! Unfortunately, they were flying deeper into enemy territory into Cambodia, and they were more than 30 minutes away from home base. Despite everything and most importantly, *the aircraft was in level-flight and under control of a 19 year-old PFC!*

1LT Moushegian gave the PFC instructions to swing the aircraft around for a new, specific heading to return home, find an undamaged radio (out of four), dial-up the specific home-frequency, notify flight operations that they were returning home and were requesting immediate medical attention upon arrival. To the young man's credit, he *maintained the new heading* until he started recognizing familiar terrain and found the correct helipad for a *good landing!* The PFC saved both of their lives and was not seriously wounded, the pilot received immediate medical attention, and the aircraft never flew again because of severe battle-damage.

Prologue: 1LT Stephen Moushegian was a highly-decorated, combat veteran who earned numerous awards for valor to include the Silver Star (the 3rd highest award behind the Medal of Honor), *four* Distinguished Flying Crosses, and the Vietnamese Cross of Gallantry awarded by the Vietnamese government as an *individual award to him* (not a unit award) complete with a citation written in Vietnamese and translated into English. Of course, he was awarded the Purple Heart for wounds in a combat zone due to enemy action. He was medically retired from the Army as a Captain, the Veterans Administration repaired his physical wounds and gave him a glass eye, and he passed away on March 22, 2019. His immediate family is waiting for instructions for his below-ground interment in Arlington National Cemetery in Virginia at a future date.

BILL GINN

U.S. Army — Infantry Officer

Dates of Military Service: 1968 to 1970

Unit Served with in Vietnam: Alpha Company & HQ Company, 1st Bn, 12th Cavalry, 1st Cav Div.

Dates Served in Vietnam: Feb 1969 to Feb 1970

Highest Rank Held: Captain

Place of Birth and Year: Atlanta, GA — 1945

ALONE WE WENT

On February 8, 1969, I checked into the Processing Center at Long Binh Logistic Base. As a 2nd Lt. with 10 months of training at Fort Benning through the Officer Basic Course, Ranger School, and Jump School, I had no real experience as an Infantry platoon leader. But, that was about to change as I was being assigned to the 1st Cavalry Division.

I flew into South Vietnam with about 200 strangers to join a unit where I knew no one. I was welcomed with a blanket and a pillow and directed to the hut where I would stay until I could get my combat gear the next day. I casually observed bunkers near the hut as I went to chow that night. Several hours after dark, the alarm sounded and I rushed to the bunker along with the other unarmed strangers. We learned in the morning that the VC had breached the wire and the Base Commander and his driver plugged the hole with his jeep while firing an M60 machine gun to thwart the attack. The Colonel had been badly wounded. Thus started the probing of Bien Hoa and Long Binh that would turn into a major battle of Tet 1969 several weeks later. What a beginning to my two-week orientation to sunny Vietnam.

After training at the 1st Cav orientation center north of Phuoc Vinh, I boarded a Huey along with a young soldier returning from the hospital and headed to LZ Cindy to join the 1st Battalion, 12th Cav. Upon arrival on the LZ, I was notified that I was now the 3rd Platoon Leader of A Company — the Platoon the young surfer from California was rejoining. This young man had been in a reserve unit that got activated and was now a grunt in Vietnam. Two days later, he and I got on a resupply chopper and joined A Company on a make-shift LZ north of Bien Hoa.

On my third day in the bush, A Company moved out of our overnight defensive position in single file with 1st Platoon taking point. About fifteen minutes out, we were ambushed and the Point Man was killed instantly. I was ordered to move 3rd Platoon into a flanking position. But, as we did, Big Ed, our point man, was wounded. My young California surfer soldier died that day trying to rescue his friend Big Ed. Welcome to Vietnam.

The next several months passed with infrequent contact after the remnants of the VC who attacked Long Binh and who ambushed us on my third day in the bush were neutralized. We spent many long, hot days hacking our way through bamboo, sleeping in the mud, and drinking water out of B52 bomb craters. On my birthday in June, we were working in an area about one klick from Cambodia. My Platoon drew the short straw and were assigned to escort an Armor Platoon as they cleared a path through a stretch of jungle that had been defoliated by Agent Orange. We had been working in the area several weeks, but didn't discover a company-sized, freshly dug, bunker complex until we worked with the Armor Platoon. The night that followed was the longest of my tour as we set up an ambush on a trail leading out of the complex. Even though we heard noises during the night and called in artillery, when the sun came up the next morning we found nothing or no one in the complex.

Upon hearing that news, our Battalion Commander ordered A Co to move about 10 kilometers east to a large meadow for an extraction. Because the jungle was thick and time was of the essence, the decision was made to follow a well-worn trail that would get us close to the designated LZ on time.

Several hours out we took fire and A Company Commander, Capt. D'Angelo, was wounded. 1st Lt. Mike Brightman, our Forward Observer, called in artillery and after about 30 minutes we moved to the LZ. When we got to the LZ, the trailing Platoon failed to post an OP to watch the trail. They were ordered to do so immediately. As the two young soldiers returned to the trail, they were ambushed and killed.

My Platoon immediately went into Fire and Movement to flank the attack and neutralize it. Captain D was extracted and another Captain came to prepare us for an extraction the next day. During the flanking movement, I hit the knuckle of my right hand where the middle finger joins the hand. This became infected days later after the doctor on LZ Grant "wisely" decided to drain the bruise and I was awarded a two-week vacation at the Cu Chi Field Hospital. When I returned to the Battalion, I learned a crop of West Point 1st Lieutenants had come in and I was reassigned to be the Support Platoon Leader on LZ Grant.

The saddest day of my tour came in November when I was returning from R&R. As I was in the Battalion rear area on Tay Ninh Combat Base, we learned that the Huey carrying our Battalion Commander, Battalion Command Sergeant Major, Captain D'Angelo, and Captain Mike Brightman had gone down and all were presumed dead. As a young inexperienced platoon leader, I needed Mike Brightman and Captain D as role models. I am convinced that I am alive today by the grace of God and the influence of these two men.

HUGH D. PENN

US Air Force — Personnel

Dates of Military Service (Active Duty and Reserves Combined): 1966 to 1972

Unit Served with in Vietnam: 633rd Combat Support Group

Dates in Vietnam: Feb 1969 to Jan 1970

Highest Rank Held: Staff Sergeant, E-5

Place of Birth and Year: Birmingham, Alabama — 1945

SO, THAT'S INCOMING!

I was born September 1945 in Birmingham, AL. I lived there until 1952 when we moved to East Point, GA where I lived with an aunt and uncle. My father was very sick and the best doctors were in Atlanta at the time. He died and was buried on Easter Sunday, 1952. We (Mother, sister and

I) lived with an aunt and uncle until Mother got a job with Lockheed and worked the night shift.

In 1966, I got an invitation to join the Army. Instead, I enlisted in the USAF. I did this on Feb 5, 1966, came home told my Mother and reported to the induction center on Feb 6, 1966. My Mother was relieved I was not going in the Army. I was inducted, checked out physically, and given the oath and then I was on my way to Lackland AF base, San Antonio, TX. Winter in San Antonio could be 94 degrees one day and snowing the next. Six weeks of basic training transformed me from a civilian to a basic airman. I learned how to follow instructions, be a team member, be punctual, make a bed, polish shoes, fold my laundry, stand in line for hours, and learned basic marksmanship, which I just barely qualified for.

I also learned how to pull KP and peel potatoes with a butter knife, wash bulk loads of dishes, and mop floors. In basic, I learned personal responsibilities that still serve me to this day. Every young man or woman who goes through this is better because of it. How much better is up to the individual.

In April 1966 came PCS orders for Amarillo AFB, TX for training. Supply school was backed up for three months so we were given an opportunity to change career fields. One choice was Personnel and OJT right there in Amarillo. As luck would have it, the Personnel school was also there at Amarillo, so I was assigned work duty in the afternoon and tech school in the morning.

Since I was permanent party (PP), I did not have to march to school, I was now a real live Airman and expected to act like it. My main assignment was to type Morning Reports. That was a form for accounting of all personnel in the unit. I was a terrible typist because the form was of rows and column of numbers and only two errors per page were allowed. Finally, the Morning Report was phased out and I was then assigned to the Machine Room.

The Machine Room was a section where paper records were converted to data and entered on IBM punch cards (do not fold, spindle or mutilate). I worked the day shift, the night shift, and enjoyed the work of coding, sorting, filing, and programming the machines. Amarillo AFB was shared by the Air Training Command and SAC. There were a wing of nuclear armed B-52s and they were deadly serious about their mission. They were still under the influence of General Curtis LeMay. I enjoyed my time at Amarillo, in spite of the Texas heat, the Texas wind, and the Texas landscape, did I mention the Texas cold?

After approximately 18 months, I got orders. PCS to Craig AFB, Selma, AL. It was a pilot training base and about 30 miles from Montgomery and 160 miles to my home in Tucker, GA. At this time I was an E-3 and shortly thereafter I was promoted to E-4 and called a Buck SGT. While at Craig I played a lot of handball in my off hours and sometimes with the wing and base commanders. I think this paid off when time came for the next promotion.

Craig was a great duty assignment, but then with 14 months left on my enlistment I received orders for Pleiku AB, Republic of South Vietnam. Before leaving for Vietnam, I had to qualify again with the M16, this time I qualified expert. I was a little more motivated. I took a leave before shipping out. I wondered what lay ahead, what would it be like, what would be expected of me? The anticipating was the worst part.

Vietnam in 1968-1969 was a very active place to be. I said goodbye to my family wearing my Class A blues, and left to report to Pleiku AB via Chicago O'Hare and Tacoma and McChord AFB, Washington. We flew over the Cascade mountains and had a stop in Japan with a good view of Mount Fujiyama, man was it cold! Next stop—VIETNAM!! We landed at Cam Ranh Bay on the coast of the South China Sea. Got off the plane in my winter Class a wool blues and **BAM,** I stepped into the heat and humid furnace, **Welcome to Vietnam!**

I changed into fatigues and was on my way to the central highlands and Pleiku AB. I got there midday and reported to in/out processing and the CBPO (Central Base Personnel Office). The clerk looked at my records, looked up at me and said, "Wait here." Great, I thought, I have been on base 30 minutes and already I'm in trouble. He returned a few minutes later and WHOA, here comes SSgt Donald J. Casey, a friend from Amarillo AFB. He was expecting me. SSgt Casey would become my NCOIC and close friend. We shared a cubicle in the barracks, I had the upper, he the lower bunk. There was another friend from Craig AFB. So with Casey and Sgt Harris, I felt a little more at ease and the transition to Vietnam was fairly smooth.

We worked 8 am to 9 pm M-F and a half day on Saturday and usually had Sunday off. For the first week or so, we would spend time on the bunker watching the war at night from afar. The Cobras, the C-47 gunships, and small arms tracers. We were close to artillery hill and I had trouble sleeping thru artillery fire. I would say, "Casey is that incoming?" He would say, "Outgoing." I'd say, "How do you know?" He would just reply, "You'll know."

Then one morning I went down to the community toilets and showers to get ready for work. **KA-WHAM!!** We took an incoming 122mm rocket and off to the bunkers I ran in my flip-flops, trailing a towel behind me. When I got to the bunker, Casey was there with my flack vest and helmet. He very calmly said, "Incoming." After that I could sleep thru all the outgoing artillery and like Casey said, "You'll know."

We worked long and hard hours in the office, and I was still pulling KP and filling sand bags. Pleiku AB had a decent chow hall, hot water, potable water. Pleiku AB was in the middle of a complex including the 4th ID Transportation, 71st Evac Hospital, Camp Holloway, MACV II HQ. The AF had some pretty good amenities and we knew it from the Army grunts who would come on base, grab some hot food, take a hot shower, sometimes with their uniforms on.

Most time was mundane, filled with work, however, we did get incoming 122 mm rocket attacks, usually in twos or threes. Fired off bamboo tripods, they were not too accurate, but were very destructive if they hit something. When under assault, we grabbed our flack vest and helmets and went to the nearest bunker.

Only a few days really stand out as far as enemy action is concerned. One day we took multiple rockets attacks with only moderate damage. That night a gale force storm blew through and tore up many buildings and a hangar or two. That was the most damage the base had.

Another was the night a VC recon patrol was spotted trying to infiltrate at MACV HQ, an old French plantation home. They fought a firefight and 13 VC were killed. That night was the only night I remember having to pick up my weapon and go to an assigned defense position.

The 633rd Combat Support Group at Pleiku AB mainly flew the Douglas A1E Skyraider. It was built for WWII but got to the show late. It was, however, perfect in the central highlands of Vietnam. The Navy flew it early in the Vietnam War, and one of our AVVBA members even shot down a MiG-17 in a Navy A1E. The AF used it for SAR (Search and Rescue), interdiction in the mountains. It could carry a lot of ordnance, linger over station a long time, and had great maneuverability. The officers who flew these planes were a very brave group of warriors and they were very dedicated.

The Air Security forces who guarded the base provided us with security 24 hours every day. They sprayed Agent Orange on the perimeter to defoliate trees and brush. We also had pilots who flew the C-47 gunships from dusk till dawn. They flew them in a three mile circle over the base, keeping watch. Most nights nothing happened, but if there was suspected VC action, a flare would be dropped to illuminate the area, if contact a red flare was dropped. The gunships, called either "Spooky" or "Puff the Magic Dragon," were armed with Gatling guns that were very effective.

The time at Pleiku was filled with long work hours, heat and humidity and rain. The monsoons were a season of rain and I saw it rain for a week solid of drizzle to driving rain. So many people who served in Vietnam have their memories of rain, heat, rain and more heat along with the engagements with the communists. (I hate commies)

I came home and had often wondered just what did I contribute to the war effort, I did what I was ordered to do which was work in an air conditioned office punching, sorting, and filing IBM cards full of information on airmen and officers in support of the war effort....but still?

Now for the rest of the story. After high school as I was attending technical school I worked for my stepfather in a small machine shop part time. The shop specialized in government contracts of different things. One of my jobs was to work on a wrench, it was closed box end on one side and open end on the other. There was a swivel reamer attached to it along with a standard screwdriver blade on it. The job required doing the same monotonous actions one at a time over thousands and thousands of these wrenches. All I knew at the time was it was a wrench for NATO forces. It was only after Vietnam that I learned that wrench was part of the maintenance tools for the M60 machine gun. That boring, monotonous seeming nothing job was actually part of something bigger and more important. I look back now and hope that work I did to assemble that wrench saved some Marine, Soldier, or Airman's life.

I left the USAF in January 1970 and went on with my life, only following the war with a distant curiosity and it was apparent the USA had no stomach for reaching out and taking the victory the military worked so hard and sacrificed so much for. The victory was ours so many times and each time we refused to see it, TET and later Linebacker II were victories for us and for South Vietnam, but when those were at hand, all that the American people wanted was just getting out. Bring our troops and POW's home and forget about why we were there in the first place.

In the end, the Soviets and the Chinese Communist were better allies to North Vietnam than the U.S. Congress was to South Vietnam.

DONALD H. NAU

US Army — Armor/Infantry Officer
Dates of Military Service (Active Duty and Reserves Combined): 1967 to 1974
Unit Served with in Vietnam: HQ Company, 3rd Combat Brigade, 25th Inf Div
Dates in Vietnam: Aug 1969 to Aug 1970
Highest Rank Held: 1LT
Place of Birth and Year: East Cleveland, Ohio — 1944

"THE WORLD" & VIETNAM — THE LORD IS KIND TO YOU TODAY

The only observable difference between the wet season and the dry season in Vietnam was — it just didn't rain as much in the dry season. More accurately, there was a wet season and a wetter season.

There was no time between lightning flashes and the deafening concussion of the thunder. At night, the continual conflict of blinding light to absolute darkness made one expect to see Satan emerge after every ear pounding crash.

I was always reminded of our basic training Drill Instructor who, it appeared, was in love with the harshness and raw existence afforded to the Infantry. On one particularly hard march in a relatively light rain, through mud, carrying full gear, wearing our ponchos, we were sweating profusely, being soaked on the inside as well as on the outside.

As we double-timed, we also sang cadence as if we were really pleased to be there. When we thought it couldn't get worse — it, of course, did! As my father would say while we watched a magnificent summer thunderstorm from the old farmhouse, "Son, the heavens have opened up!"

Upon the command of "PLATOON HALT!!," the DI bellowed, "You troops are privileged today! You people are dirty! You people smell! You people are lacking mental direction! You people are wanting the rain to STOP! But, what you do not understand is this rain is washing the mud off you! This rain is washing the sweat off you! This rain is refreshing you!"

But then, even louder — "THE LORD IS BEING KIND TO YOU TODAY! AND YOU DO NOT KNOW THAT! AND YOU DO NOT APPRECIATE THAT! SO WE WILL STAND HERE AT ATTENTION... AND THANK THE LORD FOR THINKING ABOUT US ON THIS BEAUTIFUL DAY!"

We remained there during the downpour — Thanking the Lord.

At first light after one particularly fierce monsoon night storm around Cu Chi, Vietnam, the DI proved to be right. We took no casualties — no incoming rockets or mortar rounds — no choppers shot down and there were no sappers in the perimeter wire.

The Lord *was* kind to us — and, we were Thankful!

CHRISTMAS TIME

The cease-fire was fitting although seemingly ridiculous when first ordered. After all, weren't we fighting for "Peace on Earth—Good Will Toward Men?"

That Christmas, so long ago and so very far away, is still remembered in vivid and prolonged detail though I cannot recall the exact date—no matter.

There were no families, no wreaths, no stockings hung on the chimney with care—no chimneys, no hot chocolate, no presents nor anything else that remotely resembled Christmas. The Chaplains did have a "Think Snow" sign near their hooch.

For to be without the surroundings of the Holiday season brings forth from within one's true spiritual beliefs. And so it was for us. We did have it all; we had prayers, we had promises and we had faith.

Throughout the years, I have, at times, become so occupied with life's demands that prayers would be pushed back only to reappear even stronger with the true meaning of Christmas.

Christmas with the 25th Infantry Division in the Republic of Vietnam did not have the sound of bells. Even the temporary chapel did not have bells. There is something about bells at Christmas. We DID have the snapping sound of helicopters on the perimeter circling like hawks—not reindeer. We were not angry with them for they gave us Faith. So we prayed and we made promises.

Bob Hope and his show—and we—laughed and laughed and laughed. Then we sang Silent Night and went back to the war.

That indelible Christmas, demanding and rewarding, is not longed to be repeated, for the best Christmas will happen this year.

At least, I can promise—I can Pray—I have Faith.

AUGUST 1969 *(imagine a split tv screen or two simultaneous radio broadcasts)*

Woodstock : got a ticket man peace live wherever man
25th INFANTRY DIVISON: YOUR ORDERS, SIR—REPORT AT 0500 25th INFANTRY CU CHI
the crowd is heavy man all around us man groovy
CHARLIE IS AT THE PERIMETER!

hey man they've pushed down the fence — cool

CHARLIE AT THE WIRE REPEAT VC AT THE WIRE!!!

They want us to pay man

THEY'RE IN THE WIRE THEY'RE INSIDE DAMN THIS PLACE!!

wow man someone's passing out up there cool stuff

I'M HIT!! I'M HIT!! I'M HIT!!

man can we get someone to look at her

MEDIC!! DOC!! DOC!! MEDIC!! DAMN!!

listen to the sounds man the music the vibes

WHERE ARE — WHAT IS GOING ON I CANT SEE I CANT HEAR!!!

cool man she needs to go to maybe like to a tent

THIS IS 6 NEED DUSTOFF! DUSTOFF! LZ CLOUD! HOT!! HOT!!

can she go to first aid or something man

RED SMOKE OUT!! WE'LL COVER!! WE'LL COVER!! COME ON!!

where'd she go we can't find her listen my favorite song man

DUSTOFF — 12th EVAC!! GOT EM!! TAKING FIRE!! GONE- OUT!!

man there are dudes stuck in the fence

THEYRE IN THE WIRE!! FIRE!! FIRE!! FIRE!! FIRE!!

hey man maybe we can help tomorrow be cool man

FIRST LIGHT CHECK THE WIRE WATCH FOR GRENADES

i guess she was ok saw her dancing far out

SEARCH AND CLEAR WATCH FOR SATCHELS BE ALERT

those who were here will have their own album

THOSE WHO WERE HERE WILL HAVE THEIR OWN WALL

THE PARADE

I am a Veteran in the parade

Why am I here and not afraid

Either one isn't clear
As people cheer
The Veteran in the parade
As kids we watched the parade
And, our parents found a spot in the shade
We stayed in the sun
Having so much fun
As those Veterans marched in the parade
Now we are so young
Barely twenty one
As we follow the high school band
Behind is the queen
Who remains unseen
But we continue to march in the parade
We passed city hall
That started it all
When we raised our hands to pledge
Our allegiance was pure and it was for sure
We would never just *watch* a parade
We are not here for us
But for Barnes, Green, and Gus
And those we never knew
We lost some friends and memories won't fade
Of those — not in the parade

JACK HORVATH

Branch of Service and Job: US Army—Armor and Transportation Corps
Dates of Military Service (Active Duty and Reserves Combined): 1961 to 1988
Unit Served with in Vietnam: 64TH Transportation Co, 8th TC Group; 54th TC
Bn, and 124th TC Bn, 8th TC Group
Dates in Vietnam: Jul 1966 to May 1967; Jul 1969 to Jul 1970
Highest Rank Held: Lt. Colonel
Place of Birth and Year: Cleveland, OH—1939

GUN TRUCKS PROTECT OUR CONVOY DRIVERS

My first Vietnam tour had been in 66-67, as the Company Commander
of the 64th tractor trailer company, in the 27th Truck Battalion, of the
8th Group, near Qui Nhon. My second tour was in 69-70, as a Battalion
Executive Officer for the 54th Truck Battalion also near Qui Nhon, and
as Executive Officer for the 124th Truck Battalion in Pleiku.

Having the 124th Truck battalion at Pleiku provided needed facilities for the Qui Nhon Highway 19 convoys which had not previously
existed. We now had maintenance shops to fix any problems which had
occurred on the road. We had a base of operations from which to serve
the supply depots, the Pleiku US Air Force base, the 4th Infantry Division base at Camp Enari, the Camp Holloway US Army Helicopter air
center, the artillery firebases, Kontum, and Dak To cargo customers, and
the highlands in general. Now when our drivers had to remain overnight
in Pleiku, they had a temporary visitor barracks in which to sleep and
had the four truck company mess halls in which to eat.

There were a number of notable changes in the two years between
1967 and 1969. The enemy had conducted intense ambushes along
Route 19 from September 1967 through September 1968. Therefore,
immediate truck unit security escort for the cargo truck mission was
needed. The 8th Group Commander, Colonel Joseph Bellino, encouraged the building of a new type of convoy escort security vehicle, which
soon became known as the "Gun truck."

Volunteer crews from each truck company built up each of their gun trucks from a normal cargo vehicle, the M54 five ton truck. Each gun truck had a crew of four volunteers, a crew chief, and two gunners in the back gun box, and a driver. There were several machine guns in the box, a crew radio link between the chief and the driver. There was all manner of other gun hardware, ammunition, and supplies carried on each gun truck.

Although none of the gun truck equipment was "authorized," each crew managed to be fully operational before going on to the road. For two years from 1965 to 1967 we had run convoys with about one hundred trucks and three jeeps, now things were completely changed. Convoys became thirty cargo trucks, with three gun trucks to each convoy. Our 8th Truck Group had 12 companies, each company had 60 vehicles authorized, and each company had five gun trucks. This meant that this was about a ten percent loss of load hauling capacity, but there was really no alternative. Although the ambushes continued after September of 1968, their number was very much reduced from what had been true in the previous year.

The building and use of gun trucks spread quickly throughout Vietnam during the years from 1967 to 1972. Over 500 trucks were known to have been built by almost every type of unit. They were used for area defense and convoy defense and anywhere that quick and dependable firepower was needed.

As a result of the success of the gun trucks, the 8th Group became the "showcase" transportation unit in Vietnam. This meant that my 54th Truck Battalion, near Qui Nhon, was chosen to host an 8th Group breakfast and command briefing for the visiting military and civilian dignitaries on a regular basis. These VIPs would have breakfast at 0600, complete with china and linen on the tables. After breakfast, an 8th Group representative with flip charts would explain the Group's entire day of convoy activities. The visitors would then go out to the nearby convoy truck marshalling area and listen to a convoy commander's driver briefing, and watch the gun trucks test fire their weapons at the range beside the road. Soon the convoy trucks would roar out to the west on Highway 19 on their way to Pleiku.

The VIPs would then visit the trailer transfer point operations office, and also see the activities at the Group consolidated trailer maintenance building. All of this represented visibility at the highest levels. The convoy gun trucks had gained a successful and widely known reputation.

Timothy J. Kutta wrote the book "Gun Trucks," and James Lyles wrote the "Hard Ride: Vietnam Gun Trucks."

Timothy Kutta noted that the 8th Truck Group statistics for an undefined period was as follows:

Ambushes: 36
Mining incidents: 65
Sniper incidents: 65
Bridges blown: 18
Other incidents: 39
US KIA: 38
US WIA: 203
Enemy KIA: 104
Enemy WIA: 10
Enemy POW: 5
Vehicles damaged/destroyed: 287

In January of 1968, on Highway 19, convoy commander 1st Lt David R. Wilson was involved in an ambush where he died when an enemy mortar round hit his jeep. He was later awarded the Silver Star. In the midst of heavy enemy fire he was maneuvering in the kill zone to protect his cargo drivers and trucks.

In February of 1971, on Highway 19, PFC Larry Dahl, a crewman on the gun truck "Brutus," was involved in an ambush. He was later awarded the Medal of Honor. He saved his other crew members who were in the gun box when he threw himself on to an enemy grenade which had been tossed into the gun box.

In the spring of 1971, a major two and a half month operation named Dewey Canyon/Lam Son 719 took place in the area of Khe Sanh. Fourteen US Army truck companies hauled cargo to support the Army of

Vietnam with 5,000 loads of supplies. US gun trucks were the only convoy security available on this extremely dangerous route. There were 23 ambushes, 12 killed, 35 wounded, 40 trucks damaged or destroyed. All of the 12 who were killed were gun truck crewmen, and no cargo truck crewman was killed.

After the drawdown of US forces, the enemy chose the An Khe Pass area to close down the use of Highway 19 to US cargo.

I finished my second Vietnam tour in the middle of 1970 in the highlands as the Executive officer of the 124th Truck Battalion. There was a large operation across the Cambodian border by the 4th Infantry Division. I had a two week assignment as the 8th Group truck control officer at the 4th Division forward logistic support base. With a jeep, a driver, a small tent, some pierced steel planking, and a large Transportation Corps flag, I handled the turnaround point for the dozens of trucks which daily brought in supplies for the operation. I set up a field shower by mounting a small mess hall water tank trailer on to the bed of one of our flatbed trailers. It was very satisfying to be able to ensure that every one of our trucks, trailers, and drivers were able to deliver the cargo to the customer and to return to the truck battalion every day.

Two remarkable gun trucks have survived and are now on display at the Ft Eustis, VA, Transportation Corps Museum.

The Eve of Destruction, an 8th Transportation Group vehicle, was the only original Vietnam gun truck which came to the United States. She was brought back by ship. In a humorous scenario, when she arrived at Norfolk, she was offloaded on to the dock. She had her full load of ammo cans which formed the normal floor of her gun box. These ammo cans were filled with LIVE BELTS of machine gun ammunition, a situation which was absolutely prohibited for vehicles during ship transit.

Another gun truck on display at the Transportation Corps museum is the Ace of Spades. She was brought back from Iraq. She was named after a well-known 8th Transportation Group Vietnam gun truck. She was developed in 2004, with a special gun box which was requested by Congressman Duncan Hunter from the Livermore Laboratories. Thirty of these new gun trucks had modern Livermore anti-ballistic gun boxes.

They were very effective in Iraq, but no gun truck design was adopted for regular use.

MIKE HAMER

US Marine Corps — Squad Leader; US Army — Infantry Officer
Dates of Military Service: 1956 to 2011
Unit Served with in Vietnam: B Company, 3rd Battalion, 12th Infantry Regiment, 4th Inf Div; Ops Advisor to I Corps Vietnamese Training Center
Dates Served in Vietnam: Sep 1966 to Sep 1967; Dec 1968 to Dec 1969
Highest rank: Major
Place and Date of Birth: Chicago, IL — 1938

"LET'S BUILD A SWIMMING POOL!"

During all of 1969, I was a member of Military Assistance Command (MACV) Team 5 in Phu Bai, about 10 miles south of Hue. Phu Bai was 24th Corps Headquarters and all its elements, about 5,000 troops. MACV Team 5 lived in a French-built house just outside the north gate to Phu Bai on Dong Da National Training Center, a Vietnamese installation for basic training, about 2,000 Vietnamese soldiers.

I was the Operations Advisor to the Vietnamese organization staff. One day in our Team 5 Lounge, the Operations Sergeant and I were enjoying a beer and one of us; I cannot remember who, said, "Let's build a swimming pool!"

We talked with our Logistics NCO, a master scrounger, and drew a plan on a napkin. Pool would be 40' by 20', at one end 3' deep, the other 12' feet deep. We had a tennis court in our big yard and plenty of room for a pool. The water table was very high here and we knew we could dig a shallow hole and use the dirt to shore up the sides of the mostly above ground pool.

The 101st Airborne Airmobile Division was at Camp Eagle about four miles north of Phu Bai.

We flagged down a 101st low boy truck on Hwy 1 with a front loader and asked the driver if he'd go for a hot shower and meal. He responded, "Hell yeah, what do I have to do?" We had a hole! It took the kid 20 minutes to do what we needed.

Over the next six weeks we scrounged enough plywood to build the pool walls. Phu Bai had two Seabee plants, one for asphalt; the other for cement. We made a deal with their officers; they could play tennis at our place and they'd give us concrete not up-to-building-standards. They poured for us two or three times a week until we had it all in. Our Team had a Vietnamese mason who smoothed the floor and walls.

We scrounged the packing material that came with various military equipment, stuffed a 55-gallon drum and built a drain in the pool. We hooked up a small generator and presto: we had a pool filter! When ready, we tore down the plywood and filled the pool from a nearby small water spigot, which took five days. Before we finished filling, a Captain from Phu Bai Water Works came by and said, "Looks like I found our water drainage source. No problem, we have plenty of water, just tell us next time." We told him he could use the pool anytime.

We then had a Captain from Preventive Maintenance come out and put in chlorine, measure bacteria, and tell us he'd come out weekly. He was also invited anytime. The Seabees built a concrete patio at one end and sidewalks all around. We enjoyed that pool immensely, particularly after work in the evening when the temperature went down.

Our house was situated at the north end of the Phu Bai Air Field's 5,000 foot runway which was very busy daily. We'd watch aircraft on their final approaches about 300 feet above and always we saw the wings waggle when the pilots saw our pool

I understand in 1975 the NVA took over the area, drained the pool, and stored POL in it. What a waste!

PACHYDERMS!

At the airfield near our MACV Advisory Team was the 101st Airborne Airmobile Assault Support Helicopter Company, containing CH-47

Chinooks. Their nickname was Pachyderms, each CH-47 had an elephant painted on its nose!

Our Team 5 house had a great lounge (Bar) where one day the Pachyderm company commander and I were having a beer and he said to me, "You MACV guys say you can scrounge anything; we need a baby elephant for a mascot, can you do that?" I said I'd work on it. I had a dear friend who was an advisor with a Vietnamese unit near Pleiku who had told me earlier when we were in Saigon together he knew of an Air Force sergeant in Thailand who was selling baby elephants. Did I need one? I said, "What would I do with an elephant?"

Later, The Pachyderm Company Commander was in our Team 5 lounge having a beer when I told him I thought I could get him a baby elephant. Did he still want one?

He said, "Absolutely!" I called my buddy in Pleiku and was told you can pick it up at an airbase in Thailand. We went into Phu Bai to the veterinarian's office; however, on the way we ran into my brother-in-law, Captain Johnny Herrington who had served in Korea in the late fifties, was now in USAR and had been called up for Vietnam, trained as an Ordnance Officer and was now OIC Of the Phu Bai Ammo dump, second largest next to Long Binh. I saw him every week until he DEROSED.

The Pachyderm Company CO and I then talked with a young Captain in the veterinarian's office who said, "Sure, sounds like an adventure!"

We put him in a Chinook and flew him to Thailand where he tranquilized a baby elephant. The Air Force Sergeant was paid $200 and the elephant was slung under the aircraft and brought to Phi Bai and tethered by the operations shack near the tarmac where all the unit members had a great time with their elephant.

About a week later, the unit CO was at our house and said, "Mike, you gotta take that elephant back! I have two enlisted soldiers taking care of it all day every day! I cannot afford to give them up full Time!" So back to the veterinarian's office we go to ask the now laughing captain for a return trip. He tranquilized junior who was flown back to Pleiku where the Air Force Sergeant from Thailand was waiting. No refund! The pachyderm unit then declared, "No mascot for us!"

Sure was fun while it lasted.

Later, the Pachyderms flew support for the Hamburger Hill battle, flying artillery pieces and ammo for three days.

PETE MECCA

US Air Force — Intelligence
Dates of Military Service: 1966 to 1970
Unit Served with in Vietnam: 460th Tactical Reconnaissance Wing
Dates in Vietnam: Sep 1969 to Sep 1970
Highest Rank Held: E-4
Place of Birth and Date: Memphis, TN — 1947

I DID MY DUTY

After basic, I was sent to Denver, CO for Intelligence school. For over four months we were taught mission planning, photo Intelligence, map reading, coordinates, Communist tactics, Soviet and Chinese military hardware, code breaking, dot dot this and dit dit that.

From there I was assigned to McCoy AFB in Orlando, FL, home to the 306th Bombardment Wing of the Strategic Air Command, which meant supporting Cold War missions for the deadly, yet beautiful, B-52s. On my first night at McCoy, I walked about a mile from the barracks down to the flight line. The big bombers were undergoing maintenance tests, refueling, repositioning, and high-pitched engine run-throughs on a flight line lit-up like a ball park. With several B-52s revving all eight powerful engines, nearby structures vibrated as did my eardrums. The noise was deafening. I loved it.

Part of my responsibilities including drawing radar predictions for B-52 low-level terrain avoidance missions. I could do so with special gizmos that predicted the shadows and/or radar returns using the topography of classified maps. The radar scope predictions looked a lot like bowls of spaghetti. The neat thing was, to become familiar with radar scopes plus compare my predictions to the real depictions, I was authorized to fly as an observer on the bombers. Flying on B-52s, authorized to wear a cool flight jacket on base, and toting a misspelled fake I.D. in my wallet at nineteen years of age… it just don't get no better!

Flying aboard a B-52 is beyond awesome; the noise, the majestic power and size, the almost unrestricted capabilities. I sat directly behind and between the pilot and copilot in the V.I.P. seat. The pilots implied that I was indeed a V.I.P.—Vaguely Intelligent Person. The shortest route to the Soviet Union is over the North Pole. We probed their defenses and they tested ours; we feigned crossing their border and they intercepted us, such was the Cold War. We practiced terrain avoidance bomb runs over Canada with, I assumed, permission from the Canadian government. American fighters would 'intercept' us as we struggled to 'evade' them over American airspace. The fighters always won. Then orders finally came down for Southeast Asia. No big surprise, I had volunteered.

Nakhom Phanon (NKP) Royal Thai Airbase was hot, dusty, tropical, yet noticeably improved by the American military. Positioned in extreme northeastern Thailand, a thin bottleneck of Laotian territory separated us from North Vietnam. Slow-as-Christmas Cessna O-1 Bird Dogs and OV-10 Broncos provided eyes in the sky for A1E Skyraiders and the WWII vintage A-26 Invaders, plus relayed strike coordinates for the fast-movers (jets) from other locations. Huge Jolly Green Giant and Huey choppers were utilized for rescue or special ops missions in Laos. Laos was a war that didn't exist, a clandestine interdiction campaign denied by the USA and North Vietnam. Men died in this war that didn't exist, they evaded capture when shot down, they suffered wounds, became POWS, saw a buddy perish in a ball of flame while bombing a worthless target. These men were the unknown heroes of the Vietnam War. Yet they remained silent, loyal, and soldiered on. The Paris Peace Accords DID NOT account for any American POW held in Laos.

The North Vietnamese infiltration network was called the Ho Chi Minh Trail. In reality, the 'trail' was a well-developed road system with truck parks, POL (petroleum, oil, and lubricant) dumps; fuel pipelines made of bamboo, ammo dumps, and rest areas. Although primitive by modern standards, the dirt roads were highly efficient and almost impossible to interdict due to our crippling rules of engagement.

I worked out of a Top Secret location identified by several goofy names which most likely confused non-essential personnel. We weren't James Bond but in a weird sort of way did have a license to kill. Flash Gordon, Batman, and the folks from the House of Cards would have been proud. I never discussed our location or duties until a few years ago, but that short-lived slip of the tongue came to a halt when I discovered via new satellite imagery that the location is still active. The Ho Chi Minh Trail is now an international tourist trap and of no military significance, so any explanation of the current activity is beyond my paygrade.

After 18 months at NKP, my next port-of-call was Tan Son Nhut AFB in Saigon, Vietnam. No big surprise; I had volunteered. I was scheduled for an early morning shuttle flight to Vietnam aboard a C-130, but I was up most of the night playing poker, fell asleep, overslept, and missed the flight by five minutes. NKP did not have a big terminal, but it did have a floor; I had my duffle bag which became my pillow, and the floor became my bed. Falling fast asleep, I put faith in the scheduler to wake me if another 'hop' made a pit stop at NKP.

Someone started kicking the soles of my combat boots. Impolite S.O.B.! I sprang to my feet to confront the 'boot kicker' and came face to face with my boss, Brigadier General Willie McBride. He was not the least bit intimidated by my agitated expression. "What are you doing here, Mecca?" he asked. After explaining I was en route to Tan Son Nhut but had missed my flight, the General asked, "Why?" The excuse of an all-night poker game was not the answer he sought. "I'm not talking about missing your flight, sergeant," McBride shot back. "Why are you going to Vietnam?" That I had volunteered brought forth a laxer tone. "Have you been home to see your folks?" A 'yes, sir' gained me a ride into the war. "Well, grab your gear, Mecca. I'm on my way to MACV in Saigon. You can ride with me." Cool. Instead of a lumbering C-130, I boarded General McBride's personal Lear Jet.

I was the sole occupant of a nicely-appointed cabin. McBride and his pilot sat behind the controls in the Lear's cockpit. Flying to war via a Lear Jet piloted by a one-star General and his pilot was a war story to tell future Rug Rats and Grand-Rug Rats, which I've now done, but two generations of family Rug Rats still think Poppa Pete is full of it. The fondest memory I have of the flight was General McBride coming back into the cabin. He sat with me for at least 20 minutes, asking questions about my family, my plans for future education, if I planned to stay in the military, offered advice, and if ever requested, offered a written recommendation for any endeavor. I will always remember the fatherly General who spent time with the lowly sergeant.

Tan Son Nhut AFB in Saigon, Vietnam—460th Tactical Reconnaissance Wing—I was put in charge of map layouts and plotting of every mission flown by the 460th. I was given the unrestricted freedom to correct a major problem with the incompetence and apathetic attitudes within the plotting team. Easy enough. I cleaned house and replaced the ineptness with fresh faces and better mindsets. My team made no mistakes for the entire year I was stationed at Tan Son Nhut. We lost one pilot due to catastrophic engine failure, not to a plotting mistake. In mid-1970, my team stayed up three days and two nights planning and plotting missions for the Cambodian Invasion. Enemy rockets or mortars hit Tan Son Nhut ever so often just to keep us on our toes, but we remained relatively secure in one hell of a nasty war.

When my tour was up in September of 1970, the Air Force tried to pin a medal on me but I refused to attend the ceremony. I was told to show up, or else! I chose the 'or else' option. The next day a young captain I worked with came in and slammed the medal box on the plotting table. "Not one word, Mecca, not one word!" So I took the medal I didn't want. I opened the box 10 years later to see what I had tried to refuse.

My hard-headed attitude concerning the medal was a personal choice based on what I had done and survived versus what other guys went through and paid such a heavy price. I lost school chums, I lost friends; I'd given a left-handed hand-shake because a buddy lost his right arm. Barry lost his right eye, Mike lost use of his left arm, a sniper's bullet paralyzed Steve from the neck down, a few lost their faith, a few lost fingers, a few lost toes, a few lost their minds. Dan has never lost his thousand yard stare, Joe relives Nam every night in night sweats and nightmares, and Agent Orange keeps on killing. Our countrymen said we were baby-killers; now they call us heroes. We are neither. We were just another generation of patriots who answered the call to duty but had to do their jobs under impossible rules of engagement, died for ground given back to the enemy after the battles were over, or blown out of the sky hitting insignificant, but politically correct, targets. That, was our war.

I write with humor. Perhaps that's my method of affirming that after Vietnam, every day is gravy. There are 58,267 names on a long black wall in Washington, D.C.; those are the heroes, those are the men and women who deserve medals. But the rest of us? We were just lucky.

And that's my story.

ROGER SOISET

US Army — Infantry Rifle Platoon Leader

Dates of service: 1968 to 1974

Unit Served with in Vietnam: Bravo Company, 3-7 Infantry, 199th Light Infantry Brigade

Dates Served in Vietnam: Aug 1968 to Aug 1969

Highest rank: 1st Lieutenant

Place of Birth and Date: Santa Monica, CA — 1946

A LONG TIME COMING

After graduating from The Citadel in 1968, I arrived in Vietnam in August 1969 and was assigned to the 199th Light Infantry Brigade. As a second lieutenant, I found myself as 1st Platoon Leader with Bravo Co., 3/7 Infantry Battalion. Those four months as a rifle platoon leader were the most intense of my life.

A curious feeling of joy and guilt hit me when I was given a "rear" job on January 1, 1970 as the Company XO. Eleven days later, I wondered if a rear job was all that safe, having to run convoys on the average of twice a week and having been slightly wounded in a rocket attack on the Brigade Main Base at Long Binh. And just as quickly, I realized that anything in the rear was better than humping in the jungle as I had done the last months of 1969.

But then tragedy struck "my" old platoon on January 17; my replacement as Platoon Leader tried to flush out some enemy troops with a "cloverleaf" maneuver, sending half the platoon out in opposite directions with instructions to execute right angle turns every fifty meters until either they came back together or found the enemy who was nearby—this all started when that enemy force was heard making breakfast around 0600. One element was ambushed by the enemy force, having no doubt been heard moving through the jungle. Result was one man fatally wounded—PFC Lynwood Thornton, aka "the Bear"; and another badly wounded—SP4 James Thonen. Thonen was pulled up to a hovering medevac via jungle penetrator but died before reaching a hospital. A second medevac refused to pick up Thornton, saying they only picked up wounded.

The new Lieutenant had become catatonic and the RTO, PFC Harold Jantz effectively took over running the 16-man platoon (now down to 14), calling in artillery, and arranging for a pickup for the whole platoon at an LZ about four kilometers away. According to a 1st Cav gunship, enemy troops were swarming their way and they needed to be extracted. Jantz was told by the Company Commander to leave Thornton for later pickup, and he refused that order. He handed off his rucksack, rifle, and radio and proceeded to carry Thornton in a fireman's carry. This exhausted him after a few hundred meters and the platoon made a carrier out of ponchos and branches to carry Thornton to the LZ.

Fast forward to 2013.

Richard Hearell was one of the founders of "Bravo Buddies," a veterans' group composed of men who had served in Bravo 3/7 over the four years the brigade had been in Vietnam. He was the originator of a collection of studies of significant firefights the company had been involved in while in-country, and I was gathering statements from men I had found since 1993 when I published a book on my year in Vietnam.

I had tried to find the men I had served with to give them a copy, and had found about a dozen of them. In the course of this project for Rich Hearell (the collection is stored at Texas Tech University in Lubbock, Texas) I came across an amazing statement from one of my men — that Lynwood Thornton had been awarded a Purple Heart but had not been given a Bronze Star; the other KIA that day, James Thonen, had received both. The circumstances of their deaths were almost identical, and it was a fact that in my experience, infantrymen killed in action in 1969/70 were always given a Bronze Star for Valor. So I began my quest to correct this shortcoming.

DD-638 submitted to DoD was rejected due to errors (I half expected this as I had found the forms inscrutable) but it was my hope that the rejection would guide me in how to do it correctly. Wrong. So I sought help from "lifers" who were still in the service, and tried again. Rejected again. Tried sending it through my U.S. senator. He died. Tried the other senator. He did very little other than transmit my DD-638 to DoD. Turned down again... and again. Tried to do it through my U.S. representative; he declined to help because while I was in his district, the deceased was not.

So I tried the representative for South Georgia where the Thorntons live; he turned me down because I was not in his district. At this point, five years had gone by due to the time involved in getting statements, signatures, waiting on a response from DoD, etc. I said to hell with it and told the Thorntons it did not look like it would happen, and found a great resin figurine of a soldier carrying another soldier as to embody the spirit of "Leave no man behind" and sent it to Harold ("Harry") Jantz.

And then I got a surprising email from Jantz's sister Paula who was married to a sergeant-major. She had never heard what her brother had done in January 1970, and my figurine had opened that door to the unpleasant past. Paula Beckman and her husband Tom started working with yet another senator, Lindsey Graham. Things started to actually look promising, and in May 2019, the Bronze Star for Achievement was finally awarded to Thornton and Jantz was given the Bronze Star for Valor.

Lindsey Graham personally pinned the medal on Jantz in the Senate Office Building on July 12 and I had the honor to tell the Thorntons that their son/brother/uncle/great uncle had finally been decorated. On July 27 in Thomasville, GA, four members of 1st platoon (RTO Harry Jantz, rifleman Robert Kenney, medic Richard "Doc" Uhler, and myself) joined the Thorntons, their friends, the local American Legion Post 285, the local VFW Post 4995, area veterans of the 199th LIB, and a bugler from Bugles Across America for ceremonies at the gravesite and in the Trinity Baptist Church. A flag-folding was conducted and "Taps" was played.

It is not possible to express how meaningful the above has been to me; it literally feels like a mountain has been lifted from my shoulders. As I told one of the Thornton relatives, the guilt feeling that started 49 years ago has been reduced. Not eliminated, but my regrets have been cut way back. The other members of 1st Platoon share that sense of relief, and from all the hugs and tears we received in Thomasville, I know the Thorntons feel the same. They finally know what happened on that day, and they know Lynwood died with men who loved him and still call him their brother. RIP, "Bear." The brother of Lynwood informed me that his "big" brother weighed a very solid 235 pounds, so carrying him was no mean feat.

BRYAN C.W. TATE

US Army — Infantry Platoon Sergeant

Dates of Military Service (Active Duty and Reserves): 1968 to 1974

Unit Served with in Vietnam: C Company, 2nd Bn/28th Inf Regt, 1st Inf Div;
HHC 173rd Airborne Brigade (Separate)

Dates in Vietnam: Sep 1969 to Jul 1970

Highest Rank Held: Staff Sergeant E-6

Place of Birth: Toronto, Ontario, Canada (U.S. Citizen by Birthright) — 1945

MY STORY — "AN AMERICAN SOLDIER"

(Call sign Charlie 1-5)

I grew up in Dayton Ohio and attended Ohio University. I graduated in March 1968 with a Bachelor of Science in Industrial Engineering. I received my draft notice in April, 1968.

I was very surprised that the Government could re-classify me from 2-S to 1-A so quickly, however with the TET Offensive in February 1968, it was obvious that the Army needed replacements. After receiving my Draft notice, I tried to join the Navy, Air Force and the Marines, but none would talk to me.

I reported for Active duty on July 19th to the basement of the Knott Building (Dayton) at 5:30 AM. My Mother was crying, and my Dad told me not to volunteer for anything. The Salvation Army was there with coffee and donuts, along with about 50 other draftees from Dayton. They put us on buses and drove us to Cincinnati (Federal Building) with two Ohio State Patrol Police escorts (I guess they did not want us to get off the bus).

After a very long day of in-processing and taking "Two Steps Forward" they flew us to Atlanta where we were transported by bus to Ft. Benning, GA. We arrived at 2:30 AM and were introduced to our Drill Sergeants who immediately made us do a "Police Call."

Georgia has 3-inch cockroaches and they told us to pick them up and put them in our pockets. I wanted to go home!

In our first formation the following morning, the Drill Sergeant asked for volunteers.

Remembering my Dad's advice, I stood silent. The guy next to me, Sweatt (everything being alphabetical) raised his hand. I tried to warn him about volunteering, but he was selected to be the Company Commander's Jeep driver, and would be driving the Jeep leading our "forced marches" in the deep sand. After Basic Training at Sand Hill in July and August, we went to Ft. Polk (who knew it could be worse than Ft. Benning?). Back to Benning for Jump school, Ranger Training and NCO School. I finished first in my NCO Class (300 candidates) and was promoted to Staff Sergeant E-6 and received orders for Vietnam.

As I was departing for Vietnam, my parents drove me to the Dayton airport. My Grandfather (Mom's Dad) came down from Canada to see me off. He told me I had not spent much time with him, and why didn't I come back to Canada with him until the War was over? It was a sincere gesture of love, but as a very patriotic young man, I thanked him and told him my Country had called and I was answering. Mom was crying and Dad shook my hand.

I landed in Bien Hoa at 12:00 noon (very hot and steamy after a torrential "monsoon" rain) and was sent to 90th Replacement for assignment. We were very "green" in our new jungle fatigues, and saw soldiers getting ready to return home after a year in RVN.

Their fatigues were faded almost white, and one guy had a pet iguana on his shoulder. I had the feelings of fear and despair as they called us "twinks," FNGs and other not-so-endearing names. My basic training buddy a couple of weeks ahead of me was assigned to the 1st Infantry Division, so I told the Placement Officer I would like to be assigned with him. I was assigned to Charlie Company, 2nd Battalion, 28th Infantry, 1st Infantry Division (Black Lions of Cantigny), a very storied Army unit going back to WW1.

I never did see my basic training buddy. They issued me a steel pot, flak jacket and a used M16. Since I was an E-6, they put me in charge of a convoy going from Saigon to Lai Khe through not-so-friendly jungle. Thankfully, we made it without enemy contact.

While sleeping in the NCO tent in Lai Khe, we got rocketed in the middle of the night. I got under my cot with my helmet on when the guy next to me said it was "outgoing."

How did they know? I would soon learn to tell the difference between "incoming" and "outgoing."

I arrived in Dau Tieng (2nd/28th Brigade rear) and was transported to Charlie Company at Fire Support Base (FSB) Kien. My Company Commander assigned me to 1st Platoon with a PFC (E-3) in charge (no Platoon Leader and now I was in charge). My first night we went on ambush patrol, put out Claymores, and slept on the ground during a pissing" rain. Welcome to Vietnam!

Our routine was one of Search and Destroy, later to be called Ambush Patrols (AP) which sounded better back home. We used "Eagle Flights" (I made over 50 flights which earned me two Air Medals) where my Platoon was picked up by four Hueys, flown around for about 20 minutes, and then placed in a designated location, from where we would "riff" for 5 to 10 kilometers through the jungle to our next AP location. We had to accurately report our locations (with a map and compass and no GPS) since the CP would plot all friendly locations and then begin shooting H&Is (Harassing and Interdicting) mortar and rocket fire at night.

My Company had three major Areas of Operation (AOs); the French Fort (an abandoned old location where there used to be French buildings; the Michelin Rubber Plantation (very beautiful but very dangerous— filled with "gooks"); and the Saigon River where we would support the US Navy PBRs (brown water Navy) trying to catch NVA infiltrators crossing the river at night. Agent Orange was sprayed on us on a regular basis to defoliate the jungle. It was mildly cooling when the mist came down from the helicopters. Forty years later I would be treated for an Agent Orange related cancer.

We would make contact with the Viet Cong (VC) about once a week, while less often with NVA (except for the Saigon river crossing experience). Our daily jungle experience was not just making contact with the enemy. It was red ants, spiders, scorpions, snakes and mosquitos (the size of B-52s), and "foul mouthed" lizards. We took malaria pills daily and the big red pills on Monday (that was the only time you knew what day it was).

Red ants would be on tree leaves and drop on you when you walked by. They would get down your neck and absolutely stop the War. Ruck sacks and web gear would come flying off and stripping down to get them off of you. Occasionally, we would be confronted with a "Chu Hoi" (a VC surrendering to an Army Platoon) who would jump up with his hands up, out of the elephant grass in front of our "point man" and scream "Chu Hoi."

An experience almost as scary as making enemy contact.

Our AO was also filled with VC tunnels. Although the 1st Infantry Division had "Tunnel Rats," none where assigned to our Company. When we discovered a tunnel complex, it was my job to "blow it up." We would receive a case of C-4 and I would go down into the tunnel to set the charges. My Company Commander issued me an Army .45 pistol to take into the tunnel since there was not room to maneuver with an M16. Fortunately, I did not engage "Gooks" in the tunnel, but I did see massive bugs, spiders, and snakes.

After many months of patrolling our AO, my Company Commander (CPT Ronald Stocker) told me that our Battalion Commander (LTC Richard Hobbs) had submitted my name for a "Battlefield Commission." He said the Division needed 2nd Lt. Platoon

Leaders and he had been impressed with my leadership. He said the Commission would take six weeks to get through Congress and then I would have a one-year commitment in Vietnam. Since I only had a few months to go in-country, I declined. I often regret that decision, but if I had accepted, I might have wound up as a name on the "Wall." There was a reason they needed 2nd Lieutenants.

Several months later, while on another AP, the Company Commander called to tell us we were going to get picked up the following morning at 0700 by a Chinook. He gave me no explanation but said it would be a "Hot Extraction." It was Charlie Company's protocol to let the three Rifle Platoons (1st, 2nd and 3rd) take turns returning to the

FSB for re-supply. This was so the first group back would get one of the few Cokes to drink instead of Winks and Grape Nehi (never any beer), and maybe find a container of water that had been sitting in the sun to take a hot shower. It was my Platoon's turn to go back in first. At 0700 I received a call from the helicopter pilot requesting that we pop smoke.

A Chinook is a big, dual prop helicopter that could carry my whole platoon (18 to 20 soldiers). We loaded up and took off. I asked the pilot where we were going, thinking it was going to be bad. Instead we landed at Lai Khe on December 22nd, to see the Bob Hope show. We were there all day just coming out of the field with no showers. The show was fantastic, and we got to see some "round eye" girls Bob had with him (the Gold Diggers). Too bad for the 2nd and 3rd Platoons!

Christmas 1969, we were on patrol when the CP called and asked if we wanted a ride to our next AP location. We had been walking all day, so I said yes. We got picked up by a Platoon of APCs and watched dozens of red and green flares going up in the distance by other Army units celebrating Christmas. I was really homesick!

January 18, 1970 was one of the worst days of my life. My Platoon was patrolling early in the morning. We had been assigned a patrol dog and handler, who I put on our right flank. Either my Point Man tripped a wire (unlikely since the dog and handler were very experienced on trip wires) or it was a command-detonated ambush. The explosion came from our right and killed the dog, the dog handler, rifleman Mark Tonti, and my Squad Leader, Edward "Corky" Witek, who died in my arms as our Medic ("Doc") tried to save his life. My Point Man, Bobby Jones was badly wounded and was "dusted off" to a MASH Hospital in Cu Chi, just across the Saigon River. We later visited him and learned he had lost both legs. He died a week later.

A few hours later, my Medic and I were filling out casualty reports when the Battalion Executive Officer (XO) came in on a helicopter to survey the scene. As he hovered about 25 feet above us, an enemy booby trap (155 mm howitzer round) exploded, and the chopper was hit and crashed within two feet from where we were sitting. The helicopter hit the ground so hard it knocked the Door Gunner's machine gun off its mount. The XO was killed. The Battalion Commander called in F-4 jets to engage the enemy. They came in behind us in a deep, screaming dive and it looked like they were going to hit us.

They dropped their ordnance (500 lb. bombs) with the "splash" going forward toward the enemy and large pieces of shrapnel "whizzing" above us cutting down trees.

A few weeks after the ambush, the 1st Infantry Division announced it was returning to Ft. Riley, Kansas. LTC Hobbs put me on the list of 30 soldiers (out of 1600) to go back with the Colors. I was very excited and wrote my Mom and Dad that I was going to come home early. Unfortunately, the 1st Division had a Change of Command, and the new Commanding General decided that only those with less than 30 days left in-country could go back with the Colors. I had to call my parents on a MARS line ("Working, Working") telling them that I would not be coming home early. Mom cried!

Since I was Paratrooper Qualified, and the 173rd was the only Airborne unit that was on "Jump Status" (requiring 90% of Battalion personnel to be "Jump Qualified"), I was reassigned to the 173rd Airborne Brigade (Separate) located in northern II Corps at Brigade HQ "Bong Song." I reported to the 173rd Airborne Brigade personnel officer who reviewed my 201 file and saw my combat experience including several medals (air medals and Bronze Star for Valor, Purple Hearts) and assigned me to Headquarters Company where I would become the NCOIC.

Although getting "out of the Field" was a great opportunity, life in the "rear" was marked by 122 mm rockets almost every night and "fraggings" by disillusioned soldiers. When our First Sergeant was fragged by a soldier, I was selected to escort (under arms) the perpetrator to Long Binh Jail (LBJ).

I returned to the United States (the "World"; the "big PX in the Sky"; the "Land of Milk and Honey") on July 18, 1970 on a Flying Tigers airplane out of Long Binh. As we waited to get on the plane, the "gooks" rocketed the airfield. We waited hours and finally boarded the plane in the middle of the night. As we sat on the plane waiting to take off, we all looked at the ceiling waiting for a rocket to hit us. Finally, the plane started to roll and when it lifted off, there was a very loud cheer! The same cheer would occur when we landed in Seattle-Tacoma airport.

Ft. Lewis warned us to get out of our military clothes because of the negative sentiment about the Vietnam War. Unfortunately, I did not have civilian clothes and had to endure the "cold treatment" on my way back to Dayton. It was great to get back home and enjoy Mom's cooking and the sound of flushing toilets! My Dad took me to downtown Dayton.

While we were waiting to cross the street, someone shot off a firecracker. I instinctively hit the sidewalk and felt very embarrassed when I got up.

These comments are a small fraction of my combat memories in Vietnam. The Military trains you in Leadership, Responsibility, Discipline, Confidence, Organization, Survival, Resourcefulness, and Mission (get the job done). I treasure my military experience and attribute my Military Training to much of my 49 year business success.

To this day, I stand Duty, Honor, Country and my GOD!

JAMES TORBERT

US Army – Aviator
Dates of Military Service: July 1967 to July 1970
Unit Served with in Vietnam: 281st Assault Helicopter Company
Dates in Vietnam: Nov 1968 to Nov 1969
Highest Rank Held: Captain
Place of Birth and Year: St. Louis, MO – 1944

I'LL FLY IT BACK, IF YOU WILL GO WITH ME

It was August of 1969 and I had been in country for nine months. I was assigned to the 281st Assault Helicopter Company and our primary mission was flight support for the 5th Special Forces Group. We were on a mission with Detachment B-52 (project Delta) with our operating base at Mai Loc in northern I Corps. We were inserting teams into the mountains along the Laotian border beyond the abandoned Marine fire base at Khe Sanh. I was the senior RLO (real live officer) so was responsible for flying the C & C helicopter with the Delta commander in the back. This was important so that if any decisions had to be made it could be done right there, on the spot.

I was flying overhead and watching closely as the team was being inserted. I did notice a slight flash, more like a change of color from dark to light from the tail section of the helicopter #460 in the confined LZ as the Special Forces team was jumping off and sprinting into the jungle. I radioed to ask if everything was okay and CWO Jim Baker responded, "Yes, the team is off, and we are coming out." Several seconds later I got a radio call from Baker saying they did not want to rejoin the fight and wanted to return to Vandergrift Fire Base to check out the aircraft. I did not like the idea of one aircraft leaving and flying back alone, but knew that Jim Baker had been in country for a long time, was a great pilot, and would know what to do if he needed any additional support. I agreed to his request and said that on our return we were planning on stopping at Vandergrift for fuel and if he needed help we could assist at that time.

We finished the rest of our mission and were on our way back to Vandergrift to stop for fuel. As we approached the POL area I was surprised to see aircraft #460 shut down on the pad. Vandergrift also called to say that they were experiencing incoming fire from the Viet Cong, probably because there was a helicopter shut down and that we might want to fly on to the next available POL location. We needed fuel and I said we will stop, but not shut down, and be on our way very quickly.

As the crew chief was refueling our aircraft, I ran across the pad to #460 where Baker and the rest of the crew were looking up at the tail rotor. I got a quick boost and stood up on the stinger to get a better view. It was easy to see, one of the tail blades had a puncture in the blade. I jumped down and was trying to make a decision on the best course of action when Vandergrift once again began experiencing incoming rounds. I quickly jumped back up on the stinger to take another look at the blade, maybe it was the incoming rounds but this time the puncture did not seem to look that bad.

I said to Jim Baker, "I'll fly it back if you will go with me." He agreed, maybe he wasn't as smart as I thought he was. I told the other pilot, John Korsbeck, and the crew to get into the C & C bird and get the heck out of there. Korsbeck grabbed his flight gear and ran like hell to the other helicopter. The crew chief wanted to stay with his helicopter but I sent him on and he very reluctantly went to the C & C bird and climbed in as they were pulling pitch and heading out of there.

I cranked up #460 and took off as the Marines waved us good bye with the famous one finger salute. My training had included being the Unit Recovery Officer and the Unit Test Pilot, so I had experienced some pretty bad vibrations. We were shaking pretty good all the way back. I decided to land at Mai Loc instead of flying all the way to Quang Tri just to get a runway. Baker was with me all the way as we approached the pad, calling power, fuel, and then warning lights. Just as I landed, the tail rotor gave out. I did a hovering autorotation and put #460 down right there. Mission complete, team inserted, and everyone home safe and sound.

After we returned from Vietnam we all kind of went our different ways. Like most of us I kind of drifted away and kept all of these memories buried in the back of my mind. After about 30 years, I went searching on the web and found a 281st AHC web site. I spent some time looking over the site, had some memories and signed the visitor page. Several weeks later I came home one evening and my wife told me that she had spent over an hour on the phone with a guy from Montana, who said that I had saved his life in Vietnam.

He had seen my name as a visitor to the site and was calling to say, "thank you." I asked my wife, "Who was he?" She said, "John Korsbeck." I said I do not know that name. She repeated the details of that day and the story that he had told her. Well, of course I remembered the story, I just did not remember John as being the new guy pilot in #460. What a treat it was to get reunited with John. For many years I have gotten to enjoy his phone calls (at all hours of the day and night) and his many stories and pictures on the web. Thank you, John Korsbeck, you will never be forgotten.

WITHDRAWAL PHASE

(1970-1972)

May 4, 1970 — Kent State shooting. Students were protesting government policies in the Vietnam war when National Guardsmen fired on them. Four were killed and nine wounded.

June 24, 1970 — United States Senate votes to repeal the Gulf of Tonkin resolution.

June 13, 1971 — Pentagon Papers were first published by the New York Times. They revealed that Washington had broadened the scope of the war but had failed to tell the American people or news media about the decision. This deepened the distrust between the American people and their government.

June 22, 1971 — President Nixon signs the 26th Amendment, lowering the voting age to 18.

March 30, 1972 — Easter Offensive launched by North Vietnamese on South Vietnam.

May 8, 1972 — President Nixon launches Operation Linebacker I, which includes massive bombing of enemy forces in South Vietnam, resuming bombing of North Vietnam, and mining of Haiphong Harbor.

December 18, 1972 — President Nixon orders Operation Linebacker II, a massive new bombing of North Vietnam to force Hanoi back to the bargaining table and to impress the South Vietnam government with American resolve.

Source: www.vvmf.org/VietnamWar/Timeline

JOHN BUTLER

US Army — Field Artillery Officer

Dates of Military Service (Active Duty and Reserves): 1968 to 1974

Unit Served with in Vietnam: 11th Armored Cavalry Regiment, 2/94th Heavy Artillery

Dates in Vietnam: A Troop, 1st Battalion, 11th Armored Cavalry Regiment (11th ACR), Oct 1970 to Feb 1971; 2nd Battalion, 94th Heavy Artillery, Feb 1971 to Oct 1971

Highest Rank Held: 1LT

Place of Birth: Pittsburg, Kansas — 1945

BLOWING STUFF UP

One of our operations while I served with A Troop, 1st Battalion, 11th Armored Cavalry Regiment was protecting a company of combat engineers clearing jungle with Rome Plows. Both company-sized units, about 300 of us, lived together in relatively tight quarters within different Night Defensive Positions in the jungle throughout this operation that lasted 6 — 10 weeks, as I recall.

Let me tell you, serving with a company of combat engineers with Rome Plows was never dull. These guys had welding equipment and an overactive imagination. As a result, they came up with some very peculiar and fascinating combinations. One example is they mounted dual mini-guns above the cab of a M548 cargo carrier, an open-bed tracked vehicle based on the M113 armored personnel carrier and used it like a pick-up truck. There was no "mission value" for such a combination, they did it just for fun.

As our time together progressed, it become more obvious that a lot of the equipment they had and used fell well outside of the standard TO&E (table of organization and equipment) for their unit ... or ANY unit, for that matter. That was no problem until our operation in the remote jungle approached completion and the time approached for them to return to their base camp under the scrutiny of their brass.

To make matters worse, they learned they would be having an IG inspection when they returned. One of the functions of an IG (Inspector General) inspection is to ensure that each unit has exactly the type and quantity of equipment called for in that unit's TO&E. Ooops!

We called the First-Sergeant of the engineering company, "Top" of course, and he and I became pretty good friends. We flew "the cut" together regularly in a Hughes OH-6 Light Observation Helicopter (called a "Loach"). My role was reconnaissance. His role was to check on the progress of the cut and to start fires!

The Rome Plows would push cleared jungle brush into large piles in the middle of the cut to be burned. Top decided the best way to burn the piles of brush was to drop a special, custom made, "lighter" into each pile. His lighter was a white phosphorous incendiary grenade wrapped with a stick of C-4 plastic explosive.

Let me pause to say that I'm pretty sure Top was a certified pyromaniac. He LOVED to start fires, and loved even more to BLOW STUFF UP, as you'll read later. Much to my dismay, however, his method of delivery of his custom-made lighters was via helicopter, even when I was with him! He would carry several of these "lighters" with him, have the pilot hover over a pile of brush, pull the pin on the grenade, and drop it out of the side of the chopper onto the pile. Fortunately, this worked every single time. If it had failed to work even one time, you would not be reading this story!

As we neared the end of our operation and the time approached for Top and his company to go back for their IG inspection, he began to assemble all of the maverick items and stuff them onto a light aluminum tracked trailer, we called a "Mule." I have no idea how they had come up with that Mule, but I'm confident it was NOT included in their TO&E, because Top was planning to blow it up.

We were located on the top of Hill 652 with a dirt berm all around our position. The side of the hill was cleared all the way down to the jungle at the bottom of the hill.

Top jammed every imaginable type and quantity of non-authorized equipment and supplies onto the mule and wrapped all of it with enormous quantities of white detonation cord ("det cord") connected to vast quantities of C4 plastic explosive. He even added a couple of 5-gallon cans of diesel fuel to the mix. Why? Because diesel fuel burns!

The plan was to put the Mule on the side of the hill, use a rope and stake to hold it about a quarter of the way down the hill, light a fuse connected to the det cord, cut the rope, and watch the loaded mule travel to the edge of the jungle and blow up in a spectacular explosion! Of course, there would be an audience, and I would be part of it.

With some help, Top got the mule in place and tied down on the side of our hill. Then with as large an audience as could be spared from other duties, Top went down the hill to the mule, lit the fuse, cut the rope, and ran back up the hill to join us behind the berm. Sure enough, the mule went careening down the hill and stopped at the edge of the jungle.

We waited. We waited some more. Someone asked Top, "How much fuse did you use?" Finally, we heard "pfffft" and nothing more.

We didn't know whether to laugh or cry. The mule was sitting at the edge of the jungle waiting, NEEDING, to be blown up! It was FULL of explosives, all READY to blow up!

After a long discussion and some serious soul-searching, Top decided that only he should go back down to the Mule, see what happened, and try again to blow it up. Of course, this meant he would have to get back up the hill—this time the entire distance—before the explosion took place.

It was about 100 yards from the bottom of the hill to the berm at the top of the hill and was a fairly steep slope. Top would have to use a LOT of fuse to allow plenty of time to get all the way from the bottom to the top.

We watched as he walked down the hill, spent what seemed like an hour working all around the mule, then signaled that he was about to light the fuse. He lit the fuse. We all watched breathlessly as he started up the hill, each of us hoping he wouldn't fall or run out of wind before joining us at the top.

He made it! He HAD indeed provided plenty of fuse, so he was able to catch his breath and watch from behind the safety of the berm with all of his fans! KABLOOM! A HUGE explosion shook the earth, pieces of burning metal went everywhere, some even made it up to the berm. Cheers, applause, relief!

After the smoke cleared, Top went back down, this time with a few volunteers, to make sure nothing of value was still intact. They confirmed that everything was destroyed, even though some large twisted pieces of the mule remained.

I used it as target practice for my .45 pistol. Every now and then I would hear a clink that told me I actually hit it. Just blind luck, of course. I don't know of anyone (or any .45) that could accurately hit a target 100 yards away.

LUKE DOLLAR

Army — Infantry Scout Dog Handler
Dates of Military Service: 1968 to 1970
Unit Served with in Vietnam: First Cavalry Division, 37th IPSD Infantry Platoon Scout Dog
Dates you were in Vietnam: Oct 1969 to Sep 1970
Highest Rank Held: SP4
Place of Birth and Year: Atlanta, GA — 1948

G.I. BE MY VALENTINE

As a Scout Dog Handler, I walked point for many different infantry platoons in Vietnam, which was both a blessing and curse. When they would have "Mail Call" I never got mail since I was just assigned to the different companies and platoons. I would receive my mail between missions when I was at the fire base getting my new mission.

However, on Valentines' Day 1970 I was shocked when they yelled out "Dog Man" you have mail. People from all over the country sent Valentines to soldiers all over Vietnam and every Valentine was addressed

GI BE MY VALENTINE. It was a HUGE blessing for me. I got mail in the bush and as I would later realize, it was a God thing.

The Valentine was from a wonderful lady who just happened to be from Marietta, Georgia. (coincidence... I don't think so). I wrote that lady telling her how much it meant. She continued to write and send me care packages while I was in country.

When I returned to the world, she had a Welcome Home dinner for me and asked only one thing ever from me. She asked me to attend church one Sunday with her and her family. I must admit I was not excited about it, but I would have done just about anything for her.

At the end of the service they called out three names, mine being one, to please stand up. When we did, the congregation gave us a standing ovation and a banner that said Welcome Home!! The unselfish acts this lady did for me are still with me fifty years later.

It is funny how one Valentine from a total stranger would be the only letter I ever received in the bush and then the same lady takes me to a church where I knew no one, but got my only "Welcome Home." I have a feeling she will be waiting at the gates to Heaven welcoming me home one last time.

NATHAN CRUTCHFIELD

US Army — Infantry Officer and Armored Cavalry Platoon Leader and XO
Dates of Military Service (Active Duty and Reserves Combined): 1968 to 1971
Unit Served with in Vietnam: 11th Armored Cavalry Regiment (Blackhorse)
Dates you were in Vietnam: Dec 1969 to Dec 1970
Highest Rank Held: First Lieutenant
Place of Birth and Year: Atlanta, GA — 1947

THE LETTER

The resupply Chinook had departed after dropping off the usual fuel, parts, food, people, mail. I was handed a letter with a return address I didn't recognize. We'd been out of Cambodia about a month and the letter was from the fiancée of a staff sergeant we'd lost in a mortar attack. He was our commo sergeant and ran the troop command track. A really good man, well liked.

The letter told me they had been childhood sweethearts and planned to marry on his return. The Army had not provided much in details and she wanted to know what had really happened. He had told her in a letter I seemed to be a good guy and, as I was a platoon leader, she thought I might be able to tell her more.

Back in June 1970, the Regiment was still in Cambodia. Our time there was winding down and we were preparing to leave that country. We were in a large Squadron-sized perimeter. I was at the troop command track waiting for the captain's evening meeting where we got orders for the next day and assessed what was done that day.

The attack came suddenly at sunset. It was pretty obvious we had been targeted with the intent of catching troops lined up for their supper at the mess tent and at the nearby command track with its distinct shape and antennas. As explosions began all around us, the command track crew started yelling for us to jump inside the track as they were raising up its back ramp. I had just turned to start towards them when a round exploded, picked me up and somersaulted me into the track and to safety. Our commo sergeant was caught by a blast before he could get into the track.

I was in a dilemma. Should I respond? What could I say without causing more harm or heartache? Was I violating some protocol? I got my stationary out, wrote, tore up several attempts until finally feeling ok about what I had written. I mailed the letter and hoped I'd done right.

Many, many years later, I was surfing the Internet and went to a web site on the Vietnam Memorial Wall. I searched for names I knew and was directed to a site dedicated to the sergeant killed that evening long ago. The site had several email addresses of the site managers. I selected one address and sent a message that I had known him and about my letter.

The next day I received a lengthy email telling me the story of the website. It was from the fiancée whose email address I had arbitrarily selected. She told me how she had gone on with her life, married, and had a family. Then one day all the suppressed grief came back. It was thought that an effort must be made to get as much of her grief out as possible. She and the family developed the site in honor and memory of our sergeant, her fiancé. This brought home the personal loss of war and its impact on families and how grief never really goes away.

At the end of her email, she told me she still had my letter.

LOU (GRAUL) EISENBRANDT

US Army Nurse Corps
Dates of Military Service: 1967 to 1970
Unit Served with in Vietnam: 91st Evac Hospital — Chu Lai
Dates Served in Vietnam: Oct 1969 to Oct 1970
Highest Rank Held: 1st Lieutenant
Place of Birth: Mascoutah, IL

MENDING AND REMEMBERING

Having made the decision to join the Army to "See the world" after graduation from nursing school in Alton, Illinois, I did not give much thought to the war in Vietnam. I completed school, attained my RN diploma and drove to Ft Sam Houston, Texas for Officers Basic Training in November 1968. New Year's Day 1969 found me arriving at Ft Dix, New Jersey for my first duty assignment, med-surg nursing (which means that you are qualified to care for most anyone) at Walson Army Hospital. From returning wounded soldiers, to sick prisoners from the stockade, to recruits who needed to either lose or gain weight before they were deemed fit to fight, I cared for a wide variety of patients. It was a comprehensive lesson in the varied fields of my chosen profession but paled in comparison to the challenges that lay ahead of me. In September, I received a manila envelope containing orders for Vietnam. I would quickly learn about a whole different aspect of caring for patients.

My first impression of Vietnam was felt the moment that the plane landed and the cabin door was opened. The blast of hot humid air was overwhelming. Being one of only 13 women on the flight, I was clearly not dressed for the tropical weather or combat setting. I said goodbye to my "cords"—skirt, top, hose and heels—and changed to the uniform of the war, green fatigues and combat boots. Definitely not fashionable but certainly practical and utilitarian. Within five days after touching down at Ben Hoa airbase, I was headed to Chu Lai and the 91st Evac Hospital.

The Quonset hut structure was perched on a cliff overlooking the South China Sea. At first sight, the setting was breathtaking, turquoise waters, expansive white sand beaches, clear blue skies. It took only a few minutes, however, to absorb the presence of signs of war, lookout posts, sandbagged bunkers, Army jeeps, and Huey helicopters bringing in the wounded.

My first couple weeks were filled with briefings, one at the "combat center" where we were warned about the evils of marijuana. A reefer was actually passed around as a sort of "show-and-tell." We were also given a "Pocket Guide to Vietnam," the sort of thing that one would find helpful if he/she were vacationing in the country.

By week three, I was settled into my 9' X 12' BOQ room with the bright orange door. It was furnished sparingly with a metal single bed, green metal Army-issued wardrobe, and my footlocker which had been my father's during WWII. A much-needed oscillating fan provided a bit of a respite from the 100+ degree weather. The women's latrine was in a separate building with showers, when power and water were available.

My first nursing assignment was to wards 5 and 8 which were meant for anyone who needed hospitalization but had not been wounded. Malaria was a constant threat due to the tropical environment that mosquitoes thrive on. We were given malaria pills, but, occasionally, soldiers did not take them because they turned your skin a bit orange and gave you wild dreams. Hepatitis was an issue for those who did not always have access to clean water. We saw intestinal parasites, also the result of eating contaminated food and drinking untreated water. Perhaps the most common concern was from men who came into the hospital to be treated for jungle rot. Known by a variety of names, this condition resulted from GIs walking for days in swamps or crossing rivers and not being able to change their socks and boots. The result was large sores on the ankle and legs which then easily became infected. Most of the patients in these two wards were treated and then released back to their units in the field.

On 20 February 1970, I was offered the opportunity to transfer to the emergency room. For the next eight months, I would do some of the most intense nursing ever. I was just 22 years old.

The injuries that those of us in the emergency room experienced on a daily basis varied from shrapnel wounds that could be treated under local anesthesia to massive head trauma for which there was no possible treatment except to ease the pain. My second day in the emergency room I was confronted with a patient I shall long remember. He came in by helicopter having lost both of his legs at the knees. I had never seen a bilateral amputee before. The young man had stepped on a landmine which exploded, sending shrapnel, dirt, and his body into the air. By the time a chopper was able to reach him, blood had already dried on his many wounds and on the two tourniquets applied to his stumps by a medic in the field. Fortunately, the aidman had also given him a shot of morphine to somewhat ease the pain. After cutting off his fatigues, which was standard procedure to assess the extent of his injuries, he was whisked into the operating room where doctors would attempt to repair the damage. Eventually he was returned to the States and his family, minus the legs that he had when he arrived.

Not all damage was caused by land mines. Gunshot victims filled the emergency room as well. There also were injuries from punji pits, grenades, and booby traps. Having made four return trips to Vietnam after the war, I have seen examples in a museum in Cu Chi of the many types of traps that were designed to cripple the unsuspecting soldier.

In addition to treating GIs, we treated ARVN, civilian Vietnamese individuals, even VC and NVA who were then taken as prisoners of war. On one occasion, a village had been attacked and we treated 99 patients in a 24 hour period. Not speaking much Vietnamese added to the challenge of caring for so many wounded. We learned basic terms for "doctor," "water," "pain," etc and asked questions while pointing to parts of the body. We did have a couple of Vietnamese medical helpers who provided a tremendous service by translating what we couldn't understand.

After being on duty for 12 hours a day, six days a week, I sought comfort in joining friends at the Officers' Club, playing guitar and singing, or a rare dinner out at a nearby Vietnamese restaurant. If your time off was during the day, there was a tiny beach at the bottom of the cliff that you could escape to for a couple hours. Some of us actually availed ourselves of a boat with water skis (it was there when I arrived) to glide over the waves, when the sharks were not around. There were gatherings in various hooches and occasionally visits to other compounds for special events and a change of scenery. Needless to say, alcohol was a central element of all of these gatherings.

My year in Vietnam ended in October 1970. I returned to "the world" via SeaTac airport in Seattle. I married, raised a family and got on with life. For many years, I have been speaking to students and other groups about my time in Vietnam. Thanks to a carefully kept journal and numerous slides, I have been able to recollect many of the details of my year in the war. Eventually that led to the publishing of my book "*Vietnam Nurse: Mending and Remembering*" in 2015. The story of women's involvement in wartime needs to be told. We must never forget that women also served!

Editor's note: Lou's book can be ordered from the book store at www. deedspublishing.com or from Amazon or any book store—in paperback or Kindle format.

CLYDE ROMERO

US Army— Scout Helicopter Pilot

Dates of Military Service: 1969 to 2005

Unit Served with in Vietnam: 25th Avn Co; 101st ABN Div, C Troop 2/17 Cav

Dates Served in Vietnam: Apr 1970 to May 1971

Highest Rank Held: Colonel

Place of Birth: Bronx, NY—1950

SHOT DOWN AT THE BASE OF TIGER MOUNTAIN

In August of 1970, I was re-assigned to the 101st Airborne Division, C Troop 2/17 Cav at Hue Phu Bai, Vietnam.

I was a scout pilot flying an OH-6 in a Cav Troop. A typical scout mission, called a Pink team, consisted of a OH-6 Little Bird commonly called a LOACH, 2 Cobras AH-1G, and a Command and Control ship which was a UH-1H that had a squad of infantry on it, along with an artillery forward observer. Our area of operation was the Ashau Valley. It was an avenue of approach into the southern part of South Vietnam and fed off the Ho Chi Minh Trail. The Ashau Valley was right up against the Laotian border where North Vietnamese would infiltrate supplies and personnel. The level of vehicle traffic and anti-aircraft weapons in the Ashau Valley had significantly intensified. The North Vietnamese, who were operating in the Ashau, were not fighting a guerilla war. They withstood Arc Light bombing missions from B-52s, and regular air strikes from F-4 fighters, along with artillery and anything else we could throw at them, but they persisted; they accepted casualties while intensifying their operation in the Ashau. The nature of the air cavalry mission frequently resulted in flight crews reacting to the NVA's initiative, simply meaning, the NVA shot at us first! This story is about my shoot down in the Ashau Valley.

We received information that day that there was a mass formation of NVA pushing through the Valley and to get out there right away with a Pink team. Other assets would follow and to contact the FAC (forward air controller) and get fighters on station. We decided to come from the south into the Valley and fly into Laos and come from behind them through a saddle just on the other side of Hamburger Hill FSB. We all knew where that was! It was a good plan, or so we thought, the weather was typical overcast but good visibility down low and en route we could see into the valley floor out our right side looking north. We stayed high so they would not hear us coming, or so we thought.

The Little Bird runs the mission. I peeled down out of altitude as fast as I could towards the valley floor, hugging the western edge of the Valley, trying to limit myself to anti-aircraft fire, oh did I mention they had already engaged us! The Cobras were taking fire, and then I got hit hard by a .50 cal. several times while I was still relatively high altitude for a scout. I was about 2,000 feet AGL when I was hit, so I had time to pick my spot where I would crash! Please keep in mind I usually fly at tree top level. The valley floor is flat, and very exposed, so I picked the valley edge, the base of Tiger Mountain. It was called Tiger Mountain because the face of the mountain had been bombed away to cause landslides onto the road the NVA used to transport vehicles and personnel.

I did a picture-perfect crash landing into high elephant grass at the base of the mountain. Everyone was ok and I had radio contact with the Cobras and CC ship. But we had landed not far from an NVA company hiding in the same grass! As we egressed the helicopter, we could hear the NVA close by. I got on the radio and told the Cobras to suppress fire off to my right and to shoot as close as you could stand it (danger close). I threw out a white phosphorous grenade to mark our position and left the helicopter. For those of you who have never been on the ground while a Cobra is firing its 17-pound warhead rockets around you, let me tell you it is a religious experience, to say the least.

But we still had to get to the base of Tiger mountain to get picked up by the C&C ship, which by the way was taking fire as they approached to pick us up. It was slow going through the elephant grass but we made good time to the base of the mountain which was around 25 meters away. There were three of us that had to get on that helicopter alive that day, and we did. The C&C bird did a RAMBO approach, all the while the door gunners firing their M60s above our head as we were pulled aboard the helicopter. As the helicopter took off from the base of the mountain, we could see the pith helmets of the NVA troops working their way towards my LOACH, all the while taking casualties from the Cobra suppressive fire.

The rest was uneventful, we returned to Hue Phu Bai with just some bruises and one less helicopter to maintain. This all happened before 0900, subsequently I was asked if I would go back out there to spot the wreckage and assist in blowing it in place, since there was no way in hell we were going to recover that LOACH ever! I did go out and mark it, but this time I was in the front seat of a Cobra and I had F-4 overhead and we were able to destroy the LOACH and lay waste to the NVA around the area as well. That was a very long day...

LARRY J. GARLAND

US Army — Military Police Officer
Dates of Military Service: 1969 to 1997
Unit Served with in Vietnam: B Co, 720th MP Battalion
Dates Served in Vietnam: Aug 1970 to Jun 1971
Highest Rank Held: Lt. Colonel
Place of Birth and Year: Murray, Kentucky — 1947

BEST FRIEND

Ronnie came over to the house to visit one spring day. He had just finished his Advanced Individual Training at Ft. Campbell, KY and had an assignment to the 1st Air Calvary Division in Vietnam. Ronnie was drafted after dropping his course studies at Murray State University and I was awaiting completion of my degree and a commission from the ROTC.

Ronnie was a neighbor and best friend. We had managed to share growing up in a small town, participated in local sports, passed our drivers license tests and graduated together from Murray High School in 1965. Ronnie had great concern about his orders to Vietnam and wanted to know what he could do. Said he just didn't feel good about his assignment to 'Nam. After discussing the lame suggestion that he could tour Canada, we could not come up with any suggestions to relieve his apprehension.

Four months into his tour, Ronald C. Colson took a direct mortar hit in a night attack. His death was a great shared loss by family and friends. His death resolved my determination to prepare for the certainty of my own personal tour in Vietnam.

VIETNAM BOUND

I'm 22, completed MP Officer Basic at Fort Gordon, Georgia, served as a Provost Marshal and Platoon Leader at Yuma Proving Ground, Arizona and now I have my orders for Vietnam. Twelve months in the RVN seemed on the face of it — a long time.

What kind of place was the 90th Replacement Center? How long would I stay there before filling a position slot with an established unit? My overseas flight was diverted from Anchorage, Alaska to Hawaii because of a typhoon in Asia. My first time to Hawaii. What a break. I'd never been to Hawaii. Flowery shirts, hula girls, and island scenery. Didn't matter. I couldn't enjoy it. Three days and my apprehension of my 'Nam assignment was all consuming.

Two days at the 90th Replacement and I made my appearance to the HQs of the 720th MP Battalion. Now we are moving. I drew a platoon leader slot in B Company in Tan An to learn that my first assignment would be located in the Delta city of Can Tho. That's IV Corps delta south. A discipline law and order assignment (DLO) in the bar district of Can Tho.

NIGHT DUTY

Night clubs and bars like the 007, American Bar, and Little Saigon were the soldier's entertainment and stress relief in the Can Tho bar district. Upon one of my visits to a popular bar, I noticed a glass jar on the end of the bar that contained coins and currency bills. Upon asking my fellow MP if this was the tip jar, he replied with a smile, "No, that's the baby jar. If one of the girls gets pregnant, she gets the money to help her get started in her family life." Upon reflection I wrote the occurrence off as a touch of humanity in a war zone. Not the best solution, but an attempt at thoughtfulness.

Drunkenness, drugs, weed smoking, bar fights, and curfew violations was the order of the night. Lieutenant Folsom and I had weekend competition on who could obtain the most apprehensions and violations of the Uniform Code of Military Justice. Occasionally we wrote our own rules of engagement. We would make an apprehension (the military term for arrest) and place the person(s) in a conex container while the initial paper work was processed for flow to the owning unit. After three weekends of this friendly competition, the IV Provost, Major Nix, called his lieutenants into his office and politely told us to stand down on our weekend policing activity. The admin section could not keep up with the paperwork!

FORTY YEARS LATER

My friend Rodney was giving me a tour of the Cross Pointe church at the Duluth campus. After touring the auditorium we exited the main building and headed to the lakeside administrative building. Upon entering the building, Rod turned to me and said, "You know you arrested me one night a long time ago." I hesitated and asked, "What on earth are you talking about?" Rod explained, "Can Tho, I was an aviation officer and you shut me up in a conex container." At that point I knew he was correct, but I had no recall of the event. We continue to be good friends and to this day I am reminded that the military foot print can follow you around a very long time.

CONVOY DUTY

Arrived at main gate to see six dead bodies stacked up and awaiting removal. The air had begun to sour. Events here could be life threatening.

Sappers who had tried to penetrate the perimeter were unsuccessful. I did not ask about friendly casualties.

Dong Tam was an RVN basic training facility augmented with a Navy Mobile Group, an Army Military Advisory Command, one Australian, and one American aviation company. I was now the OIC of convoy escorts in the region, police enforcement, and Viet community relations. All things considered, it was not a bad assignment. We experienced just enough ground and mortar attacks to dampen the interest of Company and Battalion HQs visits. I kept the supply chain busy with damage requests for tire and glass replacement parts for our vehicles.

My unit equipment had five assigned V-100 (M706) vehicles. They were light armored, big tired vehicles that could slosh around in the rice paddy country and ford small streams. We managed to keep about three of them in operational condition. Without exception, every Battalion Commander had to make a good impression on his Officer Efficiency Report and one way to do that was to have every vehicle repainted OD green, complete with white unit ID numbers and a white star. I lived through three command changes. Why not have bright white IDs to move around with; after all it was a combat zone!

I was ordered to find the OD paint for the jeeps and V-100s and get them painted. My local supply channel was dry. No paint could be located anywhere. After searching with no results, I contacted my local Vietnamese counterparts. They agreed to paint three of our V-100s in camouflage for ten cases of beer. We gave them Black Label beer as it was not a favorite of our troops.

I must say, it was different. A great camouflage design with a black, two browns and two greens. No unit IDs, no U.S. symbols, just camouflage. The paint job could not have been better. You could hardly see the vehicle moving through foliage. I sent one of the vehicles on a mail run to company HQ the very next day. The Vietnam land line phone system was one of the worst, but being my lucky day, the Company Commander had me on the line within five minutes of that newly painted V-100 arriving at company HQ. He asked, "Lt. Garland, do you have any more of those Disney creations on site?" "Yes sir; two more." "Well send them up to us ASAP and we will repaint them for you."—-—And that is how you get your vehicles repainted for a Change of Command at Battalion.

My closest potential encounter with an opportunity to become an in-house resident at Ft. Leavenworth (Army penitentiary) came with another V-100 issue. SGT Hammer was taking the vehicles to the Navy bay with some frequency under the pretense of washing the vehicles. Through an overheard conversation, I learned that washing the vehicles included floating the vehicles in the bay and using them as a diving and swimming platform. I cannot imagine how I would have written the report of losing a V-100, actually sinking it, in the bay. Needless to say, SGT Hammer received more than a safety briefing and was instructed to never be closer to water than crossing over a bridge in the V!

MORE THAN ONCE

Description: QL4 was like the Interstate highway for the Delta. Yes it was only two lanes for most of its mileage, but it was paved and relatively free of major pot holes that could destroy a tire or damage the structure of a vehicle.

Having received a concern from Company HQ that there were too many non-combat deaths occurring on QL4, I selected an impressive looking MP and we motored over to QL4 to apply some selective enforcement. After being on the highway for about ten minutes a five ton dump truck came roaring by at a maxed out speed. No questions here; speeding and reckless. After following the truck until we could confront the driver at its next stop, we proceeded to issue a citation. Showing my best knowledge of the law and the process I explained to the driver that the disposition of the citation would route through to his Commanding Officer who would then decide what penalty to impose. At this point the driver shrugged his shoulders and gruffly said, "WHAT ARE THEY GOING TO DO? SEND ME TO VIETNAM?" It was my first time and I walked into it.

ALAN C GRAVEL

US Air Force — Pilot

Dates of Military Service: 1969 to 1974

Unit Served with in Vietnam: 536th Tactical Airlift Squadron of the 483rd Tactical Airlift Wing; 4102nd Aerial Refueling Squadron

Dates in Vietnam: Sep 1970 to Sep 1971; May 1972 to Dec 1972

Highest Rank Held: Captain

Place of Birth: Alexandria, LA — 1945

ARTILLERY NEAR-MISS AT LOC NINH

When new Caribou aircrew arrived in country, part of the orientation was to spend some time at Squadron Operations reviewing a package of information about Caribou operations in Vietnam. One of the documents was the photograph of a Caribou on short final that had just had its tail section blown off by friendly artillery. The fuselage was pointed straight at the ground, about 50 feet up. You had to wonder what the pilot was thinking as he faced certain death. There is a photo of this incident posted on the C-7A Caribou Association web site. I remember a slightly different photo, but the image is unmistakably the same.

One day when we were flying the Bien Hoa stage, we were headed to Loc Ninh, about 50 miles north of Bien Hoa. As we flew along, the co-pilot would check with various artillery sectors and firebases to find out whether they were firing and if so, in which direction and to what altitude. We routinely had to adjust our course to avoid active artillery. As we approached Loc Ninh from the south for landing, we checked with the Artillery Controller at Loc Ninh and he said they were firing to the west. The firebase was set up just west of the runway so we set up our downwind on the east side and were going to make a left hand final turn to the south to land.

Loc Ninh had been a rubber plantation and just west of the runway, on the northern half of the runway was a mansion with swimming pool and tennis court. The place was abandoned and in disrepair, but you could tell that in some former time, life had been very good for someone in Loc Ninh. As I was looking out the open window at this mansion, the aircraft shuddered pretty violently, a feeling somewhere between a vibration and the jolt you would get from heavy turbulence.

I looked back under our left wing and saw to my horror the long barrel of an artillery piece pointed right across our tail, smoke rising from the business end. By the time we realized what had happened, it was over and it was obvious we were ok.

When we taxied into the ramp area, an US Army Lieutenant came rushing up. He apologized for the ARVN artillery mistake and tried to explain. I was too relieved to be alive to really be mad. I kept thinking about that photograph at Squadron Ops and what a fine line is often drawn in wartime between disaster and a good survival story.

LAUGH BOX AT BIEN HOA

By the middle of 1970, all the USAF C-7A Caribous in country were based in Cam Ranh Bay (CRB). We had two regular temporary duty locations in Bien Hoa and Can Tho.

For Bien Hoa, we would fly out of CRB (with clothes and personal gear for a week) in the morning, fly a full day of missions, and recover in Bien Hoa. We would then fly two days out of Bien Hoa, have a day off, and then two more days. On the seventh day, we would launch out of Bien Hoa (with our personal gear on board), fly a full day of missions, and recover in CRB. We called it the "seven-day stage." A number of Caribou crews would be in Bien Hoa at the same time, so we had quarters designated for our use.

For Can Tho, we had a similar routine but it was only three days and no day off. We lived in a civilian hotel downtown while there. The best meal I had in Vietnam was chateaubriand with avocado salad and strawberry shortcake at a small French restaurant down the street from our hotel in Can Tho.

Although these TDYs took us away from our permanent quarters and our personal stuff, they were a nice break in the routine. Cam Ranh City was closed to us. On our day off at Bien Hoa, we could take a taxi to Saigon to go shopping, eat at the Continental Hotel, and use the MARS phone at the USO to call home. In Can Tho, we usually had a few hours in the evening to walk the streets, have a nice meal, and have a drink or two.

Each morning at Bien Hoa, we would pre-flight our aircraft about dawn and taxi to the cargo ramp to pick up our first load for the day. A half dozen or so Caribous would be doing the same thing and multiple other aircraft would be starting their daily routine as well. Bien Hoa Ground Control monitored the movement of all these aircraft and you would request permission to taxi from the Caribou ramp to the cargo ramp and then to the runway for takeoff. For that reason, for a period of time each morning, all the Caribou aircraft and many others would be tuned to the Bien Hoa Ground Control frequency. By 1970-71, many air traffic control functions, particularly the less critical functions such as Ground Control, were being turned over to South Vietnamese Controllers. The operational control language was English but many of the South Vietnamese Controllers spoke English with a heavy accent.

One morning someone keyed their microphone and activated a "laugh box" on Ground Control frequency. Laugh boxes were novelty items available from many sources in country. As I remember them, they were a small plastic box with batteries and a speaker. They were usually sold with a small cloth sack with a drawstring. When activated, they played audio of continuous uproarious laughter.

That morning, each time the Vietnamese Ground Controller would transmit instructions to an aircraft, an unidentified aircraft would key the mike and transmit a few seconds of laughter. As I remember it, it went something like this.

Ground Control: "Aircraft XYZ, taxi and hold short of runway XX"
Mystery Aircraft: "Ha Ha Ha Ha Ha Ha Ha Ha Ha Ha"
For the first few times, the Ground Controller tried to ignore the laughing but at some point, he could no longer ignore it.
Ground Control: "Aircraft transmitting on Bien Hoa Ground, say call sign."

Mystery Aircraft: "Ha Ha Ha Ha Ha Ha Ha Ha Ha Ha"
Ground Control (more emphatically): "Aircraft transmitting on Bien Hoa Ground, say call sign!"
Mystery Aircraft: "Ha Ha Ha Ha Ha Ha Ha Ha Ha Ha"
Ground Control (even more emphatically): "Aircraft transmitting on Bien Hoa Ground, say call sign!!"
Mystery Aircraft: "Ha Ha Ha Ha Ha Ha Ha Ha Ha Ha"
Ground Control (now clearly losing his patience): "Aircraft transmitting on Bien Hoa Ground, SAY CALL SIGN!!!"
Mystery Aircraft: "Ha Ha Ha Ha Ha Ha Ha Ha Ha Ha"

And then from another unidentified aircraft, very dryly, "He might be crazy but he's not stupid."

It clearly was disrespectful of the Ground Controller, but it was some much-appreciated comic relief at the start of our day.

DAVID B WALLACE

USN, Officer-in-Charge, PCF/Swift boat
Dates of Military Service: 1968 to 1973
Unit Served with in Vietnam:Task Force 115/94, Coastal Squadron ONE,
Coastal Division 13
Dates in Vietnam: Sep 1969 to Sep 1970
Highest Rank Held: LT (0-3)
Place of Birth and Year: Pittsburgh, PA—1946

ANOTHER NIGHT IN PARADISE

A Friday night in March of 1970, super dark with no moon. We were patrolling in the southernmost area of the Bassac River (the main branch of the Mekong River), around Cua Lao Dung ("Dung Island"). At about 2100 we rendezvoused with the LSSC (Light Seal Support Craft) at the northern tip of the island. We had another Swift with us from the adjoining patrol area. Our job was to cover a SEAL insertion by driving with them down a small canal on the northwest corner of the island. The Swifts were noisy and the LSSC was very quiet, we would fake a probe into the canal bumbling around pretending to be investigating something while the LSSC sneaked the six man team into the bank. The SEAL mission that night was to grab a VC or NVA officer passing across the island.

We were in this 300 yard long canal for a good 45 minutes, making noise and maneuvering unexpectedly. The Swift's twin super-charged 12V71s had aluminum "clappers" on the four exhaust ports that made so much noise we suspected that any self-respecting Victor-Charles would be waking up, furious with us. The SEALs departed and the LSSC made its way out past us noise makers. When they had cleared the canal, they called and told us it was OK to make our way out which we did post haste. The limited maneuvering areas in the dark had lost their appeal to us months ago.

The second Swift was sent back to our regular patrol area. To further the ruse that it was just another night on the river we simulated anchoring and setting up watches on the boat. After drifting for an hour we resumed a very quiet patrol. We made the operation's target the center of our AO keeping within 2000 yards and running at bare steerageway on the port engine only.

The SEALs clicked twice on the operational push so we'd know they were still around; we'd become their link to the outside world. If anything went wrong on our end we'd pop a couple of illumination rounds over an area south of them. The SEALs would see them but not be lit up. Everything was quiet from then on. I finally assumed a horizontal position about 0300, leaving our Leading Petty Officer in charge.

By 0615 I was looking for coffee which successfully eluded me when I heard the radio break static and the SEAL Chief saying, "We have a problem here. I need all the 81 mm you can throw in here **RIGHT NOW!**" and gave the coordinates for the rounds. I asked if he wanted a spotting round and he came back, "Shoot every *COLORFUL-NA-VY-SPEAK* round you have right now!" The coordinates were right where we estimated the team would be and he angrily confirmed this and suggested that we do it or we would need to come in and pick up their bodies. In total, this whole exchange took less one minute. Within another minute, we had a firing solution, set the propellant charges, matched up the magnetic compass, and had rounds in the air. We moved during the initial mortar shooting to get the bow of the boat onto the beach to give more stability. This also opened us up to Chuck or his friends tossing something nasty onto the boat—we had a crewman manning an M60 on the bow.

The Chief told us that he had at least five automatic weapons emplacements surrounding him and was in a world of shit. They were dug into shallow trenches and hunkered down. Our mortar rounds were keeping Chuck's head down so he couldn't get a good shot at our guys. There were several things wrong with this as a long-term strategy:

- One of our well-intentioned rounds might hit our guys
- A Swift typically carries only about 120 81mm mortar rounds (some were white phosphorous, some illumination, some direct-fire flechette rounds—leaving us with 80 high explosive rounds)
- Within not-too-long Chuck could set up his own mortars to use on the team or bring some nasty rockets to bear on us
- As we were shooting 7+ rounds per minute, something had to give pretty quickly

In combat, things never seem to happen one thing at a time. While we were busy scrambling helicopter gunship support, the Chief called back to report that he had a wounded Victor November November (Vietnamese Naval person), his interpreter. This is all going on while we are calling the Sea Wolves (Navy UH1-Ds), Black Ponies (Navy OV-10s), Army Cobras and Hueys and, as a last resort, the Air Force. The Sea Wolves and Black Ponies were tied up on another mission. The Army—bless their hearts—was scrambling at least three gunships with an ETA of 15 minutes, but a station keeping time of less than 20 minutes (this was due to limited fuel and the heavy load of ordnance they could carry).

We told them to get everything and anything they had ready to get in the air and to be prepared to keep something(s) on station for the next hour or so. Their simple response, "Roger that!"

We hadn't forgotten about the wounded guy, it's just that if we weren't able to at least stabilize the situation, they would all need medical help—and that had to be our first priority. We normally liked to use the Army Dustoff choppers for medical evacuations (the famed MEDEVACs) as they were familiar with the boats and we with them. In the past, they had even rested the fronts of their skids on the mortar boxes of boats to evacuate wounded. Naturally, we called the nearest medical evacuation facility for some hurry-up help; the Air Force had an operation about 25 clicks away.

When I finally got them on their push, they asked if I had their checkoff list. I responded in the negative. He said, "Very well. Line one is …" Keep in mind that we were still firing mortar rounds right on the top of our own guys, had Huey and Cobra Gunships arriving on station, and just had the LSSC tie up alongside to see if they could help. I exploded at the Air Force guy telling him to get a bird in the air, we have a wounded man on the ground. When he asked if the LZ was HOT, I let fly with my best string of quaint Navy invective questioning his ancestry, wife, and his parents. I "hung up" on this guy. Next call was to the Army to see if they could help. Just then, I noticed one of the LSSC crew about to drop a second mortar round into the tube of our already loaded mortar (naval mortars could be fired by "dropping" the round onto a firing pin or by trigger). This would have caused a rather nasty explosion on the stern of the boat that would have been likely to set off the other fifty or so rounds in the open mortar box or laying on the deck. A similar accident had killed two on another Swift and had sunk the boat. Dropping the radio mic I raced to the stern of the boat, tackling this well-intentioned guy, screaming all the while. I threw him into the LSSC and said I'd probably kill him if he came back on the boat.

By now the Army Dustoff guy thinks we've been sunk or something else awful. I reassured him that we "only" needed a dustoff for a HOT

LZ! He double-checked to coordinates and again that wonderful reply, "Roger that!"

So now the gunships are blasting away and we're shooting mortars when there are no gunships in the area, and our guys are actually beginning to return some fire on the ground. A guy driving a Cobra says he sees some guys (not his actual description) shooting at a couple of piles of mud on the edge of a clearing. I reply that that's probably our guys and the pilot asks for smoke. It's pretty routine for a gunship to ask for smoke to be popped when shooting close to friendlies and he relayed this request to me. The SEALs only had access to one push at a time and had camped out on the Navy push, meaning they didn't have direct communications with the gunship guys. I relayed this immediately to the SEALs and the Chief came back that it would be easier to throw a smoke grenade into the bad guys and have the gunships shoot them. It took only minutes to convince the pilots that this was a good plan and they identified green smoke, I confirmed that back through the Chief, and they blew up one of the bad guy locations. Only four more to go!

At this point the Dustoff arrived. It was a regular "slick," an unarmed Huey used for transporting troops and other things. The Army didn't have a regular dustoff chopper available so they sent what they had. The pilot confirmed through me that there was a wounded guy on the ground and asked if the LZ was as hot as it looked. "That's affirm," I replied. I had visions of him fleeing the scene, claiming some mechanical problem—I wouldn't have blamed him. He yakked it up with the circling gunships and then did one of the wildest things I saw in my entire tour. From a height of about 500 feet he turned the slick so the nose was pointed straight down and dove into the bushes. By this time, we had two Cobras and two Huey gunships overhead and they shot off every round they had and were ready to start throwing seats, binoculars, and anything else they had on board at the bad guys. We couldn't shoot anything because we couldn't see anything and the risk of hitting the slick was just too great. The radios went quiet as no one was talking, we were all holding our breath. The slick was probably on the ground for about two minutes, but it seemed like they were there for a couple of hours.

Because of the shooting, we couldn't hear anything that sounded like a chopper taking off. The noise from four attacking gunships and everything else that was going on made it hard to even think. All of the sudden, this slick just popped out of the trees and took off for the hospital. We could see some battle damage to the aircraft and later learned that they had taken (this means had been hit by) over a hundred rounds. For months we tried to find out who the pilot and copilot were without success. The crew of PCF-32 must have consumed the output of Pabst Brewing Company for a year, buying drinks for Army helicopter pilots, and I'm sure the SEALs did as well. This was an incredible rescue, even by the standards of these guys who made their livings doing stuff like that.

By this time it's approaching noon and the tide is going out rapidly. The tidal drops in Southeast Asia are like nowhere else in the world. A tidal range of 12-14 feet was considered normal. Our charts showed we'd get 20 feet this day. The highest guns on a Swift are at 14 feet; with a 20- foot drop, we would be shooting up at the banks if we had to go into the canal of the previous night and get these guys out. It also meant there was a possibility we could get hung up on a sandbar in the shallow water or—joy of joys—that Chuck could hide along the bank of this 40 foot wide canal and simply drop or lob all manner of ugly, exploding things into our boat. If we didn't get something done in the next two hours, we'd have to wait for the tide to come back up and the SEALs didn't have enough bullets for this kind of lollygagging. Without ground support there was zero chance of getting these guys into a slick and evacuating them by air.

It was now four machine gun nests against our five SEALS. If not for being almost out of bullets and out of hand grenades, it was probably an even fight. I'd gotten the gunships, the SEALs, and our boat on the same push. Try to imagine a discussion of alternatives when all three "principals" are directing weapons firing at bad guys or trying to keep everyone's head down and find a way to escape. There were two avenues of escape that seemed feasible: go west to the narrow canal or east to the main channel of the Bassac River. The canal was approximately 300 yards away and the river was over 1200 yards. We decided to have the team make a break for the canal and be picked up by either the Swift or the LSSC. We only had .50 caliber and 81mm flechette rounds left as we had shot everything else onto the SEAL position. A little-known use for .50 caliber machine guns is in land clearing. Because the round is very heavy and travels at approximately a mile each second, it will take down most medium-size and smaller trees quite effectively. Larger trees might take two or three rounds.

The plan was to have the Swift pull up to the bank of the canal nearest the SEALs and to "hose down" (a technical Navy expression) a path from the SEALs to our boat. While we were making this path the SEALs would have to stay close to the ground and avoid the falling vegetation. The LSSC would pull into the bank about 100 yards south of us and the escaping team could also go there. Since the smaller boat was decidedly below the bank, they would have no weapons to use on this operation and would have to wait for some guys to jump out of the bushes into the boat. This is not really a great plan, but it seemed the best we could do in a drastically shortened planning time. At least our twin .50 gunner would be able to see down onto the bank and perhaps give some warning if guys in the wrong uniforms (or wearing their black PJs) started running toward our boat. He could not see the bank in front of the LSSC, so these guys had to kind of "gut it out."

As mentioned before, things rarely go the way the best and brightest minds have planned. The textbook explanation for this phenomenon goes along the lines, "The most brilliant plan rarely survives the first contact with the enemy." The Swift could not make it into the canal because of the tidal drop. We hit sandbars on both ends and, rather than add to the list of American assets at risk by trying to get into the canal, had to back out.

Another unplanned event took place at about this time: a Cobra pilot reported seeing five blobs of mud "beating feet" (technical military term for running) toward the canal. He wanted to know if I thought he should "waste" them. Before I could respond, a gravelly voice came up and said, "Those are five **friendlies** beating feet two-seven-zero." (Navy term for heading west). For whatever reason, there had been a break in Chuck's attack on the guys who were dug in and they responded by making their break toward the canal. Another of those awful silences ensued that seemed to last about six hours. We knew the LSSC couldn't have gotten into position in this short a time and the thought of the SEALs coming to the bank of the canal with nowhere to go except into the water, where they'd be sitting ducks. The LSSC crew heard all this going on both on their radio and probably noticed the marked increase in gunship activity, not to mention the increase in bad guy shooting.

A remarkable boat, the LSSC is a gasoline powered, water-jet propelled, screaming machine. It is incredibly quiet and can operate in about two teaspoons of water. Though never tested, to the best of my knowledge, it could probably pull a couple dozen water skiers up to speed in somewhere around four seconds. This little critter pulled up alongside the bank of the canal just as the "beating feet" SEALs reached the edge. The team simply rolled over the edge of the bank and into the boat.

The last we saw of them, they were headed north in the main channel of the Bassac River at over 50 knots with a rooster tail that would make any off-shore racer proud. It was the only time any of us had ever seen an LSSC in the daylight and the last one we saw in the Bassac. The LSSC crew reported that the SEALs were grumbling about losing the prisoner they had gone in for and were trying to get information about their wounded team member.

To this day, an Army helicopter pilot cannot buy a drink around me. No one, to the best of my knowledge, ever received an award for their heroics in this action or even thought about writing it up. It was another day of "Going up the River"—less pleasant than many, but still spent with the best people ever collected in one small spot for any reason.

That's my story and I'm sticking with it. We never did get any coffee made that day.

RICHARD L MENSON

US Army—Judge Advocate General Corp

Dates of Military Service (Active Duty and Reserves): 1965 to 1973

Units Served with in Vietnam: USARV, Engineer Troops-Vietnam (Staff JAG Office) Jan 1969 to Jun 1969; Headquarters, 4th Inf Div (Staff JAG Office, Chief Trial Counsel, Administrative Head of Court Martial Convening Authority) Jul 1969 to Feb 1970; Staff Judge Advocate—Cam Rahn Bay Support Command Feb 1970 to Aug 1970

Highest Rank Held: Captain

Place and date of birth: Chicago, IL—1943

WHAT DID A LAWYER DO IN VIETNAM

I was in ROTC at Ripon College in Ripon, WI, graduating in June of 1965. I had applied for participation in the excess leave program for law students who were in the Army. I went to Northwestern Law School in Chicago where I was attached to 5th Army Headquarters. Upon Graduation from Northwestern and taking the bar examination, I was asked where I wanted to go for my first duty assignment and I volunteered to go to Vietnam.

During October and November I attend the basic course at JAG School on the campus of the University of Virginia. While there I took my arms training, I think at Fort Belvoir. I received my orders upon graduation that I was going to the 1st Infantry Division which was located outside of Saigon. On January 6, 1969, I arrived at Ton San Nhut Air Base. We were taken to the Replacement Depot. After getting all our equipment and uniforms, I was called in to meet two JAG officers, not from the 1st Division but from Engineer Troops Vietnam. They explained to me that they were promised because one of their lawyers was rotating back to the States, that they would get the first JAG officer who arrived in country. That was me.

They took me to Long Binh, home of USARV Headquarters. Engineer Troops Vietnam was headquartered there and was responsible for all the separate Engineer Battalions within Vietnam that did not support a combat unit. They made the roads, strung the telephone line, constructed headquarters, and did what normal engineers do. The battalions were located from the DMZ to the tip of the Delta.

I was going to be the 'newbie' in the office, with no experience. I had to get to know what to do from all different directions. I was duty officer from time to time, responsible for a certain part of the perimeter since we had a portion of the perimeter to guard at night. I would get calls from the perimeter, staffed by clerks from Engineer Troops Headquarters that they had seen some VC and could they fire at them. I wasn't trained for it.

When I was shown my billet, the first thing pointed out was where the bunkers were since my billet was near the wire. During my time at Long Binh, we were hit with 22 attacks from January through May 1969.

I became involved heavily in traveling around the country interviewing witnesses for courts martial that I was assigned to, both for the prosecution and the defense. I also went to our battalions to assist them with various legal issues pertaining to their soldiers. I also had a unique experience. With as many troops as we had there, the big issue was the use of money that was supposed to be used in the PX or others facilities within the camps. We were not allowed to carry and use US Currency. What we had was called Military Payment Certificates (MPC). It was colored a certain color in units of cash from a one dollar MPC to a 25 dollar MPC. Every so often to prevent the MPCs being used on the black market, the war stopped and you had to turn in the MPC for a different color MPC.

In my first month in country, my boss said, "Go pack for a week's trip to Saigon." He didn't tell me what it was about. I got to Saigon and met up with other junior officers and we were told that our job was to count the MPCs and then burn them. So the first day we were very efficient in counting each MPC. After we realized that if we continued on that way it would take months, we started stacking and comparing stacks and putting on the paperwork the amount each day before we burned the MPCs. The most interesting thing was when I looked at who made the boilers we were burning the MPCs in, they were made by the company for which my Dad worked.

After about three months and having taken over some Court martial cases and assisting on others, I was traveling all over Vietnam to spend time on both my prosecution cases and defense cases, interviewing witnesses, seeing the area where the offense happened, and seeing the scenery. On one case I had to travel to an area just short of the DMZ where a soldier had fired upon Vietnamese workers. In another case I had to go visit our Battalion that was deep in the Delta.

I went by planes, helicopters, and other types of transportation. I remember that one time I was in the Delta at the 9th Division trying to get back to Saigon when we were informed that we were not taking off since the VC were on the other side of the river shooting at whatever took off from the runway. I also remember going to a Forward Base in a Caribou. We got to about one thousand feet and the pilot said, "Hold on, make sure you are strapped in tight." The next thing that happened was he put us in a full diving circle on the way down and only pulled out of it when we were landing. He said if he came in for a regular landing, the VC would try and shoot the plane down.

At the middle of May I was called into a meeting with USARV's Staff Judge Advocate and informed I was being transferred to the 4th Infantry Division, effective at the end of June. I had a talk with him and he said I was going. He thought that if I was going to make the Army a career, that it would be good to have on my record a divisional tour in a combat situation. Even though I had been at Engineer Troops fresh off the plane, what most JAG officers did was to spend the first six months in the field and the second six months at a staff job, but I was going to the 4th Infantry Division for another field job.

Before I went to the 4th I was sent to Saigon with Gary Hullquist, another JAG Officer in the office, to investigate a civilian contractor who was being charged with shooting at the Military Police. We determined that he was not the shooter. Since he was not subject to the Uniform Code of Military Justice, we had to attend the trial in the Vietnamese

Courts to make sure the trial was in the accordance with the agreement we had with the Vietnamese.

One comment about the weather. When I got to Vietnam, we were in a dry period before the monsoon season, which came in April and May. It was beautiful flying in that type of weather. I made a trip to Ban Me Thuot in a U-21. We also flew from Binh Hoa to Pleiku, to Cam Rahn Bay, and to my final destination. I had a meeting with witnesses for three hours. The pilots stayed with me. On the way back in an empty plane, they put me on the radio as we flew in and out of the clouds and thunderstorms at the beginning of monsoon season. The winds and the rain would blow so hard that out of our office window in the midst of the monsoon season, I watched a helicopter get blown off the USARV heliport and slid over in front of our building.

The 4th was located in the Central Highlands outside of Pleiku, close to the Cambodian border. Pleiku also contained other commands which supported the 4th which we were responsible for in the JAG office. Our JAG office had approximately 35 personnel, including some Vietnamese and was located next to the Commanding General's office.

I was assigned two jobs. The first job I was assigned was lead trial counsel at General Courts-Martial for the Division. My second job was to establish a new Division-wide Special courts-martial Jurisdiction that would combine all Special Courts-Martial for the entire Division under one roof. We moved a portion of the JAG Office to a new building which would be responsible for processing all Courts Martial within the Division. There are two different types of Courts Martial, a General Court-Martial and a Special Court Martial. To show you the enormity of the number of Cases within the Division when we established the new jurisdiction, we had about 10 General Courts-Martial being handled, but we had 47 Special Courts-Martial. So keeping track of them and making sure all of the cases were handled in the proper way was enormous. Each case required a lot of administrative time along with the trial time if needed. General Courts-Martial required a written trial transcript, had a military judge presiding, and a jury. Special

Courts-Martial until January 1970 had no military judge and could only have punishment of six months confinement.

Since we were within a certain area, my traveling was curtailed. Most of my travel was within the Central Highlands. In September 1969, I tried a number of General Courts-Martial. I had a number of guilty plea cases, but I had one soldier who I was representing turn down a deal. He was accused of pointing a weapon at a superior officer and threating him. He and I had long discussions about what he did. He decided that he wanted a trial so we went to a trial. He was found guilty but his sentence was forfeitures of certain pay for a year with no confinement. I learned a lesson. You cannot let personal feeling influence you. You need the defend someone to the 'Best of Your Ability." I was really learning to be a lawyer.

We had a Vietnamese lady, the wife of an ARVN officer, in the office. Her main job was to review claims which the local people had filed for something that we had destroyed like water buffalos or fields where they were growing food. Under the International Claims Act, we were responsible for their loss. She processed them and then someone would take the claim and pay it.

Outside of Pleiku was a Montagnard village. I would tag along from time to time with the group which would go out. At the village we were welcomed and sat down around a pot in the middle of us. In the pot was homemade rice wine. Our interpreter explained as part of the ceremony we had to drink some rice wine with the people sitting around the pot. Someone had filled the pot up and had left it overnight so that the rice could ferment. They then brought out a stick with another stick pointing down into the pot. You were expected to drink the rice wine until the stick pointing down didn't touch the top of the rice wine. Before the next person would drink, they put more rice wine in the pot. Everybody enjoyed themselves, the Montagnards got their claim paid, and we went back to our living quarters, after drinking rice wine all afternoon.

About this time I felt that I was growing up day to day, which is one of the reasons I went to Vietnam. I was considering extending my tour of duty for at least four months, and had volunteered to become a Military

Judge for Special Courts-Martial while I stayed in Vietnam. In discussions about staying and becoming a Military Judge, I was informed that I would have to come back to the States for training and then go back to Vietnam. I turned that down, but I said that I would stay for the time I had extended and until August of 1970, so I could go to another command. I was informed that after the beginning of the year I would be transferred to Cam Rahn Bay Support Command as the Staff Judge Advocate.

During my time with the 4th, I participated in 30 General Courts-martial, with most as the Trial Counsel and a few for the defense.

Cam Rahn Bay Support Command was on a peninsula which bordered the South China Sea. I was the Staff Judge Advocate for the Support Command, arriving in February 1970. I had an office with about 25 personnel. Since I was on the Commanding Officer's staff, whenever rockets were fired at us from the other shore and the sirens went off, I had to report to the Command Bunker with sidearm and full combat gear, including my helmet. One time I forgot my helmet and the Deputy Commanding General took me to task. But why did they need me in the Command Bunker?

The whole peninsula was occupied by the three branches—Army, Air Force, Navy. It was sometimes questionable whose jurisdiction you were in. The hills were large sand dunes. The water was sparkling blue. Everybody worked hard to keep the supplies moving along. The docks were busy 24 hours a day in order to supply all of the units in central Vietnam. People worked and got in trouble. It was another part of my schooling and helped me understand why we were there.

I left Vietnam from the Airbase at Cam Rahn Bay. I arrived in Vietnam as a "Newbie." After 20 months in country, I was leaving Vietnam as an "Old Man. I learned a lot while I was there about what a lawyer was expected to do, but now I was going home to see my family. The countdown time was zero. And that is what this lawyer did in Vietnam.

KURT MUELLER

US Army — Infantry Officer, Rotor Wing Aviator

Dates of Military Service (Active Duty and Reserves Combined): 1968 to 1972

Unit Served with in Vietnam: C Troop. 3/17 Air Cav; D Troop 3/5 Cav.

Dates in Vietnam: 1970 to 1971

Highest Rank Held: Captain

Place of Birth and Year: Tulsa, OK — 1947

CAMBODIAN INCURSION: OBLITERATION & DEATH

By the end of May, 1970, President Nixon authorized us to go into Cambodia to attack the North Vietnamese Army (NVA) who was trying to migrate into South Vietnam through Laos and then Cambodia. We were restricted from going no further than six kilometers or about 3.7 miles inside Cambodia or as we called it, "The Tricky Dicky Line." As we found out, the NVA knew this and would go just past the "The Tricky Dicky Line" to hide or take a rest. Of course, the restricted six kilometers limit did not stop some of us from accidently going too far into Cambodia.

It was about the second week in June 1970 and I was the lead on a mission with our C Troop 3/17 Air Cav, call sign "Charlie Horse" flying slicks (Hueys UH-1H) as part of our hunter killer team consisting of two Cobras (AH-1G) and one Scout (OH-58) to search for NVA in Cambodia.

On that particular day, cloud cover was very low and with only a 1:100,000 map to guide us, we set out on our mission flying above the clouds. After what I suspected was an adequate amount of time, we dropped down below cloud cover to find a schoolhouse and a large soccer type field to be completely vacant of any children playing outside around the lunch hour. At one end of the soccer type field there was a large one story building with a grassed thatched-roof. Our scout pilot, Larry Brown, flew low to that building and caused several NVA troops to run out of the building of which he and his crew quickly neutralized. Larry then flew back toward the building to the back side to be surprised by a number of NVA armed with AK-47s cranking up two deuce and a half type trucks which were equipped with quad fifties in each. He quickly exited the area and radioed for all of us to get out of the area and get as low as possible so we could not be targets.

As we moved out, the two Cobras flew low level back around and to come in on the two deuce and a halfs. As they came up on the targets, one of the Cobra pilots accidentally jettisoned all his flechette rockets, hitting both trucks dead center, obliterating the trucks and the NVA solders... a very bad day for the NVA. Those NVA were just about eight kilometers inside Cambodia.

My buddy and flight school roommate, Cal Binder, arrived about the 1st of June and was assigned to a sister company, A Troop 3/17th Air Cav, call sign "Silver Spurs." When he arrived, he starting flying missions in Cambodia as well. On June 29, 1970, the last official day of the Cambodia operation, Cal's Cobra, in which he was flying front seat, was shot down. From what I understand, his pilot, Jim Elkin, was shot and killed. The aircraft crashed and killed Cal when the aircraft hit the ground.

A few days later, my commanding officer called me in and gave me orders, at the request of Cal's parents, to escort his body back to the United States. I had to go through an orientation as to the proper protocol required in escorting his body back to the US. This was going to be a particularly tough situation for his parents because the Army had decided that it should not be an open casket funeral.

I left Vietnam on July 7, 1970, on a jet transport with Cal's body in a casket in the baggage area. I finally arrived in Iowa where his funeral was going to be held and where he would be buried. It's an understatement to say that this event made quite an impression on me, especially being the fact that it was a closed casket and his parents were never able to see his body before they buried him.

The process of escorting Cal's body back turned out to be a traumatic event for me. Part of the protocol of escorting an individual's body was to stand at attention and hold a salute as the visible casket was being moved anytime during the transportation back to their home city. As I was changing planes in St Louis, Cal's casket was being moved down the conveyor system out of the aircraft's baggage holding area. There were two baggage handlers watching the casket come out of the aircraft on the conveyor when it became obvious the casket could fall off the conveyor onto the ground, as it was getting sideways on the conveyor system. The casket suddenly fell off the conveyor, hitting the ground and the two baggage handlers stood there looking at the casket and started laughing.

I was really extremely angry, unfortunately I wasn't much in a position to do anything other than complain to an airline representative who was at the check-in counter. The airline representative expressed his sincere apologies for what happened. The disrespect some people showed for our fallen during this war was unimaginable.

I arrived in Rembrandt, Iowa with Cal's casket and got off the airplane to meet Mr. and Mrs. Binder and their other three children, Marilyn, John and Scott. Sally Blitch, Cal's girlfriend from Savannah, also came to Iowa to join the Binders for Cal's funeral. I don't remember a lot about the funeral but I do remember specifically the Binders asked me a number of times how did they know that it was Cal's body in the casket, since the Army requested it was a closed casket funeral.

They also asked me several times what I remember about his death and all I could tell them was that his Cobra was shot down and he and his pilot were killed instantly. I also assured them that there was no mistake that Cal's body was in the casket and the Army did not make mistakes like sending the wrong body to the wrong family. I never explained to his parents what actually caused his death as it would have only made the loss of their son even worse.

After the funeral, I was able to take some time off before going back to Vietnam and I went back to my home in Tulsa to see my parents and my soon-to-be fiancée, Marilynn Hood, who came in from Savannah. Once I arrived back in country, our unit was alerted that we were going to be moving to a new operation base near the DMZ called Quang Tri.

MAX W. TORRENCE

US Army — Field Artillery, Heavy Lift Helicopter Pilot

Dates of Military Service: 1966 to 1988

Unit Served with in Vietnam: 273rd Aviation Co (Skycranes); HQ, 12th Aviation Group

Dates in Vietnam: May 1970 to May 1971

Highest Rank Held: Lt. Colonel

Place of Birth and Date: Bellefountaine, OH — 1946

FIRST MISSION IN VIETNAM — INTO CAMBODIA!

Superhook 1-6

I arrived in Vietnam at the end of May 1970, fresh from helicopter flight school and a lucky transition into the CH-54 Skycrane helicopter.

I quickly passed through the 90th Replacement Battalion and was transported with several other aviators to the 12th Aviation Group headquarters at a nearby base known as Plantation. We had an office call and quick pep talk with the Group Commander and finally got our in-country assignments. As a first tour aviator, I was lucky again to get assigned to the 273rd Aviation Company (Skycranes) at Long Binh. I was directed out of the headquarters to a waiting UH-1 for a short flight to Sanford Army Airfield at Long Binh.

The next week was spent in-processing to the unit, drawing flight gear, jungle fatigues, and TA-50. The last two days of in-processing was a mandatory Vietnam Orientation course taught at Binh Hoa airbase nearby. I learned a lot, but mostly I realized how fortunate that I was to be assigned as a helicopter pilot and not as a grunt. In my meeting with the unit commander, I learned that I was only the second Captain assigned to his unit that was authorized five Captains. Most of his officers were senior Warrants, CW3 and CW4. I became the 1st Platoon Commander, call sign Superhook 1-6.

On June 3rd I was assigned to my first flight mission, but not into South Vietnam. US Forces had been sent into Cambodia beginning on May 1st to join South Vietnamese units in clearing out the enemy stockpiles of weapons, rice, and supplies along the border. My unit had been flying missions in support of the operation for the past four weeks to move bridge sections, artillery, ammunition, and engineer equipment into place. Our mission that day was to continue to provide support to US Forces in Cambodia.

My Aircraft Commander (AC) was a third tour CW4 and I'm sure that he was really thrilled to have a new, first-tour helicopter pilot as his co-pilot. We departed Long Binh and flew direct to Tay Ninh to pick up our first load, a 155 mm howitzer going into the Fish Hook area of Cambodia. On heavy loads, we often cut back on fuel to give us the additional power needed. Not to worry though, our Operations has coordinated for a 500 gallon bladder of jet fuel to be dropped at our destination to give us enough fuel to get to the fuel point at Song Be.

As we arrived at our destination, the AC overflew the fire base and asked the Flight Engineer to visually confirm that a black fuel bladder was there. He confirmed and we made our approach, put the howitzer in place, and hovered over to the bladder. By this time, our low fuel warning lights were on so we shut down #2 engine and pulled #1 to flight idle. The two crew members jumped out to begin hooking up the refuel line. Almost immediately, the Flight Engineer was back on the intercom to inform us that the bladder he saw was 500 gallons of water and not jet fuel. The CW4 immediately shut down the aircraft, jumped out without saying a word and quickly headed to the far end of the fire base. I exited the aircraft and walked with the two other crew members to where the CW4 was standing.

When the three of us caught up with him, I asked why he had so quickly moved to the far side of the fire base. He said, "Captain, our Skycrane has now become the biggest target for any bad guys in the area. I suggest that you stay here with me and look for a place to go if the mortar rounds begin to drop." The voice of experience got my attention!

In about 30 minutes, we heard the unique sound of a CH-47 in the distance. As he got closer, we could see that his load was a 500 gallon bladder that we hoped was our fuel. Since our aircraft had blocked one end of the fire base, the CH-47 pilot dropped the bladder at the far end and departed. It was our jet fuel. With the help of several of the artillery troops and the battery commander's jeep, we pulled the bladder across the firebase to our aircraft. We took turns on a manual 15 GPM pump to load the fuel in record time! We were off and headed to Song Be within about half an hour!

So, my first mission of my tour in Vietnam included boots on the ground in Cambodia!

DAN HOLTZ

U.S. Air Force — Healthcare Administration
Dates of Military Service (Active Duty and Reserves Combined): 25 years
Unit Served with in Vietnam: HQ MACV/CORDS, Military Provincial Health Assistance Program, Ninh Thanh Province (Phan Rang, RVN)
Dates you were in Vietnam: Nov 1969 to Nov 1970
Highest Rank Held: Colonel
Place of Birth and Year: Indianapolis, IN — 1943

FRIDAY NIGHT AT THE MOVIES

During my tour in Vietnam as an Air Force Medical Service Corps Captain from November 1969 to November 1970, I was assigned as health advisor to MACV/CORDS in Phan Rang in Ninh Thanh Province. I was part of a medical advisory team of Air Force medical personnel working at the Province Hospital in Phan Rang. Our mission was to advise and assist Vietnamese civilian and military medical personnel to provide healthcare to the citizens of Ninh Thanh Province. Our 15-person team of advisors consisted of two surgeons, an internal medicine physician, a nurse anesthetist, a healthcare administrator (me), and a cadre of enlisted personnel including three medical service technicians, two public health technicians, two radiology technicians, two clinical laboratory technicians, and a medical administrative technician.

Phan Rang was basically a quiet village located near the South China Sea about 40 miles south of Cam Ranh Bay and 55 miles south of Nha

Trang. Most of the war at that time was ongoing to the west of our location. Just six miles down the road was Phan Rang Air Base and because of the several missions located there, F-4s, C-130s, C-47s, etc., it was the prime target of the Viet Cong (VC) forces who routinely lobbed artillery shells into the base. The city of Phan Rang was left reasonably untouched by attacks from the VC.

The American forces in the city were housed in three compounds. The MACV compound housed the Army troops who were military advisors to the Vietnamese provincial forces and included the Deputy Province Senior Advisor, an Army LTC and the other senior Army leadership team. The CORDS compound was primarily comprised of civilian advisors in the province including the Province Senior Advisor who was a Foreign Service Officer—3 (O—6 equivalent). The third compound housed the "spooks." The MACV compound was a typical Army compound with few individual amenities. On the other hand, the CORDS compound was built for occupancy by civilians and had private rooms with private baths, showers—otherwise they were first-class accommodations. One surgeon, the internist, nurse anesthetist, and I resided there to provide medical capability as needed.

All the compounds had provisions to show movies sent over from the States. At Phan Rang, we shared movies among the three compounds and at the CORDS compound we watched them on the patio in the open air and enjoyed the breezes that occasionally moved inland from the coast.

One Friday evening during the American action into Cambodia, the war along the seaside had basically come to a total halt because the enemy supply lines had been cut. The aircraft based at Phan Rang were all deployed to the western part of Vietnam to support troop actions taking place there. Things were quiet and two other residents (a Psychological Ops Officer and a female Foreign Service Officer) and I at the CORDS compound were relaxing on the patio as the sun went down at about 2000 hours. The movie (I do not recall the title) started and had run for about 15—20 minutes when we noticed tracer rounds going back and forth on the south side of Phan Rang in a small hamlet about eight miles to our south. The Army Psychological Ops Officer ran to his room to get his radio, so we could listen to the traffic to find out what was happening.

Most of the conversation was coming from the U.S. Tactical Operations Center (TOC) in downtown Phan Rang, located on the grounds of the Province Headquarters and 8—12 feet below the surface in a large bunker. The U.S. soldier manning the TOC was working with Vietnamese counterparts, including an interpreter. The U.S. soldier was briefing the Deputy Senior Advisor about a VC "action arrow team" that had invaded the hamlet looking for food and other supplies and who were surprised by a government Regional Forces (RF) patrol in the hamlet. A firefight had ensued in the dark and neither side knew exactly what they were shooting at.

The Vietnamese RF commander was requesting aerial support from an American gunship, Spooky or Stinger, both of which were based at Phan Rang air base. The TOC sent a request up the channel and got back a negative response because all those aircraft were in action elsewhere in country. The firefight continued for about 15 minutes when the TOC reported the higher headquarters had offered assistance of a Haiti's aircraft that had no guns, but had flares attached to small parachutes that could be dropped from the plane to light up a field of battle to provide "daytime" visibility. The RF commander accepted the offer of the Haiti's aircraft, but was told he would have to wait about 30—45 minutes before it could arrive on station. Meanwhile, those of us at the CORDS compound continued to watch the movie and the tracer rounds in the distance from several miles away.

About 35 minutes later, the Haiti's aircraft arrived on-station and the aircraft commander ID'd himself to the TOC. From the CORDS compound we could see the aircraft's lights in the sky and heard the pilot ask the TOC for a slow count, so he could home-in the aircraft onto the TOC's location. Here was the problem with that request; the TOC was about six miles away from the hamlet where the firefight was taking place and it was underground. Dutifully, the TOC began the slow count and the Haiti's aircraft made its approach from the south heading north homing in on the slow count. The Haiti's aircraft began releasing about

10 flares on parachutes. The pilot called the TOC on the radio and asked if the light they had provided had done the job, as the aircraft began a 180 degree turn to the left.

The U.S. soldier in the TOC told the pilot he would have to come up and out of the bunker to see and when he did he rushed back to the radio and exclaimed, "You just lit up all of 'Papa Romeo'. You are nowhere near where we need the light!" The Deputy Senior Advisor located at the MACV compound had no eyes on the situation because that compound was located on the north side of Phan Rang and the ability to observe what was happening south of the town was impaired by trees. However, the CORDS compound was located on higher ground and our visibility to the south was fairly clear because there really was nothing between our compound and the hamlet but several rice paddies and farms.

At this point, you must remember, "medics" are non-combatants in accordance with the Geneva Conventions. For whatever reason, the Psychological Ops Officer was reluctant to get involved in the activity because of the distance from the action and the female Foreign Service Officer did not have any training or experience in military operations. So, what to do—do we sit there, or does someone take some action? Well, being somewhat the assertive type, the medic, me, who had a radio callsign, picked up the handset, called the TOC, and confirmed to the TOC the fact Papa Romeo looked like it was high noon, not 2100 hours. I advised the TOC the Haiti's aircraft had missed the real target by several miles.

At that point, the pilot of the plane came on the ratio to the TOC and said if there was someone on the ground who could see where the plane needed to go, to please ask that individual if he could guide the plane to where the light was needed. The TOC asked me if I could provide guidance to the plane to get to the hamlet and light up the right location—I agreed to do my best. At this point the aircraft had done a second 180 turn and was headed back north over Phan Rang. When I could see the plane was coming even to my location, I told the pilot to make a 90 degree turn to the right, which he did and then about a minute later I told him to make another 90 degree right turn to put him on a southerly heading and going for the hamlet. Then, I told the pilot to light up the terrain below him, which he did and daylight came to the hamlet and the Vietnamese RF force was able to put eyes on the VC and drive them out.

The rest of Friday night was spent around Phan Rang by all forces remaining on high alert in case contact was made with any VC forces left over from the firefight at the hamlet, or from any reinforcements. All was quiet.

The next night, Saturday, the female Foreign Service Officer went to the Than Hai District Headquarters and compound to attend a birthday party for several Vietnamese and American officers. While sitting in the "bar," she overheard several members of the American Advisory Team that rode around in a jeep with a .50 caliber machine gun mounted on the rear, who were complaining about that "d—" medic who sent the Haiti's aircraft right over their position and exposed them and their Vietnamese comrades to possible loss of life from enemy fire because their positions were exposed. Unfortunately, this Advisory Team had been trying to contact the TOC to advise them of their situation all along; BUT, the battery on their radio had apparently died before they could transmit their information. The Vietnamese RF force was too busy fighting and apparently unable to contact their highers to give information to them about requesting the needed support to relay to the Americans through the TOC.

The good news from the firefight was, there were no Americans injured or wounded, nor were any of the Vietnamese RF personnel. The VC were run out of town!

Noncombatant medics can step in and provide assistance when needed.

THE MILPHAP MISSION IN PHAN RANG, RVN

I was assigned to the Military Provincial Health Assistance Program (MILPHAP) Team-14 located in Phan Rang, Ninh Thanh Province, Republic of Vietnam. MILPHAP was composed of 21 medical teams made up of seven teams from the Army around Saigon and south into the Delta, seven teams from the Air Force along the coast from II Corps into I Corps and seven teams from the Navy located in I Corps.

Our team's mission was to assist the provincial health system in Phan Rang. We were based at the provincial hospital located next door to the local Province Medical Chief's house and the Vietnamese Provincial Governor's compound in downtown Phan Rang. The local MACV/ CORDS Headquarters was located across the street in a converted office building, where the U.S. Province Senior Advisor and the Deputy, a U.S. Army LTC, had their offices.

The hospital was built by the French back in the early 20th century and was composed of several buildings that included a surgery suite with two operating rooms with open windows for fresh air where our two surgeons and nurse anesthetist performed general surgery procedures on patients and an intensive care bed unit for recovery and intense medical treatment manned by our internist. Another building housed an open ward that could accommodate up to 30—40 patients. There was a small clinical laboratory with two Vietnamese laboratory technicians, assisted by two U.S. Air Force laboratory technicians. There was an x-ray room that could handle routine x-ray examinations. We had two preventive medicine technicians, whose jobs were to visit hamlets in the province and assist local public health officials in projects to improve the health of the citizens of the hamlets. About six months before my arrival at Phan Rang, the Army of the Republic of Vietnam had merged their medical mission of care for the Vietnamese forces with the provincial health mission of caring for the local civilian population.

A major focus of our public health efforts was sanitation and water purification to prevent the spread of disease. The local public health officials worked with the hamlet residents to educate them on how to better handle their daily bodily functions to prevent the spread of diseases. As for water purification, the effort centered on education and covering the local wells, which usually were about six feet in diameter in the center of the hamlets. Covering wells was an interesting effort because we had to make the covers out of reinforced cement on the ground next to the wells, manually lift them up and place them on top of the open wells and install hand pumps in the tops, so the water could be drawn by the *mama-sans*. These well covers were one of the biggest frustrations to our team members because we would assist making them, installing them, and teaching the locals how to use them and then we would come back a week later to follow up and find the cover had either been removed or pushed to one side so the water could be drawn by the old method of rope and bucket.

We understood the goal of our mission in Ninh Thanh Province was to be part of the larger pacification of the local population mission of the MACV/CORDS Advisory Team, to which we were attached, by showing the local population their lives would be better under the South Vietnamese government, as opposed to the Viet Cong and North Vietnamese. Our mission was complicated by traditions which in some cases went back centuries, not just years and decades. The resistance to change was an ongoing issue in just about everything we did.

Teaching the local health folks they needed to change needles between patients when they gave vaccinations and inoculations was an ever-present issue. Part of it was driven from ignorance, but the bigger driver seemed to be the cost of using new sterile needles *when the one they had just used looked perfectly clean*. The members of our Advisory Team had to constantly observe what the locals were doing and be ready to intervene quickly to correct reverting to old bad habits.

One of the biggest satisfactions we had was a visiting ophthalmologist from the Air Force hospital at Cam Ranh Air Base. This guy was a life saver for many local residents who suffered from cataracts and other eye diseases that impaired their abilities to see. This ophthalmologist came to visit us once a month on a Saturday and would perform 10 to 15 cataract surgeries on each visit, sometimes on both eyes at the same time. Back in 1970, cataract surgery was accomplished by freezing the lens with an instrument that resembled a pellet gun compressed air tube. When the lens was frozen to the instrument, it was lifted out of the eye and replaced by a clear prosthetic lens that enabled the patient to see clearly. For many patients this was for the first time in years they were able to see. The smiles on the faces of the patients who could now clearly see family members, new babies, and blue sky was so rewarding.

I remember taking pictures of the surgical procedure with my Pentax SLR camera and sending the photos back to my wife with explanations of what was being done, Wow. Wow was my reaction and Linda was sort of OK with them; however, *yuck* was the reaction of my family and my wife's family.

Was our mission a success? That is an unanswered question. Personally, I believe it was. Part of the reason I believe this is, we knew we were treating Viet Cong fighters as part of the patients who came to the hospital for care and who were in the hamlets benefiting from our preventive medicine activities. The local population was either poor or was living in housing that appeared to be poor, or just not well kept up, but they were basically healthy.

I am convinced we did good things while we were there. The fighting of the war was generally not close to Phan Rang, so it was mostly a peaceful place. The locals were doing everything they could to live productive lives for their families and for themselves—they were trying to be normal (whatever the definition of normal was).

The countryside along the coast was a beautiful place. Had the war not been happening, it might have been a good place for people to visit and to vacation. The people of Phan Rang were friendly to the Americans who were there to help them and I believe they appreciated what we were doing for and with them. Unfortunately, when the North Vietnamese took over, many of the locals were sent to re-education camps and may have never returned.

My year there was an interesting time. I hope I am right about my assessment regarding the success of our mission. I hope the people who are there now are some of the ones who were there when I was and they remember fondly the good things we brought to them.

CLIFF PENROSE

US Army—Armor Officer, Aviator-Fixed Wing and Rotary Wing

Dates of Military Service (Active Duty and Reserves Combined): 1967 to 1978

Units Served In Vietnam: 335th Trans, Aircraft Direct Support, DivArty Aviation, Americal Div

Dates Served In Vietnam: May 1970 to May 1971

Highest Rank Held: Captain

Place of Birth: Henderson, NV—1943

IT'S A SMALL WORLD

I arrived in country in May 1970 and was immediately assigned to the 335th Transportation Company, Aircraft Direct Support. The 335th was assigned to the 'Americal Division', 23rd Infantry Division located in the southern part of I Corps. I was met at the airfield by an old classmate who took me to the company area to settle in and meet a few of key people in the company. I was then introduced to the In-Country Orientation which consisted of training on the basic operational issues, weapons, booby traps, the indigenous people, and the ever-present Live Fire Course! During the live fire session I was the senior officer present so I carried the white flag which I could raise to stop the fire, if necessary.

Well, during our movement through the course, the sergeant in front of me came crawling back and said, "Captain, there is an alligator in the path in front of us." I proceeded to move in that direction and as I turned the corner, there it was! It sure looked like an alligator to me so I decided to raise the flag to stop the firing. The training officers were not happy but they did check out our observations and explained to us that it was just a large lizard and they had it removed. We finished the course and passed the orientation.

Our mission was to perform all 2nd and 3rd level aircraft maintenance in southern I Corps. We also had aircraft recovery responsibility. This included rigging and lifting small aircraft out with our Hueys and rigging larger aircraft for lifting by larger helicopters. This recovery and extraction operation was not an easy task as it had to be done quickly and efficiently, sometimes with snipers in the area. It was a great unit with a bunch of well qualified and dedicated people.

In early September of 1970, my request for a Regular Army transfer to Armor was approved and I was able to look for units more suited to my branch. There were no Cav slots available in the Americal Division but I was able to land a spot at DivArty Aviation as XO. We were the largest aviation section in the Army and had 18 OH6A aircraft. I was immediately sent to Vung Tau for transition into that beautiful little aircraft. After returning to Chu Lai I quickly got acclimated to the unit and to our various missions. Our missions on a daily basis included VR (visual recon), artillery support, some C and C (command and control). Most of our missions were geared toward support of the fire bases in central and southern I Corps. Some of our other missions included flying various Special Forces missions and scouting for the river boat crews.

Periodically we had to supply an aircraft to fly 'Donut Dollies' to remote locations. This was always a great break for some of the guys. I also had operations responsibilities although SSgt Cooper handled everything without a hitch. I made it a point to fly every type mission we were assigned so I was familiar with what the guys would be experiencing.

It was early October and the fire bases and LRRP teams out toward Laos had been observing a significant amount of build up by NVA forces coming across the border. We had been searching for activity and calling in artillery fire all day and it was getting late. On the way back to Chu Lai we monitored an open call for any rotary aircraft in the vicinity of such and such coordinates. We were close and I was ready to respond when my door gunner came alert in a flash and asked me to be cautious with my response. Please note that in a Loach we many times flew with just a door gunner. I think he was thinking that I was real new to the unit and was probably concerned as to how I would handle an unusual situation. Needless to say, I answered the call and found out that a Chaplain had been stuck with an infantry company that had been experiencing heavy contact. They had broken contact and said they had a 'cold LZ' available for an easy extraction. As my door gunner was quick to explain to me, there was no such thing as a cold LZ.

I made contact with the unit to find out where they were and proceeded to the LZ. As we approached and made visual contact, they popped smoke which I confirmed and they led me in to a fairly tight spot. A LT came running out of the heavy foliage to talk to me and said he had a Chaplain that needed to get back to Chu Lai. It was late in the day and dusk was approaching fast. I told him to get him out to aircraft ASAP so I could get going. He then told me they had four body bags and wanted me to take them. I had to explain that we were in a loach not a Huey! I told him I could only take one due to the weight. As we were lifting off we received some light fire but did not take any rounds.

We headed back to Chu Lai, first to the morgue and then on to Ky Ha where I told the Chaplain that we could get him a ride back to his area. He told me that he probably owed me a beer and I was quick to point out that it was more than one. He did buy me more than one over the next few months and he even talked me into attending Mass a few times as well as Christmas Eve. We had a great friendship for four months until he rotated out of country. In my letters home to Linda, I often mentioned Father Nick and that I had wished we had kept in better contact after he left country. I served my remaining time in Vietnam experiencing many exciting times flying the Loach!!

After Vietnam, I attended the Armor Officer Advanced Course at Ft. Knox KY and was assigned to the 8th Infantry Division in Germany. Initially to Manheim but then immediately sent to Mainz to be part of a new battalion assigned to the 1st Brigade Airborne. A tank battalion in an airborne brigade did not make much sense to me but that was the way it happened. We formed up rapidly and I was assigned as Company Commander of A Company. I was soon to discover that the Battalion Commander was not fond of aviators for some strange reason but I managed to overcome that issue.

Linda and the boys were not able to join me yet as there were no quarters available but it was not going to be long before we could work something out. No worries as I was working very long hours. After one normal long day, I was leaving the company area on the way to my car. The weather was turning nasty and the wind was blowing hard and it was snowing heavily. As I was getting close to where my car was parked, I lifted my head enough to acknowledge another officer coming toward me, and as we passed each other something seemed familiar. As I turned to ask he also turned and called out, "Cliff!" I immediately said, "Nick!" It was Father Nick. I could not believe that it was actually Nick. We had some great times catching up on our Vietnam experiences. When Linda and the boys arrived in Germany, we all got together many times. Nick loved Linda and the boys.

What a small world!

LINDA PENROSE
Cliff's wife

EXPENSIVE PHONE CALLS

It was a very long year, May 31, 1970 to May 31, 1971. Right before Cliff left for Vietnam, I moved back to Wichita to be close to my parents. I tried to keep very busy, to not think about what he was doing half way around the world. Our two boys were four and two. With my Mom's help taking care of the boys, I went back to Wichita State to work on my degree.

I did meet some Air Force wives whose husbands were also in Vietnam, so that helped. We played bridge once a week. We would always compare our phone calls with our husbands. In 1970-71, long distance phone charges were accomplished with the help of Ham Operators who relayed the call from Vietnam to the States, and would get as close as they could to the city where the spouse was. The further the phone call, the more the phone call would cost. So I asked the other wives, "How close did your phone call get?" They would tell me, Kansas City, Denver, or Omaha. Mine was always Oregon, California, or Alaska. They enjoyed a big laugh over my expensive phone calls.

It was a treat for the boys and myself to get to spend time with

Chaplain Nick in Germany in late 1972 and early 1973. We had many great times together.

DANIEL O HYDRICK

US Army — Engineer / Engineer / Weapons
Dates of Military Service (Active and Reserves): 1969 to 1975
Unit Served with in Vietnam: Americal Div "Southern Cross — 23rd Infantry — 26th Engineers
Dates Served in Vietnam: Jan 1970 to Dec 1970
Highest Rank Held: E-5 Buck Sergeant
Place of Birth: Wynne, Arkansas — 1947

BEHIND THE CURTAIN OF WAR...

1969 was a rough year for America, but history will tell that tale.

After graduating in May of 1969 from Harlem High School at 17 years old, 75% of my male classmates and I were given a choice: draft notices. I would turn 18 in August, and that's all the time I had to decide what I would do.

I didn't want to be drafted; for some reason it didn't sit right with me. As a child of the '60s, I knew something was wrong with that war and I had already seen many returning childhood friends on our block who came back torn and emotionally unstable.

My choice was clear: with my father being a military veteran and hero (a WWII Sergeant with the Hell on Wheels Division under General Patton, and awarded the Silver Star after being reviewed for the Medal of Honor), and my brother and best friend currently serving in Vietnam with the 101st Airborne, it was my turn. I took my chances and "volunteered for the draft" before I received my official notice.

At the Army enlistment office, I flipped through the big book of career opportunities to find something in the W's that might keep me out of harm's way. I wasn't a coward; I was just playing the odds to beat the system. "Water Purification Specialist" ... who shoots at that guy?

After enrolling, I graduated at the top of my class and was recognized as an Honor Graduate of United States Army Engineering School. I thought I would be sent to some safe haven like Germany or Korea, or maybe stay in the States to teach. It was not to be.

After a seven day leave to visit and say goodbye to my family, I would be on a plane. I was only given seven days instead of the normal 30 days because of some extra special training by the Army that they felt would help me to be a better soldier. The real reason was that I qualified as an expert in firing several types of weapons (even receiving an Expert Badge in throwing hand grenades). I was given orders for RVN (Republic of Vietnam). In fact, I believe my time was delayed, waiting on my brother to get out of Vietnam so the Army wouldn't have two sons in combat at the same time.

I landed in Long Bien, Vietnam, courtesy of Continental Airlines. We didn't have long to realize where we were because as we were landing, we came under mortar attack and were told that when the door opens, RUN!!! Welcome to Vietnam.

As I stepped up to the Colonel in what was called "The Combat Center" for orientation, I was asked what my MOS was (Military Occupational Specialty). I proudly stated, "WATER PURIFICATION, SIR!" Confused, he looked at me as said, "What the hell is that?" But before I could explain he said, "Never mind." And off I was sent.

The Americal Division, 23rd Infantry, 26th Engineers, in a place called Chu Lai, this was to be my first home. It was a division only activated in war time in the Southern Pacific, but that is another story.

I took my first helicopter ride out to a place God forgot, called LZ Wrong Hole. The LZ (Landing Zone) wasn't big enough to have a real name. We were sent as a part of President Nixon's Pacification Program, to win the hearts and minds of four villages, about 7-10 clicks (kilometers) outside the city of Quang Nai. A place they called, "Step-and-a-Half Valley," the reason being was that this was your life expectancy, with all the "booby traps" the Viet Cong continually set up for soldiers. Booby traps were everywhere we walked, buried pits, trip wires, daisy chains (a string of explosives in a line), and even in the trees.

We set up a NDP (Night Defense Perimeter) to begin work on becoming friends of the people of My Lai (4). This place would become infamous because of the massacre and destruction of one of the four hamlets just a year earlier. We didn't know this until we got there, and let's just say, the villagers were still angry and hostile.

One of my many jobs was to teach the villagers about clean water and how to treat it and store it, etc. Now coming from an 18-year-old, the old men were not too impressed. So began my next nine months, trying to help.

The place we lived was all underground bunkers and we lived like rats in a cave, but somehow, we called it home. Every place you went on this little compound was buried to protect against mortars. We didn't have a lot of support because we were so far out. We knew what the situation was, and as young as we all were, we accepted it and banded together in the darkness to keep each other amused with stories of where we were from and our families.

My plan to avoid heavy combat by choosing an educated job in in the Army had failed. The Army did find a good use of my ability to fire all those weapons and even throwing grenades, but on a brighter note, I was now living just outside of a village full of old men, women, and many, many children to protect. I found peace with that, in between moments of trying to stay alive, helping those that only wanted to live in peace made sense.

The children and I were inseparable. They called me by the name the guys gave me, Goliath. This was not because of my size, because I was small. They called me Goliath because of my big feet. My brothers in the platoon felt at ease if I walked point (the lead soldier advancing through hostile territory), because they would step in my foot prints and feel safe that they wouldn't step on any booby traps. At least they didn't call me Big Foot.

I in turn, gave the children American names. I had two special young orphan lads about 10 or 11 years old (best guess), who followed me everywhere I went in the village, to "watch my back." Tommy would carry my M16 and stay close at hand, and his little brother Danny would carry my machete. We made a hell of a team, even though they knew that just being with me could cost them their lives.

My water purification days were numbered, as you can imagine. The mortar rounds would fall and destroy the equipment, we would repair it, and they would blow it up again. We were constantly attacked as we made our way to the watering hole. Besides, I was needed to handle other duties more suited for my weapons capabilities than my chemical science abilities. On a side note, the bomb crater that we used to pump our water from was sprayed with Agent Orange, and I came to realize that wasn't good.

Time passed and I was promoted from Corporal to Buck Sergeant and given added duties. We went out more and more on Recon and ambush work. Our presence in this place was tenuous at best, and the Viet Cong were hardened to the fact that we could not stay.

We tried our best to protect the villagers from the night raids of the Viet Cong. You see, just by being there we brought more destruction to these peaceful villages. If the villagers helped us, the Viet Cong would take it out on them. If we provided aid to the villagers, the Viet Cong would come at night and take it. We did what we could to let the villagers know we were sorry and only meant to help.

To distract the Viet Cong's attention away from the villages, the Army decided to build a road and a bridge to help the fishing villagers get to the outer island, "for their own good." In my opinion, if the villagers wanted a road and bridge, they would have already built it a couple of centuries ago.

You see, the irony of this road and bridge became a "bone of contention" with the Viet Cong. Our plan was working. We would work on it during the day and they would blow it up at night. So, we had a mission; build it faster than they could blow it up, and also keep them away from the villages.

Well, we won. The bridge was beautiful… well, as beautiful as Third World country bridges go… and the road was elevated some eight feet. As we took the time to marvel over our dedication and perseverance, the rains started. This would be the announcement and the beginning of "The Monsoons."

Someone should have known better. You see, LZ Wrong Hole was closer to the coast of the South China Sea than the villages. These were fishing villages and they were set further inland. They were smarter than us.

Within the month, we were underwater and had to abandon LZ Wrong Hole. Our underground bunkers filled with water. We had to sleep above ground, out on the berm, and abandon our equipment, eventually walking out in chest deep water. The monsoon rains and the coastal tides took out the road and left our bridge sitting out in the ocean as a testament to American ingenuity.

I left Vietnam a hardened, young 19-year-old. I had seen things that I wished I had never seen and had done things that an 18-year-old shouldn't have had to do. I came home to a very different place.

I needed a break, so after my time was done with the Army I took off to the high country of Colorado. I wanted to pay respect to a family of a friend who didn't make it back, and then I decided to stay.

This was a place that was peaceful, quiet, and let me slip back into "the World." I even brought my brother out. He still lives there today, somewhat at peace.

Maybe it was me, or my father's words of his time, "You do what's right in your heart; they can't take that away from you." I look at all my pictures now some 50 years later and smile. You see, they are of the people and mostly of the children. They give meaning to my time there. I see me and feel my heart was in the right place, just not the right cause.

Of all the stories about my time in hell, I want people to know that I fought hard, I went the extra mile to do a good job, and I helped the children.

In the end, history has dealt the blow… but that is another story.

RICK LESTER

US Army — Aviator, Armor Officer

Dates of Military Service: 1967 to 1994

Unit Served with in Vietnam: 10th Combat Aviation Battalion (1st tour); 48th Assault Helicopter Company (2nd tour)

Dates in Vietnam: 1969, 1970 to 1971

Highest Rank Held: Lt. Colonel

Place of Birth and Year: Marietta, GA — 1948

REMEMBERING ED BILBREY — "JOKER 07"

It had been many years since Ed Bilbrey was killed in action on Lam Son 719, when a fellow pilot forwarded a note to me saying that his widow, Karen, had signed our website's guest book expressing a desire to track down those of us who knew and served with Ed. Following is his story as I have recorded it in my personal journal…

Karen asked me about the mess officer duties and I gave her some background to explain, but I thought I'd share this story, especially since it seems so appropriate for the holiday season. Looking through my journal, reviewing the period of September — November 1970, one entry I noted was about the time we had an officer's call in the club. This meeting was, just coincidentally, on a day we had come back from a long and 'shitty' mission to find that, for the fourth time in five days, the chow hall was serving for dinner some form of roast beef!

If you recall, some days it was roast beef and gravy, sometimes it was gravy and roast beef, beef "tips" and gravy, sliced roast beef, beef stew or beef with onions and gravy. Many times, returning late from missions, we would find that the chow which had been saved for us was stale sandwiches with…you guessed it, SLICED ROAST BEEF!

Operations gave us the heads up about dinner as our fire team was hovering out of POL to the gun revetments and we couldn't believe it! Ed Bilbrey and I were pissed and decided to skip the mystery meat, commonly referred to as "Blue Star special," retreated to my hootch, ate C-rations and had a "few" shots of tequila which we chased with some of that 'great' Crown Beer the ROKs had given us. By the time officer's call started, Ed and I had worked ourselves into a lather bitching about how 'the troops' deserved better, how all the f****** roast beef had probably come from LBJ's Texas ranch and how we weren't going to put up with this crap any longer!

As Officer's call began and the Executive Officer was briefing all the requisite administrative crap, Ed stood up, interrupted him and said, "The hell with all this Military Pay Certificate control BS! We need to talk about a serious morale issue, we're all getting sick of having this f****** roast beef almost every night!" Most of the pilots chimed up in support and Ed looked over at the rest of the Jokers, smiling and nodding his head.

The Commander, who was seated in the front row, now stood up and, to everyone's surprise, voiced HIS support, saying he was also getting sick of all that roast beef. Ed proudly raised his hands giving a double 'thumbs up' as everyone cheered. The CDR then added, ". . . and Lt Bilbrey, I believe YOU are just the guy who can correct this serious morale problem!" He then turned to the XO and said, "Place Lt Bilbrey on orders as the company's new mess officer, effectively immediately!"

That comment kind of cut through our tequila fog and as Ed sat down he looked at me and, even over the noise of everyone's laughter and cheers, you could hear him expressing how he felt about his new 'additional duty.' As it finally got quiet and everyone was looking back toward the XO, Beau Newton stood up and said, "Hey, Lt Bilbrey, what are you serving for chow tomorrow night?."..and everybody broke up laughing!

Ed took a lot of grief over the next few weeks and the roast beef stayed off the menu for the most part, but when we did have roast beef, Ed had it rough. We all took notice as he really became serious about improving the mess operations and soon, he and the mess sergeant became very close. When anyone complained about the food, Ed would 'return fire' and tell them about how he thought all the cooks and maintenance guys worked their butts off, but never got any credit.

Almost every time you saw Ed he would be reviewing the Army

manuals about mess operations and management of the Army's "master menu." He reviewed all documents related to food procurement, equipment, and the military occupational specialties of required personnel and decided the Table of Organization and Equipment (TO&E) needed to be modified. The TO&E being used only addressed equipment and staffing for an assault company, minus our attached support, so he and the XO figured out a way to increase the number of personnel assigned to the mess hall.

Things were starting to improve in the mess hall and in early November Ed told me that he wanted to make the 48th's Thanksgiving dinner the best ever and asked me to help him out. I had met an Army veterinarian at Cam Ranh Bay after he noticed me carrying an AK-47. The Vet was fascinated with the gun and asked me if I would be willing to trade it. I asked him what he had to trade and he responded that he was 'a Vet.' I told him I wasn't in the market for any animals and we didn't have any which would require his services. He said, "You don't understand, don't you know what my job is here? I have to inspect ALL Class I (rations) which arrive in country via the port of Cam Ranh Bay."

It didn't take long to realize what a valuable contact this guy was and, after gladly relinquishing my AK-47, I was soon loading cases of steaks and BOTTLED beer into the admin bird.

I told Ed this contact could be a big help to him, so we took my aircraft on a' test flight' to Cam Ranh Bay so I could introduce him to the Vet 'gun collector'. Over the next few weeks, Ed did a lot of networking with this guy and a bunch of his Air Force friends. I don't know what all he and the Mess Sergeant actually gathered from those guys, but I know that every time we could, we picked up weapons captured by the ground units we supported throughout our area of operations and gave them to Ed. He would head off to Cam Ranh Bay on a "business trip" and soon the mess hall became a lot more popular! It even became almost routine for the Mess Sergeant to walk into the Officer or EM clubs late in the evening with trays of fried chicken, hamburgers, hot dogs, fresh fruit or cookies and cake. That was quite a welcomed change.

In my journal I noted on 16 November 1970, "Ed says he's in a bind with a Thanksgiving deal he's working with the Vet at CRB...needs four AK-47s and two M-2 carbines. The deal will get us cases of fresh fruit, whole hams, wine, (other than that Mateus Rose shit), prepared (?) pies...and real LIVE turkeys! Ed's a great guy, but I'm betting against him on the live turkey deal. I think the Air Force guys are feeding him some BS...so I told him not to promise the old man on that one."

We heard through the rumor mill that the CDR is planning a 'health and welfare' inspection of our hooches and the enlisted barracks within the next week, so we developed a plan to gather some extra trading material for Ed to use in his efforts to close the deal with the Air Force. The SOP for the health and welfare inspections calls for all illegal/contraband weapons collected to be destroyed. The 48th's method of destroying the weapons was to have one of our aircraft fly off the coast and 'deep six' them. We had to somehow make sure that one of the Joker birds got that mission.

To make a long story short, the health and welfare inspection was a goldmine. And, yes, the Jokers flew the mission to "destroy" the contraband weapons and even "properly" certified the destruction paperwork.

Ed was very quiet about what he was up to, but finally on 20 November, I noted, "OPS got a call from an inbound slick and relayed a message from Ed Bilbrey telling me to meet him at the command pad. I wasn't sure what was going on until I walked to the flight line and saw Lt Ed Bilbrey, gun pilot and Mess Officer Extraordinaire, sitting in the back of a slick grinning from ear to ear while covered in feathers and turkey shit! He was wrestling with this 'brace' or 'gaggle' of live turkeys who, one or two at a time, would escape from their makeshift cages and thrash around inside the aircraft.

This event drew a lot of attention from almost everyone on the compound and after a rowdy welcome to this excited group of turkeys, who had just experienced their first helicopter ride, Ed's prized turkeys were finally escorted to their new digs up by the mess hall.

Every day you'd find a group of guys, acting like little kids, checking on the turkeys and playing with them. Ed's "big score" was a true hit with everyone and seemed to take our minds off the war. It also seemed like everyone was getting into the holiday spirit and looking forward to the big feast. Ed and I flew a mission together two days before Thanksgiving and I told him the Mess Sergeant had been bragging about him, telling me how much he had done for all his guys.

Ed was really humble and didn't say much, but he was really proud of what he and the Mess Sergeant had been able to accomplish. I asked him when they were going to prep and cook the turkeys and he said, "Well there's been kind of a change in plans, no one has the heart to kill the turkeys."

He said most of the cooks as well as the rest of the unit had become "kind of attached to them," and he was even referring to THEM, the turkeys, BY NAME as he told me the Mess Sergeant had promised everyone that they wouldn't kill and cook them. I asked him what we were going to have for our Thanksgiving dinner and he said they were "…working on that."

He and his mess team came through in spades and we ended up having all kinds of great food. The mess hall remained open all day and for once, you could have as much to eat as you wanted and because the food was so good, we all ate too much, but the only turkey we had came from a can. The "guests of honor," Ed's turkeys, stayed in our company area for a while after Thanksgiving but, one or two at a time, they "escaped" and as Ed said, "were probably over there hiding…IN THE ROKs!" as he pointed to the Korean Whitehorse Division compound.

I can't believe that was so many years ago. This year, I'll be celebrating Thanksgiving in my family's traditional way. I'll be amazed, as I am every year, by the variety and abundance of unbelievably delicious fare, as it is lovingly prepared and presented in the finest Southern tradition. I will give thanks for every blessing God has graciously rendered me.

After attending the funeral for the crew of Blue Star 811, who had been missing since 1967 before being recovered and buried at Arlington National Cemetery, and spending time with many of their family members, I will be thinking about what the families who have lost loved ones or have loved ones missing in action, have had to endure all these years. I can only imagine what it has been like to suffer the loss of someone you cherished so much or, for so many years, to cope with the anguish of wondering about the fate of those still missing. I have witnessed the pain of many of our friends who still face their personal demons from their Vietnam experience and I will pray they let that experience temper them, make them stronger, and enable them to find their peace.

I will try hard to recapture the feelings I experienced walking with fellow Blue Stars and family members across that hallowed ground at Arlington. Being there among those who have gone before us, viewing the granite headstones whose etchings not only reflect letters of a name, but illuminate memories of those with whom we shared life. Those men whose strength of commitment, in a most difficult war during one of the most trying times in our country's history, cost them their future. I will always remember their courage and selflessness.

In the excitement of the holiday's activities, I will do as I have done every year since 1971 and seek a place where I can be away from others who may not understand. I will cover the ice in my old canteen cup with scotch and, in that peaceful time of sunset, toast my Blue Star brothers who gave their all. While smiling through my tears, I will also remember a special warrior named Edmond David Bilbrey, who made his last Thanksgiving... my most memorable.

Captain Edmond David Bilbrey is honored on Panel 4W, Line 36 of the Vietnam Memorial Wall in Washington, DC. A documentary which highlights the heroic efforts of the 48th Assault Helicopter Company and provides details of Ed Bilbrey and all the unit's losses can be viewed by going on "YouTube," selecting "Battlefield Diaries" and then "LAMSON 719."

DON COWAN

United States Marine Corps — Aviation Maintenance

Dates of Military Service: 1968 to 1972

Unit Served with in Vietnam: VMGR-152 Sub Unit

Dates you were in Vietnam: Mar 1970 to Nov 1970

Highest Rank Held: Sergeant

Place of Birth and Year: Long Island NY — 1947

THANK YOU FOR YOUR SERVICE

I knew I was going to be drafted because I had gone for the draft physical. I didn't hear anything from the Air Force, got a letter from the Army, a call from the Navy, but the Marine recruiters came to my house one evening. They had a large book of all the military occupational specialties (MOS) and OCS (Officer Candidate School). They happened to flip the book open and one of the recruiters said, "Well, now if you enlist for four years, we guarantee to put you in the aviation field." And I thought, "They protect airplanes."

I enlisted in the Marine Corps in April of 1968. Got through Paris Island and Camp Geiger up in North Carolina. After that I was in a holding platoon for a while waiting to go to Avionics School in Tennessee. When I got to Millington, Tennessee for school I spent time in another holding platoon.

Probably the one thing that affected me most during my time in service happened there in Tennessee. Here I was fresh out of boot camp. I'd been in for four or five months, and I was now on burial detail. I was one of the pallbearers for Marines KIA in Vietnam. I did about six funerals. As a PFC, I was the ranking person among this set of raw recruit pallbearers. In each funeral we folded the flag, and then I would hand the flag to the body escort. We didn't do the slow deliberate salute back then, we did a very crisp salute. The body escort would then wheel and deliver the flag to the family, with those words, "On behalf of a grateful nation." I thought those Marines who paid the ultimate price deserved better.

I graduated from Avionics School in February 1969 and was assigned the C-130 Squadron VMGR-252 at Cherry Point, North Caro-lina. Served there until October 1969.

In October I got word that I was going overseas. After some leave at home I went to Camp Pendleton, California. Thanksgiving of 1969, left California headed overseas and landed in Okinawa. I stayed in Okinawa in the C-130 squadron VMGR-152, commonly known as Ichi Go NI, Japanese for 152. In April of 1970, because we had planes on the ground at Da Nang at all times, we had to have maintenance people there to work on the planes. The small maintenance detachment was affection-ately known as the 152nd High Altitude Slave Labor Battalion. The squadron insignia was a caricature of a ruptured duck holding an eight ball and a can of beer.

Da Nang was an Air Force Base. Our squadron tarmac was on the west side of the base. We Marines were the interlopers, if you will. We were just a C-130 squadron. You could look across the runways and see the Air Force barracks with window A/C units handing out. We were lucky to have a fan.

Most days were just the same. Some nights the sirens went off and then you heard the explosions. Other times you heard the explosion and the sirens went off. We were lucky not to be in a targeted area. You're bored to death waiting for the planes to come in to have any work done on them. When the plane taxied in, all the maintenance people jumped on as fast as they could because the plane was air conditioned. You got on, closed the doors, and did all your work. Everybody got off the plane at the same time. There was no going in and out and opening the doors.

One day we heard the sound of a plane engine failing during a take-off. You knew something was wrong. The plane crashed. It was an AD-4, which was a single engine prop with wings joined under the belly of the plane. It crashed between the two runways, slid and caught on fire. We were maybe just forty yards or so from it. The EOD, Explosive Ordi-nance Disposal jeep, came flying down the runway, circled the plane and took off. I don't know if they got the pilot out or not. From our vantage point we couldn't tell. We watched the plane burn and burn and burn. Finally the MAG firefighters come up in their truck and start to foam the plane. They ran out of water so they couldn't continue foaming the plane.

One of the Marines in his asbestos suit jumped off the truck with his hand-held fire extinguisher. He just about put all the flames out except for where the wings met under the belly of the plane. He was underneath the plane with a hand-held extinguisher putting the flames out when the EOD people ordered him away from the plane. The fire fighters came over to us cussing and complaining. The plane sat there and smoldered for a while, and then it blew up. Not a huge explosion, just a dull thud and the plane dropped into two pieces. The Air Force firetruck came and started foaming from forty yards out, inching closer to a smoldering nothing. We all thought, "The guy in the asbestos suit, we want him on our crash crew when they come out." It's just the way Marines do it.

I was in Da Nang for about four or five weeks before I came down with hepatitis. I didn't know what was wrong with me. I just knew I couldn't hold down solid food. I didn't even want to eat. I was living on Coca Cola, five, six, seven a day. I could hold down ice cream if I could get it. Availability of ice cream in Da Nang? I was sent to Freedom Hill Hospital where I stayed for a few days before I went to the Air Force Evac Hospital at Da Nang. I spent the night there and was loaded on an evac plane the next morning. Some people getting on that plane had serious injuries. The plane was due to leave Vietnam and go, if I remember correctly, to the Philippines, Guam, California, Texas, and then Washington, DC, dropping people off all along the way. I still didn't know what I had, but I had an idea it wasn't good because they put me in the back of the plane, and strapped my stretcher to the wall. The nurse came by, picked up my records, and went, "Hmmm" put on gloves and a mask and pulled a curtain around me. I thought, "This isn't good."

I was put off in Guam where I got my first taste of Navy hospital life. Because hepatitis is contagious, I was supposed to be kept in isolation. The admitting ward idea of isolation was to put me in the last bed at the end of the ward. Everybody else was up at the center of the ward, where they had fans blowing on them all night. I woke up at three the next morning, literally covered in mosquito bites, because I didn't have a fan blowing on me. I was just down there at one end of the hall. I started to get into a fight with a Navy Corpsman about taking a fan and putting it down there. The next morning I went through admitting and was sent to the ward I was assigned. Because it was an old ward-style hospital, their idea of isolation was, "You people will live on the sun porch. You have one head (head meaning bathroom in naval lingo) that you could use. You don't go anywhere else or don't do anything else."

Every Thursday the Navy had Field Day so the nurse came in and said, "Clean this area here, and clean the head then go clean the rec room." I looked at her and I said, "We're not going to clean the rec room. We don't get to go in there. We're in isolation. We'll clean this area; we'll clean our head. Thank you very much." And from there my relationship with the nurse just blossomed. Nurse Ratchet as I called her.

While I was there, I had a dream one night that my mother had made me a turkey sandwich on gummy white bread with cranberry sauce. The next morning I felt like I wanted to have breakfast. Over the course of the seven weeks I had lost 36 pounds. I'm six foot one and weighed less than 145 pounds. You know, if you don't eat for five or six weeks you tend to lose weight. A few more weeks went by before I was discharged. I went from Guam back to Okinawa where I was put in a holding platoon to go through training. I talked to the platoon sergeant and said, "I work on planes that are across the other side of the island." A few phone calls, and about an hour or two later, a jeep came by to pick me up and take me back to my squadron. I was there for a few weeks then back to Vietnam.

I was in Vietnam just in time for the summer monsoon season, followed by the heat. We were always scouring the PX for mini refrigerators or fans, because if you had a refrigerator or a fan, you were golden. We would go up to Freedom Hill because there was a large PX up that way. To get from the airbase to Freedom Hill, you went outside the base and went through a place we called Dogpatch, which was just a two-lane road. On one side was rice paddies. On the other side was a row of 8x8 wooden crates that had a piece of fabric for a door. You saw how the other half lived. How people in the direst straits in that war zone were just living hand to mouth.

Of course, when you rotated back out again, your fan and refrigerator went up for auction to the highest bidder because without a fan you were eaten alive!

I was in Vietnam until late November. When I got back to Okinawa I was due to rotate back home. Luckily for me I knew the Squadron Clerk in Okinawa. I said, "Scotty, if you keep me here for a couple more weeks by losing the paperwork, I'll be able to spend Christmas and New Year at home, before I go to my next duty station." The Gunnery Sergeant in charge of the Avionics Shop just said, "Make sure you show up some time each day so we know you are alive, but other than that..." I pretty much had free run for a couple of weeks.

Another odd thing, the Air Force didn't have any C-130 maintenance people on the ground at Da Nang because they didn't have any C-130s based there. They could get parts, but they had no one to work on their planes so occasionally we would be called to fix something on one of their planes to get things going. I remember one time I went over to work on a piece of gear, and the plane was full of Vietnamese civilians. The plane didn't have the bench seats pulled down. These civilians were just literally in the belly of the plane sitting on the floor. I got there to work on the plane and the crew made all of them get off the plane.

Now this is summertime in Da Nang. There was one lady there with an infant just a few months old in her arms. These people were not in my way. I could have worked around them on the plane. Now this lady and her baby were standing in the sun and I thought, "Couldn't you have gotten some shade for these people? Couldn't they stand under the wings of the plane or something? Why have you decided that these people don't deserve to have any kindness whatsoever?" They had to get out of the way and then get back on the plane. Herded cattle. Something about it just didn't seem to fit with what we were trying to do.

I've never been back to Vietnam nor Parris Island. I do see fellow Marines occasionally. I had a great Christmas about two years ago. I had gone to a UPS store to send a package to my son and grandsons. There was a long line that sort of circled about halfway around the store. Across the room from me was a black gentleman wearing a well-worn field jacket that had a big Marine Corps emblem on it. I said, "Semper Fi, Marine. How are you doing?" And he gave me a Semper Fi reply. I asked what he did in the Corps. He was a mortar man, a ground-pounding Marine. I asked him if he'd been to the Nam and he nodded back. We conversed back and forth before my turn at the counter came up and my packages were off.

As I started out of the store, this gentleman stopped me on the way out and said, "Thank you. I really appreciate being recognized." And I said, "It's what Marines do. We look out for each other."

He said, "I hope you have a Merry Christmas." I said, "I think we just did."

Simultaneously, as we turned to leave, we said, "Thank you for your service." Then we both said, "Semper Fi!" That also is what Marines do.

FRANK CRANFORD

US Army — Military Policeman
Dates of Military Service: 1970 to 1973
Unit Served with in Vietnam: 188th MP Company, 504th MP Battalion, 18th MP Brigade
Dates in Vietnam: Dec 1970 to Dec 1971
Highest Rank Held: SP4 (E-4)
Place of Birth: Macon, GA — 1948

MY FIRST DAY ON DUTY — HOW I GOT HERE

Exactly a year before my first duty day, I had been a senior at the University of Georgia, enjoying life except for the ultra-low number that had been drawn for me in the first draft lottery. I figured I still had a deferment until June to figure a way out of it.

In July, I reported and was sent to Fort Jackson, South Carolina for Basic and then on to Fort Gordon, Georgia for Military Police School, thinking I might go to Germany or even stateside with such an MOS.

In December, after 18 days of leave, I departed for RVN via Oakland

Army Terminal, Anchorage, Alaska, Japan and in to Tan San Nhut Air Base, still having no idea of which unit or location I was headed for. I do know that I filled enough sand bags to acquire a secondary MOS while I waited in the Replacement Station. Apparently, the Army can never have enough sand bags.

Finally we started north by Army transport aircraft, landing seemingly everywhere the pilot saw an airfield big enough to set down on. Some of the replacements had their names called and deplaned, and others took their places in the slings provided for seating.

We finally arrived at Da Nang Air Base, where several of us were met by the driver of a deuce and a half, who transported us about five miles to the HQ of the 504th MP Battalion. There we found temporary quarters while the S1 folks figured out where to put us. I was told to take a walk up to HQ where an E-6 took my file and asked if I could type. I really didn't want to do that so I declined and was sent back down the hill.

WAITING FOR AN ASSIGNMENT

Luckily for me, I had a few days down time to settle, when one of the platoon sergeants came by and told a few of us that there was a deuce and a half headed to the USO Bob Hope Show and to go sign up if we wanted to go. Several of us were all over that, so off we went to the Freedom Hill Amphitheater which was a hill with a stage set up at the bottom and room for several thousand troops to sit on the ground and watch the show. I honestly don't remember much about the show, except it was hot as hell and there was no shade from the sun. I'm not the brightest guy in the world, but I had enough sense to keep my shirt on, unlike some of the other FNGs who were in attendance. One of the guys who came over with me got absolutely blistered and was out on sick call for another week.

Sometime just after Christmas, I was assigned to the First Platoon of the 188th MP Company which was located right where I was at Camp Land. There was a Marine unit there of MPs and dogs, a stockade, and Battalion HQ.

All in all, not too bad of an assignment.

GOING TO WORK

I received my initial equipment issue, most of it well worn. Steel pot and cover, black helmet liner with "MP" painted on the front and "504" on the side, brassard, web belt, ammo pouches, flak jacket, all the normal stuff MPs need to function, I suppose. Also an M16 to be kept in the hooch and carried at night and a Colt 1911 to be kept in the armory and checked in and out were also assigned. Other weapons could be drawn out as needed.

We would work 12-hour shifts from 0600 to 1800 hours daily, changing to the overnight after a month or so then back to days. First Platoon happened to be on days when I arrived. Except for the Bob Hope Show, I had not been off the compound since I arrived a week or so earlier. I had managed to find the EM Club and the Day Room, so all was not lost.

I made my first guard mount, which was held in the company area at 0500. I was paired up with a Coloradan named Buck, who looked to me like an MP should look: Fairly tall, solidly built, good tan, with a properly fitting uniform without wrinkles; he wore mirrored teardrop sunglasses to complete the look.

From there we travelled by convoy of around ten jeeps containing Army and Marine MPs to our AO, the City of Da Nang. The route took us back through the air base and about ten kilometers or so further, crossing a broad river into the city. I probably don't need to tell you about the culture shock I felt witnessing life in the city. There were few cars, but seemingly thousands of motorcycles and scooters and Lambros and the occasional cyclo being pedaled by an old papasan. All motorcycles, regardless of make or power, were called Hondas. They all drove like bats out of hell and I did not look forward to navigating in what appeared to me to be total chaos. Buck seemed to handle it fairly calmly.

Our convoy arrived at the downtown Combined Police Station after about a thirty minute trip. The desk inside the old, dingy green and white building was manned by a Desk Sergeant, a Radio Operator, and a Clerk. The ARVN, the Koreans, and the Vietnamese National Police also operated from this location. There was a raggedy D Cell attached which was attended to by the turnkey, or jailer. Being the turnkey was crappy duty that I would try to avoid as long as we operated from this location.

ON THE ROAD

At the CPS, there was another guard mount where we received intel which meant nothing to me at the time, and were assigned an AO. Buck and I moved out, he making small talk, advising me to keep my arms inside the jeep at the risk of losing a watch, and me on high alert because it's my first day among the locals who I don't know and don't trust.

Not too long after we began our patrol, Buck pointed out a Honda among all the traffic that had an African American in the green army jungle fatigues riding as a passenger behind a local driver. Buck told me in so many words that riding on civilian transportation was unauthorized and that we would be stopping them to check out the GI as he was possibly AWOL, and at the least was in an off-limits area. The entire city was off-limits at this time.

We pulled behind them and Buck began to blow the horn and flash his headlights, signaling the Honda driver to pull over, which he did. We stopped behind them and since I was fresh out of MP school, I knew exactly what I was supposed to do. I exited our vehicle and walked to a point to the right and behind the bike. Buck approached from the left. As we approached, the GI got off the bike and reached into his pocket for what I thought was his identification. Turned out it was a six-inch blued .38 revolver that he pointed directly at my face. I knew it was loaded because I could see the rounds in the cylinder. Time slowed down for me as I looked down the barrel of the biggest .38 I had ever seen in my life and I eased my hand toward my sidearm. The GI was highly agitated and shouted, "Why y'all fuckin wit' me?!", along with some other expletives as he waved the weapon around.

At that point, both Buck and I told him we didn't intend to hurt him and kept our hands away from our sides, attempting to diffuse the situation and stay un-shot in the process. He quickly hopped on the bike, put the gun to the Vietnamese man's head and they rode away into the traffic. We remounted and Buck got on the radio as we followed the GI and his now-hostage Vietnamese driver. Buck radioed in that we were in pursuit and that deadly force might be necessary. He looked over at me and told me to lock and load, so I racked the slide back on the .45 and chambered a round.

The bike continued to dodge between all the other traffic, and we followed, blowing the horn and flashing lights trying to warn other drivers and pedestrians out of the way. We were soon joined by some other units and of course I had no idea where we were, since this was my first day in the city. I was also holding a locked and loaded .45, unsure if I was going to have to use it.

As we were following, the bike took a sudden turn into an alley and into the ville, a neighborhood of small buildings crowded together on each side of a very narrow street, just wide enough for one jeep. We had to slow considerably in order to follow, but we were able to see the GI jump off the bike and run into a hooch while the Vietnamese bike driver continued on, much to his relief, I imagine.

By this time there were four or five units involved and stopping in the alley. Eight to ten MPs were now on foot and moving toward the hooch with M16s and .45s drawn, locked and loaded. You could hear the radios squawking in the background. To add to the confusion, Vietnamese were yelling and running and some of them pointing toward the hooch. After a short time, the GI exited the hooch, however he was now wearing papasan clothes that looked like black pajamas, a conical coolie hat, and waving a machete about eighteen inches long over his head. He was yelling that he was the "Black Moses" and had come to free his people.

Once he looked around and realized he was probably overmatched, he dropped his machete and put his hands in the air. He was quickly put into hand irons and placed in the back seat of our jeep along with another MP for our trip back to the CPS.

BACK AT THE STATION

Back at the station, we turned him over to the Turnkey, and he was placed in the D Cell, still ranting as he had been since he had been apprehended, about being Black Moses and his mission. The D Cell was pretty ramshackle, and had to be accessed from outside the station, even though it was attached. Somehow, the Turnkey and another couple of guys from inside the station convinced him it was in his best interest to sit down and shut up.

He did have an ID with him, so Buck took it and we went around and into the station to complete the DA 19-32, which is a form all MPs are familiar with since it is used to document any incident, no matter how large or small.

We sat at a small desk and Buck began filling out the report. After several minutes, I noticed Buck seemed to be having trouble with the report, and had asked me how to spell some relatively simple words. I glanced over at the report and could hardly read his writing, but I saw enough to know his spelling and grammar were terrible.

Even though it was my first day, I could tell he was relieved when I offered to write the report for us. Being fresh out of training, I was thoroughly familiar with the Form DA 19-32, having spent a couple of days learning to write one. It also didn't hurt that I had very precise handwriting that was easily read. The form had to be printed, and I had spent one summer working at my Dad's engineering company doing some drawings, so my printing was neat and legible.

As a matter of fact, after the Desk Sergeant reviewed it, he asked who wrote it. I admitted I had done it and he just said, "Nice" and went back to whatever he was doing.

Just goes to show that no matter what your talent is, the Army could find a way to use it.

So even though it was an exciting day, I found myself hoping that they weren't all going to be like that. And they weren't. Most were better and a few were worse. I spent the next weeks learning all I could from my more experienced partners (even the Marines) that I rode with.

But I do have to say that my first day was memorable and not at all what I expected. I learned a lot. Including the meaning of the phrase "dinky dow."

JOHN BUTLER

US Army — Field Artillery Officer
Dates of Military Service (Active Duty and Reserves): 1968 to 1974
Unit Served with in Vietnam: 11th Armored Cavalry Regiment, 2/94th Heavy Artillery Dates in Vietnam: A Troop, 1st Battalion, 11th Armored Cavalry Regiment (11th ACR), Oct 1970 to Feb 1971; 2nd Battalion, 94th Heavy Artillery, Feb 1971 to Oct 1971
Highest Rank Held: 1LT
Place of Birth: Pittsburg, Kansas — 1945

CHRISTMAS ON HILL 652

One of our missions when I was an artillery forward observer (FO) with the 11th Armored Cavalry Regiment (11th ACR) was protecting a company of Army Engineers utilizing "Rome Plows" to clear jungle. Rome plows were D-7 and D-9 Caterpillar bulldozers with special land-clearing blades manufactured in Rome, Georgia.

Our mission was to clear a path 100—200 yards wide along what on the map showed as being a road. The road was completely grown over so that we ended up "defining" the road where the map said it was supposed to be.

The mission lasted about six weeks. We set up night defensive positions (NDPs) periodically from which we operated every day, relocating to a new NDP about once each week to keep close to where the plows were operating. Each NDP was an area cleared by the Rome Plows and surrounded by a dirt berm pushed up by the plows. The NDP was large enough for two company-sized units consisting of about 300 soldiers with about 60 tracked vehicles. The 30 or so 11th ACR tracked vehicles would set up just inside the berm facing outward all around the perimeter to provide protection for the NDP.

A few days before Christmas of 1970, our NDP happened to be located on the top of Hill 652 (652 meters above sea level), the highest point in any direction for many miles. When we occupied and set up that NDP, we discovered that it had been recently occupied by a large NVA unit. Plenty of equipment had been left behind, along with countless 20 MM shell casings from attacking US war birds.

As soon as we set up our NDP, our guys decided that we should create a huge Christmas tree in the sky using hand-held flares. Our plan was to make it happen at midnight on Christmas Eve. We scrounged and saved as many flares as we could get our hands on, especially red, white, and green ones. We even managed to have some brought out to us with other supplies on our daily supply chopper.

These hand-held devices were about 1 ½" in diameter and 10" in length. When the cap was removed and repositioned on the opposite end, then struck hard with the palm of a hand, a colored ball would be launched 500—600 feet in the air and burst into a very bright colored display.

By Christmas Eve, we had a few dozen of them ready to go. We radioed our supply base back in Song Be to let them know to be looking our way at midnight. Sure enough, at the stroke of midnight, we lit up the sky with a huge, bright display of red, white, green and amber light. Our guys back in Song Be said that sure enough it looked like a giant Christmas tree!

We thought we were so cool and tough to pull this off to help celebrate Christmas. Then as the adrenaline from our little adventure wore off and we were just individuals, alone in the dark night, I heard a familiar song playing on someone's radio on Armed Forces Radio Network (AFVN), and I wasn't so tough or cool anymore.

Every time I hear that song, I remember Christmas on Hill 652. The song: "I'll Be Home for Christmas."

CHARLES LOUIS SINGLETON

US Army – Combat Intelligence and Operations

Dates of Military Service (Active Duty and Reserves Combined): 1969 to 1975

Unit Served with in Vietnam: 1st Cavalry Division, Airmobile, 2nd Battalion, 12th Cavalry

Dates you were in Vietnam: May 1970 to Apr 1971

Highest Rank Held: Sergeant E-5

Place of Birth: Summerville, SC – 1947

A VIETNAM SOLDIER'S STORY, BODY BAGS

When I arrived in the Republic of Vietnam at the Bien Hoa Military Airport on May 25, 1970, the very first thing that caught my attention was the large number of body bags (looked similar to black garbage bags) that were stacked on the Bien Hoa Military Airport tarmac. Immediately, I said to myself, "I hope and pray that I will not go home back to the United States in a body bag." And, I further felt that this would hurt my parents too much to learn of my death and being placed in what looked like a garbage bag.

Thus, during my tour of duty in South Vietnam, I survived several "firefights" and three ambushes. Looking back, mentally, one reason I survived several near-death experiences in the jungles of South Vietnam was my determination not to be killed in combat! Thus was the case, on October 7, 1970, after being wounded by a Chicom grenade. During this grenade episode, while I was being treated, I kept saying to myself, "I hope and pray that I will not go home back to the United States in a body bag."

A few months later, while we were on a reconnaissance patrol, a North Vietnamese radio code in English was sent to us and requested our location. One soldier in my unit said to our RTO (Radiotelephone Operator) and Lieutenant, "Don't SHACKLE: 10 letter code words to map co-ordinates!" This was a North Vietnamese RTO trying to locate us, so that his mortar unit can fire on our location. Shortly afterwards, they fired mortars trying to get us to call in artillery support from a neighboring US Army Firebase. While this was going on, I once again kept saying to myself, "I hope and pray that I will not go home to the United States in a body bag."

That day, my fellow squad member's timely advisement helped to secure our location and saved our lives.

HUBERT "HUGH" BELL

US Army — Aviator
Dates of Military Service (Active Duty and Reserves Combined): 1966 to 1997
Unit Served with in Vietnam: Co A, 25 AVN BN, 25 Inf Div (first tour); 73rd Surveillance Airplane
Co (second tour)
Dates in Vietnam: Mar 1967 to Mar 1978; Jul 1971 — May 1972
Highest Rank Held: Lt. Colonel
Place of Birth and Year: Elberton, GA — 1942

A SICK AIRPLANE BUT A FORTUNATE END

On my second tour in Southeast Asia I was assigned to the 73d Surveillance Airplane Company based at Long Thanh North, located southeast of Saigon just off the highway to Vung Tau. I arrived in Vietnam on this tour in July 1971. The 73d flew SLAR and IR ("infrared camera") missions each night all around the III and IV Corps area. There I was given command of the SLAR (side-looking airborne radar) platoon and in addition, I was the unit instrument instructor pilot and the unit instrument flight examiner. I had recently commanded the Instrument Flight Examiner School at Fort Rucker before completing the Armor Officer Advanced Course at Fort Knox, Kentucky, and the Army Fixed Wing Qualification Course and the OV-1 Mohawk transition course.

After a few months in RVN, 73d was ordered to take over the Mohawk detachment at Udorn Royal Thai AFB in northern Thailand. This detachment flew SLAR and IR missions in the Plain of Jars and the Bolovens Plateau in Laos. I was selected to command this detachment, probably because I was the third most senior officer in the 73d after the company commander and the operations officer. I went to Udorn on TDY to command the airplane unit. I had about 30 people in my detachment, including pilots, technical observers, crew chiefs, aviation maintenance people, and imagery interpretation people, and other support personnel. I had five OV-1 Mohawk aircraft, both SLAR and IR models.

All our missions were fragged out of the Army attaché office at the US Embassy in Vientiane, Laos, 40 miles north of Udorn. We were OPCON to the embassy in Vientiane. I was a member of the Embassy country team which included representatives of the Air Force, CIA, State Department, USAID, and other alphabet soup agencies. I attended monthly meetings at the Embassy but was not allowed to fly into Vientiane in my own aircraft to maintain the fiction that there were no US troops on the ground in Laos. The Embassy would send an Air America airplane, (a Volpar converted Beech 18 twin engine airplane with tricycle gear and PT-6 turbine engines) to pick me up for the 20-minute ride across the border to Vientiane.

My boss was still at Long Thanh North in Vietnam, and we were in contact daily by telephone. Although the US Air Force 432d Tactical Reconnaissance Wing was the principal US tenant at Udorn and provided inter-service support for my detachment, any maintenance on my aircraft beyond the crew chief level had to be performed back at Long Thanh.

One of my aircraft required some maintenance, and I decided to fly it across Thailand, Laos, and Cambodia to Vietnam for the necessary service. I was alone in the aircraft; no T.O. with me. I flew southeast to Long Thanh early in the morning, got the work done on the aircraft, and departed Long Thanh North airfield just at dark. I was travelling northwest across Cambodia to northern Thailand and had been airborne an hour or a little less.

It was completely dark, few lights on the ground, about 2100 hours. Suddenly the annunciator panel lighted up. The engine oil chip light for the number one engine came on. This meant a bearing in that Lycoming turbine engine was disintegrating. In a moment the low oil pressure light for the number one engine came on. That engine was in serious trouble.

Although the engine was continuing to run, it could not last long. Therefore, in order to avoid engine damage beyond repair I elected to shut down that engine. I shut off the fuel supply to that engine, feathered the propeller, and turned back east toward Saigon.

I came up on the guard channel on my UHF radio, 243.0 mhz., and transmitted Pan, Pan, Pan which is the internationally recognized aviation distress call indicating a problem not quite yet an emergency, which would be signified by transmitting MAYDAY. I still had one engine operating normally and the aircraft was quite flyable.

I radioed Saigon Approach Control. The controller asked me to squawk 7700, the transponder code for an emergency so that my aircraft would be given special identification on the radar screens. I requested radar flight following and vectors to an ILS approach to Tan Son Nhut. I wanted 12,000 feet of runway and the best firefighting equipment for a single engine night landing.

Saigon gave me vectors for a straight-in approach and, although the weather was good and skies were clear, I shot a localizer ILS approach to the long runway just to make sure I didn't screw up the alignment with the runway while I was dealing with other things.

I still had hydraulic pressure which, in the Mohawk, operated so many systems, including the landing gear, flaps, nose wheel steering, brakes, etc.

At the proper point in my descent to landing, I dropped 20 degrees of flaps and then to begin the final descent to the runway I lowered the gear. I was very glad to see the three green lights indicating that my main and nose wheel gear were down and locked and I still had hydraulic pressure for steering and braking. I then lowered the flaps to 40 degrees, switched on the landing lights, reduced to landing airspeed, and made a normal landing. The only difference with single-engine landing with all other systems operational is that reverse thrust is not available to slow the aircraft on the runway. Therefore, I reduced the thrust lever for the good engine back to flight idle and relied only on the hydraulic brakes to slow the landing roll.

Ground control directed me to a parking spot near a maintenance hangar. I contacted 73d Operations by telephone and asked them to send out a couple of maintenance people with another aircraft the next morning. They did and the next day I returned to Udorn in a different aircraft.

BILL MCRAE

US Army – Armor Officer, CH-47 Helicopter Pilot
Dates of Military Service: 1969 to 1973
Unit Served With in Vietnam: 132nd ASHC, 14th CAB, 16th CAG, American Division
Dates Served in Vietnam: Oct 1970 to Oct 1971
Highest Rank Held: Captain
Place of Birth: Gainesville, GA – 1947

PHU BAI FIASCO

In early January 1971, my unit, the 132nd Assault Support Helicopter Company, was ordered to relocate from Chu Lai to Phu Bai, and we were to take all flyable aircraft with us. We took off in flights of two and basically followed Highway 1 north to Phu Bai. We were to take part in the largest Army aviation operation of the entire Vietnam War. Within a few weeks we had all sixteen of our aircraft on the ramp at Phu Bai. The mission was called Lam Son 719.

Approximately 44 Army aviation units were to assist in the re-establishment of Khe Sanh as a forward operating base and to fly full support for the South Vietnamese Armed Forces, when they hopped the fence and went after the North Vietnamese Army in eastern Laos. The general direction of the offensive was westward along Highway 9, from the border southwest of Khe Sanh, all the way to Tchepone, Laos. The 132nd lost one aircraft, shot down and destroyed in Laos. The five man crew was successfully rescued. To my knowledge they are the only crew in history to successfully survive a dual hydraulics failure in a CH-47. Their survival story was written up in the "Stars and Stripes."

The 132nd ASHC was a medium lift helicopter unit. We flew the CH-47B, Chinook. Most of our loads were external type, consisting mainly of artillery pieces, ammunition, fuel, water, and food. We also recovered downed aircraft in the area. This work went on all day long, every day, seven days a week. In the evenings we flew back to Phu Bai to pull maintenance on the aircraft and to prepare for the missions planned for the next day. Any downed aircraft would also be hauled back to Phu Bai for repairs.

On one day I was part of a two ship mission to return two damaged aircraft to Phu Bai. We completed the mission and returned to our company ramp area to park our aircraft and shut down for the night. It was well after dark by the time we got everything done, so we could call it a day. The company area was relatively close to the airfield, but too far to walk. So, our Flight Operations had a 2½ ton truck to shuttle crew members to and from the flight line. A clerk, Private Carter, from Flight Operations was the driver.

Each Aircraft Commander had two classified items that they had to account for and turn in at the end of each day. One was the SOI with all the radio frequencies and call signs, and the other was the secure radio, called a KY-28. These items were all turned in, and we were ready to get back to the company area, grab something to eat, and hit the rack, so we could do it all over again tomorrow. We normally flew from sunup to sundown, or until the missions were complete.

There were ten of us in the two crews waiting to get a lift over to the company area. The truck is in the usual parking spot, but the driver was missing. We yelled and called for Carter all around the Operations hooch, but we got no response. Of the four pilots standing there, I was the ranking man, as a 1st Lieutenant. So, I took charge and asked if any of the enlisted crew members could crank and drive the truck. A SP4 Edwards said he could, but that he did not have a license. I told him I did not care about the license. I told him we needed to get everybody back to the company area, and we did not want to wait for somebody, somewhere, to find another driver and get him over to the flight line. Everybody was tired and ready to go. So, I told Edwards to get in the cab and crank the truck. I got in with him. The other eight guys climbed in the back. The diesel engine started right away. I held my flashlight so we could figure out how to work the headlight switch. The truck ran for a little while before we finally figured out that switch.

It was probably about 2100 hours when Edwards finally backed the truck out into the road and headed down the empty dirt road which paralleled the runway. We were doing just fine for about 300 yards, when all of a sudden the guys in the back started pounding on the cab and yelling for us to stop. Edwards started slowing down, and I looked at him and said, "I don't know what is going on, but one thing is for sure; you did not run over anybody!" When the truck stopped, the guys in the back started jumping out and running back down the road. I stepped out on to the running board and looked in that direction. To my amazement, there appeared to be a body lying in the road. I jumped down and ran back to the crowd gathered in the road.

There was in fact a dirty, nasty looking, guy lying in the road, covered in dust and dirt from head to toe. It was Carter! He had gotten high on something and went to sleep under the truck. When we backed out, he grabbed the axle, and we dragged him down the road, until he finally let go. Carter had some scrapes and bruises, but he was not seriously hurt. I really did not care. I was pissed. I hoped his minor injuries would hurt like hell for a long time. He still could not drive.

That's when my attempt to get us back to the company area got complicated. The MPs showed up! We were all American Division troops, and the vehicle had Americal markings. They were 101st Airborne Division. We were visitors in their neighborhood. I explained to the E-5 MP what had happened and that we would take care of Carter. The MP figured out that our driver did not have a license. That's when he informed me that he would be issuing a traffic ticket to Edwards. I told the MP that I was aware that Edwards did not have a license when I told him to drive the vehicle. I told him that Edwards did exactly what I told him to do, and that he did not deserve any punishment. I told him we had been flying out of Khe Sanh all day, and we just wanted to get back to our company area, so we could do it all over again tomorrow. I further informed the MP that this is a war zone and not a stateside post. I told him that if he felt like a ticket was justified, he should issue the ticket to me and not to Edwards. He said the ticket has to go to the driver. I said the only reason Edwards is the driver is because I made him the driver. It was clear we were not getting anywhere, and Edwards got the ticket.

A few days later, I got back to my hooch and found a note saying the CO wanted to see me in his office at 1900 hours that night. I reported as instructed. Already in the room were the CO, the XO, and SP4 Edwards. The CO informed me that he intended to give Edwards an Article 15 for driving without a license. I explained to the CO everything that had happened. I told him that Edwards did exactly what I told him to do and that he did not deserve any punishment. I suggested that Edwards should get a gold star for obeying my instructions. I also suggested that the CO punish me instead of Edwards, if he felt some kind of punishment was warranted. The CO said, "Lieutenant, I intend to do just that. I'm putting a Letter of Reprimand in your 201 File!" I said, "That's fine sir, if you feel you need to do that. Just don't punish Edwards for doing exactly what I told him to do." He said the Article 15 was a done deal. I responded by telling the CO that I would go to bat for Edwards if he wants to fight it.

About a week or so later, Edwards and I went over to the JAG Office at the 101st Airborne Division. We met with a JAG lawyer, who agreed with our position on the Article 15. It was dismissed and removed from his record. I don't know if I got the Letter of Reprimand in my 201 File or not.

All these years I have blamed myself for what happened that night in Phu Bai. It just occurred to me. I did not cause all the problems. It was Carter! It was Carter who caused the scene that attracted the MPs. Otherwise, nothing would have ever happened. That's my story, and I'm sticking to it!

RICK MAROTTE

US Army—Medical Service Corps Officer
Dates of Military Service: 1969 to 1973
Units Serviced With in Vietnam: 571st Medical Detachment/237th Medical
Detachment
Dates Served in Vietnam: Jul 1970 to Jul 1971
Highest Rank Held: Captain
Place of Birth: Denver, Colorado

LAM SON 719 AS A DUSTOFF PILOT

I served as a Commissioned Officer Medical Evacuation (Dustoff) Helicopter Pilot in Vietnam from July 1970 to July 1971. In late January 1971, I returned to Vietnam after spending a week in the US on leave with my family. When I left the US, there was a news blackout for Vietnam. There was something going on they didn't want to tell to the rest of the world. When I got to Vietnam, in-country flights had been cancelled except for those in direct support of combat efforts. The action seemed to be in the north, close to the DMZ, my area of operation as a Dustoff Pilot.

So how did I end up in Vietnam flying an unarmed UH1-H (Huey) helicopter into combat with a big red cross painted on the side as an enemy target? It started with college ROTC. I had decided to become a helicopter pilot during our summer camp at Ft. Riley, Kansas. After six weeks training in the hot Kansas sun, I decided flying was better than walking, and a dry bed in a home base was better than sleeping in the jungle on the ground. My senior year in college I was accepted into the Army flight training program and was given fixed-wing flying lessons (I guess to make sure I wouldn't chicken out). On June 19, 1969, I received my BS Degree in Civil Engineering and a Commission as a Second Lieutenant in the US Army Medical Service Corps. After flight school, I went to Vietnam and was assigned to the 571st/237th Medical Detachment in Northern I Corps.

Lam Son 719 was the operational name for the invasion of Laos by the South Vietnam Army (ARVN). The operation was part of President Nixon's Vietnamization program and was made top secret to keep information away from the North Vietnamese Army (NVA). The Lam Son 719 mission was to cut off the Ho Chi Minh trail, a major food and arms supply route from North Vietnam to South Vietnam. The ARVN executed this operation with Air Support from the US, including medical evacuation (me). On February 8, 1971, the ARVN invaded Laos.

Four days after returning to Vietnam, I was at our base camp in Khe Sanh, the former Marine base that endured a long siege in 1968. I was living in half of a corrugated culvert pipe covered with sand bags, my bunker. Many missions were deep into NVA-controlled Laos. During this operation, aircraft losses and crew casualties were at the highest level of the war. As a pilot, I was called upon to go to the front line to rescue wounded soldiers and pick up the dead while under fire. Communications with the ground troops was limited to the radio operator who spoke broken English and soldiers that understood sign language to say, "hurry up," "get down," "run" and "bye bye."

I recall my most dangerous mission was the rescue of injured Special Forces troops that were in heavy contact with the NVA. Trees and jungle made the use of a rescue hoist necessary. Hoist missions must be executed while maintaining a high stationary hover while individual soldiers are lifted to the helicopter. Adding enemy fire to the situation made this the most dangerous mission of my tour. We were fortunate to have two Cobra Gunships for our protection. I had the Cobras saturate the enemy with mini-guns and rockets while we flew to the pickup location and while still under enemy fire, hoisted four casualties into my helicopter. We flew away from the site, flying low level in the tree tops, until we were clear of enemy fire then flew to the closest aid station and waited for the next mission.

The closest I came to death's door was deep in Laos on our way to Fire Base Sophia, when we took fire from a 51-caliber, anti-aircraft machine gun. We knew there were anti-aircraft weapons in the area, so we flew at 1500 feet and made lazy "S" turns. We hoped the NVA wouldn't waste radar guided fire on a Dustoff Helicopter, so we expected our altitude and evasive maneuvers to make us a small target. **WRONG!!** The NVA let loose on us. I heard the noise of gunfire then as I felt a sudden jolt to my controls, my crew chief yelled, "We've been hit." I knew he was right. The aircraft was still flying, all the instruments were in the green, so we decided to risk an aircraft failure and try to save the casualties. We made the pickup and flew back to Vietnam but this time on the other side of the valley, away from the NVA gun. After landing to drop off the patients, we shut down to inspect the helicopter. There was a hole in one rotor blade and one in the tail boom. We thought we were still flyable enough to make it back to the airfield for repairs. We didn't know what caused the jolt but we thought we were OK.

The next morning, the maintenance sergeant presented me with a gift, the front half of an armor piercing 51-caliber anti-aircraft bullet. The bullet was lodged in the split cones, just below the "Jesus" nut, the nut that holds the rotors to the rest of the helicopter. It is called the Jesus nut because, when flying, it is the only thing between you and Jesus. He said if it had hit a few millimeters over in any direction, we would have probably lost our rotor and learned first-hand about the name of the nut. But, we were safe and knew what had caused the jolt. Needless to say, I had an extra few drinks at the officer's club that night.

The most difficult mission and the most emotional occurred on the night of February 20, 1971. Several Dustoff Helicopters were commissioned to make a large-scale extraction deep in Laos. The mission was to be conducted at night. Unfortunately, fog settled into the area making flying conditions difficult at best. Shortly after refueling, one of our Dustoff Helicopters went out of control in the fog, crashed near the Khe Sanh runway, then burst into flames. The pilot and crew chief were immediately killed, and the co-pilot and medic seriously injured and burned. We had just refueled when we saw the crash. We hovered to the burning wreckage where my crew attempted a rescue. In the process, I used rotor wash to blow heat and flames away from my crew. We managed to recover the co-pilot and medic. In the meantime, heavy fog had moved in, what we pilots referred to as pea soup. There was no time to analyze the situation, figure out options, or discuss what we were to do next. I called air traffic control, declared an emergency, and did an instrument takeoff into the fog. I was familiar with the mountains surrounding Khe Sanh and knew the safe direction to fly. I flew on instruments to the hospital and delivered the patients to the emergency room of the 18th Evac Hospital (MASH) for care. The crew chief and medic were treated, unfortunately, they did not survive.

There were many other noteworthy missions. One involving intentionally chopping down trees with the rotor blades to clear a landing zone. The Huey blades could easily cut down small trees less than six inches in diameter. Another was where I followed a parachute flare through fog into an LZ only to have the flare drift almost into our rotor blades. It was scary but we were able to dodge the flare, pick up the casualty, and return to base.

Finally, we made perhaps a Guinness Book of Records extraction of over 25 ARVN soldiers that stormed our helicopter to catch a ride home. The ARVN soldiers dropped the casualties on the way to the helicopter and jumped on board. Towards the end of the operation, the ARVN were aware that the last soldier out of Laos would probably be killed or captured. They saw the Dustoff Helicopter as their ride home. On this mission, we had so many ARVNs on board that we didn't have enough power to take off or even lift off the ground. My crew chief and medic pushed, kicked, and punched ARVN soldiers to get them off until we could finally break ground and make a running take off through some bushes and trees. We made it back and I don't think anyone fell out. After that mission, the crews put grease on the skids to make it easier to manage too many passengers.

We saw many seriously wounded soldiers, some friendly and some enemy. After each pickup, I would check out the condition of our patients and usually look into their eyes and give them a smile and thumbs up. In return I saw and felt their fear and pain. Sometimes they would smile back, I don't know if it was from shock or from just being happy their war was over, and they would be going home. Sometimes we saw the enemy as the guy with a gun shooting at us and sometimes, we would pick them up as patients. We treated them as just any other wounded soldier. The enemy soldiers feared what might happen to him in the hands of the Americans. That fear would quickly go away when they saw the great care given to them by our medics and the doctors and nurses at the US Hospitals.

Dustoff pilots became very personally involved with their patients. We made personal commitments to ourselves to do everything we could to save a life. "To Save A Life" was the motto of our unit.

I flew probably 250 days during the year I was in Vietnam, completed over 1,000 missions and carried over 1,500 wounded. Many of the missions were flown into active battles. My helicopter was shot up many times, I was shot down once and had my helicopter riddled with anti-aircraft machine gun fire. I lost two roommates to combat, but fortunately, none of my crew were ever hit.

I have talked little about my experience in Vietnam until recently. My year in Vietnam, with two months flying during Lam Son 719, have been the most major experience of my life. The best part is knowing how many lives I saved. From time to time, I receive a phone call from someone who was rescued and remembered my call sign, Dustoff 502.

JOHN BUTLER

US Army—Field Artillery, Forward Observer with 11th ACR; Battalion Fire Direction Officer with 2/94th Heavy Artillery
Dates of Military Service (Active Duty and Reserves): 1968 to 1974
Unit Served with in Vietnam: 11th Armored Cavalry Regiment, 2/94th Heavy Artillery Dates in Vietnam: A Troop, 1st Battalion, 11th Armored Cavalry Regiment (11th ACR), Oct 1970 to Feb 1971; 2nd Battalion, 94th Heavy Artillery, Feb 1971 to Oct 1971
Highest Rank Held: 1LT
Place of Birth: Pittsburg, Kansas—1945

LIFE ON A FIRE BASE DURING LAM SON 719

After serving as a field artillery Forward Observer with the 11th Armored Cavalry Regiment (A Troop, 1st Battalion) for about six months,

I was transferred to a heavy artillery battalion headquartered in Da Nang. Our Battalion S1 assured me that I would have concrete sidewalks and hot showers at my new duty station ... it would be posh assignment!

After living, working and sleeping in our mechanized vehicles in the jungle for six months, I was looking forward to safer and more comfortable duty. When I arrived at the headquarters of the 2nd of the 94th in Da Nang, sure enough, I saw lots of plywood buildings, concrete sidewalks, and a large outdoor shower area. So far so good. But I didn't see any soldiers!

I found the HQ building and was introduced to our battalion S3, a major. I asked him where everyone was, so he showed me a huge map on the wall of I Corps (the northernmost portion of South Vietnam. The map included the DMZ on the north, the Laotian border on the West and the Gulf of Tonkin on the East. He pointed to Khe Sanh, about five miles from Laos and ten miles south of the DMZ.

He said, we're operating "right about here," and pointed to a spot *west* of Khe Sanh on Hwy 9 about an inch (on the map) from the Laotian border. Since the roughly 60 miles of I Corps was shown on a map several feet wide, that told me we were spitting distance from Laos and within artillery range of North Vietnam. Oh S*** ... I was not thrilled with this news.

Our mission was to support an ARVN (Army of the Republic of Vietnam) insertion into Laos to disrupt the constant flow of men and equipment down the Ho Chi Minh trail. The closer we could be to the Laotian border, the further into Laos we could shoot to support our Vietnamese brothers. The operation was named "Lam Son 719" for reasons I still do not know.

I was assigned to be one of two Battalion Fire Direction Officers, and I left immediately for our battalion's field headquarters which was located in the center of a fire base "on" the Laotian border. We had three other firebases located near the border as well, but I didn't visit any of them. The battalion consisted of four firing batteries, each with four 28-ton self-propelled 175 mm guns.

They were designated "guns" rather than "howitzers" due to the length of the barrel (more than 35 feet) and range of each piece. Each gun fired a 175-pound projectile filled with TNT up to nearly 40 kilometers (about 25 miles). The powder required to launch each round such a long distance weighed nearly 100 pounds and came in a metal sleeve about 4 feet long. After use, the empty sleeves were often filled with dirt and used for protection around our sleeping quarters. The stress on the barrel of each gun was so severe that it had to be replaced after firing between 300 and 400 rounds.

I would be working at the battalion's Fire Direction Center (FDC), which was a radio van mounted on a 5-ton truck. A dozer had dug a huge trench deep enough for the top of our FDC to be below ground level when at the bottom of the trench. Huge wooden timbers were positioned across the opening of the trench above our truck and sandbags were laid on the timbers. We created flat areas on the otherwise sloping ends of the trench where we could sit.

I had to provide (build) my own sleeping quarters, so I initially stacked ammo sleeves on two sides and one end of an area just wide enough for my bunk bed (something new I obtained from supply in Da Nang). I found and used a few pieces of curved corrugated metal for my roof: comfy, but a bit too exposed for my nervous self, especially because we received incoming from NVA artillery periodically.

So another Lieutenant (LT) and I built ourselves a trench home with separate sleeping rooms. We dug a "Y" shaped trench for an entrance, a tiny common area, and two bedrooms. Each bedroom was just wide enough for a bunk and deep enough to allow about 2—3 feet above our heads when we slept. The entrance had steps down to the floor of our cozy little home and was at an angle to each bed room.

We covered each bedroom with dirt-filled ammo cannisters, topped with several layers of sandbags, and eventually large pieces of plastic in case it rained. It was ugly, but comfortable and pretty safe.

Since our firebase also included our field battalion headquarters, we were provided with four squad-sized units of air defense artillery protection positioned in strategic spots around our perimeter. There were two "Quad-50" crews and two "Duster" crews. The Dusters were tracked hulled vehicles with twin 40 mm rapid-firing cannons that fired 240 rounds per minute, each with an explosive projectile. Dusters were designed for air defense, but they were very effective ground defense weapons.

The quad-50s consisted of 4 synchronized 50-caliber machine guns mounted on the bed of a 5-ton truck, facing toward the back of the truck. It was capable of firing 800 rounds per minute when all 4 guns were in operation. Like the Duster, the Quad-50 was originally designed and intended to be used as an anti-aircraft weapon, but it was an amazingly effective weapon for ground defenses on our perimeter.

For some reason, the four crews assigned to augment our perimeter had been racially segregated, so that each crew consisted of men of the same race. Whatever the rationale for that decision, the result was that one of the Quad-50 crews became openly antagonistic to the rest of us on the firebase, especially officers.

I had quite an unpleasant encounter with that crew while walking our perimeter one night when it was my turn to serve as officer of the day. As instructed, the crew had positioned the truck facing toward the middle of the perimeter with the guns facing outward. The truck was positioned at the end of a narrow point of terrain with relatively high elevation. As I approached the truck, I could see that all of the crew members were sitting in the cab of the truck instead of being on guard. I guessed they were smoking weed. It had become all too common, even while on duty, but that's a whole different topic. One of the crew members climbed out of the cab and took a few steps toward me with his M16 in his hands. I heard the distinctive sound of a round being loaded into the chamber, and I called to him that I was the officer of the day. He claimed that he didn't know who I was. Right.

A few nights later, an M-79 grenade exploded in the sand bags surrounding our battalion commander's tent. The round had been fired from the direction of that crew's Quad-50 truck. That crew was transferred from our area immediately. I have no idea where they went or what happened to them.

I shared the job of Battalion Fire Direction Officer with another

LT. We worked 12 on and 12 off, and would trade days for nights every week or two just to keep things fair. The guns fired day and night, nearly always four guns firing within seconds of each other. Each time even one gun fired, the ground would shake. The sound was tremendous. Our gun crews wore ear protection. Our FDC crew each kept radio headphones on while they were on duty. I tried to keep ear plugs in my ears all the time, but all too often I would take at least one of them out to talk with someone visiting the FDC. I now wish I had been more careful with my hearing.

Occasionally, we would receive target coordinates from one of our firing batteries requesting clearance for that (or those) coordinates. We knew what that meant. It meant that one or more of the guns had accidently fired the wrong power charge (called Zones), and the battery had calculated where the rounds would have landed based on that incorrect charge. Sometimes we got clearance, and sometimes we didn't. I've always wondered if civilians or friendly forces were injured or killed when these accidents occurred.

One night while sitting on the sloped area outside the FDC fan, we experienced a huge explosion ... it was different than the sound of one or more guns firing. It also was different than the sound of an incoming artillery round. It was kind of a combination of both. Also, the area where I was sitting exploded with dirt flying everywhere.

We assumed it was incoming until we heard cries for a medic coming from one of our gun crews. We learned that the breach of one of the guns had blown up, that two of the gun crew were killed, and that two others were wounded. A few hours later after dealing with that awful disaster, I returned to the FDC and to where I had been sitting. I saw a large hole in our sandbag wall between where I had been sitting and the gun that had blown up. I looked some more and found a large metal piece embedded in the sandbag wall on the other side of where I had been sitting. It was the breach bolt from that gun. It had missed my head by only an inch or two. I still have it.

One final thought: My mom and other members of my family prayed for me every day while I was in Vietnam. I have no doubt that their prayers were primarily responsible for my survival and ability to return home in one piece.

CLYDE ROMERO

US Army— Scout Helicopter Pilot
Dates of Military Service: 1969 to 2005
Unit Served with in Vietnam: 25th Avn Co; 101st ABN Div, C Troop 2/17 Cav
Dates Served in Vietnam: Apr 1970 to May 1971
Highest Rank Held: Colonel
Place of Birth: Bronx, NY—1950

RESCUE OF KANE AND CASHER DURING LAM SON 719

In 1971 while US ground forces were prohibited from crossing the Laotian border, a South Vietnamese Army corps, with US air support, launched the largest airmobile operation in the history of warfare, Lam Son 719. The objective was to sever the North Vietnamese main logistical artery, the Ho Chi Minh Trail and its hub Tchepone in Laos. It was an operation that according to General Abrams could have been the decisive battle of the war, speeding up the withdrawal of U.S. troops and ensuring the survival of South Vietnam. The end result was defeat of the South Vietnamese army and extremely heavy losses of U.S. helicopters and aircrews.

The operation lasted from 30 JAN thru 24 MAR 1971. According to the U.S. Army Center for Military History, the total count of aviators killed is 215 with 1,149 wounded (I was one of them). There were 94 extractions under fire, which resulted in the rescue of 347 crew members. Nine attempts were unsuccessful, leading to 30 crew members being listed as MIA. There were 632 incidents of battle damage involving 441 aircraft. Every aircraft that flew in that operation got hit. 197 helicopters were listed as "irrecoverably lost."

A little background is needed here, I was due to DEROS on 17 April 1971. The policy was if you are within 60 days of going home, especially having been a scout pilot and haven been shot down previously, you were pulled off the schedule and did not have to fly. Lam Son 719 broke that rule big time. This story is about the rescue of a Cobra crew from my unit, C troop 2/17 Cav 101st Airborne Division (AMBL) during the debacle called Lam Son 719.

It was mid-February, around the 10th I believe, and everyone was getting hit. The NVA were caught initially with their pants down, but they were calling in reinforcement quickly. Please keep in mind from Khe Sahn combat base to Tchepone was only 35 miles. And all the action was with in five miles of Highway 9, which was the main road into Tchepone. Numerous landing zones were to the north and south of Highway 9 to support the movement of troops on the highway into Tchepone. The main action was around LZ Hope and Sophia. Our area of responsibility was north of Tchepone, right in the heart of Indian country. Usually scouts have to find the enemy, not during Lam Son 719, they found you, and engaged you like I had never seen before. They had no fear of TAC Air, which consisted of F-4 and numerous B-52 strikes. Even after all that bombing, the anti-aircraft fire was the most intense any of us had ever encountered.

While we were suppressing the anti-craft fire north of Tchepone, Jim Kane's Cobra got hit badly. He was able to crash land it on a piece of terrain not far from where I was. The only good thing so far was that he didn't roll over during the crash, Cobras love to do that. But God was with us that day. The other Cobra gunship was taking fire as well but managed to escape. The NVA was using triangle fire to shoot Cobras. In other words, one gun pit would fire at the Cobra to draw its attention while two other gun pits fired on the attacking Cobra gun ship from its sides, where the Cobra was most vulnerable. The NVA were using .51 cal and 37 mm anti-aircraft weapons. We as scouts would engage those gun pits at low level. It was a cat and mouse game, also there was always considerable small arms fire to contend with from random troops on the ground.

We were able to locate Jim's Cobra relatively quickly considering the amount of enemy fire everyone was taking. The C&C ship was close by but didn't see where they went down, so we had to put his eyes on the downed bird, along with everything else that was going on. Can you say Chinese Fire Drill?

The only thing we had going for us was we had plenty of station time; a bonus... you can hang around longer around the enemy who knows how to shoot down helicopters! Once the C&C ship got eyes on the downed Cobra, the issue was the terrain surrounding Kane's Cobra. There were small trees that prevented a landing, unless you want to use your rotor blades like a weed eater to chop down the trees. Needless to say, that is exactly what happened. The UH-1 C&C ship hovered close by the downed crew, while myself and another scout drew fire for them. Kane and Casher where pulled aboard and we flew low level back to Khe Sahn. Upon arrival, both were taken to the medical tent. Kane was sent home and Casher after a couple days was returned to the cockpit for more fun and games with the NVA. As for myself, it was just another day at the Rock. We had taken a couple of rounds in the ship, but nothing major and no one on my ship was hit. I consider myself lucky in that regard, but my luck would run out not soon after.

But that's a whole story upon itself.

MY PURPLE HEART

And here's another story from Clyde Romero, the rest of the above story, as he told it to a news reporter...

On February 13, 1971, just five days into the operation, the Condors suffered their first loss. Two Cobras were downed that day, including one that went in with the crew, killing them both. Romero felt lucky thus far, but sensed it was just a matter of time before he'd be killed as well. He wasn't fatalistic. The odds just didn't look good for him to make it home if he kept flying into such intense anti-aircraft fire.

The next day the first Loach in the unit was shot down, but the pilot made it back safely. The day after that two more aircraft were shot up and knocked out of service. Barely a week into the battle the Condors were running out of serviceable aircraft. Pilots started flying whatever helicopter was available instead of the bird and crew normally assigned to them.

February 18 was another bad day. The Condors had recently received a new Loach pilot, and Romero was tapped to quickly train him. Clyde concluded that the new guy was not cut out to be a scout pilot. He wasn't a bad pilot, he just didn't have the instinct for scouts, which was something that in Romero's experience couldn't be taught. Romero lobbied the higher-ups to take the guy out of Loaches. "He's going to kill himself, Sir. Send him back to flying Hueys, or at least let him train some more." Nothing was done in response to Romero's pleas. With Lam Son 719 in full swing, the Cav was short on Loach pilots and couldn't afford anyone extra training time. The new guy had performed poorly the day before and asked Romero to let him take Romero's mission the morning of February 18 for himself. Romero consented. "Ok, man, but keep up your speed. You can't hang around out there. Keep moving."

With more missions scheduled than crews to fly them, Romero ended up flying another mission anyway after he gave his up to the new guy. Romero was working an area nearby when the new guy's helicopter went down. He raced to the scene to provide help if needed. There was nothing he could do when he got there. The helicopter was obliterated. The new guy was dead before completing his seventh Loach mission.

As more and more aircraft succumbed to battle damage, the nature of the missions over Laos began to change. One Loach and one Cobra on a mission was referred to as a "Pink Team," since scouts were designated "white" and Cobras "red." The hunter/killer team configuration with the heaviest punch was one Loach and two or more Cobras, a "Purple Team." As the supply of serviceable Cobras during Lam Son 719 shrunk, Romero increasingly found himself as the lone helicopter on a mission, a single ship "White Team." With no protective Cobra overhead, the only firepower at his disposal was what he brought with him. An observer sat in the front left seat with an M16 rifle and grenades. In the back behind the pilot, another observer was tied in with a "monkey strap" that allowed him to dangle out the rear door and fire a light machine gun from the right side of the aircraft. The main purpose of the two armed observers in the event of contact with the enemy was to make enough racket to keep the enemy's head down while Romero made a hasty retreat to get the crew out of range of the ground fire. If they had the luxury of a Cobra circling overhead, one of the crew would drop a smoke grenade to mark the spot for retribution gun runs. The Cobras did most of the killing. Romero's job was to sniff out the enemy and get out of the way. When there were no Cobras to be had, he still had to sniff out the enemy. There just wouldn't be any retribution as he was leaving.

On occasion the Loach crews did claim kills of their own. On the morning of February 23, Romero and his lightly armed three-man crew stumbled upon a large enemy encampment. The observer in the back opened up on the NVA soldiers with his machine gun as they scampered about and killed two of them. Romero managed to get the Loach away without suffering much damage from the surprised enemy.

Later that afternoon he and his crew ran into trouble again. He never felt anything hit his right shoulder, but the splattering blood blowing back onto the rear crewmember as they sped away from the scene made it clear that something had happened. Romero examined his shoulder once they were out range of the ground fire. It was clearly cut, but everything was working. He felt fine. He flew back to base, bandaged himself up and thought nothing of it. A week later a piece of shrapnel popped out of his shoulder. He never knew it was there. Someone found out about his injury and put him in for a Purple Heart.

Lam Son 719 raged on. One day blended into the next for Clyde and the Condors. A Cobra lost, then a Loach, sometimes two aircraft in a day were destroyed in combat. Charlie was active on both sides of the border, with crews drawing enemy fire from takeoff to landing. The aircrews lived in tents on a compound on the South Vietnam side where they were constantly harassed by NVA mortar attacks. The crews were exhausted and dirty. They soldiered on, grimly doing their duty.

Toward the end of February his platoon leader, a newly-arrived Lieutenant, tried to take Romero off flight status due to his short-time status. It was a general courtesy within Vietnam helicopter units to take a pilot off flight status if he'd survived for most of his tour. It was best if there wasn't an official last mission. No need to jinx the guy. Without warning, tell him he's no longer on flight status. Such was the official pronouncement that he'd made it through his Vietnam tour alive.

Romero didn't care much for his platoon leader. He was a new guy that Romero didn't think was anymore cut out to be a scout pilot than the other new guy who had just killed himself. In fact, Romero had tried to get the Lieutenant out of Loaches too. Romero appreciated the offer to stop flying, but turned him down. He felt like the mission wasn't done yet. There were still bad guys out there shooting down Americans. He wanted to fly until the operation was complete, regardless of his approaching ticket home.

The next day the Lieutenant was killed short of completing his fateful seventh mission. The front seat observer managed to fly the aircraft back to base with the platoon leaders blood and guts plastered about the cockpit. Romero took the Lieutenant's death hard. Two inexperienced scout pilots had died in a matter of weeks. He'd tried to get them both grounded, but the need for pilots during Lam Son was too great for the commander to consider it. If only he'd protested louder, maybe one or both would still be alive. It wasn't until someone was killed sometimes that you recognized what a good person they were. Romero saw that now about his platoon leader. If only he'd protested louder.

Operation Lam Son 719 ended on March 25, 1971. It was a disaster for the US and their South Vietnamese ally. South Vietnam's ARVN was seriously mauled, retreating from increasing pressure from the North Vietnamese without achieving their objective of destroying major enemy supply depots or seriously impeding the future use of the Ho Chi Minh trail. What was worse, some of the South Vietnamese military's most elite forces, including rangers and marines, had been decimated by being fed into the battle piecemeal with little follow on reinforcement from other ARVN units that remained in the relative safety of their home country. When Romero learned that even the vaunted ARVN Ranger units had been defeated, he felt like the war was lost.

Romero survived Lam Son 719 with but a scar on his shoulder. During his one-year tour in Vietnam, Bro Clyde was shot down nine times and wounded three times, although he was only recognized with one Purple Heart for the shoulder wound.

Romero finished his tour and returned to the US. He learned when he got home that he'd become somewhat of a local celebrity. The military publication *Stars and Stripes* ran a story on Clyde during the height of Lam Son 719, publishing excerpts of a diary he kept. The *New York Times* picked up the story as well, which made Clyde's mom proud. A kid from the Bronx made the *New York Times*!

ALAN C GRAVEL

US Air Force — Pilot

Dates of Service: 1969 to 1974

Unit Served with in Vietnam: 536th Tactical Airlift Squadron of the 483rd Tactical Airlift Wing; 4102nd Aerial Refueling Squadron

Dates in Vietnam: Sep 1970 to Sep 1971; May 1972 to Dec 1972

Highest Rank Held: Captain

Place of Birth: Alexandria, LA — 1945

LANDING PROCEDURE AT DAK PEK

All of the airfields in Vietnam were categorized as Type I, II or III. Tan Son Nhut, Bien Hoa, Cam Ranh Bay, and similar major bases were Type IIIs. Smaller fields with 3,000 — 6,000 feet or so of runway might be Type IIs, although the width, surface type, surface condition, and other factors entered into the classification. Type I fields were those with the shortest, narrowest, most difficult runways. The classification would vary for aircraft type — the same runway might be a Type I for a C-123 and a Type II for the Caribou.

We operated into a number of fields that were also used by the C-123s. Only once did I see a C-130 land at a forward field. At Caribou Type I fields, we were generally the only fixed wing cargo aircraft operating there.

Beyond Type I there were Type I Restricted fields. To fly into these fields, the aircraft commander had to have previously been checked out at that particular field by an instructor pilot. Typically, they had unique non-standard landing procedures or challenging conditions. Dak Pek was a good example.

The airfield was surrounded by small hills that were very close to the runway. The Tactical Aerodrome Directory says, "C-7 wing will clear hill if wheels on runway." With a cross wind, this was particularly exciting as the hills would alternately block and concentrate the winds. On both ends of the runway, the approach was over a small river with an abrupt upslope to the end of the runway. The runway was Peneprime treated laterite (oiled dirt), 1,500 feet long and 60 feet wide. There were other problems, but you get the idea.

As I remember it, the landing procedure I was taught was as follows:

Fly the downwind above a small valley that was more or less parallel to the runway. From this downwind you only got occasional glimpses between the mountains of the runway off to the side.

When your left wing comes abeam of a certain large dead tree on the hill, turn 45 degrees left between two small hills and begin the descent. You cannot see the runway from here.

When you reach a small river, turn 90 degrees left. You are now 45 degrees from the runway heading and continue to descend. On this leg you will begin to see the end of the runway.

When you intercept the extended runway centerline, turn the final 45 degrees left and land. In making this final turn, you have just enough time to level the wings and the runway is under you.

During this procedure, you were in a constant descent, hoping to be on glide slope when the runway appeared. The abrupt upslope just off the end of the runway distorted the normal landing "picture" so you had to be careful not to be too high or too low.

WM. JOSEPH BRUCKNER

US Army
Dates of Military Service: 1970 to 1971
Unit Served With in Vietnam: MACV Advisory Team 26
Dates in Vietnam: Dec 1970 to Oct 1971
Highest Rank Held: Captain
Place of Birth: Atlanta, GA—1944

MY UNFORGETTABLE YEAR

After graduating from The Citadel with an Army ROTC commission and The University of Georgia Law School, I entered the Army in February 1970 as a 1st Lieutenant. My Dad served in the Army during WW II and Korea, and my grandfathers and uncles also served. I always knew that I would eventually join the Army.

After completing Armor School at Ft. Knox, Kentucky and Military Intelligence School at Ft. Holabird, Maryland, I was stationed at Ft. Holabird until I left for Vietnam ("VN") just after Christmas 1970.

On the trip over I made my first visit to Alaska, where I experienced my first sub-zero degree temperature. There was a lot of conversation on the plane. But as we started our approach into Saigon, everyone got very quiet.

My first impression upon exiting the plane was the contrast between the heat there and the winter temperatures I left at home. Because I arrived as an individual, as opposed to being with a unit, I had no idea where I'd be or what I'd be doing.

After a few days, I was flown to Nha Trang where I spent a few days, including New Years. I remember listening to the Sugar Bowl on Armed Forces Radio and realizing that I had traveled a long way from home in a short time. I was then notified that I would be sent to Dalat in II Corps and be the Asst. S-2 on a Military Assistance Command Vietnam Advisory Team ("MACV").

As we approached Dalat, I was struck by the beautiful trees and hills that reminded me somewhat of North Georgia. I was met at Cam Ly

Airport by Jeff Rhodes, who became one my best friends on our team. He drove me to our Tactical Operations Center ("TOC").

I couldn't believe what I was seeing as we drove to town. I was already impressed with the relatively cool weather. But as we traveled through lush forests, I saw some residences that had architecture similar to those in European Alpine areas. Dalat was where Europeans living in VN and some well-to-do Vietnamese went to escape the heat of Saigon and other warmer areas. There was a golf course no longer in use that still had foxholes dug by the VC during Tet of 1968. I eventually saw a photo in the house of our Chief Province Advisor of Teddy Roosevelt with a Water Buffalo he'd killed nearby.

MACV Team 26 consisted of 10-11 members. We provided support to our South Vietnamese counterparts. Captain Long was the Vietnamese S-2 and became my counterpart when our S-2 was evacuated after putting a bullet through his hand in the middle of the night during my first month. Because he was not replaced, I became our S-2 and inherited his Jeep! I also had a Vietnamese Interpreter who was invaluable to me. I had been promoted to Captain by the time I arrived.

I worked closely with my counterpart, and we got along very well. Each morning I would present an intelligence briefing to our team based on information we received from our sources and Capt. Long's sources. On many days, I'd be involved in one of two types of air missions. I would participate in either a "sniffer mission" or an L-19 recon.

During a sniffer mission, I would ride in a door gunner seat of a Huey with my weapon and a grenade launcher. The seat opposite me was occupied by the operator of a machine that could detect human (and sometimes animal) presence. These were Free Fire zones, meaning that anyone there was a "bad guy." In the event activity was detected visibly or by the machine, the pilots would notify gun ship pilots flying above us. The gunships would move in and do their work. We would occasionally take fire, but it was always small arms fire and not very accurate. Occasionally a round would ding the chopper, but no damage was done. I've often said that those Huey pilots were more skilled at their profession than any group of people I've ever encountered.

The L-19 recon missions were flown in a fixed wing two seat L-19 with a pilot in front and me behind him. These were observation missions looking for signs of enemy presence. We flew into river valleys and other areas where the enemy would often be. The most significant memory of these flights was the beauty of a country that most people associated with rice paddies and remote fire bases.

I realized how inexperienced I was my first week in Dalat when after eating a decent sized breakfast I went on my first L-19 recon mission. As we were weaving through a river valley I completely lost my breakfast. As soon as we landed, the pilot turned and asked, "How was your breakfast, Captain?" That was my initiation, and I never made that mistake again.

While I could mention other memorable mission oriented experiences, I can't go without relating something that made my time in VN somewhat unique. My wife spent several months in Dalat during my tour. It's a very long story that I'll try to make short. A former classmate at The Citadel and UGA had deployed, and his wife went to Bangkok to teach at the American School. She contacted my wife, Lucy, to tell her that there were teacher openings and that she should come over. The friend said that she and her roommate would periodically fly to Saigon and visit their husbands.

Right after I left, my wife sold our car and did everything necessary to secure the job and make it to Thailand. The only problem was that her friends' husbands were in Saigon, and I was well north of there. About a month after she arrived in Thailand, I hopped on an Army transport, she got a flight, and we spent a weekend in Saigon. Our Province Advisor, who was a civilian and a great guy, urged me to have her visit Dalat. I think he felt this would be a sign to others that Tuyen Duc was relatively pacified.

She did visit, and he even had a party for her and invited some local Vietnamese and members of our team. She got to know some American missionaries who worked with the Montagnards, the indigenous mountain people who had no use for the Communists.

After Lucy returned to Bangkok, her roommates' husbands received "early outs." She wasn't going to stay there alone, and the missionaries said that since she had medical training (veterinarian), she could help them in their work with the Montagnards. It was on a volunteer, no pay, basis but was a way for her to stay. We were young, fearless, and probably a little crazy. If I was out overnight, she stayed at a house where a couple of the Police Advisors (State Department) that I worked with lived.

The Vietnamese Secretary who worked at our TOC invited us to a Sunday lunch at her home with her husband and young son. Very few of us got to have such an experience. We also occasionally taught English to local kids at an area grammar school. I was very fortunate to have a personal relationship with so many Vietnamese.

My wife and I returned to VN in 1995. This was before many American tourists started arriving. The trip reminded me of what a beautiful country it is but also made me sad for the plight of those in the South. We saw almost no Americans other than a family that lived in Doraville returning for the first time since being "boat people." I also met a small group of American soldiers in civilian clothes in Da Nang. They were part of a search team working with the Vietnamese trying to locate the remains of American KIAs and POWs. Based on info provided by local Vietnamese, they had found the remains of one soldier and were looking for others.

Unfortunately, I'd been told not to ask about Vietnamese I'd worked with because it could endanger them. One sad experience on my return trip occurred in Saigon when a lady walked by and quietly said, "We miss the Americans so much."

The bond the members of Team 26 had was truly special. Plus, I was able to learn about another culture at a young age while at the same realizing how lucky I was to live in the US. I'm very proud to have served in the Army. And although I did nothing heroic, I'm particularly proud of my service in Vietnam.

WAYNE LARUE

US Air Force — AC-130A Pilot

Dates of Military Service (Active and Reserve): 1968 to 1989

Unit Served with in Vietnam: 16th Special Operations Squadron, Ubon Ratchathani Royal Air Base

Dates in Vietnam: Jan 1971 to Oct 1971 (Permanent Change of Station) CCK

Taiwan; Aug 1972 to Dec 1972 (Temporary Duty)

Highest Rank Held: Lt. Colonel

Place of Birth and Year: Monroe, NC — 1946

VETERANS / HEROES

During five years active duty in the USAF one of my assignments was
to the 16th Special Operations Squadron from 18 Jan 71 — 25 Oct 71
at Ubon Ratchathani Royal Thai Air Base, Thailand flying AC-130A
gunships. I logged 280 combat hours that year on 60 plus sorties.

Our primary mission was to destroy trucks carrying weapons, ammo,
fuel, food, PAVN (People's Army of Vietnam) troops, and other items
needed to resupply the Viet Cong in South Vietnam, Laos, and Cambodia. Viet Cong were Vietnamese forces fighting the South Vietnamese
Government and the United States. These materials were supplied by
the Soviet Union, People's Republic of China, and North Korea over-
land and via ships using Haiphong Harbor in North Vietnam. Trucks
would depart southwestern North Vietnam through mountainous passes Ban Karai and Mu Gia under the cover of darkness southward on the
Ho Chi Minh Trail, eventually unloading in Laos and western South
Vietnam.

Each night AC-130As (call sign "Spectre") flew to a safe area firing
on a practice target for the purpose of tweaking or bore sighting each
gun for accuracy. Then continue flying to their assigned combat area. Two
consecutive shifts would cover that area from first dark until relieved by a
second gunship, exiting prior to sunrise. Operating well within range of
ground based North Vietnamese Russian or Chinese anti-aircraft 37 &
57 millimeter guns was fatal to AC-130s so easily spotted during day-

light. Almost 100% of the Specter missions were flown under the cover of darkness. The distinctive C-130 drone noise could be heard for miles, night or day. Specter could only be effective in a permissive airspace; i.e., no Surface to Air Missiles (SAMS) No MIGs (Russian Fighter Jets).

Each "A" model gunship was equipped with two 20 mm Vulcan (Gatling) guns fixed mounted on the left/port side of the aircraft. Firing 20mm guns simultaneously for three seconds would put a shell in every square foot of an area the size of a football field. Heavy duty snow shovels were used to fill 55 gallon steel drums with spent 20 mm brass shells. "A" model gunships also had two Bofors 40 mm single shot cannons that could go through an engine block from two miles away. Late in 1971, M105mm howitzers were added to the AC 130 gunships.

Now AC 130 gunships are armed with 105 mm M102 howitzers and GAU-12 Equalizer cannons in place of the two 20mm cannons. AC 130 gunships have been in operation for 50 years. The AC-130A model gunship was the first in a line of continuously updated AC-130 models.

Each gunship flying from Ubon Air Base Thailand over the Ho Chi Minh trail in Laos was supported by three F4 Phantoms tasked with destroying anti-aircraft guns stationed along the trail and defeating any Russian MIG fighter jets threatening the gunship from bases in North Vietnam. One F-4 would be refueling at a tanker. A second F-4 was en route from tanker to the combat area as the third F-4 remained on target protecting the gunship. Time on target for each F-4 was 30—40 minutes.

Fighters flew two rotations each for a total of three to three and a half hours total support on target. The Spectre shifts were five hours, comprising one hour each way to and from target area and a total of three plus hours on target destroying trucks.

During the spring of 1971, our flight crew met at the intelligence shop for weather, scheduled air activity namely ARC Light flights AKA B-52 bombing runs, and threats affecting our mission. The first aircraft covering our assigned area near Tchepone, Laos (now known as Xepon or Sepon), reported heavy anti-aircraft ground fire, including rockets. Intelligent officers believed they might have been unguided "point and shoot" rockets. No guided SAMS (Surface to Air Missiles) had been used in Laos, and it was unlikely that the roads or trails were in shape to handle all the motorized vans, gear, and missiles necessary to fire "guided" missiles.

At 0300 we were flying large counterclockwise circles around the trail/road near Tchepone, destroying trucks. All our BDA (battle damage assessment) was recorded by low light video recorders to be graded by specialists on the ground at Ubon, housed in a portable van. As we fired on vehicles in the area, we received 37-millimeter ground fire, usually seven shot bursts by multiple guns, forming an arc bracketing us. During missions, an average of 400 rounds of 37mm shells per night were targeting gunships working Tchepone.

On board we had pilots, navigators, sensor operators, loading crews, and an IO assessing enemy gunfire. The IO was strapped onto the ramp with half his body in and half out of the aircraft looking underneath and assessing ground fire (accurate or no threat) and calling out any evasive maneuvers deemed necessary. The IO initials came from the original mission of "Illuminator Operator." Originally the AF had the bright idea of having huge lights that would roll to the rear of the aircraft on rails bolted to the ramp and light up an area the size of a football field so that ground and air forces could see the enemy at night. Unfortunately the light source could easily be seen from the ground and render the gunship the main attraction at the turkey shoot. A very brief usage indeed. It was used occasionally to light up nightly performances for Bob Hope shows and such. Total crew complement was 14.

The evening of our encounter, IO shouted out that two rockets were fired directly at us. Our Electronic Warfare operator simultaneously said we had been locked on by enemy radar. Our fighter yelled, "SAMS FIRED! Evade!, Evade!" We immediately rolled into 120 degrees of bank, steeply diving for the ground to break radar guidance — inverted flight was a new maneuver for us and definitely outside the operating envelope of a 17 year old C-130! The flying light poles (surface to SAM II air missiles) flew high and right of our position. Our good fortune enabled us to stay **Veterans, not Heroes.** We flew low until we were out the threat area and back to Ubon.

It was determined the missiles were SA-2s and had been launched so close to us that they were not taking guidance yet from ground base radar type units until they were past us. We were blessed to escape the SAMs. Operating outside the limits of the aircraft resulted in hundreds of metal fasteners found underneath the aircraft and rippled sheet metal covering the aircraft skin. After debriefing Intel, a two-ship flight of F-105 Thunder Chiefs Hunter Killers confirmed the SA-2 sites, including radar vans, and destroyed them after daybreak.

Many more of the aircraft I flew sustained damage from 37mm and 57mm guns, including a four-foot by three-foot hole in an elevator one night, but nothing was as harrowing as that night over Tchepone. That night was the first reported use of SA-2s outside North Vietnam. Our crew received the Distinguished Flying Cross, and the opportunity to continue living as **Veterans.**

I've been living on borrowed time since that night! I'm proud and blessed to be a Veteran. There are many more **Heroes,** including **POWs;** however, these AC-130 gunship crewmembers gave the ultimate sacrifice that each person who signs up for military duty agrees to give the USA. **God Bless the USA!**

HEROES

On 29 March, 1972, Major Irving B. Ramsower II, aircraft commander; Capt. Curtis D. Miller, pilot; 1st Lt. Charles J. Wanzel III, pilot; then Major Henry P. Brauner, navigator; Capt. Richard Castillo, infrared sensor operator; Major Howard D. Stephenson, electronic warfare officer; Capt. Barclay B. Young, fire control officer; Capt. Richard C. Halpin, low light TV senior operator; SSgt. James K. Caniford, illuminator operator; SSgt. Merlyn Paulson, flight engineer; SSgt. Edward D. Smith, Jr., aerial gunner; SSgt. Edwin Pearce, aerial gunner; AFC William A. Todd, aerial gunner and AFC Robert E. Simmons, aerial gunner; comprised the crew of an AC130A gunship named "Prometheus," tail number 55-0044, call sign "Spectre 13."

They departed Ubon Air Base, Thailand on an armed reconnaissance mission with F4D fighter escorts over Laos to interdict North Vietnamese supplies moving south into the acknowledged war zone, then return to Ubon. At 0300 hours, the F4D's aircrew saw a surface to air missile lift off the ground. Before the AC130A could take evasive action, the SAM hit Specter 13. A few seconds later, the AC130A impacted the ground on the east side of a jungle covered mountain, followed by secondary explosions.

All Search and Rescue (SAR) efforts were terminated at 1830 hours on 30 March 1972 when no trace of the downed crew was found. Because of the heavy enemy activity in the area, including numerous anti-aircraft artillery (AAA) and surface-to-air SAM sites, as well as a large concentration of NVA forces, it was believed any surviving crewmen would have been undoubtedly captured by then. All 14 crewmen were listed as Missing in Action (MIA).

In 1984, William Todd's remains consisting of five small bone fragments were provided to US officials by a Lao refugee as well as AFC Todd's dog tag. Also Curtis Miller's ring with the inscription, "Forever Sue," was found by a reporter and returned to his family. Other human remains were found but were rejected by the families of Robert Simmons and Edwin Pearce. Likewise based on claims made by CIL-HI (Central Identification Laboratory-Hawaii) forensic personnel, Richard Halpin, Richard Castillo, Irving Ramsower, Charles Wanxel, Merlyn Paulson, and Edward Smith were identified and remains accepted by their families.

On 27 May, 2008, the Department of Defense POW/Missing Personnel Office (DPMO) announced that remains of four of the Specter 13 crewmen have been identified and would be returned to their families for burial with full military honors. They are Maj Barclay B Young of Hartford Conn, and Senior Master Sgt. James K. Caniford of Brunswick, Md. Remains of two other crewmen are being withheld at the request of their families. Remains that could not be individually identified are included in a group which were to be buried together in Arlington National Cemetery.

Vietnam Veterans Memorial "The Wall" Panel 2W Row 121-124

Further Reading: Steeljawscribe.com Search the Site: Airmen Missing

Nearly 600 Americans were lost in Laos during the Vietnam War, and many were known to have survived their loss incident. However, the U.S. did not negotiate with Laos for these men, and consequently, not one American held in Laos has ever been released. These men were never negotiated for either by direct negotiation between our countries or through the Paris Peace accords which ended the War in Vietnam since Laos was not a party to that agreement.

HEROES

30 MAR 1972 AC-130E C/N 4345 First "E" model gunship attrition Shot down over The Ho Minh Trail, Laos Second AC-130 loss in three days

18 June 1972 AC-130A tail number 55-0043 Shot down by SA-7 SAM in the A Shau Valley, SW of Hue, South Vietnam

21 Dec 1972 AC-130A tail number 56-0490 named "Thor" shot down 40km NE Pakse, Laos East of Ubon, Thailand

JON P. BIRD

US Army — Signal Corps Officer

Dates of Military Service (Active and Reserve): 1968 to 2000

Unit Served with in Vietnam: 221st Signal Company

Dates in Vietnam: Nov 1970 to Nov 1971

Highest Rank Held: Colonel

Place of Birth and Year: Syracuse, NY — 1944

IT WAS A DARK AND STORMY NIGHT

I was sitting around the Officers' compound, 221st Sig Co, on Long Binh, when an enlisted man ran in, saying, "Captain Bird, you have a phone call from Major January." I put down my beer, one of several I had, and went to answer the phone.

"How can I help you, Sir?" I asked. Major January replied, "Captain Bird, I need to borrow your jeep." I replied, "What's wrong with your jeep, sir?" January said: "My jeep is in the motor pool and I need to go off-post." Then I reminded the Major, "Of course it's OK, but you know it's our policy to have a driver when you go off-post." Major January noted, "I'm not going to be gone too long, so I'll just come on over to get your key to the jeep lock."

The Major came down, got the keys, and off he went. Meantime, I continued to drink beers with my junior officer buddies. Enlisted men often drove officers around, according to 1st Signal Brigade or US Army Vietnam policy — I don't know which — but that also meant you'd have different drivers and you'd never know what you might find in the jeep.

Time passed, as did the beers, and I got another phone call from the good Major. "Captain Bird, you need to come down to the MP station and get me," he said. "The MPs stopped me because I was driving alone and they found a cookie tin in the tool box with marijuana in it!"

Unfortunately, I had to answer, "I'm sorry, Major January, but I can't come and get you!" To the immediate response of, "Well, why not?" I had to reply: "Well, first, Sir, I've had too much to drink and can't come to get you, and second, I don't have a driver tonight and you have my jeep!"

PATCHING THINGS UP

It was a dark and dreary night when Captain Jon P. Bird held a weekly staff meeting with his combat photographers. One major topic of discussion was whether the men of the 221st Signal Company (Pictorial) could have a patch on their uniform to represent their function. The patch would go on the shoulder above the unit patch and would be black letters on an olive drab background.

My job was to decide if this patch should say U.S. Army Official Photographer or Official U.S. Army Photographer. I think we even toyed with the idea of adding Correspondent to the patch, but ended up with Official U.S. Army Photographer. A week later, the patches were produced and then handed out. Hearing laughter from the rear of the room, I asked "What's so funny?" The answer came, "Sir, we wondered if you would like one of our "special" patches?"

Well, there it was, Official U.S. Army Pornographer. Ha, ha, I thought, but then, black letters on an OD background, who would care or even know? I stuffed my five patches in my pocket for our hooch maid to sew on and promptly forgot about it. Little did I know . . .

The lab technicians surfaced another issue that required a solution. For the awards and departing dignitaries, we often provided color photos. While we had a machine that processed 8x10 color photos automatically for a very low cost, larger 11x14 photos had to be done manually. The cost per photo was more and the chemistry was injurious to our lab technician's hands. My goal was somehow to get the 1st Signal brigade to standardize on the 8x10 size.

My solution was to get a cherry picker crane and have one of our best photographers take photos of the 1st Sig Bde's headquarters building with the flags flying and a nice blue sky with white clouds. We made 8x10 prints and then raising the enlarger, made 11x14 prints as well. The only difference between the prints, other than size, was that the 8x10 had much richer colors than the 11x14.

Then I set up an appointment to see the commanding general, MG Hugh Foster, through my friend, who was his adjutant. I got a haircut, put on my best uniform and boots and armed with my photo samples, some medical reports, and my quick wit, I headed up to see the Big Man. I arrived a bit early and talked with my friend before my time with the general. At the appointed moment, I knocked, walked in, and saluted General Foster. He returned my salute and I proceeded with my plan. I told him my visit was two-fold, first to present him with photos of his headquarters that I had heard he wanted, and next to relate this issue to a problem within the command. He was sitting down and I was standing in front of his desk. I gave him the 11x14 photo as a token of appreciation from the men of the 221st Sig Co. He was impressed with the quality, so then I gave him the 8x10, which was so much more colorful.

I explained to him the cost factors and the medical ramifications and laid down the medical reports, then asked him for a policy on the "grip and grin" photos. As he was looking at all the paperwork, he looked up at me and gazed at my shoulder, saying "Captain Bird, that doesn't say what I think it says, does it?" I looked at my shoulder patch and among the five uniforms that I had, I had selected the one that my hootch maid had sewed on the special patch that read *Official US Army Pornographer.* Looking the general straight in the eye and swallowing, I said, "No sir, it doesn't say that." Well, my meeting was over and I went back to my Unit. The policy was settled and I never again heard about the "special" patch.

GEORGE S. PEARL

US Navy—Aviation Electronics Technician
Dates of Military Service: 1968 to 1974
Unit Served with in Vietnam: Tactical Air Support Squadron 13 (TACRON 13)
Dates in Vietnam: May 1971 to Sep 1971
Highest Rank Held: E5 (Petty Officer 2nd Class)
Place of Birth and Year: Vicksburg, MS—1949

I WAS A FLIGHT LINE GOD!

I went to Vietnam in a Strike Force of ships running the Tonkin Gulf in operations, but before that for two years I was in support of the Navy's training of fighter jet pilots who were heading to Vietnam. The following is just a little of what I was doing in support of the training of these pilots.

After finishing electronics school in 1968, which back then cost the government around $80,000 to send me through, I was sent to Beeville, Texas, and a place called USN Chase Field. I was assigned to Training Squadron VT-24 which during the Vietnam War was a training base for fighter jet pilots. Beeville was a very small town and the base was about 10 miles outside of town. It was located about 60 miles northwest of Corpus Christy, Texas, out in the desert. I spent 25 months there repairing F-9 Cougar Jets that were in use for the training of new fighter jet pilots.

The first couple of months I worked out of the hanger electronics shop repairing aircraft, but they must have found me too able since they sent me to rocket seat ejector school and high-pressure chamber testing so that I could fly. What the squadron needed was an electronics technician who was qualified to fly in fighter jets and knew the operation of the electronic equipment in use on board the craft. After my getting certified, I was the only tech qualified to fly when there was a problem with some piece of electronic equipment in the air that we could not figure out the remedy for or even repeat the problem being complained about on the ground.

I actually prayed for things to break down! Flying in fighter jets doing wing overs and rolls was such was a thrill. I was able to tell the pilot to do all sorts of maneuver's flying around and pulling Gs to see if I could get the equipment to fail. Sometimes I found that it must be in the wiring to an antenna or a power connector for the unit to start switching on and off by itself. Flying was a great 'gift' to me actually. It wasn't a job doing that. At the time, I could not be a pilot due to my near-sighted vision and glasses, but this was allowable for me being the AT tech on the aircraft.

After several months in the shop, I was transferred to VT-24 Troubleshooters out on the flight line. I was in troubleshooters for my duration of time at Chase Field. The troubleshooters were an elite band of technicians from each different system on the aircraft. The idea was that when groups of aircraft were going out on a 'mission' if something didn't work right when the aircraft was started, then the plane captain would call for a troubleshooter to come inspect it and quickly fix the problem so that that aircraft could fly out with the rest of the grouping.

The squadron had jets flying out and coming in 24—7 every day of the week. We operated around the clock day and night. Rain or shine. In troubleshooters, some of us could learn the craft of another technician in a different system than ours. No other technicians could figure out electronics, but I could understand hydraulics, some power plant issues, some structural issues, and such to cover for another technician if they were working on another 'Bird' and yet another request for that kind of technical help came in to trouble shooters.

We got to drive around in little yellow vans with all our needed equipment inside. When driving around on the flight line there were always aircraft 'turning up' that you would have to watch out for. The exhaust could blow your vehicle over if you got too close and they were revving up to pull out of a spot on the ramp. There was noise from jet engines and we all had "Ears" that we had on our heads all day long except in the troubleshooter's shack. These were plastic protective ear phones that had foam and sound deadening inside to allow us to work around loud noise. We also called them Mickey Mouse Ears because they were sort of large like that and went around your whole ear. Mine were red. When I didn't have those on, I had a radio headset and mike boom that I could hook up inside the cockpit to run radios and talk to the flight tower or other people from the aircraft.

The old F-9 Cougar Jets being used for training were rugged. The Grumman F9F/F-9 Cougar was an aircraft carrier-based fighter aircraft for the United States Navy and United States Marine Corps. It was the first supersonic carrier-based fighter for the Navy, and was used extensively in the Korean War. I never got the opportunity to fly out to the carrier and land there like our student pilots did, but I found that even though these were old aircraft that were replaced at that time by F-4

Phantom Jets, they still were amazing for what they could do. So much so that one afternoon the structural troubleshooter was busy and they called for one to do an overstress G-Force Inspection on one of our aircraft. I would be mostly looking at cracks on the wings and the bolts that secured them. The head of troubleshooters pushed me out the door and told me to go pull a G-Force inspection on the plane.

When I got out there, I first looked in the cockpit to see just how many Gs the Bird had pulled. I was floored looking at the gage, the needle was all the way to the right to the maximum. Wow, I could not understand how anyone could have pulled that many Gs and still recovered to fly the plane back. I can't remember how far up the gage went, but it was like 8 or 9 Gs or more, I think. I called the ready room and asked to speak with the student pilot about his flight. When asked, he said "Oh, that was nothing…I did that on LANDING!!!" In other words, he had a controlled crash into the ground with his plane. I then had to really check the landing gear and the wing bolts of that plane to allow it to fly again. I could find no stress related problems and so I let it fly again. Those planes were extremely strong.

The troubleshooters were **GOD** on the flight line. If there was a systems problem with an aircraft and we were called to inspect it, sometimes we found a really big or bigger problem than the pilot or plane captain called us on, and we would DOWN the aircraft. That meant that for the mechanical reason we found unrepairable at that moment, we found the aircraft unfit to fly and the pilot would need to shut the plane down and get out. Sometimes there was time for him to walk over to another plane and take that one instead, but more than not he would need to reschedule his flight training for another time.

Once I was working days and one afternoon was called out for a hydraulic question by a plane captain who was about to send it out with a group of jets on a training mission. I drove my little yellow van and pulled up next to the Bird. The plane captain pointed at the front actuating cylinder with red fluid leaking out of it. Sometimes I will let a Bird fly if it is just a few drops now and then, but this one was actually bleeding. I knew that if this one flew out that it might very well have the front landing gear collapse upon landing with such a leaking cylinder. I **DOWNED** his aircraft! The **Word of GOD had spoken,** but for some reason this pilot rejected the Word of GOD, being a big know-it-all himself, and he commanded his plane captain to remove the chocks to his wheels and off he went...

I had already pulled out with my van, but happened to look around and saw this aircraft pulling out and rolling rapidly down the ramp to the runway to take off! I went NUTS! I had DOWNED that plane and we all had an attachment to our planes that we loved each and every one of them and didn't want those pilots screwing them up. I gunned the van and drove it as fast as it could go to the runway. The plane was there at the end of the runway about to take off so I just drove my van right in front of him and just sat there giving him the cut engine sign. I called the tower and the troubleshooters and had a tow truck come out to the runway to tow his plane back to the ramp. I got the guy out of the plane and drove him over to the ready room in the van where I called up to and reported what he had just done rejecting the Word of GOD! That was his last day in the Navy.

I went on to a tour in Vietnam and when I returned was assigned to repair F-4 Phantoms with the Top Gun squadron VF 121 at Miramar, California. I was with Top Gun VF 121 until I left the service in 1972.

VINCENT C. CORICA

US Army — Field Artillery Officer
Dates of military service: 1969 to 1978
Vietnam service with: 3rd Brigade (Separate), 1st Cavalry Division (Airmobile)
Dates in Vietnam: Jul 1971 to May 1972
Highest Rank Held: Captain
Place of Birth: Johnstown, PA 1947

FIELD ARTILLERY COMBAT GUN SLINGER!

I was commissioned a 2LT in the Field Artillery on 4 June 1969 upon graduation from West Point. I then completed the Field Artillery Officer Basic Course, Jump School, and Ranger School. After a very educational 16-month airborne assignment with C Battery, 1st Battalion, 319th Field Artillery in the 82d Airborne Division at Fort Bragg, NC, I was off to Vietnam. I arrived in Vietnam a freshly-promoted Captain on 1 July 1971. I left Vietnam 11 months later on 26 May 1972.

My unit in Vietnam was the 3rd Brigade (Separate), 1st Cavalry Division (Airmobile), APO San Francisco 96490. Our Brigade Commander was BG James F. Hamlet, a revered, heroically decorated aviator and combat commander. The Brigade operated out of III Corps, in the vicinity of Bien Hoa.

(On 29 April 1971, the bulk of the 1st Cavalry Division was withdrawn to Fort Hood, Texas, but the 3rd Brigade remained as one of the final two major US ground combat units in Vietnam, departing 29 June 1972. The 3rd Brigade's 1st Battalion, 7th Cavalry, Task Force Garry Owen, remained another two months.)

BATTALION AIRMOBILE ARTILLERY LIAISON OFFICER (LNO) OCTOBER 1971 – JANUARY 1972

After my four-month assignment as Brigade Assistant S-3 for Operations in the 3rd Brigade TOC, I was now getting right into the action as an Artillery LNO, "firing shots in anger" every single day.

Second only to commanding an artillery battery (which I got to do in my final RVN assignment), Battalion Airmobile Artillery LNO was THE badass job for an Artillery officer in Vietnam. Why? Because the LNO was a gunslinger, but with Artillery weapons and helicopter gunships instead of pistols!

I was part of a two-officer Command and Control Team who climbed into a helicopter crewed by two pilots and two door-gunners day or night, flying over combat operations in the jungle. Our mission: to protect and aid our troops with fire support as needed and to make sure they knew they weren't there alone. Can you feel how exciting that would be?

The Battalion Commander I was honored to support was Lieutenant Colonel Charlie Hodges, Commander of the 1st Battalion, 7th Cavalry, Task Force Garry Owen.

LTC Hodges was a 6'3" muscular guy, proud of his Florida heritage and his classy Southern accent. He had a John Wayne walk, a big voice and a ready smile for all "his boys." He was about 10 years older than I. We tried not to stare as he folded his substantial body into and out of the back seat in our Bell UH-1 Iroquois "Huey" Command and Control helicopter. LTC Hodges was the kind of leader who drew people to him, and we all wanted to be a part of anything he did. LTC Hodges' favorite C&C chopper pilot was CW3 Joe Aubey. Joe had several different co-pilots in our time together, but no one else ever commanded our aircraft other than Joe.

Because LTC Hodges loved commanding Soldiers in combat, he spent very little time in the Battalion TOC. He wanted to be on the ground with his troops or orbiting overhead. He called me his "Artillery Shadow." I never left his side. When troops need fire support, Artillery is THE sustainable source. Day or night, rain or shine, a constantly-available provider of steel on the target.

I rode side-by-side with LTC Hodges in the back seat of the C&C chopper, each of us talking on our respective radio network. He would be talking to his ground commander or the battalion or brigade TOC on the Command Net. I would rapidly switch networks between the Command Net and my Arty Net. I needed to know the situation on the ground as well as speak with one or more Artillery batteries and/or one or more Cobra gunships, coordinating fire support for our troops.

The science of it all in 25 words or less? "To know precisely the map coordinates of our troops and to coordinate and control accurate fire support to protect and aid those troops."

I have no idea how many hours LTC Hodges and I spent airborne. We never took a day off, and in our four months as a team, we spent a minimum of 10 hours in the air on every day which was suitable for flying. Performing these missions, always ferociously eager to protect "our boys," is how I earned my precious 11 Air Medals.

The "US Army Veterans Medals Home Page" says the following about Air Medals:

"The Air Medal is awarded to any person who, while serving in any capacity in the armed forces of the United States, shall have distinguished themselves by meritorious achievement while participating in aerial flight. Award of the Air Medal is primarily intended to recognize those personnel who are on current crew member or non-crew member flying status which requires them to participate in aerial flight on a regular and frequent basis in the performance of their primary duties. Examples of personnel whose combat duties require them to fly include those in the attack elements of units involved in air-land assaults against an armed enemy and those directly involved in airborne command and control of combat operations."

This is a perfect description of what LTC Hodges and I did together every day for four months.

One of my Air Medals includes a "V device for Valor."

Here's an excerpt from the Department of the Army General Orders published by 3rd Brigade 1st Cavalry Headquarters on 2 March 1972. It's framed in my office. I am not embarrassed to say I look at this citation almost every day and thank God I am here to read it.

"The following award is announced.

Corica, Vincent C, Captain, Field Artillery, United States Army, Battery B, 1st Battalion 21st Artillery.

Date of Action: 8 December 1971

Air Medal with "V" Device.

Theater: Republic of Vietnam

For heroism while participating in aerial flight in the Republic of Vietnam. Captain Vincent C. Corica distinguished himself by heroism in flight on 8 December 1971. Disregarding his own safety, he courageously exposed himself to the dangers inherent in aerial flight as he directed his efforts toward neutralizing the enemy threat. His heroic and valiant actions contributed materially to the successful accomplishment of the United States mission in the Republic of Vietnam and were characterized by a great concern for the welfare of his comrades. His heroic actions are in keeping with the highest traditions of the military service and reflect great credit upon himself, his unit, and the United States Army."

What exactly did I do that day? I honestly am not sure! What???

No one told me they were submitting me for this award (as is often the case in a war zone, in my experience.) I didn't keep a journal of my days in Vietnam, so I don't know for certain exactly what I was doing on December 8, 1971, other than riding in the chopper with LTC Hodges.

That said, I do have a very strong idea of what combat action that day caused someone (almost certainly LTC Hodges) to submit me for the "V" device Air Medal award. I didn't know I had earned 11 Air Medals and the Valor award until I ended my deployment in Vietnam! My Field Artillery Battalion Commander at the time, LTC Jack B. Keaton, an extraordinary officer, pinned my combat awards on my chest in my last few days in-country. I was stunned to see the 11 Air Medals and the V. I not ashamed to say I wept with humility and pride as I looked down at the medals.

Now...I will describe what LTC Hodges and I did three times during our time together because it's extremely likely that one of these was what occurred on December 8, 1971.

On three occasions, Charlie Hodges ordered our C&C ship to land and pick up crew members of downed Hughes OH-6 Cayuse Light Observation Helicopters...which everyone called "Loaches." We'd perform these rescues without thinking. It was just what you did... you left no man behind. We knew others would do the same for us. The creed of fighting soldiers.

These rescues all followed pretty much the same sequence. We'd be in the air, engaged in supporting ground troops. The Command Net would light up with a Mayday call of a helicopter shot down in our AO. Usually the call came from the downed bird itself, which made answering these calls instantly with a confident voice very important to the downed crew. These calls were immediate alerts for any helicopters near the downed bird to rush to that location.

LTC Hodges loved these opportunities. He would respond to the

Mayday in his loud, confident voice and direct CW3 Aubey to get above the downed bird as fast as possible. We were lucky in that all three of our rescued crews had managed to get their birds into usably-open pieces of jungle. The door gunners on both sides of our ship would fire their M60s into the perimeter of our rescue-LZs. Charlie Hodges and I were the only two people on board our ship without a specific job to do in these instances, so if the crew of the Loach couldn't walk, he designated himself and me as "Rescue Critters." Charlie directed recurring dry-runs of this action and we each had our respective roles mastered..."Land, dismount, retrieve, bug-out."

In two of our rescues, all Charlie and I had to do was change our seats to make room for the two new passengers, perhaps also giving them a hand up from the ground to the skid and into our ship. We'd rush back to base and drop off the rescued men into the hands of medics if needed. Charlie would "order" the rescued guys to "medicate with cold beer" immediately. We'd refuel and get back in the air. We never knew our rescued guys' names. They never knew ours. The radio never stopped when a ship was down; there just wasn't time to make introductions. Quick "Thank You"; quick "You're Welcome" and off we went.

Here's what I think happened on December 8, 1971. On one of our three rescue events, our two guys to be rescued were injured. We could hear the Pilot loud and clear on the Command Net. They weren't wounded, but the Pilot told us they couldn't walk. Charlie and I dismounted, ran to them approximately 50 yards away, put them into a fireman's carry just as we'd rehearsed, and scooted back to our bird. Charlie and I had to take our flight helmets off to leave our ship, of course. We then had no contact with anyone except each other and that was just by shouting and hand signals. Get this... on this occasion, when Charlie and I jumped out of our respective sides of the bird, he shouted, "Rangers lead the way" and started laughing! Therefore, so did I! When I looked over at him as he grabbed the Pilot and I grabbed the Observer, he was laughing and whooping. What a guy.

On the day Charlie told me his tour was ending, he also told me he had arranged to have me designated as Battery Commander of B

Battery, 1st Battalion, 21st Field Artillery, right here in my beloved 3rd Brigade, 1st Cavalry Division. He smiled a giant smile. He told me the last day of his command would be the first day of my command. He had arranged it so that when he and I ended our partnership, neither of us had to "fall in love" with another counterpart. Can you imagine my joy, my admiration, my gratitude?

Charlie and I lost touch after he returned to the States and I went on to my Battery Command assignment. A few years ago, I tried to find him via social media. I was saddened to see that "Charles Hodges" had died several years earlier in Florida, which was Charlie's home state. The obit said he was a Vietnam veteran. The photo did not look like what I imagined Charlie would look like then, but I last saw him more than 40 years ago. What I do remember is a man bigger than life. A soldier through and through, a leader, a large man whose heart was as big as his body.

Charlie Hodges, a man who taught me so much without even trying to.

RALPH H. BELL "ROTORBRAKE"

US Air Force — CSAR (Combat Search and Rescue)
Dates of Military Service: 1966 to 1991
Unit Served with in Vietnam: 3d Group, 38th ARRS, Detachment 7 (DaNang)
Dates in Vietnam: May 1971 to Dec 1972 (19 ½ months)
Highest Rank Held: Colonel
Place of Birth and Year: Texas City, Texas — 1943

"THE BOX" MEMORY OF BOB SWENCK, AS WRITTEN BY HIS DAUGHTER

I served in the Vietnam War, and I would like to remember my friend, Bob Swenck, and let you read a letter of love and remembrance written by his daughter, Stacy. Her letter could be the reflection of any American Soldier through the eyes of his children on the fateful day when Soldiers come to personally notify your next-of kin "on behalf of a grateful nation, that you were killed by enemy action." BTW, all Vietnam Soldiers are listed on a website www.virtualwall.org by name or state and comments by friends and family are posted there. I knew Major Bob Swenck as we served together in a Helicopter Rescue Squadron at DaNang AB Vietnam. Bob was easy going and always wore blue jeans when not flying… most of the rest of us continued to wear our flight suits day and night. Bob and I talked on our days off about life in the Air Force, the seemingly slow promotion rates for helicopter pilots, and also about our families. He proudly showed me photos of his wife and three little girls.

He was anxious to get back to them, but one day he was assigned a mission down South near Saigon and Bien Hoa, which seemed to be a safer mission than up North where we were at DaNang and Quang Tri. Then I was shocked to hear the news that Bob and his entire crew: copilot, Flight Engineer, and three PJs flew into the Song Nha Be River, in bad weather. We all prayed for survivors and as most pilots fear, the dreaded cause of the accident. Had Bob flown too low? It was bad weather after all and overcast and heavy rain. Only two crewmen survived, one crewman was never found, and bodies of the pilot, copilot, and one PJ were recovered. The cause of the crash was what we called "the golden bb." Some sniper on the river bank managed to time his shot to enter the helicopter pilot's window and went underneath Bob's body armor and into his neck. Bob was killed instantly and likely unknown to the copilot until they hit the water. Sadly it all happened on Thanksgiving Day. One of his daughter's, Stacy Swenck, has posted her Memories on the website alongside her father's biography.

24 OCT 2002 – MEMORIES OF MY FATHER, A VIETNAM WAR HERO BY STACY SWENCK

How strange I felt reading that obituary! At age twelve, what had I survived? I still lived in our little brick house in Fern Creek, Kentucky, I still went to Fern Creek High School, and I still slept in my own bed at night. My father had died in Vietnam, his chopper shot down by a sniper on Thanksgiving Day in 1971. I didn't understand then, as I do now, thirty-one years later, what I had survived.

The morning after Thanksgiving that year, as we were eating donuts and listening to music, an unmarked car pulled up in the driveway. Two men in dress blues stepped out. All military wives know what that means, and the men had only spoken a few words when my mother collapsed on the front porch. They carried her crying to the couch, and we were told that my father's chopper had crashed in the river, and he was missing in action. Mom sobbed; I felt shocked and numb. I took my two little sisters, ages six and nine, downstairs to the basement and kept them there. We stayed quiet and listened, forgotten in the confusion. For two days we made sandwiches and stood apart, watching. Dad's body was recovered two days later. Mom cried some more. People brought us food for a few days, and then we were left alone. Mom stopped crying and walked around the house with a distant stare. A dark cloud was in our house for a long time.

For months I imagined my father showing up in my classroom door to take my hand and walk me home. Even now, I still wonder if his spirit will ever visit me in a dream, giving me guidance I still long for.

Soon after his death, I remember running into Mom in a dark hall of the house and her sharp intake of breath as she said, "You look just like your Dad." She searched my face and saw every resemblance; my cheeks, eyes, nose, and frown stung her heart. I couldn't help being a reminder of sorrow for my Mom. Did we remind everyone we met of Death, our faces omens of the grief Death brings?

We had been mainly sheltered from television images of the war. Besides, those men wearing camouflage running around on the ground were not my Dad. My Dad was a pilot; he wore a plain flight suit when he went to work. He was doing something Top Secret and the return address on his letters was fake. As far as I knew, Dad was at work on a very long mission. I didn't understand what war was. Then he was killed. As I grew older, I came to hate the Vietnam War and blamed the government, even though I had no idea about the war's causes and avoided any mention of it. I became a rebellious teenager, transforming hurt into toughness. I grew bitter in my ignorance. Hating the war helped when I had to tell someone my father died fighting there. I couldn't say he has been drafted unwillingly; he was a career soldier who served two tours in Vietnam. Joining in the belief it was a bad war made it easier to get along with others who hated it, which was most everyone I met. I did not know then that my father had been a hero.

I found out about that from the Trunk. On Thanksgiving Day, 1996, the 25th anniversary of his death, Mom appeared and deposited a heavy blue trunk in my living room. She firmly announced that these were my father's things and were now mine to keep. She was finished with them. I wasn't so sure I wanted the Trunk, and I put it in the closet. Many months later, I opened it and found my father's blue uniform, several medals, and a large stack of letters.

I had seen the medals as a child, set out on the bookcase in the basement. He had earned them during his first tour in 1969. Now I saw what they were: a Silver Star (one of the top medals a soldier can earn), three Distinguished Flying Crosses, and an Airman's Medal for Valor (he was most proud of this one), eight Air Medals and a Purple Heart. I read over 100 letters he had written to Mom and to us. I found out that Dad was a highly skilled Air Force chopper pilot for the 20th Special Operations Squadron (the Green Hornets), who were flying "sensitive and classified missions" on the real, secret front lines of the war. The 20th SOS's mission was to rescue long-range reconnaissance patrols on the ground in Cambodia during a time when both the United States and North Vietnam were denying any involvement there. He repeatedly flew extremely hazardous missions head-on into gunfire with frequent disregard for his own life. One time he and his men ran to a crashed and burning chopper and lifted it enough to free a man who had been pinned underneath. Reading the letter he wrote to my mother took my breath away.

My search for information has intensified since Sept 11, 2001. Seeing the faces of those who had lost loved ones made me know I could not wait any longer to tell my family's story. I wanted those children to know someone understood their loss. I recently made several internet contacts with men who flew with my father. They are telling me stories about his bravery and skill, and about his ever-present wit. Everyone respected him. The day he died, he had flown in the rain to rescue fourteen Navy men from battle. After sharing a turkey sandwich with them at their home base, he was flying back low under the rain clouds when a sniper's bullet shot him dead in midair.

My own memories have surfaced. I remember him being tall and having feet so big I could sit on one and get a ride around the house. He and Mom had lots of friends and gave big parties. He would kiss Mom in the kitchen or squeeze her knee during road trips in the car. He affectionately rubbed my head until my hair shook, and he challenged me to always do my best in school. He taught us proper manners in a restaurant. His shadowy presence looming in the doorway was enough to make my sister and me stop our nighttime giggling and lay quiet as mice in our beds. He was always telling jokes and could make anything sound funny. He told us he had earned the Silver Star by flying a general to use a real latrine.

I am proud to be Dad's daughter. My Mother deserves a medal of her own. She is an original Super Mom who worked full-time and raised three young girls up into happy, successful adults. The strength of both parents lives on in us. My sisters still don't like to talk about Dad, but one day our children will be glad to hear about him. Bob Swenck never saw his daughters graduate college (all three of us!) or get married; he can't kiss his grandchildren or hold his wife's hand as they grow old. Heroes historically have fought for a good cause. My father was a hero because he believed he was defending the freedom of his family and country. He could not know he was jousting against the windmills of Communism. The Viet Cong was a real enough enemy, vicious in battle. He fought hard and bravely and gave his life for his fellow man. He deserves the same honor as any soldier. He did not die for nothing. He died for his belief in freedom, because he was a Soldier doing his job. I write to give him that honor.

GLENN PEYTON CARR

US Army — Aviator

Dates of Military Service: 1958 to 1986

Unit Served with in Vietnam: 213th Assault Support Helicopter Company (Chinook) B Troop 7th Squadron 17th Cavalry Regiment (Air)

Dates Served in Vietnam: XO & CO 213th — May 1967 to May 1968; CO B Troop 7/17, XO 52nd Combat Aviation Battalion — 1971

Highest Rank Held: Lt. Colonel

Place of Birth: Shawnee, OK — 1934

TREMBLING FRIGHT DELAYED

While in command of B Troop 7th Squadron 17th Air Cavalry, Camp Holloway, Pleiku RVN, we had a mission to take a large portion of the troop aircraft, UH-1H Hueys and AH-1G Cobras, way up to the northeast corner of II Corp area. The time was early 1971 after TET, the lunar new year. I was flying the CnC aircraft.

Now for a brief explanation of military terms and acronyms:

UH-1H The infamous Huey helicopter of Vietnam fame designed to carry nine infantrymen into combat.

AH-1G Referred to as Cobra or gunship. The first helicopter designed from the start as a gun platform, the basic aircraft being only 36 inches wide. Armament consisted of a mix of 20mm cannon, 2.75 folding fin rockets, 30 caliber mini-gun, 40 mm grenade launcher and later TOW and Hellfire missiles.

Max Ord The highest altitude an artillery round would reach. This depends on the elevation of the howitzer (gun tube) and the powder charge used to reach the desired target.

105 Shells 105 mm artillery ammunition. The projectile that flies to the target.

Super Quick A term describing the fuse setting. Super quick means the round will explode upon impact. Fuse delay allows the round to penetrate structures or burrow into the ground before exploding.

CnC Ship The Huey designated for the unit commander to ride in and control the battle. It has multiple radios and a pair of 30 cal machine guns on the right side and a 50 caliber machine gun on the left side.

Radio call sign containing 6. In normal army radio procedure, a call sign with 6 designates the unit commander, 5 for the executive officer and 3 for the operations officer. My call sign was Embalmer 6.

FDC Fire direction Center and/or Flight Operations Center. These two can when asked give artillery firing advisories to aircraft flying in their area.

ARVN Army of the Republic of Vietnam.

Battery The name of an Artillery unit. In this case, the Battery consisted of six 105 mm howitzers.

So, here we are early on a sunny morning enjoying the beautiful Vietnam countryside, passing NE of Dak To. I had just called the artillery FDC for any artillery firing along or crossing my flight path. I was quickly given "GUNS COLD" meaning no guns were firing in their area. I confirmed that by responding "Understand ALL guns are cold, is that correct." The FDC responded, "That is correct, sir."

I then said, "This is Embalmer 6, please advise me on this frequency if you have any new firing activity in the next three zero minutes. Can Do."

"Roger sir, we'll keep you posted."

So, we just relaxed, navigated, and enjoyed the countryside. Vietnam from the air is a very beautiful country. Cruising along at 3,000 feet Above Ground Level, after about ten or fifteen minutes all of a sudden I caught a glimpse of movement to the left. A very quick jerk to the left looking out the left window where I was seeing two 105 artillery rounds passing about ten to fifteen feet below the skids of my aircraft. What is now being described lasted only about three to four seconds. I was screaming on the intercom a multitude of four letter words, phrases, exclamations, and proclamations when my pilot Capt. Butch Cleveland, God rest his soul, (he passed away three months ago), announced in his crisp voice, witty humor and sarcasm, "Sir, I didn't get a good look at 'em, but I think they were set on super quick."

The crew chief said he saw them, followed by the gunner and passenger on the right side saying they were OD in color with yellow markings. Confirming they were in fact artillery rounds and not a pair of super-fast hawks, wings folded going into a dive. Upon first sight of what I knew to be artillery everything went into slow motion.

Those rounds were so graceful and smooth flying, it put your mind at ease, mesmerizing to say the least. Seems I watched those rounds for at least ten minutes. Butch broke my trance and said he wondered if those two rounds were the first call for fire spotting rounds. If so, a fire for effect could be on the way with who knows how many rounds. I immediately ordered all aircraft to execute an emergency climb to 4,000 feet, just in case. We saw no more rounds.

Now comes the day of reckoning. "FDC, this is EMBALMER 6."

"This is FDC..."

"EMBALMER 6, didn't you give me a guns cold a few minutes ago?"

"FDC, that is correct, sir."

"THIS is Embalmer 6, That is B*** Sh*t. I just had two 105 rounds pass ten to fifteen feet below my aircraft and five folks on board saw them. My Crew Chief even said they were OD with yellow markings."

"FDC, Sir standby, we are checking the Batteries now."

After several conversations with FDC resulted in no resolution, a deep mature sounding voice came on the radio saying, "Embalmer 6, this is Artillery 6." I don't recall his actual call sign except he was a 6. Therefore, I'll call him Artillery 6.

Then he followed by saying, "I'm rechecking all MY Battery's now, standby."

During that pause I realized the FDC is normally a battalion section, and he said he was rechecking HIS Batteries. The only guy I know of that commands numerous Batteries is a Battalion Commander—a LT Colonel.

He returned to the radio saying, "Embalmer 6, I confirm all my Batteries are cold and have not fired a shot for almost two hours. However, I have an ARVN Battery attached to me and they were firing and failed to report their fire mission to FDC and, Embalmer 6, I humbly regret and apologize for that mistake."

I said, "Artillery 6, apologies accepted, and I judge by your call sign that you have the wherewithal to correct that situation."

He responded, "I sure do, and I have already ordered my helicopter."

I responded, "Good. I have one last request, 6. DO NOT GIVE ME THE LOCATION OF THAT ARVN BATTERY because I have ten loaded Cobra gunships behind me, and I'm mad as hell."

He responded, "Yes sir, I'll see that doesn't happen."

With that bit of excitement behind us, we resumed navigating and enjoying the countryside. The landing at destination was uneventful, but when the five of us got out of the aircraft, we all began uncontrollable shaking and almost had slurred speech as the reality of our close encounter with being blown out of the sky hit us like a ton of bricks.

The two platoon leaders came up to join us and even being a hundred and two hundred yards behind us they also had the shakes from hearing all the radio traffic. After about fifteen minutes, we regained our composure and went to the briefing for the next mission.

There is an old army aviator story that says flying a helicopter is hours and hours of shaking and boredom, interrupted by a moment of stark terror. I can now subscribe to that.

RICHARD "RICK" LAWSON LEAKE, JR

US Army, Accounting Specialist
Dates of Military Service: 1970 to 1971
Unit Served with in Vietnam: 7th Finance, 22nd Finance Section
Dates in Vietnam: Nov 1970 to Nov 1971
Highest Rank Held: Specialist 4th Class
Place of Birth and Year: Atlanta, GA — 1948

PAYROLLS, MPC'S, AND MAIMED CHILDREN

My story is an "our story" that begins June 8, 1969, in Oglethorpe Presbyterian Church Chapel. Rebecca "Becky" Allen became my wife. She had finished her freshman year at the University of Georgia, and I had finished my Junior year. We took a quick honeymoon to Jekyll Island and returned to Athens for summer school. We had a life plan. I was to finish early, start a career, and she would finish college, start a career, and we would eventually start a family. Before fall quarter started, we visited her aunt and uncle in Delaware. He was a research chemist at DuPont. He arranged a job interview for me. I was called to Chattanooga, Tennessee for an interview and was eventually offered a job upon graduation in March of 1970 as a cost analyst at the Chattanooga DuPont plant. Our plans were in motion.

Then came the draft lottery on December 1, 1969. It did not take long to find out that January 15th, my birthday, was draft pick 17. No doubt, after graduation, I was being drafted. I investigated joining all services. I had two years of Air Force ROTC at Georgia but did not like the options the recruiters presented. I tried enlisting in the Army Finance Corps but there were no openings there.

I started working for DuPont in March. I was called to report for induction in June. My wife insisted on taking me alone to the Army induction center on Ponce de Leon Avenue in Atlanta. With tears, we kissed, and she drove off, one year and ten days after our wedding. We had no idea where I was going or when she would hear from me again. She headed back to Sandy Springs where her parents lived. I went into the building for THE medical inspections and THE OATH.

Later that day I was placed on a bus to Fort Jackson, SC. Meanwhile, my young bride was experiencing her first military wife crisis after leaving me. Her car overheated on I-285, the engine seizing and leaving her stranded on the side of the highway. Someone stopped to help, she called her dad and the car was towed. She found employment and bought a rebuilt engine and began to wait for my return.

Basic training at Fort Jackson was not hard for me. As I filled out the paperwork, for skills, I emphasized how much I liked classical music, reading, and so forth, and I knew nothing about camping and guns. As a Southern boy, reality was quite the opposite. I paid close attention in all our training. I was in fair shape when training began, better when it was over. My wife drove from Atlanta to Columbia, SC for quick visits when permitted. When it came time to get our next assignment, our Sergeant had never heard of my MOS. Accounting Specialist school, Ft. Benjamin Harrison, Indiana. I was elated.

Accounting Specialist school was much like college classes in accounting and auditing. About 18 were in the class that was taught maybe once every year or so. Most of the attendees were guys who had elected to leave OCS and were given their choice of assignments. I think about six of us were draftees that had been placed in the class. We were told we would all be assigned to embassies to audit facilities around the world and would not go to Vietnam. Across the hall, 100 men or more were being trained as finance clerks and most were expecting to go to Vietnam. When duty stations were announced, we six accounting specialists were assigned to Vietnam, but none of the clerks were.

I arrived in Cam Ranh Bay. There were no accounting specialist jobs in Vietnam. The six of us were spread out across the country and we never kept in touch. I was sent to a small unit in Di An via Saigon. This office was closing. We packed up and went to Bien Hoa Army base, to a unit in a corner of the base away from the airfield and everyone else. I was assigned to a small group that functioned as payroll calculators that computed and prepared payrolls for field units. We took each personnel voucher, reviewed it, then totaled it. The individual vouchers were then assembled into payroll units. The payroll clerks calculated how much money was needed by denomination to be disbursed. The monies were in MPC, Military Payment Currency; it looked like Monopoly money.

The payroll was picked up by the field units' disbursing officers who arrived in well-guarded jeeps. After paying their units, the records and any remaining MPCs due to any soldiers who were on leave, MIA, or KIA were brought back to us. We reconciled the records and monies. Then we started the month processing all over again. We were an efficient group accurately processing thousands of pay vouchers monthly. We labeled ourselves "Chairborne."

I had a lot of free time with not much to do. Our unit was small and was essentially divided into the guys that gathered every night to drink, be alone, or smoke pot. Some of us kept to ourselves. There was a small post exchange on the base; beer and alcohol were cheap, both in quality and price. We hardly ever saw our unit officers.

Our clothing was washed and pressed and living areas were cleaned by Vietnamese maids. They were paid by payroll deductions. Our "hooch maid" was named Hai.

I took up photography; there was a place on base that taught film processing and printing. I spent a lot of time there. We also traded tapes and copied them. I wrote home and made cassette recordings to send home and spent time listening to the tapes I received. Because we were so few, our sleeping quarters had been sub-divided and we each had private space. I had a large poster of my wife on the wall next to my bed. It looked like she was lying next to me.

Back home, my wife worked, baby-sat, wrote letters every day, and tried to avoid the hyped-up TV news with graphic war images. I know she worried a lot even though I wrote home or sent a tape almost every day.

We were responsible for guard duty at a bunker on the perimeter. Three of us would spend the night in the bunker about every 10 days. We were equipped with M16s, a .50 caliber machine gun, and an M79 grenade launcher. The bunker had flares and Claymore mines and concertina wire set up around it. During the night, we could observe the Cobra helicopters firing in the distance. Occasionally, there would be a few rockets fired at the adjacent air base. The most dangerous night we had was on New Year's Eve when the guys in the next manned bunker thought it would be fun to fire their M79 at us. We hunkered down in flak vests and helmets and called the officers. The other guys were on drugs and were removed, replaced, and court-martialed.

Christmas Day 1970 was memorable. I had been on my first guard duty the night before. Guys were headed to Long Binh for the Bob Hope Christmas Show. I went to the main gate and got a ride; the place was hot and packed. Men even had climbed high on telephone poles to get a view. I had to stand in the back, but what a sight! I will never forget seeing Bob Hope, Lola Falana, and Johnny Bench.

A pass to Saigon and stopping in at the USO there impacted my entire Vietnam experience. At the USO I saw a sign that mentioned a trip to an orphanage. I decided to go. It changed me. A small number of us went and they took us to a facility called Me Linh. The children there were missing arms and/or legs, but we were greeted with excitement and a desire for companionship. We tossed balls and Frisbees with them. When the USO stopped offering the trips, I went by myself and on some occasions I took my friend Brian Pennington. My wife sent packages filled with small toys. I took portraits and then on my next visit I would give the children their photos. I always tried to make them look their best. They were wonderful subjects and posed well. Once, some of the boys dressed in Boy Scout uniforms and posed with a picture of scout founder Lord Baden Powell.

I took a tape recorder and recorded singing and laughter. They loved to hear themselves. I was invited for lunch. Lunch was rice with a meat. It was delicious. I asked what it was. The kids all laughed and told me in Vietnamese "it is dog" trying to get me to react. I really loved several of them, especially a two or three-year-old girl we called Moonshine and a boy who had lost both legs we called Chuck. I asked my uncle, a retired Army Colonel, an Orthopedic Surgeon who had served two tours in Vietnam what did he think about attempting to adopt Chuck, and he discouraged me. I still tear up today thinking of those adorable children and their amazing spirits.

I took an R&R trip to Hawaii. My wife had saved and flew from Atlanta to Hawaii. I guess I don't need to say we had a great time that went by all too fast.

After eight months in Bien Hoa, our unit was sent to MACV Headquarters in Saigon for my final four months. It was 1971, the war was winding down. Our function as payroll calculators was being taken over by a computer. We were now verifying computer printouts. Vietnamese clerks were doing the bulk of the work. Our living conditions were worse. We were now in large open barracks sleeping on bunk beds like basic training. Boredom was constant. Much more time copying cassette tapes in a lab that was on base. The government was facilitating illegal copyright infringement.

When it came time to return stateside, I had less than 18 months in service out of the two-year draft commitment. If I had more than 18 months of service, I would have been discharged when I returned stateside. So I knew I would have a new duty assignment. I later learned the advantage of having more than 18 months active duty qualified me for VA benefits that I have been able to utilize.

When I returned, it was via San Francisco. My retired Army Colonel uncle made my conscientious objector cousin (his son) take me from the air base to the San Francisco airport to avoid the war protestors.

I got a month leave after Vietnam. I was assigned to Kirkland Army/Air Force Base Hospital, Albuquerque, New Mexico. My wife and I headed west, found an apartment with a short lease and toured the area. I reported for duty, was given a week to get winter uniforms and get settled in and report back. They did not know what to do with my specialty, so I was assigned to the reception desk handing out medical folders.

A few days after starting my new assignment, I was called into the commander's office, asked if I wanted to get out of the Army, I replied, "Yes SIR." He told me to take a week off to get my affairs ready, report back for discharge. On December 23, 1971, my Army experience had come to an end. In January, I reported back to work at DuPont. My wife reentered college at the University of Tennessee at Chattanooga; two years later, she had graduated, our first child was on the way, a house had been purchased with a VA loan and our life plan was being executed.

At the time I was drafted, I was not happy at being called. Serving in the military was derided and was not part of our life plan, but I felt I had to follow in my father's and ancestors' footsteps and serve as my duty. I was not going to take steps to avoid service as some classmates or relatives did. I always was glad I had served but now I am even more appreciative of what a privilege and an honor it was to serve my country.

HAM HENSON

US Army — Infantry Officer
Dates of Service: 1969 to 1978
Unit in Vietnam: Charlie Company 1/502 Infantry 101st Airborne (Airmobile);
8th Radio Research Phu Bai
Dates served in Vietnam: May 1971 to May 1972
Highest rank held: Captain O-3
Place of Birth: Savannah, GA

THE RED CROSS RUINED MY R&R

I was a platoon leader with C/1/502 Infantry—101st ABN in the fall of
1971. We were operating along a river NW of Hue in platoon-size elements. My RTO called me over and said we have an order to prepare an
LZ for immediate extraction due to an emergency Red Cross message to
evacuate back to the States for a family emergency. While I understood
that you did not argue with these orders, I did grumble that there was no
LZ close by and we would have to blow a couple of trees to fit in a Huey
so I asked my Platoon Sergeant to proceed with securing a perimeter
and setting up the LZ.

When I went over to the RTO to confirm that we were complying
and give the coordinates for the LZ he said, "LT, there is a problem. The
line number (everyone had a numerical line number so we could identify
individuals in a clear radio transmissions without saying their names in
the clear since only Company Commanders and FOs had secure radios)—is YOURS!" After more discussion, I confirmed it was really me
being ordered out. I pulled over my Platoon Sergeant and Squad Leaders and told them what was going on. Everyone responded professionally
and personally and I gave up my maps and code books and distributed
my munitions per SOP. It was then that I realized I had one more item
to give out. I had my ticket to Australia for the last R&R to go there. I
had lobbied hard for that seat as I had more time in the field than any
other Platoon Leader in the battalion (or brigade—I forget which). It
was the land of beaches with blond and blue-eyed girls. It was a dream

come true now being taken away from me. I pulled out the ticket and gave it to my Platoon Sergeant with instructions that he should award it to the most deserving man when he saw fit.

I made a speedy trip to the States via Da Nang. The Red Cross did not give details for an evacuation order until you reached the US. I called upon landing and learned my father was in the hospital and proceeded to Macon GA. I found that the emergency that had ruined my R&R wasn't really much of an emergency. So, after three days, I decided to have my own R&R anyway, called some friends in Atlanta and stayed with them, enjoying the city.

It was nice, but here is the good part and the point of this story that doesn't talk about my specific experience, similar to many others in this book. My unique experience is that my friends arranged a blind date with a blond blue-eyed girl for me! It wasn't Australia but it turned out to be infinitely better. I fell head over heels in love with that girl and we were married when I returned to Ft. Benning. We will celebrate our 47th wedding anniversary this year. My "ruined" R&R turned out to be the best thing that ever happened to me.

Oh, and yes…when I returned to my Company 30 days later…they were at Eagle Beach in Da Nang. The kid who had my ticket came over to me and said he had a great time and appreciated the chance to go. I didn't know it yet but my reward was much greater than his.

RICK MAROTTE

US Army—Medical Service Corps Officer

Dates of Military Service: 1969 to 1973

Units Serviced With in Vietnam: 571st Medical Detachment/237th Medical Detachment

Dates Served in Vietnam: Jul 1970 to Jul 1971

Highest Rank Held: Captain

Place of Birth: Denver, Colorado

THE NIGHT I DISOBEYED A BRIGADIER GENERAL

I served as a Commissioned Officer Medical Evacuation (Dustoff) Helicopter Pilot from July 1970 to July 1971. During my tour in Vietnam, I served with brave and dedicated pilots and crews that put their lives on the line to rescue wounded soldiers that were putting their lives on the line serving their country. This is a mission that saved lives, but I drew a line in the sand when ordered to risk my crew to evacuate healthy soldiers under very dangerous conditions.

In the northern most operating area in Vietnam, I Corps, summers are hot and humid and winters cool and rainy with low cloud cover and fog. Starting around October, the winter monsoons challenge pilots with heavy rain, low ceilings, and fog. Poor weather conditions caused more Dustoff casualties in I Corps than enemy action. Clear skies would suddenly change to fog and rain, creating dangerous flying conditions for poorly instrumented helicopters and pilots with little hands-on instrument flight experience.

Under conditions where it was impossible to land, we used rescue hoists to lift patients to the helicopter. Most of our helicopters were fitted with a hoist that was operated by the crew medic. The pilot could also operate the hoist from the joy stick. If we had an emergency with the hoist cable hung up or tied to a tree, we could flip a switch and a 45-caliber bullet would severe the hoist cable.

At 2230 hours on December 18, 1970, the DMZ Dustoff radio shack received an urgent mission request to pick up a squad of Rangers that walked into a mine field in the DMZ, well north of the most Northern American Fire Support Base Alpha 2 (sometimes call the Alamo). The Rangers had found themselves in a field of "Bouncing Betty" mines (when triggered, these mines pop out of the ground then explode). They had four wounded soldiers, the rest of the team was not injured. The Bouncing Betty mines would be triggered by my helicopter rotor wash if I tried to land. To make things worse, there was heavy fog over the pickup site. I was told that the Commanding General was at the fire base and wanted me to land there before the pickup to plan and coordinate the mission. After landing and strategizing the mission, we agreed to a plan. The plan was to position tanks around the perimeter of the mine field for security and to use their high intensity spot lights to light up the field and allow me to see the ground while hovering.

I hovered the helicopter into the DMZ at about 50 feet above the ground towards the lighted area. We hovered over the mine field using the tank spot lights for reference and maintained a stationary hover while all four wounded soldiers were hoisted onto the helicopter. Before continuing the mission to pick up the remaining soldiers, I elected to hover away from the pickup to relieve the tension in my crew and me. I soon learned that the fog had grown much thicker when I lost my ground reference. The sudden instrument flight conditions created temporary disorientation and the helicopter entered a steep left bank. As all pilots know, when things go bad, you can't have too much air below or too little sky above, so I increased power and started to climb as I regained control. We climbed out of the fog to clear skies. I flew back to the 18th Surgical Hospital (MASH) in Quang Tri and dropped off the wounded. There were still soldiers at an uncomfortable location on the ground, but we had extracted all the wounded and they were safe.

I radioed Alpha 2 and had them tell the General that conditions were currently unsafe for a rescue and I would pick up the remaining team after the weather cleared. I was asked to return to the fire base while we waited for the weather to clear.

When we landed, the General radioed me and said he wanted me to pull his troops out of the mine field that night. I reiterated that flight conditions were not safe, and it would not be a good idea to risk my helicopter and crew to rescue soldiers now. He said, "Lieutenant, I order you to pick up my troops." I responded to the General respectfully that I was in command of my helicopter and I would not risk my crew on this mission. I also informed our radio shack of the situation with the General and asked the radio operator to contact our commander and tell him of our situation. Word came back to me that I was not to risk my helicopter and crew regardless of what the General ordered. My commander then contacted the Vietnam Medical Command (MEDCOM) to inform them of the situation. He radioed me back that we had MEDCOM's full support.

I remember the General walking from the Alpha 2 bunker to the helipad and while standing on my aircraft skid he said, "Lieutenant, this is a direct order, you are to fly back to the pickup site, rescue my troops and either take them to the hospital or bring them back here." I again respectfully said, "No sir." The General told me to exit the helicopter, his crew would complete the mission but he needed our rescue hoist. I radioed my commander and told him the situation. He said under no condition should I give my helicopter to the General.

Good grief, I was a lieutenant in the crossfire between a General and my commander with contradictory direct orders. I told the General that my commander had ordered me not to give up my helicopter. In response, the General unholstered his 45 pistol and said again, "I order you to give me your helicopter." Of course, I said, "Yes sir."

He had his pilots take over the aircraft. The new crew were not familiar with the rescue hoist so my medic volunteered to go on the mission.

The General's crew hovered into the mine field and entered heavy fog. They too became disoriented and had to regain control of the helicopter, climb out of the fog, and return to base. They were not able to complete the mission either.

I spent the rest of the night with my pilot and crew chief in the command room at Alpha 2. The General was with us but there was little conversation. However, I learned that he was the key speaker at my flight school graduation and he pinned on my flight wings at graduation, a bit of irony.

The next morning, the General's crew returned my helicopter to the fire base. I completed the rescue with my helicopter, returned the uninjured soldiers to the fire base, then flew to my home base.

Later in the day, the General contacted my commander and asked for him to send me back to Alpha 2 the following day. He wanted to award me an Army Commendation Medal for Valor for rescuing the wounded troops. Considering that I did not follow the General's orders, I didn't think I should accept the medal and respectfully declined the offer.

This almost ends the story except a few days later I received a note from a good friend and fellow Dustoff Pilot stationed in Saigon. He said he overheard some senior officers at the officer's club talking about a defiant Dustoff Pilot up North that had a run-in with a General. He asked if I knew anything about it. It seemed to have created quite a stir with the brass in Saigon. I told him, "Yes, I had heard of it."

ROBERT REESE

US Army — Infantryman
Dates of Military Service: 1968 to 1971
Unit Served with in Vietnam: B Company, 1st Battalion, 26th Infantry, 1st Infantry Division
Dates Served in Vietnam: Dec 1968 to Nov 1971 (two tours)
Highest Rank Held: Sergeant E-5
Place and Date of Birth: Atlanta, GA — 1949

LIVING IN A GRAVE

Home sweet home to B Co. was a fire base. A fire base is a semi-permanent position built in a clearing in the forest or the edge of a rubber plantation. The concept was that infantry companies would be located there as a launching point for combat operations in the surrounding countryside. One or two companies would occupy the firebase for 2-3 days while other companies of the battalion were on operations then rotate out on their own missions while the other units came in for a few days 'rest'.

You can think of a typical firebase as about the size of a major league baseball field — without the nice infield and outfield grass of course. Just red dirt and dust. Think about living and working there. Not just working during that day then going home to your house or apartment... but literally living there 24 hours a day.

In the middle of the base were a few fortified bunkers made of sandbags and timbers where the command folks — officers and communication people — hung out. Sometimes in the center was a wooden watch tower with some communication antennae on top.

In a circle outside that were pits where artillery batteries were placed, usually 105 mm howitzers. These batteries were called upon by units 4-5 miles away out in the forest when needed to assist in fighting off the enemy. The artillerymen were permanent residents of the base.

In a larger circle outside the artillery positions were a series of simpler bunkers constructed mostly from sand bags. The infantry company soldiers were disbursed there — four to five soldiers to a bunker. These represented the last manned outer defenses of the base. There were maybe 15 or so of these that were the outer defense circle. Outside the perimeter bunkers were coils of concertina wire, a razor-sharp version of barbed wire. Outside that was the forest and whatever was out there.

The perimeter bunker was home for us squad members. When we were at the firebase, you hung out at these all day. The bunker was a hole in the ground with overhead cover built out of sandbags. There was room for maybe three people inside if there were an attack. Nobody would ever go inside otherwise. It was really just a dank hole, usually with vermin such as stinging insects, leaches, snakes, or scorpions inside. If you wanted to — and had permission — you could take a short stroll around the base, but it all looked the same — dusty, hot, filthy.

The daily routine was to sit on top of the bunker or in a low uncovered pit behind it in the broiling sun — just shooting the breeze, cleaning your weapon, listening to music on a transistor radio, writing letters, reading the *Stars and Stripes* (an excellent Army published-daily newspaper complete with comprehensive sports page), eating primitive meals, and mostly just wishing you were somewhere else. Time did not exactly fly by.

I heard it once described as like living in an open grave—which was a pretty accurate description of what it was like.

For meals in a fire base, most times you would get a hot breakfast—scrambled eggs, bacon on a paper plate, coffee. Lunch/dinner you may or may not get something hot from the mess tent. Kool-Aid—the Army was big on Kool-Aid. So you get your paper plate meal, go back to your bunker to eat sitting in the sun. I remember, more than one time, a supply chopper picks that time to arrive, blowing huge clouds of dust all over the fire base, including over your eggs and bacon.

When you are at the fire base, you get to take a shower maybe one time in the 2-3 days you are there. To take a shower was a great luxury and takes some advance planning. First you go to the water trailer (hauled in by chopper—because there usually are no roads to the base from the outside.) You fill up a 5-gallon canvas bag that has a shower fixture on the bottom that you screw to on-off positions. You hang that bag up in the sun for 2-3 hours so it becomes—presto—hot water. Then you hang it up on a pole (no shower stall usually…just out in the open). Open the shower head and gravity does the rest, a hot shower. Soap up and rinse quickly though. This is one of the important luxuries of a two-week period—one shower. Hot, dusty, work-filled two weeks … I often think about it when I have a shower at night in my well-appointed suburban bathroom.

One wore the same clothes for a week or more at a time. Every 10 days or so a chopper would bring in a big load of clean but more than slightly used jungle fatigues. They were simply dumped on the ground then we would rummage through them to find a set that fit. Those were then worn for the next week or so—on patrol, sitting around the firebase, sleeping, whatever—24x7. I doubt that many in our mainstream society would ever have such an experience these days, except maybe a homeless street person.

If you need to use the toilet, to pee you go to a big tube (a used casing for an artillery shell—universally called a 'piss tube') embedded in a hole with a wheelbarrow of chlorine pellets for sanitation. For a bowel movement, there is a wooden throne out near the barbed wire where you go. Like an outhouse but without the house... just out in the open field. The only privacy is the 40 yards that separates you from the nearest perimeter bunker.

You try to relax as much as possible hanging around your bunker. But the Army doesn't believe in relaxing, they always want you to be doing something. So one of the details one could get assigned to was "OP" (Observation Post). This delightful assignment was for two of you to take a radio and your M16s, go out 100 yards or so outside the concertina wire perimeter and sit out there in a bomb crater on guard duty for 3-4 hours in the broiling sun. The idea is that if the enemy approaches the base, they overrun you first, thus giving an early warning to the main force inside the wire.

If you are on OP, you sit in a shell hole with your buddy and hope that nothing comes your way. It usually didn't that near to the base, but it's the idea of sitting well outside the wire in a combat zone with one other guy—well it's just not comforting. Not to mention ever-present mind-numbing boredom. There is an old saying which is quite true—being in the infantry is 98% boredom and 2% terror.

At night, except for the mosquitos, life around the bunker was slightly more pleasant—well, cooler anyway. The first part of the evening guys just sat around smoking cigarettes and shooting the breeze. Then some would get some sleep while one or two stood guard for a couple of hours, then trade off. Guard duty is the worst—you're dead tired and there is nothing to do but stare at the gloom and think about all the things you are missing "back in the world" and how much time you have left before you can go home. However, the view or the sky at night was a real spectacle. The stars and Milky Way were crystal clear, like nothing one would ever see in our light-polluted skies today. There was something very exotic, even romantic in the adventurous sense, about sitting in a jungle clearing under a moon-lit sky listening to the night sounds.

One was constantly brought back to reality though because you could look in any direction of the compass in the night sky and see far-off tracers from gunships shooting up some part of the countryside. They appeared as ghostly streams of red fire from the sky in the distance, so far away you couldn't hear them but most always observable at whichever point of the compass you gazed. This was true anywhere in country for years and years. An eerie manifestation of a war that seemed destined to go on and on without end.

But eventually, like all wars, it did end. I've traveled back to Vietnam many times since the war ended. The firebases we once defended are now overgrown patches of ground or a peanut field or maybe home to a small industrial park. Absent the memories of the men who fought there, they might never have existed at all.

JOHN D. SOURS

US Army – Judge Advocate General's Corps Officer
Dates of Military Service (Active and Reserve): 1969 to 1979
Unit Served with in Vietnam: HQ, US Army Vietnam (USARV), U.S. Army Procurement Agency Vietnam (USAPAV)
Dates in Vietnam: Jan 1971 to Dec 1971
Highest Rank Held: Major
Place of Birth and Year: Harrisburg, PA – 1944

"JUST GETTING THERE"

I served in Vietnam between 5 January and 1 December 1971 as a Captain in the U.S. Army Judge Advocate General's Corps ("JAG"). Prior to that, I had been fortunate enough to, first, be deferred from active duty for two years to finish law school after being commissioned in the Infantry; and, then, to be given a plum initial assignment to the JAG office at Fort Lewis WA, where I got to defend scores of courts-martial and spent the balance of duty time reviewing contracts and claims against the government—an activity known in military jargon as "procurement law."

I volunteered for duty in Vietnam for three reasons: curiosity about what was really going on there, amid all the largely negative accounts flooding the American media; recognition that I was surely thought by JAG personnel officials to owe the Army such service due to all the breaks I had already received; and, lastly, the hope that I would get an interesting assignment there and another one when completing my tour. As it turned out, the latter two reasons and hopes largely panned out.

I can't say that the first—my curiosity—was fulfilled simply because the scope and pace of developments was so large, the dynamic of events so varied, and the nature of one's individual perspective so dependent on where, when, and how he spent his tour. Nevertheless, I've never regretted my decision and the experiences I had, nor have I forgotten the people I met or the feeling that, once I came back home, the work assigned to me seemed so mundane and insignificant in comparison to what I had done at the staff agency in Saigon (the Army Procurement Agency—USAPAV) where I was assigned. However, in looking back over nearly 50 years, some of the most significant remembrances I retain have to do with things largely unforeseen that occurred during the time of (and shortly after) my arrival.

My orders for RVN arrived at Ft. Lewis in early December 1970, a day or so before the arrival there of a more senior JAG Captain who told me he had been assigned at the 2nd Brigade of the 101st Airborne Division headquartered at Phu Bai, south of Hue, and that I would replace him there on my arrival in-country the first week of January 1971. How he knew that was never clear. He also helpfully informed me that, since I was stationed at Ft. Lewis, I would undoubtedly leave from McChord Air Force Base next door, then fly to Honolulu, then to Guam and on to RVN, all as he had done a year earlier. All of this sounded quite logical, and so, realizing that all those places, and certainly my final destination, were hot weather spots, I packed light—TWs, field jacket without liner, etc.

The day I was to depart, I received orders to report to SeaTac Airport, not McChord, and to take a United flight to San Francisco, then surface transportation to Travis Air Force Base to begin my trip across the sea.

The flight from SeaTac is where my adventure really began. I noticed that I was the only passenger in military uniform, but thought nothing of it at first. Then I met my seatmate, an older Roman Catholic nun from Seattle, who shortly after takeoff asked me if San Francisco was my final destination and then noticeably blanched when I responded that I was actually en route to Vietnam.

She quickly produced a very lovely set of prayer beads and asked me to say the rosary with her. When I responded that I didn't know how to do that because I wasn't Catholic, she asked warily what my religion was and, when I replied that I was an Episcopalian, she breathed a small sigh of relief and said, "Well, that's close enough." She then proceeded through the entire litany, adding at the end a special and undoubtedly heartfelt prayer for my protection from harm and safe return from "this horrible war."

I thanked her and quickly changed the subject, searching for something more palatable to discuss. Surprisingly, that turned out to be baseball and our conversation lightened until we descended at dusk into San Francisco where, the plane having just barely touched the runway, we rapidly re-ascended. She immediately recalled, to my considerable discomfort, that there had been several hijackings of commercial airliners in recent weeks, mostly to Cuba, and expressed great concern that I was in uniform. I reminded her that Cuba was quite far away and that it was doubtful we could be re-routed there; and, besides, it was a Communist country where persons such as she, also being in "work uniform," may not be welcomed any more than would I. "Oh, dear," she replied, "that's right," and out came the rosary beads once again.

Fortunately, though, before she had launched deeply into another round of prayer, the calming voice of the pilot intervened, announcing over the intercom that he had been directed by flight control to abort the landing and come around again because, on his initial try, he had put the aircraft down too far to the right on the runway and was putting out all the lights on that side.

After landing successfully on the second try, we quickly gathered our belongings, disembarked and went our respective ways through SFO.

I descended to the baggage claim area, where I encountered two other Army Captains and a CWO, all pilots and all headed to Travis and on to RVN. We decided to share a cab for the nearly 50-mile ride, crammed ourselves and baggage into a dilapidated Ford taxi driven by a guy whose first language wasn't English, and headed out at a fast clip, noticing that the space on the dash where the radio would normally be found, was vacant, replaced by a rush of rapidly chilling air.

One of my fellow passengers, perhaps wanting to make other travel arrangements and knowing that we didn't have to report at Travis for nearly four hours, quickly informed the rest of us that a Santana concert was about to begin at the Fillmore Theatre on Geary Street and that we could get there for the start if we hurried. That sounded like a much better way to spend the evening than riding for an hour and a half in a cold cab, and we all quickly agreed. The concert was wonderful, though not a few other attendees gave us dirty looks as we entered and took our seats, one exception being a very lovely young woman who gave each of us a big hug when learning we were on our way to Nam.

Our joy at this very welcome respite quickly turned to something considerably less, though, when we left the Fillmore at the end of the two-hour concert and discovered that the only transportation available was the naturally air-conditioned cab that had dropped us there earlier.

Not having time to scout out alternatives, our party reluctantly piled back into the decrepit Ford and endured the trip to Travis, arriving about 20 minutes prior to the deadline for reporting. Our efforts were then rewarded with an unplanned three-hour delay until we were loaded into a Seaboard World Airlines charter around 3 am—every licensed air carrier was flying to and from Southeast Asia in those days. There followed a rapid trip down the runway, ending in an equally rapid stop near its end, followed by an announcement that the flight crew was unable to get the radio to work, so that there would be a brief delay while the problem was fixed.

The brief delay turned into a half-day, during which we were the airline's guests at an early-vintage Holiday Inn next to Solano State Prison near Vacaville, sans our baggage. I seems that the radio was irreparably damaged, so that the aircraft had to be replaced with another, from Chicago. So we regrouped back at Travis at the end of the following afternoon, January 6, and took off uneventfully, not for Honolulu as we all had expected, but rather headed to Anchorage, where we landed in 18-degree weather at about 2am on January 7.

After a planned two-hour refueling stop, which took more than six hours, we headed off again, almost immediately noticing a very loud screeching sound emanating from the Seaboard World's right side engine as it sped down the runway. This obviously bothered our pilot as well, for he fishtailed the aircraft to a stop about 25-30 yards from its end, and then announced that he didn't think those engines could generate enough power to enable takeoff, so that we would have to return to the gate. We were impressed with his maneuver, and even more so with his logic, notwithstanding that we had to evacuate the plane 100 yards or so from the lounge, there being an apparent jetway shortage as well. All of this brought to our minds the fact that, about two weeks previously, a similar airplane carrying outbound troops had crashed after takeoff into Mt. McKinley, which lies directly to the east. We then all slept soundly in the new Hilton in downtown Anchorage, to which we were bussed just in time for breakfast.

The result of all the foregoing activity (and inactivity) was that we had spent two nights and three days en route to Vietnam without yet leaving our own country. An AG-type in our traveling party somewhat gleefully reminded us that this travel time counted against our twelve-month short tour obligation, as would any further delay time prior to arrival in Vietnam. With that pleasant thought in mind, everyone looked forward to our next interim stop at Yokota Air Base near Tokyo, secure in the thought that one more breakdown would lead to one more overnight, to be spent sampling the delights to be had in the Ginza. Sadly, though, things worked like a charm and our schedule two-hour layover at Yokota lasted just two hours.

One more series of developments still awaited us, however, as we flew at altitude over the Vietnamese countryside watching tracers arcing throughout the night. Our pilot announced, around 2 am (it seemed as though a lot of things happened at that hour) that we would be soon making our descent to land at Bien Hoa Air Base, to be met and bussed to the reception and assignment center at Long Binh nearby. He advised that the descent would be steeper and faster than normal, that the plane's engines would be kept running, and that we should exit out of both the front and rear exits of the 727. These measures were necessary, he said, because another Seaboard World aircraft had been subject to a rocket and mortar attack when landing at about the same hour the previous evening. No sooner did the pilot finish the announcement than he put the plane into a steep dive, causing most of the items in the overheads to spill out and scatter throughout the cabin, and the flight attendants to fall across or into some of the passengers, an experience thoroughly enjoyed by my seatmate, a grizzled E-8 arriving for his third tour.

Having collected ourselves and our belongings, we were instructed to exit in a hurry, to locate an open-air shed with a metal roof that would be about a hundred yards away, and to sprint for cover. Everyone complied and we all began running across the tarmac, hardly able to catch our breath in the oppressive heat and humidity even of the night, until we collectively realized two things: first, there was no attack being conducted by anything but mosquitos; and, second, the shed was not 100 yards but more than a quarter mile away. Whereupon, a Colonel (actually a Lieutenant Colonel) uttered one of the most appropriate orders I ever heard: "STOP—this is silly!"

After a desultory "greeting" by another Colonel who seemed to resent having had to be awake at that hour, we were ushered onto several wire-festooned buses and taken a couple of klicks to the northwest edge of the Long Binh perimeter to await the arrival of morning and the opening of the mess halls and assignment shed. I teamed up with two other JAG Captains from my Basic Class of a year or so earlier and we spent the rest of the night talking and smoking cigarettes, which it seemed everyone did back then.

When the assignment shed opened, I ventured there and was told that, as a JAG officer I needed to call the Sergeant Major at the US-ARV SJA office, to make arrangements to meet with him as soon as possible. I dialed his number, got him on the line and told him who I was. He asked if the other two JAG officers were there with me, I said they were, and he then instructed us to gather together in front of the shed and await transportation. About ten minutes later, while some of the field grade officers were being picked up in worn and dirty jeeps and utility vehicles, a horn sounded outside the gate leading from Highway 1. When it opened, a sparkling as new Oldsmobile 98 sedan appeared, drove up to the shed and the SJA's Sergeant Major exited, introduced himself, we all saluted and, at his invitation and to the evident dismay of an array of Majors and Colonels of all types, ensconced ourselves in the air-conditioned vehicle.

We did, however, also invite the smart and practical Lieutenant Colonel who had short-circuited the arrival fiasco of the previous evening, to join us for a ride to HQ, USARV. That turned out to be a very good move, especially for me, because that 0-5 turned out in short order to be a key staff officer in the USARV G-4 shop, to which our little agency reported.

MARK STEELE

US Army — Adjutant Generals Corps Officer
Dates of Military Service: 1969 to 1971
Unit Served with in Vietnam: MACV Armed Forces Courier Service
Dates in Vietnam: Mar 1970 to Mar 1971
Highest Rank Held: 1st Lieutenant
Place of Birth and Year: Brooklyn, NY — 1947

DID HE OR DIDN'T HE? WHAT WAS IN IT?

My year in Vietnam was spent as an Courier Officer in the Armed Forces Courier Service (ARFCOS). We were responsible for the secure transport within Vietnam of all Top Secret/Crypto material for all five branches of the service, the CIA, NSA, DIA, Department of State, etc.

Our main Courier Station was in Saigon, located adjacent to the flight line at Tan Son Nhut Air Base. Smaller stations were located on the Air Bases at Cam Rahn Bay and Da Nang.

My fellow ARFCOS Courier Officers were all ROTC graduate Lieutenants with a two-year active duty commitment. None of us planned to make the Army a career, hence our Branch choice as the Adjutant General's Corps (AGC). Contrary to popular belief, we were NOT the lawyers. They were the Judge Advocate General's Corps. AGC Branch is the administrative branch of the Army. For those who remember typewriters, "We did not retreat, we backspaced!"

We were a motley bunch, similar to the characters in the television show MASH. We figured since we were already in Vietnam, anything the Army could do to us would be better than spending one year in Vietnam.

One day in the middle of my tour, I was the Officer of the Day for the Saigon Station. This meant I spent 24 hours on duty, made delivery assignments, and slept at the Station at night. All of our highly classified material was either in a vault in the Station, or if too large, was outside, guarded 24/7 by Air Force police.

It just so happened while I was on duty that day, General William Westmoreland was on his way from the Pentagon to MAC-V Headquarters (Westmoreland was Chief of Staff of the Army, not commander of Vietnam at the time). For some unknown reason, he had sent his personal briefcase on ahead through ARFCOS channels. There it was, a battered, brown leather unlocked briefcase in one of the cubby holes in our vault. It was late at night and I was the only one in the Station.

Did I look inside General Westmoreland's briefcase? What was inside it? If I tell you, I would have to kill you...

THOMAS "TOM" YEARIAN

US Army — Combat Engineer Officer

Dates of Military Service: 1968 to 1971

Unit Served with in Vietnam: 14th Combat Engineer Battalion

Dates you were in Vietnam: 1971

Highest Rank Held: Captain

Place of Birth and Year: Atlanta, GA — 1946

IT'S A SMALL WORLD

I was with the U.S. Army in Vietnam in 1971; Company Commander of a Combat Engineer Company operating out of Quang Tri Combat Base and along the DMZ. Although I saw very little actual combat, there was one incident that left its mark.

I must begin by explaining that maybe 25 or so years ago, I was a Boy Scout leader and received a call asking me to serve on a committee; the purpose of which I cannot recall. I agreed and went to a meeting at a gentleman's home in Stone Mountain, GA. Arriving early, I took note of his medals in a shadow box and particularly his Air Medals. I asked, "What did you fly in Nam?"

"I was a Cobra pilot," he replied. The conversation then went something like this, as is typical when Vietnam vets meet:

"When were you there?"

"1971"

"Me too!"

"Where were you?"

"Based on Quang Tri."

"No kidding. Me too!"

"Whoa! When my Company was on the Combat Base, we used to pull guard on the perimeter around the airfield."

"So, your guys had my six while I slept at night?"

Small world! We marveled over the coincidence. I then said, "You know what, there was only one time I needed you guys," and began to tell him the story. My company was on an NDP (Night Defensive Perimeter) on a mountain top between Khe Sanh and Firebase Vandegrift. I had gone into Quang Tri to bring the Chaplain out for a service (we had lost two men; the only ones I lost, thank God!). On the return trip via jeep, we got between five or six ARVN deuce and a half trucks. As we rounded the curve in front of The Rock Pile (two bare mountain peaks about a click or two from the road), the road was mortared. The trucks were disabled, and the ARVNs ran over the hill, leaving us (the Chaplain, his assistant, and my driver) stuck. With a sharp drop off, we were unable to maneuver around the wreckage.

Presently, we noted a North Vietnamese unit (about a squad and a half) making its way across the open area in front of The Rock Pile, headed straight for us. I radioed my Battalion Net at the Combat Base in panic, hoping to get an artillery strike. Within a few minutes, however, two Cobras came in, like angels from heaven. The command ship called me on my frequency, and although I don't recall the exact conversation, it probably went something like, "Relax, guys, we've got this." They hit that unit with their mini guns; destroyed them; and we were saved.

As I told the story, this fellow got almost ashen colored. I thought he might be having a heart attack. "Are you okay?!" I asked.

He softly replied, pointing his thumb at his chest, "I was the guy you were talking to on the radio!" We just sat a while and stared at one another, at a loss for words. You see, he was from northern Virginia, and when he left the Army, he took a job with the IRS in D.C. He was subsequently transferred to Atlanta, and there we sat… in his family room… for a Boy Scout meeting… fate having re-connected us for some unknow reason in the cosmic sphere of things.

It's a small, small world, indeed.…

RALPH H. BELL "ROTORBRAKE"

US Air Force — CSAR (Combat Search and Rescue)

Dates of Military Service: 1966 to 1991

Unit Served with in Vietnam: 3d Group, 38th ARRS, Detachment 7 (DaNang)

Dates in Vietnam: May 1971 to Dec 1972 (19 ½ months)

Highest Rank Held: Colonel

Place of Birth and Year: Texas City, TX — 1943

RESCUE AT THE BURNING BARRACKS

When I went to Vietnam in May 1971, all Combat Rescue (CSAR) pilots first reported in at the Headquarters (38 ARRS) at Tan Son Nhut AB, RVN, co-located within Saigon City. I was first subjected to a hair-raising tour of the city and surrounding areas in a HH-43F "Huskie" helicopter borrowed from the local Tan Son Nhut Detachment, and after surviving that "dollar ride," I was allowed to choose the base that I would go to for permanent basing. Since I was in Vietnam to rescue as many survivors as possible, it made sense to me to ask for the base with the most activity, translated as the "most saves." Quickly the admin clerk said DaNang and typed out my orders.

My Commander explained the system that was long established, pretty much 24 hours on Rescue Alert and 24 hours off. During your 24 hours off you were allowed to do anything, and most of the men simply stayed around the Det 7 Alert Facility which had the adjacent (built by self-help efforts) "Pedro's Cantina."

I caught a ride to the Main Compound and bed. At first sight, my barracks looked scary, run down and even the surrounding revetments and sand-bags looked like they had been stacked by the French before Dien Bien Phu (May 1954): two story, window AC rooms but no windows, hallway to community showers and commodes; laundry room near the showers and at the end, the "hooch bar and grill" for beer and steak. My room was "pick one that was not occupied and had a working AC." The barracks are another story.

Then on the night of 8 July, I was on Alert and thinking that it was my brother's birthday and remembering that I had sent him a card and maybe some souvenir from Vietnam. He was in the Air Force also, although not qualified as a pilot. Then, not unexpectedly, the rockets began a night-time thumping, on the east side, probably impacting DaNang City. So, I followed the tradition and greedily chugged my FREE Beer (crazy rule, beer was declared free during a rocket attack), and NOT free Pabst Blue Ribbon.

BLAMMMMMB, BLAMMMMMMMMMB, Phssst. All was black. The dirty bastards had hit the power station, obviously blind luck since the katyushas (Soviet-made rockets) were not guided, just random lobs... Brrrrringggg- Rrrrinnnggg- Ling went the CRASH Phone. Oh NO! My practice was to be ever ready and the Medic always took the calls as I ran to the helicopter and started her up. But wait, I had a half-full can of ice cold beer in my right hand? Wasting cold beer wasn't an option so I gulped and did not hesitate my run out to the ramp and into the pilot seat. I pushed the start button and as I always did, began strapping in as I manipulated the engine control on the collective and watched the meager cockpit instruments.

Strapped in, I released the rotor brake (coincidentally also my nickname) and the whistle of rotor blades added to the whine of the turbojet engine (HH-43 nickname was "Whistling Shithouse"). I had already launched the helicopter up and into a five-foot hover when my medic came on intercom. "Sir, it's Gunfighter Village, they hit the barracks." And probably a few other spots I thought to myself from the looks of the darkened airbase.

The control tower had emergency electrical power, but it soon looked like nobody else did. I began a flight in roughly the direction toward Gunfighter Village as the control tower responded to my request to cross the now blacked out runway with, "Pedro is cleared direct to Gunfighter Village." But, I thought, "I cannot see anything: pitch black with some weak lighting reflected on the horizon from across the river to DaNang City." I thought it was important to climb to 500 feet and avoid hitting unlit objects until I could see the barracks.

WOW, there was a BIG fire over where Gunfighter Village had been, and the closer I flew, I could start to see that one of the enlisted barracks was on fire and surrounded by those big fire trucks. I flew a low pass for reconnaissance at 100 feet, risking the darkness and realizing the bright fire was causing spatial disorientation, like a moth to a candle. Next was running the pre-landing checklist and looking for a spot to land. There was no landing pad, just the space framed between the overhead spaghetti of power lines, three fire trucks, and the barracks. So I tried my Motorola FM radio (secured with a bungee cord to the canopy beside my right leg) to call the Fire Chief as he would be the on-scene commander. No luck, and I was suspecting that I would soon be called upon to pick up numerous burn victims. So without any other reassuring voices except from my crew, I began a non-standard landing approach to a burning building. I tried not to look directly at the fire, but I also had to gauge my approach to not fly through the high pressure water stream from the fire trucks.

I landed smoothly and without a hover, just a gentle landing while staring straight ahead at 40-foot power lines with wires running in all directions. Helicopter pilots are afraid of wires and telephone poles, rightfully so. My medic and two firemen leaped from the helicopter, and the flight mechanic set up a perimeter so that we wouldn't get run over by the fire trucks and also so nobody would walk into the rotating rotor blades. I began to hear something like popcorn, then like firecrackers. Hmmm. What would THAT be? Someone from the fire department dressed in his Aluminum — Asbestos Fire Suit came up to me while I was sitting in the pilot's seat and SHOUTED: "they have live ammo and it's cooking off." Probably that's why the firefighters were all on the other side of their trucks. I also began to hear zings and bings as well as the pop — pops. I was trapped between the burning barracks and the fire trucks waiting for my medic to return, not a good feeling and of course at that time he seemed to be taking forever.

As I was contemplating taking flight again and attempting to land the helicopter and my remaining crew further away, maybe on the roadway. Obviously it would be safer on the other side of the big fire trucks protected from the projectiles from those "illegal souvenir weapons." But, at that moment, here came my medic and two litters loaded up with two surviving burn victims. I gave him the "clear to load survivors" signal and my practice was to look over my shoulder back into the cargo compartment to see the situation for myself.

The two men had totally gray-white faces which startled me. Why? Were they dead already? And we had a policy not to haul bodies unless they were in danger of being lost. But my medic reassured me they were okay for now, but we needed to take them to the 45th Army Medical Hospital, which could handle their injuries. I gave the crew "Prepare for Take-off" and did a quick scan of the scene. Years later, I realized the men had been smeared all over to cover their burns with an ointment like Zinc-Chromate or some other white salve.

It was clear to me that once I flew above and clear of the wires and away from the light from the fire trucks and the fire, it would be back to flying into a pitch-black night. I anticipated more IFR (Instrument Flight Rules) than usual night flying VFR (Visual Flight Rules). Of course, I had my own lights so I set my search light to about 60 degrees deflection for a possible autorotation (engine failure during flight) and flood lights for the early part of the takeoff before they would become ineffective (above 100 feet). As soon as I confirmed "Ready For Take-Off," I began a maximum power takeoff on a steep angle toward the top of the highest power line. Just as we broke ground and began to accelerate beyond ground effect, suddenly the co-pilot pulled back on the cyclic, turning our flight path into almost vertical climb, impossible to sustain at our heavy gross weight. I screamed "GET OFF the controls" and luckily he did. Then in an embarrassing and dangerous mid-takeoff maneuver, I pushed forward, rocking the nose down and we rocketed upward, barely clearing the top wires. And, into the darkness on the other side.

Fortunately, the survivors were heavily sedated with codeine so they didn't get airsick or worse on that rough "rocking-horse" take-off. I always wondered what the firefighters thought of my airmanship. It must have looked almost acrobatic. In the darkness I could see flying IFR was the answer so I flew straight and level climbing to 1,000 feet. By the time I leveled the helicopter, I could see the DaNang River bridge dimly lighted by vehicles and some sort of navigation lights for the river boats.

That was a good sight to see and from there I needed to only fly over the city below to White Beach between Marble Mountain and Monkey Mountain and let down over the water where we were more likely not to encounter unlighted obstructions. As they say, it was going to be a piece of cake.

When I let down to 500 feet over the ocean and past the beach, I stabilized everything onboard and reverted to a normal approach to the Army Hospital Helipad. I radioed the Hospital of the survivors or "Souls on Board" (we have two SOBs). I was trying to relax and fly smoothly as we descended with "Before Landing Checklist" and transitioned to a shallow approach.

As we approached the beachfront helipad for the hospital from the ocean side, I apparently was a little bit too low when my landing light illuminated the unexpected 8-10 foot chain link fence with three rolls of that triple concertina razor wire. Of course, the Army hospital had to have perimeter security. But it made for another rapid pull-up and dive over in my otherwise text-book approach and landing. I then safely landed the helicopter in the center of the "Maltese Cross" marked helipad and watched as my medic, firefighters, and flight mechanic performed the offload efficiently and with the care expected.

I took the opportunity to lecture my copilot once more and ask him, "What in the hell were you thinking." He said, "I'm sorry sir, but in helicopter school they taught me to establish a hover at the height of the obstacle. Then when assured I could clear it, fly straight ahead." WOW! He nearly caused us to crash. If he had had more experience, he would have known that we did not have enough power to hover at 45 feet in the air (height of the obstacle) with two survivors and a crew of six. I hope he never forgot his lesson that night; I know I haven't. When the gurneys rolled away with the two survivors and the medical attendants from the hospital gave me a "thumbs up." it was time to go back for more.

Flying back, I tried to make peace with my copilot, letting him think he was flying while I gently guarded the controls. I checked with the control tower on the radio to see if we would need to haul more survivors from the barracks. I could see the fire was out and the problem of landing again was probably worse than before: no lights. I was relieved (copilot also) when the tower relayed that the Fire Chief had said "thanks for a job well done, and no further need for helicopter support."

My copilot landed the helicopter at the Rescue Alert Pad where we had NF-2 generator/light carts to illuminate the entire area. We all helped refuel, clean up, and re-cock the Alert helicopter. We were soon back on Alert in our sweat drenched flight suits awaiting the next call. And I tried to forget that FREE beer.

In later days, I helped to modify the Det 7 tradition to provide a free soda to alert crews during rocket attacks. Beer was provided only for those not on alert, and there was always ample consumption as the all night poker games continued and the katyusha rockets kept coming.

PETE ALEXANDER

US Air Force – Pilot
Dates of Military Service: 1968 to 1979
Unit Served with in Vietnam: 309th SOS
Dates in Vietnam: Oct 1969 to Sep 1970
Highest Rank Held: Captain
Place of Birth and Year: Pittsburgh, PA – 1945

IN THE JUNGLE: A GIRL IN A POWDER BLUE DRESS

I was a C-123 Provider Pilot in Vietnam in 1969 — 1970. I was based at Phan Rang Air Base, a mostly large jet fighter base about 50 miles south of Cam Rahn Bay. Most of my missions consisted of flying military supplies to about 80 airfields throughout Vietnam, most of which were in the jungle. A few times I performed air drops at camps that were deemed too dangerous to land. I flew several medevac missions taking wounded soldiers from jungle strips to army hospitals on the coast. I could probably talk for several hours on my experiences with these missions. I, however, would like to write of a mission I flew that I, to this day, consider a very human interest mission.

My mission one morning was to fly several tubs of ice cream to Phan Thiet, a small field halfway down the coast to Saigon. Accompanying me that day was a young Red Cross girl probably about 23 years of age. She sure was cute all decked out in her pretty powder blue dress. She rode in the jump seat with my co-pilot and me. We finally landed at Phan Thiet and I proceeded to park on the small ramp. I noticed immediately that there were perhaps 100 soldiers, most of them asleep just off the ramp. I immediately thought that they were probably on an all-night mission.

The aft ramp was lowered and our pretty Red Cross girl was about to exit the aircraft from this ramp. Just before her departure I mentioned to my co-pilot what reaction these soldiers would have when they first saw her. Keep in mind that these soldiers operating in the jungle probably have not seen a pretty American girl since arriving in Vietnam. The first soldier to see this girl reminded me of a prairie dog that stuck his head out of a hole to look around. It took only a few seconds for all of these very tired soldiers to come alive and proceed to the back of the aircraft. What a wonderful sight for these soldiers to find themselves in the presence of a beautiful American girl in a pretty powder blue dress in the middle of the Vietnam jungle.

Probably not more than a half-hour later, we were once again loaded and ready to depart. Our Red Cross girl was back on the jump seat. I asked her just one question. How were you treated? She said she was treated as if she was every soldier's big sister. What a wonderful experience to bring part of America to these young soldiers.

STAN CHAMBERS

US Navy — Aviation Ordnance Chief Master
Dates of Military Service (Active Duty): 1968 to 2000
Unit Served with in Vietnam: USS Kitty Hawk (CVA-63) Bomb Assembly and
Air Missiles Divisions
Dates you were in Vietnam: Feb 1972 to Nov 1972
Highest Rank Held: Master Chief Petty Officer
Place of Birth: Atlanta, GA — 1950

RIOT AT SEA

When I was a year old, my parents and I moved to the south suburbs of
Chicago ,where my dad found a better job. I grew up through my teenage years in Harvey, IL. After graduating from high school, I decided
college was not for me and joined the U.S. Navy. I started basic training
at the Recruit Training Command (RTC) Great Lakes, IL, on December 2, 1968.

My test scores were not high enough to qualify for advanced training
of any kind, so I was given general assignment orders to my first duty
station at Naval Air Station (NAS) Cecil Field, FL, a Master Jet Base in
February 1969. Because I was an undesignated airman, I basically could
be assigned to any low-level job available at the time which could have
been working in the base galley or cleaning barracks.

When I checked-in at the administration office they gave me choices
of assignment from which to choose. Now, in my mind I'm thinking
galley or barracks duty. To my amazement, I was told that I was going to
the air station's Weapons Department. I was given two division choices,
Station Weapons or Air Missiles. Air Missiles sounded more interesting
to me, so I spent the next two years storing and assembling missiles. I
was eventually promoted to Aviation Ordnance Third Class Petty Officer (AO3).

In February 1971, I transferred to Naval Station (NS) Kodiak, AK.
Shortly after arriving I was offered an opportunity to qualify for air crew
on the NS's aircraft (C-54 Skymaster and the HU-16 Albatross sea-

plane). I qualified as a Flight Attendant/Load Master. This allowed me to fly all over the state of Alaska out to the end of the Aleutian Islands to the northern most point at Point Barrow and everywhere in between. One of our pilots was one of the last enlisted aviation pilots from WWII, AFCM (AP) Lou Drumm. I learned a great deal from this flying Master Chief Petty Officer pilot, and his leadership example served me well years later when I achieved the same rank.

Late that year, the Navy began to transfer ownership and base operations to the U. S. Coast Guard. So, I was transferred again in November 1971, to my 3rd duty station to Helicopter Combat Support Squadron Seven (HC-7), Seadevils (AKA. Big Mothers) at NAS Imperial Beach, CA. HC-7 flew Combat Search and Rescue (CSAR) missions off the coast of North Vietnam recovering downed pilots in the Gulf of Tonkin. I qualified as a plane captain on the HH-3A Sea King helicopter as well as armament maintenance on the M60 machine gun and the GAU-2B mini-gun.

After completing helicopter qualifications with HC-7, I was issued a set of TAD orders to our detachment at NAS Cubi Point, in the Philippines. It seemed only days later that my TAD orders were suddenly cancelled, and I was given yet another set of transfer orders to my 4th duty station in four years to the USS Kitty Hawk (CVA-63). I asked why and was told the USS Kitty Hawk's scheduled deployment date to Vietnam had been moved up 30 days, and they were still short of ship's company Aviation Ordnance men (AO). So being one of the more junior AOs in the squadron, even though I was trained and ready to go on detachment, the squadron transferred me out due to the needs of the Navy.

So, my Vietnam journey began as I checked aboard the ship on the morning of February 17, 1972, the day the ship left San Diego for the western Pacific. I was initially assigned to the Weapons Department G Division bomb assembly crew. We arrived in Subic Bay, PI, March 3rd for final preparations and on March 4th, we departed for our first line period to Yankee Station in the Gulf of Tonkin off the coast of North Vietnam supporting the 7th fleet's bombing missions.

For the first few months of our deployment we assembled bombs 12 plus hours a day, for the most part, during non-flight operations in order to have enough weapons ready for the next day's flight operations. Every three or four days we conducted an Under-Way Replenishment (UN-REP) of supplies, fuel, and munitions. After UNREP was completed, we went right back to our job of bomb assembly. During this time, I heard of an opening in the Air Missile (AM) Division and with my previous experience with air missiles at NAS Cecil Field, I was reassigned to AM Division. I stayed in the AM division for the remainder of the cruise.

Our line periods on Yankee Station seemed pretty routine from my point of view. The longest line period at sea during our 10 ½ month-long cruise was 52 days with monthly R&R breaks to Subic Bay, Philippines and one trip to Hong Kong. I do remember once going to General Quarters because the ship had been alerted that two North Vietnamese MiG Fighter Jets were heading our way. One of the jets was shot down by the USS Chicago, and the other one returned to home base.

Shipboard life was good for me, but that all changed the evening of October 12, 1972. We had just left a port call in Subic Bay earlier that day and was once again on our way back to Yankee Station for flight operations. An altercation in the aft galley between a black crewmember and a white mess cook turned into a fight. Things quickly got out of hand and fighting escalated between more sailors, resulting in a full-blown race riot: a mutiny if you will. As soon as the command became aware of the situation all crewmembers were ordered, through the 1MC, back to our workspaces. That order apparently fell on deaf ears to a group of very angry black sailors. I was unaware of the size of this group or their issues at the time.

The upset blacks formed several groups and started beating up any white sailors they could find. These travesties took place mostly on the 2nd deck of the ship, just below the hangar deck. They weaponized themselves with dogging wrenches or anything they could use to hurt unsuspecting white sailors and it didn't matter if they were asleep in their bunks or walked alone down passageways. They attacked in gangs. One of my shipmates who I worked with was grabbed from behind while walking up the forward mess deck ladder and was severely beaten and punched in the face. One guy even found refuge in our berthing compartment who had escaped from an apparent beating himself.

A couple of us made our way to our shop just forward of the hangar deck to make sure our guys on duty were ok and that the shop was secure. At some point, the XO, a black officer, was able to get most of the angry blacks up to the fore castle of the ship for a meeting to hopefully calm down the situation. Many of these sailors passed right by our shop and tried to open our shop door to attack us. Fortunately for them, they were unsuccessful because there would have been hell to pay. A couple of us had spent our previous port time at jungle survival school and we had our machetes from that training with us, and we would have used them in self-defense. The skipper and XO finally regained control of the situation by the next day, and we did not miss any flight operations, but the damage was already done.

This incident was captured in the news of the day and later in a book titled "Troubled Water" written by Gregory Freeman. It revealed many things that I was unaware of at the time. Personally, I never had any issues with the black sailors I worked with on the ship. We seemed to all get along well. This situation, however, changed my demeanor, and I became an angry individual for what had happened and the lack of justice not to mention those who had been severely beaten for no justifiable reason. For the remainder of this, our final line period in Yankee Station, there seemed to be a sense of uneasiness anytime black and white sailors encountered one another. The anger and bitterness that I developed from this unfortunate experience stayed with me after Vietnam and affected my personal life for many years.

We completed our final line period on November 8th and returned to San Diego on November 28, 1972. I was released from the Navy on December 1, 1972, after four years of active duty.

I returned to my childhood home got a job and got married. After a year of working two full time jobs, I decided to go back in the Navy in February 1974 in Full Time Support status in the Navy Reserve. I was initially assigned to a P-3 Orion patrol squadron and qualified as the in-flight ordnance man serving in several Patrol squadrons. During those first few years after returning to active service, I was blessed with two beautiful daughters.

I steadily rose up the ranks to Aviation Ordnance Chief Master (AOCM) working in several management positions, including serving on the staffs of Reserve Patrol Wing Pacific and Commander Naval Air Reserve Force as the weapons program manager. I completed my last five years of duty as the Command Master Chief of a C-130 squadron (VR-54) and retired with 30 years of service.

Although very successful in the Navy, the anger and bitterness that I carried with me from the incident on the Kitty Hawk had an effect on my family life and my first marriage which sadly ended in divorce. It was, by far, not the only issue that resulted in divorce but truly a contributing factor. All of that led to deep depression after retirement for which I saw no way out. But with the encouragement and prayers of family and some dear friends and the Lord's touch on my life, my life was restored. I have since remarried to my wonderful wife, gaining two more daughters in the process as well as eight precious grandchildren.

I thank God for His grace and mercy.

ALAN C GRAVEL

US Air Force — Pilot
Dates of Service: 1969 to 1974
Unit Served with in Vietnam: 536th Tactical Airlift Squadron of the 483rd Tactical Airlift Wing; 4102nd Aerial Refueling Squadron
Dates in Vietnam: Sep 1970 to Sep 1971; May 1972 to Dec 1972
Highest Rank Held: Captain
Place of Birth: Alexandria, LA — 1945

RE-FUELING THE 1972 EASTER OFFENSIVE

When we finished KC-135 tanker school at Castle AFB in Merced, CA, we reported to the 913th Air Refueling Squadron at Barksdale AFB in Bossier City, LA. We moved our mobile home from Alexandria, LA, where Sheri and our son Alan W. had lived while I was in Vietnam in Caribous to a nice mobile home park in Princeton, LA, a few miles east of Barksdale. We settled in and started training flights about the first of March 1972. I was a senior first lieutenant with almost 1,000 hours of combat time in Caribous in Vietnam, but I had only the upgrade training hours in the tanker, so I started out as a co-pilot. I was assigned to a crew led by an aircraft commander who had come to tankers right out of pilot training, accumulated the minimum required 500 hours in the tanker, and had just recently been upgraded to aircraft commander.

Early in May, virtually the entire squadron was deployed to Clark AB, Philippines to fly missions into Vietnam as part of the US response to the 1972 Easter offensive when the North Vietnamese tried to overrun South Vietnam. Unlike my other experience in Caribous, we deployed as a unit. The morning we left Barksdale was a very dramatic scene, with wives and children crying and saying their goodbyes and the aircrew and support personnel preparing to embark on a long flight across the Pacific while at the same time bidding farewell to their families.

On a typical mission from Clark we would takeoff and fly about two hours to "Purple Anchor" over the South China Sea north of Da Nang. There, we would fly a 60-mile north-south racetrack pattern waiting for aircraft to come up needing fuel. When we had either burned or off-loaded enough fuel to reach the minimum amount we needed to return to Clark, we would leave the anchor. There would usually be at least two and sometimes three or four tankers in the anchor at one time, separated by 4,000 feet of altitude. We mostly were refueling F-4s and F-105s. Missions could be as short as 6-8 hours but frequently lasted 12-14 hours.

Each day one crew would be designated as a "maintenance alert" crew. Your job was to pre-flight an airplane and taxi it down to the approach end of the runway, pull into the run-up area off to the side of the taxiway, and shutdown. We could plug a headset into the airplane and monitor the ground and tower frequencies while sitting outside on the tarmac. Most of the time, you would just wait all day there and never have anything to do. If one of the crews had a problem with their airplane that could not be quickly repaired, or if they aborted on takeoff, or if they had to return to base in the first few minutes of their flight, you would launch and take their mission.

On our day on maintenance alert, we had waited several hours and watched one tanker after another take the runway and takeoff. One of the Barksdale crews taxied by and waved as we waved back. When they were well into their takeoff roll, we heard "Abort, Abort, Abort." We jumped up, got into the airplane, started the engines, took the runway and got clearance to takeoff. Just as we lifted the nose gear off the runway at rotation speed, the outboard engine on the left side exploded. This was the upwind, outboard engine which was the most critical one for maintaining directional control. This failure at this phase of the flight was the "skull and crossbones" scenario for KC-135 tankers. They want to roll over on their back which would obviously have been fatal.

The crew that had aborted was taxiing back to the ramp on the parallel taxiway and had a front row seat to the explosion. They said flames shot 30-40 feet out the front of the engine and 60-80 feet out the back. To us it felt like someone had hit us in the left shoulder with a sledgehammer.

However, our training kicked in perfectly. The A/C and I both locked our legs on the right rudder, and we pushed the other three engines up to maximum power. We leveled off about 500-600 feet and turned about 30 degrees left to avoid the mountains ahead and stay over the rice paddies. I got the landing gear and flaps up and started dumping fuel to reduce our weight. We declared an emergency and asked air traffic control for radar vectors to return to Clark. We dumped about 100,000 pounds of fuel until the airplane could climb and be reasonably responsive to the power we had available.

As soon as things were under control, and in accordance with the emergency procedure for this situation, I notified the A/C that I was pulling the throttles back to a normal power setting for climb-out. He stopped me, saying, "No, No, we need the power." I tried several more times and each time he stopped me. He insisted on maintaining the maximum power setting until we leveled off at 10,000 feet. When we pulled the throttles back through about 82% rpm, the whole airplane shuddered, and we knew we had another problem.

While level at 10,000 feet and following the radar vectors back to Clark, we increased the power of each engine separately and determined that it was the other engine on the left side, #2, that was vibrating badly at 82% rpm. We set that engine below the vibration level, set the #4 outboard engine on the other side to idle, and used #3, the inboard engine on the right side to make all power adjustments.

We made an uneventful, albeit tense, approach and landing back at Clark. They had scrambled the fire trucks and the ambulances, and every senior officer who had a blinking light on his car was out on the tarmac to watch us bring it in. We pulled into the ramp and shut down and were immediately surrounded by a crowd. The Navigator and I were so angry at the A/C that we decided we better not say anything for fear of ending his flying career.

While the crowd was standing around and we were trying to gather ourselves, the maintenance personnel opened the cowling on the #1 engine that had exploded. About a dozen turbine blades fell out on the pavement. I picked one up and for years have kept it in my office desk drawer as a reminder of my good luck.

Later we were told that the main thrust bearing on the engine had failed and the rotating parts had moved longitudinally into the stationary parts, causing all hell to break loose inside the engine. We later learned that the other engine on that same side that had been vibrating was found to be failing in the same way, just not catastrophically.

One night in the early summer of 1972, we were returning to Clark from Vietnam. As we approached the base, the Navigator could see a large thunderstorm about 20 miles or so off the approach end tracking straight at the runway on more or less the same compass heading. We made our downwind and turned a 10-mile final between the storm and the runway, intending to land in front of it.

During the final approach the Navigator was calling out the distance from the ground and our vertical position relative to the glideslope. This was all perfectly normal for this situation. The A/C was flying the airplane. A mile or so out the Navigator called "100 feet below glide-slope." The A/C did not react appropriately. A few seconds later, "150 feet below." Still no reaction. A few more seconds, "200 feet below." By this time, we are ½ mile out and this is not looking good. I looked over at the A/C and it was as if he was in a trance, catatonic or something. I said "Do something!" Still nothing. At that point, I shoved the throttles up, pulled back on the yoke, and called, "Go Around" on the intercom and to tower.

The A/C complied with my "Go Around" and continued flying the airplane, climbing slowly back up toward traffic pattern altitude. I asked him, "What are you going to do?" and there was no answer. By now the storm was nearing the approach end of the runway and we needed to get the airplane on the ground quickly. We were over the departure end of the runway.

Somewhere from the deep crevices of my brain, I have no idea how or why, came my solution to the problem. I asked the tower for a "Left 90, Right 270 to land opposite direction." To his credit, the tower controller immediately knew what I was asking for, although extremely unconventional, and he "Approved" without hesitation. I told the A/C to turn left and he did. We drove away from the runway a few seconds and I told him to turn 270 degrees back to the runway and he did. We rolled wings level on a very short final and landed. As soon as the main landing gear touched and before the nose gear touched, we hit extremely hard rain and could see nothing out the front windscreen. He watched the white line on his side of the runway, and I watched the white line on my side as we rolled out.

The A/C never said a thing about this. Not "you were insubordinate," not "I could have made the landing," not "kiss my ass," not "thanks for saving my ass." The Navigator and I determined that we would ask for a crew change at the first opportunity.

In the first 28 days of June in 1972, it rained 98 inches at Clark AB. Up in Baguio in the mountains, it rained 176 inches in that same period. We were leaving Clark every hour with about 150,000 pounds of JP4 jet fuel which was normally supplied to Clark by pipeline from the Subic

Bay Naval Base about 50 miles to the south. The excessive rains washed out the pipeline to Subic. For a while they tried to maintain Clark's stock of fuel by truck but that was not possible. The decision was made to move the whole operation up to CCK in Taiwan.

We were there on temporary duty (TDY). USAF regulations would not allow a TDY to extend longer than six months so at something less than six months they would rotate you home for 28 days and then they could deploy you again on another TDY. The Barksdale crews had been rotating home for weeks and our crew was in the last group to rotate.

In early August of 1972, most of the tanker operation moved up to CCK but we only had a few days left before going home so we stayed at Clark until rotating home. Almost exactly in the middle of my 28 days at home, Sheri gave birth to our twin sons Mark and Ryan at Barksdale AFB. I never knew if my Squadron Commander intentionally scheduled my rotation that way or if it was just a lucky coincidence.

HAM HENSON

US Army — Infantry Officer

Dates of Service: 1969 to 1978

Unit in Vietnam: Charlie Company 1/502 Infantry 101st Airborne (Airmobile);
8th Radio Research Phu Bai

Dates served in Vietnam: May 1971 to May 1972

Highest rank held: Captain O-3

Place of Birth: Savannah, GA

SCARY REMF MOMENTS

I spent most of my tour as a Platoon Leader and Company XO with C/1/502 Infantry 101st ABN operating out of Phi Bai. The 101st displaced to Ft Campbell Ky. I was excited to be going back with them. As we were packing out the last of our gear, I was called to Brigade HQ to meet with a Lieutenant Colonel. That "conversation" went like this... "Henson, I am staying here to command the security force for 8th Radio

Research and you are going to be one of my officers. Here are the orders transferring you to my command. You are staying at Phu Bai."

I didn't know 8th RR Station existed. It was housed in an old concrete French fort that was inside 101st HQ compound. All that was bulldozed and cleared as the 101st departed. Their mission was to monitor radio traffic in SE Asia. To do that they had several subfloors below the actual fort itself. I don't know how many because I was never allowed inside. It also had a huge antenna field. The antenna field was defended by a group of former RVN soldiers. Our job was to organize the defenses around the fort and operate a TOC (Tactical Operations Center) with direct communications to First Regional Assistance Command in Da Nang.

I had spent the first seven months of my tour as a Platoon Leader in C/1/502. The overwhelming time was either operating in the field or on a firebase as a grunt. While I was scared plenty of times during that period, I want to relate how a REMF (Rear Echelon MF) can be scared as well.

Three specific incidents come to mind. The first was early in my assignment. As TOC Duty Officer, I was responsible for checking the various defensive positions. One night I was making the rounds and stopped at a bunker manned by 8th RR troops. Understand that these men were all very intelligent (their day jobs were linguists and intelligence analysts) with high IQs. But they had all been thru basic training. I dropped in on a four man team in the bunker and announced they had been selected to fire at targets in the mined field to their front using their aiming stakes. I called for an illumination round on standby from the local artillery and obtained permission to fire from the TOC. The idea was to give them confidence that they could engage predetermined targets without illumination then show them with the flare that it worked.

I said, "Fire at will in standard 3-5 round bursts." The gunner pulled the trigger and there was a distinct "thunk" as the firing pin engaged. I calmly said, "OK, now what do you do?" The kid said, "I dunno, sir.," so I explained we declare a misfire so everyone in the bunker is aware and conduct immediate action to reset the weapon. He looked at me like I was speaking Mandarin Chinese—no I take that back—they probably spoke fluent Mandarin Chinese. Anyway I said, "OK, let's learn something—take off your poncho and pick up the red flashlight." Poncho goes over our heads and we turn on the light.

As I started to explain the immediate action sequence, I looked at the M60 and saw the belt was placed in BACKWARDS! There was no more training at that point. I cleared the bunker and called the Armorer. We were lucky the round didn't detonate and blowup the receiver in the kid's face. A few minutes after leaving, I realized I wasn't surrounded by dependable combat infantrymen like I had been in the bush. If we were actually attacked, it would not be a good fight for our side. I got scared as a REMF.

The second time I got scared was a few days before the Easter Offensive. I was an observer on a Pink Team that flew out past our boundaries from the airfield at Phu Bai as a reconnaissance every day to be sure there were no enemy troops we could observe. We also flew over the land route we would have to take if we had to evacuate on foot to a LORAN station on the coast for pickup by the Navy. Granted that would be the last thing we wanted to do, but you planned for the worst.

On one of those flights we saw fresh tracks from tracked vehicles—as in armored vehicles that had likely infiltrated down the Ashau Valley to our west and come in at night on our old roads to the firebases. With limited firepower, we were not equipped to take on PT-76 tanks. But we had lots of air assets and naval gunfire if needed. Did I mention we were the northernmost US-manned facility in I Corps? There were advisors to RVN units all the way to Quang Tri but we were solely US troops for security reasons due to 8th RR's mission. I was scared as a REMF.

As an aside, when the NVA invaded across the DMZ in the Easter Offensive, I had a chance to meet several of those MACV advisors when they showed up in our TOC. One I remember in particular was Jim Avery who came in wearing boots, boxers, a t-shirt, his helmet, M16, and webgear. That's all he had time to grab on the way to the last helicopter that made it out of Quang Tri. One of us, I think me, gave him one of our uniforms to wear that day. Later, he was assigned to the Ground

Committee at the Airborne Department at Ft. Benning where I was stationed after returning. Small world—we became friends.

My last time being scared happened a few days later. Fast forward many years—I had joined the AVVBA as a result of running into Stewart Davis at a garage sale in our Sandy Springs neighborhood. He took me to my first AVVBA meeting. "Moose" Davis was a retired Army 0-6 who was a multi-tour captain when we were at 8th RR. He ran the day shift and I served on the night shift with Bob Rosa with whom I had trained before deployment. Small world. Moose died from cancer several years ago, but I had the privilege of being his sometime-driver to and from chemotherapy. We talked a lot while he was getting his meds and he related that the scaredest he had ever been was with me in the days after the Easter Offensive began. Mind you, Moose had been in heavy action on his first tour.

At 8th RR, he and I checked out a Jeep and took our weapons—he had an M16 and I was carrying an M79 and a vest of rounds. We drove up Highway One to just observe what was going on and report back. We drove north past a stream of civilians and RVN troops all headed south for a couple of miles. Then Moose said it was time to turn around. His reason was simple—there was no way we could be sure all the people were from South Vietnam. Some reports had NVA infiltrators wearing captured RVN uniforms and driving captured RVN equipment.

What a wonderful target we were—two dummies with top secret clearances driving alone in a US Jeep in US uniforms! In chemo, Moose shared that was as scared as he ever was in Vietnam. We had a good laugh remembering that day where we were both scared as REMFs.

ED DEVOS

US Army – Infantry Officer

Dates of Military Service: 1969 to 1989

Unit Served with in Vietnam: Battalion Advisor, 21st ARVN Division

Dates Served in Vietnam: Dec 1971 to Dec 1972

Highest Rank Attained: Lt. Colonel (P)

Place and Date of Birth: Grand Rapids, Michigan – 1947

THE 1972 NVA EASTER OFFENSIVE... THROUGH THE EYES OF ONE AMERICAN ADVISOR

Late March and early April 1972 has been labeled the Easter Offensive, the North Vietnamese Army's (NVA) attack into South Vietnam. The offensive lasted almost four months. Fourteen NVA divisions attacked within a day of each other, all aimed at significant South Vietnam cities: Hue, Quang Tri, Kontum, and An Loc, with the ultimate goal of taking Saigon. Because most American combat units had been withdrawn earlier in 1970 and 1971, only the South Vietnam Army (ARVN), their American advisors, some U.S. Army aviation units, and U.S. Air Force assets still in support of the South Vietnamese stood in the way of an NVA victory. The comments below are several of my experiences as one Infantry advisor with the 21st ARVN Division during this timeframe.

Within a week of the NVA making significant advances in III Corps toward An Loc in early April, 1972, the 21st ARVN Division, a tough, well-trained, highly regarded division known for its cleaning out of the VC in the U Minh Forrest, was deployed by C-130, C-123, and C-7 aircraft in two days from the Ca Mau in IV Corps to Lai Khe in III Corps. The division's mission was to attack north up Highway 13 to relieve and/or reinforce the ARVN units defending An Loc and stop the NVA advance. Intelligence estimates indicated we would be facing at least three, maybe four NVA divisions. The friendly units at An Loc, elements of the 5th ARVN Division, were under attack from several NVA divisions reinforced with tanks and a great number of indirect weapons including some U.S. made 105mm artillery pieces the NVA captured at Loc Ninh, a town northwest of An Loc, in the first few days of their offensive.

Once the NVA observed the 21st ARVN Division's movements as we begin to move north along Highway 13, they began a series of counterattacks to halt our advancement. Fights were vicious and our progress was slow despite the grit shown by our ARVN soldiers and their leaders. There were many occasions when the NVA units used civilians as shields. Each step we took was met with barrages of mortar and artillery fire which we could only counter with U.S. and VNAF airpower as we had little artillery with us. Because we paralleled the Highway 13, we were easy to track and the NVA took full advantage of this. One of my sister ARVN battalions was overrun four or five clicks from my location. The battalion advisor of that unit, Captain Hank Faldermeyer, was killed in that attack. It should be noted that at this time in the war, we were down to one battalion advisor per ARVN battalion. To make matters worse for me, my interpreter was wounded in our first fight and was never replaced.

In mid May, the ARVN battalion I was with was given the mission to make a night move to get around the NVA positions along Highway 13, going due east of the highway for three or four clicks, then turning north for twelve or so clicks before turning back to the west to occupy the village of Tan Khai, located ten clicks south of An Loc. The purpose of this move was to set up an artillery fire base to give fire support to An Loc and along Highway 13 to augment the close air support that was keeping An Loc from being overrun.

This battalion had NEVER done any night moves prior to this mission higher than squad level. We began the move east with about one hour of daylight remaining. Once it turned dark, the move became a major Charlie Foxtrot. Many voices, noises, and rattles of all kinds, flashlights, numerous stops and starts for navigation checks. It was the Ranger patrol from hell. Somehow or maybe through sheer blind luck and/or God's intervention, the NVA didn't make any moves against us. With about two hours of darkness remaining, we stumbled around and finally made the turn to the west, halting several clicks from our objective.

After quiet settled in all around us, without warning, one of our M60 machine gunners on our perimeter opened up. It was quickly discovered that there was nothing out there; the soldier just got spooked. As soon as dawn came, the ARVN Battalion Commander gathered all his soldiers around him and pistol-whipped the soldier for a minute or so as he explained to the rest of the battalion that this one man's actions could cause us all great problems because the NVA may now know where we were. His object lesson was simple. Don't give away your crew served weapons positions until there is a genuine need. And I understood—my counterpart handled this lesson in a way appropriate to his culture, and without question, the soldiers of this ARVN unit got the point their commander was making. We then saddled up and occupied Tan Khai by mid-day without a shot being fired.

Within an hour of our arrival in Tan Khai, a significant roar from the south could be heard as six CH-53 Sky Cranes with a large gunship support element came toward us at about 2,000 feet. Under each Sky Crane was a 105mm howitzer. Other helicopters in this air flotilla carried the artillery crews and ammunition. After unloading these artillery pieces, within the hour the ARVN Artillery unit began to receive fire missions from An Loc.

While the idea behind taking Tan Khai and making it a fire support base made sense, the NVA were not blind and certainly not stupid—it isn't hard to see a Sky Crane flying at two grand with an artillery tube on a sling underneath. Within two hours we began to receive effective artillery and mortar rounds from everything and anything in the NVA arsenal. We could hear what seemed like hundreds of mortar rounds coming out of the tubes as they flew toward us. We also caught hell from big guns—122s, captured 105s from Loc Ninh, and everything in between. I was told later by some intel guy that the NVA fired 5,000 rounds at us. I don't know who counted all those rounds or how they did it, but I will not argue with that number.

The battle of Tan Khai lasted four days. NVA ground attacks came throughout the day, each preceded with bugles blowing at various times from all directions of the compass as they surrounded us. Our ARVN soldiers, well dug in, stopped them every time. It was close a few times but non-stop U.S. and VNAF aircraft made the difference. Our only resupply for those days of both food and ammunition was by HALO drops by C-130s at ten thousand feet. 80% of what was intended for us fell into NVA hands. Our only food was tuna fish and dry rice—water was at a premium. We had Air Facs overhead from dawn to dusk and C-130 Specter Gunships at night—all as near as my radio. The Specters were my favorite. When the NVA attempted a night attack, I give "Specter" a target in a six-digit grid (accurate to within one hundred meters). Once he had the target spotted, he called me back to confirm the target with an eight-digit grid (accurate to within ten meters). Needless to say, the NVA attack was stopped in its tracks by a rain of fire from above. I don't believe they ever knew what hit 'em.

Once the NVA finally pulled back from Tan Khai after the beating they had received from our airstrikes, I was able to get to the site of a Cobra crash several clicks north of our perimeter during the time we were trapped in Tan Khai which was shot down by one of the first, if not the first, SA-7 shot in South Vietnam. We found one body but not the other one; no sign of that man whatsoever. During the evacuation of the one airman we found, we were literally pushing civilians off the skids of the helicopter so we could get airborne. The ultimate whereabouts of that missing airman still haunts me to this day.

Author Note: Ed DeVos is a great military historical fiction author. He has written books about the Centurion at the Cross (The Stain), The Chaplain's Cross (WWII), Revenge at King's Mountain (Revolutionary War), and Family of Warriors (WWII).

VINCENT C. CORICA

US Army — Field Artillery Officer

Dates of military service: 1969 to 1978

Vietnam service with: 3rd Brigade (Separate), 1st Cavalry Division (Airmobile)

Dates Served in Vietnam: May 1971 to May 1972

Highest Rank Held: Captain

Place of Birth: Johnstown, PA — 1947

AIN'T NUTHIN LIKE COMMAND!

At Last...Battery Command!

105 mm Field Artillery Battery — Feb 1972 — May 1972

With no disrespect for my time as a Field Artillery officer in non-command positions in Vietnam, there's NOTHING like command of American Soldiers to bring out the best in a person and to remind you to never lower your vigilance below 100%.

I took command of B Battery, 1st Battalion, 21st Field Artillery with seven very relevant months of combat-zone Field Artillery experience under my belt.

The war was winding down and we all knew it. I gave myself one mission as Battery Commander... no one under my command would be wounded nor killed in battle. My Battery consisted of as many as 100 men and as few as 70 during my four months commanding it. We operated at the full complement of six M102 105 mm howitzers, a precision-fire artillery piece with an Effective Firing Range of about 7 miles. The sustained rate of fire for the 105 is three rounds per minute. With all six pieces firing simultaneously, a 105 mm Battery can put 18 eight-pound projectiles PER MINUTE on a target! With a Kill Radius of 18 yards per round, that's a lot of hot-steel protection for our Infantry brothers.

We were the first unit to occupy "Firebase Grunt II," so-named by our extraordinary Brigade Commander, BG James Hamlet, in honor of what Infantry soldiers lovingly called themselves... "Grunts."

I was the Commander of the Firebase and the Artillery Battery.

Why do I make that distinction? Because that meant I was accountable not only for the Artillery pieces and the men, equipment, and ammunition required to safely and accurately fire those pieces. I also owned the security of the Firebase. Did we have enough Concertina wire around our perimeter? Did we have enough sensors in our surrounding jungle to warn us of person-sized weight traversing an area? Were our earthen berms high enough? Did we have enough Claymore mines emplaced? Where they aimed correctly? Did we have interlocking fields of fire from our M60 placements? What about our mutual Fire Support Plans with our sister batteries? Did we have enough sandbags protecting our ammo, our hooches, and our howitzer positions? What about my men and drugs? How would I really know if they were 100% ready to operate when needed? Were we properly maintaining our howitzers? How fast could we shift to manual fire direction processes if our new FAD-AC (Field Artillery Digital Automatic Computer) field fire direction computer went down? Every single piece of equipment, including our individual weapons…were they clean and reliable? What about our food supplies? Our water?

I was never worried about any of these things. I thought about them, yes. But I had surrounded myself with terrific Lieutenants, NCOs and enlisted soldiers. I was certain that I was not alone in my commitment to "everyone goes home healthy." I still remember many names of this 'posse' of mine, 47 years later. I have a few dozen photos of our Firebase and of my Soldiers, but I should have taken so many more.

For the last two Veterans Days, I have Posted the same content in Facebook. I am going to end this story with that Post. This Post is about emotions… about what it felt like to support our intrepid Infantry troops. Here it is:

"During my time in Vietnam, I had the honor of commanding a Field Artillery Battery. The spectacular men in my Field Artillery Battery and I fired our 105 mm howitzers day and night in support of our magnificent US Army Infantry brothers in the 3rd Brigade of the renowned US Army 1st Cavalry Division.

We protected and aided our incredibly courageous brothers as they helicopter-inserted into dangerous, hostile, jungle Landing Zones.

We put "steel on the target" before every helicopter-borne airmobile assault to make those Landing Zones safer for our troops.

We put steel on the target when our brothers reported enemy movement near their positions or as our troops advanced on enemy positions.

We put steel on the target when our brothers engaged the enemy in close combat.

We were accountable for keeping our brothers safe.

We were accountable for assuring them that they were not alone as they did their astonishingly dangerous work.

We were ACCOUNTABLE. We knew it. We felt it. We never let our guard down. Ever. We whooped and hollered and wept like babies when our brave Infantry brothers radioed their thanks back to us for 'getting them out of the shit'.

My men and I were in the middle of the jungle ourselves on our Firebase, so we too were in harm's way. We never forgot that. We improved our defensive positions and readied our artillery ammunition every minute when we weren't firing for our brothers.

All the men in my command returned home without a scratch, thanks to our constant vigilance and our in-depth defensive preparations and rehearsals. Men of every race, creed, and color. Strangers when we met; trusting and beloved comrades after one handshake. Volunteers as well as draftees, career soldiers as well as short-timers, officers as well as enlisted. "All for one and one for all." It was no more complicated than that. We WENT to Vietnam for our Country. We FOUGHT to protect ourselves and our brothers.

Other than my wife, my family and my friends, my eight years as an officer in the United States Army are the proudest, most indelibly important achievements in my life. When people thank me for my service, even 40+ years after taking my uniform off for the last time, I swell with pride, smile, stand up a little straighter, look them in the eye and reply, 'Thank YOU, it was an honor to serve'.

CARL H. "SKIP" BELL, III

U.S. Army—Armored Cavalry Officer, Aviator

Dates of Military Service (Active and Reserve): 1967 to 1998

Units Served With In Vietnam: A Troop, HHT, B Troop, 1st Squadron, 4th Cav, 1st Inf Div (First Tour); C Troop, 3rd Squadron, 17th Air Cav; 18th Corps Aviation Co; G3, HQ 1st Aviation BDE (Second Tour)

Dates in Vietnam: Feb 1969 to Feb 1970; Feb 1972 to Feb 1973

Highest Rank Held: Colonel

Place of Birth and Year: Decatur, GA—1945

CONVOY ESCORT MISSION TO PHNOM PENH

In the spring of 1972, my unit (C-3/17 Air Cavalry—callsign: Lighthorse) was given the mission to escort a convoy of ships up the Mekong River to Phnom Penh, Cambodia. At that time, Phnom Penh was under intermittent siege by Khmer Rouge forces and highway travel from the port of Sihanoukville was unreliable, so supplies for the capitol were delivered using the river. At that time, C Troop was attached to 7/1 Air Cavalry, and we were all based at Vinh Long airfield in the Mekong Delta of South Vietnam. The gun platoon callsign was Crusader.

The plan was that we would send four Crusader AH-1G Cobra gunships and a UH-1H Huey from Troop HQ as the Command and Control (C&C) aircraft. We were to orbit the convoy from the time that it entered the Cambodian portion of the Mekong River until it arrived at Phnom Penh. Two of the gunships stayed near the head of the convoy, the other two were near the rear (I don't recall how long the convoy was, but it stretched several kilometers). The C&C orbited the middle of the convoy and could direct the gunships when/if contact was made. The operation took several hours; we had a refuel point midway up the river between South Vietnam and Phnom Penh at a place called Neak Long (which was also a ferry boat crossing site).

When we refueled, we did it 'hot' (that is, we did not shut down the aircraft). The hot refuel process in the Crusaders was for the aircraft commander to exit his position and do the refueling (the refuel port was on the right side of the aircraft) while the gunner remained in his position and monitored the aircraft controls. I was newly assigned to the unit and flying in the gunner's position on that mission. The refuel pads at Neak Long were located on the side of what appeared to be a soccer field. Across the field from the refuel pads were some open-sided warehouses. I noticed that the warehouse closest to the soccer field was stacked from bottom to top with palates of Coca-Cola! I thought to myself, "here we are in the middle of nowhere, and you could still get a Coke here!" Amazing.

The trip up the Mekong River was relatively uneventful. There was one incident where an RPG (rockt-propelled grenade) was fired at one of the ships and one of the gunships went in and put fire down on the area from which the RPG was launched. Results of the engagement were unknown. It was an interesting day from the perspective of going into Cambodia and being able to see Phnom Penh, even though we did not get to land there.

We were told not to overfly the Presidential Palace, which we did (I have photographs). From the air, Phnom Penh looked in many ways like a large, sophisticated European city (wide boulevards, large buildings, parks, etc.). When I heard several years later about what the Khmer Rouge did to the city and its inhabitants (in the *killing fields*), I was saddened. What a waste.

The trip to Phnom Penh with the Air Cavalry Troop opened the door to another opportunity to visit Cambodia later on that same tour. After the Air Cavalry Troop rotated back to the States (to become D Troop, ¾ Cavalry, 25th Infantry Division in Hawaii), I transferred to a General Support Aviation Company (the 18th Corps Aviation Company) in Can Tho. That unit also had a mission to supply two aircraft to the U.S. Embassy in Phnom Penh once or twice a month. I was given the opportunity to go on that mission because of my previous trip to Phnom Penh with the Air Cavalry Troop. But that's another story . . .

RALPH H. BELL "ROTORBRAKE"

US Air Force — CSAR (Combat Search and Rescue)

Dates of Military Service: 1966 to 1991

Unit Served with in Vietnam: 3d Group, 38th ARRS, Detachment 7 (DaNang)

Dates in Vietnam: May 1971 to Dec 1972 (19 ½ months)

Highest Rank Held: Colonel

Place of Birth and Year: Texas City, Texas — 1943

ONE-HANDED RESCUE — DANANG AB SOUTH VIETNAM

I was assigned to Detachment 7 of 38th Aerospace Rescue and Recovery Squadron (ARRS) at DaNang Air Base, South Vietnam. The detachment was manned with about 45 men and 2 HH-43F ('Huskie') helicopters. The HH-43F was lightly armored and had a bit more powerful engine than the "stateside" HH-43B. We took off the doors on the back and flew with the sliding doors removed or pinned back, mostly due to the heat. The armor was some titanium metal plates strategically placed and two hinged side flaps that shielded the pilot seats from the open sides. The weight of the armor made the HH-43F a great helicopter to autorotate since when you reduced power, you descended like a rock and the two synchronized rotors immediately sped up to the 250 rotor RPM with little effort.

On one hot and humid but also normal day on alert at DaNang AB, South Vietnam, the "red alert phone" rang, and we were launched by Panama Control on a potential search and rescue mission. The victim was said to be a Vietnamese pilot lost at sea off the coast of DaNang. I remember that the rescue coordination center said he had been missing for three days but was hopefully in a life raft. I had not ever picked up a "real survivor" in a life raft before, however, we all had training which included many actual water pickups over lakes and reservoirs. At the time we were alerted, it seemed to me to be almost an impossible mission. I was thinking that a man drifting somewhere off the DaNang coast for three days in a "one-man dingy" could be almost anywhere.

I conducted my crew mission briefing and reminded everyone of all of the techniques I knew that we could use while looking for a survivor over water. We would initially fly directly to the last known position and then begin an expanding circle search pattern at 500 ft above the water. If that didn't provide results, then we would vary the search pattern or altitude. If any crew member sighted anything at all, he would have to immediately call for or launch an over-water smoke grenade. It was important (as we were taught), to NEVER take your eyes off the victim once sighted. In training, the instructors cited tragic cases where a survivor was once sighted and in the excitement of discovery, the survivor was subsequently lost from sight and not ever reacquired. I admit I was very skeptical about the success of this mission as we started out.

I lifted our HH-43 into a hover and checked the instruments and the torque, which indicated the power available. After calling DaNang tower, I flew down the taxiway and gradually accelerated with a full load of fuel. My crew consisted of pilot, copilot, flight mechanic, two firefighters, and a medic. My plan was to have lots of eyeballs busy scanning the ocean for a life raft on this over-water search mission. The HH-43 flew for about four hours, no matter the speed or altitude… so I picked up 60 knots and tried to make my shaking helicopter become a stable search platform.

This proved to be a daunting task. After I had flown an ever expanding square search pattern, beginning at a point about four miles out to sea, we were again back over the beach. Everything was beginning to look green and greener. Then, I began to fly a creeping line-search pattern beginning at the shoreline about 10 miles up and then back and forth gradually flying out to sea. I initially flew at 500 feet above the water. Then I tried another expanding square search pattern but at 1,000 feet beginning back at the point where the survivor was last predicted to be. The weather was good, with some clouds but no rain, and the visibility was great. But, the water was a dark black-green color, exactly the color of green NOMEX flight suits and also the color of olive drab one-man life rafts.

Several crewmembers began to grumble that this search was hopeless. I was keeping one eye on the fuel gage and subconsciously thinking that we were probably at risk ourselves flying a single-engine helicopter so far from shore... I sure wouldn't want to try to swim four or five miles if the engine quit and we went down. At the southern end of my now random search track and probably the greatest distance we had thus far drifted from land, I thought I saw a black dot in the water below from my right-seat door. YES?

I called excitedly on the intercom as I began an immediate descent, but most of my crewmembers were asleep or so bored they didn't respond right away. Since the training guys said, "always keep the victim in sight," I did so with a vengeance. No one else on board had seen the survivor... only me! Maybe I was seeing things? I had been about three hours flying in circles and rectangles looking at nothing but green water. Oh, oh! I remembered as we continued to descend in a tightening spiral around the dot... we HAD a checklist to follow. Hurriedly, I called for the crew to initiate the Water Pickup checklist, but due to my rapid descent, we were already below 200 feet when the first flare was tossed out the aft ramp... and at the same time I was seeing the survivor more and more clearly. I was so happy the life raft wasn't a mirage and I wouldn't be embarrassed by a false sighting and my abrupt diving spiral was justified. Still no one else saw the life raft.

Of course the text book procedures and our training had been a little bit different than the reality before us. Rescue pilots were always trained to fly a box pattern around the survivor and drop three smoke flares to form a floating triangle. As you orbited these three flares, the pilot could judge the direction of the wind and sea currents as the rest of the crew methodically prepared for the pick-up. Well, before any of this could happen, I had flown to a 20-feet hover over the water with only one flare deployed. The smoking flare that had been tossed was directly behind me, but somehow (luck maybe), I had flown my final approach into the wind and was sitting, more-or-less in a stable hover about 100-200 feet away from the small (make that tiny) life raft. I could feel my heart beat and my crew members were jubilant and almost cheering. Survivor IN SIGHT.

At this point in the mission, I regrouped control and began to proceed as we had trained; now flying as a polished and coordinated crew. My SSgt Flight Mechanic (FM) was standing in the doorway, checking the electric hoist and attaching a flotation device. On intercom, crew voices were humming with suggestions and comments. My co-pilot was carefully watching our engine gages and rotor tachometer and we all seemed to be poised again for a "text-book rescue." Then I began to hover forward in a slow and controlled motion directly toward the survivor. At about 100 feet out, I could see that the man in the life raft was unconscious... I switched on the loud speakers on the front of the helicopter and began to holler at the survivor. He was supposed to get out of the raft, into the water and be ready to reach for the rescue device. He wasn't moving. I was screaming "*chao Ong... chao Ong*" into the loud-speaker, which is Vietnamese for "Hello Man"... but still no movement. My co-pilot reminded me that we had been flying for over three hours and fuel was low... "Nice predicament," I thought... "if we went back for fuel, we would never be able to find this needle-in-the-haystack again."

I guess I was asking for suggestions as I moved slowly forward again, not really having any special plans. I would just hover overhead, lower my crewman on the hoist, and he would attach the cable to the survivor and reel him in (like a fish). No one on my crew had anything special to add. And then, very suddenly and without any warning, WHING... zit... whoosh. My rotor-wash (wind from the spinning rotor blades) caught the one-man life raft blew it away, separating cleanly from the survivor, bouncing over the water and getting smaller as it literally flew... last looking to me like a very small life-saver candy standing on edge and then gone. Meanwhile, at the same instant, the unconscious man was now sitting in the cold water. "Oh shit," I said, thinking I had just killed him. I hovered straight for him as miraculously the survivor was shocked back to consciousness by the cold water. But he was going down... He flapped his arms like a drowning bird as I moved in, faster and lower. The helicopter canopy was now covered with salt-water spray and I saw the survivor passing by my doorway with one arm raised as if a final thrust to grab us before he would sink from sight.

Then I felt my FM jump out of the cargo door and grab the extended arm of the victim… with one hand. He was hanging out the door and I had another sick feeling he was going overboard. But, he was wearing a gunner's harness around his waist which was fastened to the D-ring in the center of the floor. As I was distracted by the unexpected actions, I flew the helicopter even lower until my landing gear and "bear paws" were beginning to touch the surface of the water. This was a very BAD thing. If you break the surface tension of the water, the power needed to "unstick" the helicopter is probably too great… we were touching the water and with an "AUGH" I pulled up on the collective, righted the helicopter with the cyclic and somehow "LEAPED" us back into the air. Salt water spray was everywhere and forward vision was impossible. I used my side-door vision to climb away from the swirling water with the two men still hanging out of the helicopter on the gunner's harness and with a hand-grip. The Vietnamese man was small and light, maybe 75 pounds dripping wet (no pun)… but still quite a load.

My other crewmembers, a medic and fireman pulled the FM and his precious cargo inside as I continued to climb straight ahead… away from land. The medic began treatment, mostly for dehydration and pronounced, "he's still alive…great job, Captain Bell." I didn't want to confess to the team that we had just survived a near crash and were lucky to be alive ourselves, and I began to return to normal pilot duties.

Oh yeah, the fuel! Well, if the fuel gage was correct, we were on fumes but maybe only four—five miles from land. Obviously if the engine flamed out, I could land by autorotation on the beach, if we made it that far. But where was the nearest beach? The instruments had tumbled during the lift-off and my cockpit was sprayed with salt water. My co-pilot and I were wet all over and the back end of the helicopter was flooded also. Not exactly a text-book mission, but what was one anyway? With no sight of land, I began climbing and turning, hoping not to continue further out to sea. The horizon looked the same, even at 1,000 feet so I continued climbing. By 2,000 feet, I felt more confident and the magnetic compass seemed to be stable so I flew northwest (NW) and soon saw the beach ahead, way south of DaNang. Once over the beach, I flew up the coast and began radio communications with the world again.

"This is Pedro 61, and we have one survivor." The rest was pretty much routine. I flew to the Vietnamese Air Force (VNAF) portion of the ramp where a few Vietnamese soldiers in a jeep took our survivor amid cheers and hugs. They had obviously given up on him being recovered. I was tired and wet, shaken but happy that my crew had logged another SAVE for our Rescue Detachment scoreboard. I finished up the Search and Rescue Data Base (SARDAB) mission report which was always teletyped to 38th Air Rescue and Recovery Squadron at Tan Son Nhut as a permanent record. When our 24-hour alert ended, I drank me a few of those rusty steel cans of Pabst Blue Ribbon, the beer was free (nothing was too good for our servicemen?) Thinking nothing more about my last mission… anxious for the next one.

ROCKET ATTACK ON DET 14

I served the USAF in Vietnam from 15 May 1971—29 Dec 1972. Within the first 90 days in country, I volunteered for a six-months extension to the standard 12-month tour so total time, including a special 30 day leave, totaled 19 1/2 months.

As a CSAR (Combat Search and Rescue) helicopter pilot, I flew primarily from DaNang, Marble Mtn, Hue Phu Bai (Quang Tri), Tan Son Nhut and Bien Hoa Air Bases in South Vietnam. I was also TDY 30 days to Udorn RTAFB and six months to Utapao RTNAB, Thailand providing manpower assistance as a temporary commander, HH-43 Helicopter Flight instructor and Flight Examiner.

In Jun 1972, I was selected by Headquarters Aerospace Rescue and Recovery Service (ARRS) as Commander of Detachment 14, 40 ARRS (previously 38 ARRS), Tan Son Nhut Air Force Base (Saigon) including a Forward Operating Location (FOL) at Bien Hoa Air Base (about 30 miles NE). I was responsible for maintaining 24-hour alert with four HH-43F helicopters and over 60 personnel (over eight helicopter pilots and direct support of a Maintenance Supervisor, HM (Helicopter Mechanics), MT (Medical Technicians), FF (Fire Fighters), Administrative staff and other specialists.

Both Da Nang and Bien Hoa claimed nicknames of "Rocket City." I counted 109 rocket impacts one morning at Bien Hoa with a second barrage exactly one hour after the first, just after the base sounded the all-clear. Tan Son Nhut was located on the northern city limits of Saigon and repelled a serious ground attack as part of Tet in January 1968, however following Tet, rocket attacks were less frequent but sometimes numbered nearly 100 launches. Tan Son Nhut was also an International Airport and continues to support commercial airlines with hundreds of civilian passengers today.

On 6 Dec 1972, we suffered a direct katyusha 122 rocket attack. It heavily damaged two of my four HH-43F helicopters while killing one Helicopter Mechanic Tech Sergeant and wounding another staff sergeant while they were conducting a preflight inspection of the alert helicopter early in the morning: 0746 hrs. They were my two new maintainers transferred from Thailand and rocket attacks were new to them. When they arrived, both had told me that Thailand duty was boring, and they wanted to see more action and were anxious to qualify and earn combat pay.

Before the attack began, I was asleep in officer quarters in the Main Compound on the other side of the base from my Alert Facility. The "crash phone" in my room began loudly ringing. I had remembered hearing the "thumping sound of rockets" in the distant, but I was talking to one of my pilots who excitedly shouted, "Sir, we've taken a direct HIT," and I replied, trying to sound calm, "I'm on my way." Jumping into my flight suit and boots I ran out the door to my blue Alert pickup and found five or six Vietnamese maids begging for a ride to their on-base quarters. They wanted to check on their families; so I said okay and they all piled in.

Then I started driving, and the emergency base sirens began their loud shrieking and warbling with a loudspeaker repeatedly barking out something like, "We are under attack, take cover immediately." But, I didn't. I drove past the Vietnamese housing and slowed while the maids all jumped out. Then I turned on the truck rotating beacon and siren continuing toward the flight line Alert Facility. The flight line access was blocked and Security Police with M16s shouted at me to HALT. I rolled down my window and shouted, "I'm the Commander and I'm going through!" And I did, while in the background the base emergency siren continued shrieking.

My TSgt maintainer had been atop the alert helicopter tilting the rotor blades to his SSgt helper on the ground as part of a daily preflight inspection. At that instant there were a series of rocket impacts coming closer and closer until the last rocket hit the concrete ramp less than 10 feet from my helicopter, creating a three-foot crater and spraying concrete and shrapnel into the helicopter and both maintainers. The sergeant caught on top slumped and slid/fell from the helicopter to the ramp. My other maintainer had alerted on hearing the "Thud...Thump...Blam" sounds of incoming explosions and was running for nearby revetments surrounding the alert facility when he caught a spray of shrapnel in his back.

Personnel inside the facility hit the floor as the large plexiglas window was blown inside and the entry door was blasted off the hinges, close to a direct hit.

My alert medics quickly determined the maintainer with back wounds was less serious, and they focused attention on the maintainer that was thrown from the helicopter. They placed him on a litter and when I arrived from across the runway, they continued to perform CPR although there was evidence of a head wound. Their efforts created a false life-like groaning sound when they released pressure on the victim's chest. They loaded him into the back of my Alert truck and climbed aboard, continuing CPR. I drove the alert truck on the ramp and across the runway with siren and rotating beacon flashing disregarding any Security Police driving directly to the Base clinic. When we arrived, I ran inside and grabbed a gurney from the entrance of the clinic and my medics transferred the victim from the litter to the gurney. I loudly shouted for the on-call doctor and begged him to provide emergency aid. He looked at me calmly and said, "That man is dead." Ouch. We were all visibly shocked and upset; we wanted a miracle.

The doctor had so quickly pronounced my wounded TSgt as DOA

without even listening for a pulse. Everyone in my unit had hopes he was still alive, and I wanted to get assurances from the clinic. Sadly, I now had to return and confirm his death to the waiting members of the unit and also send a KIA casualty notification message.

I started to drive back to the Alert Facility when I saw the Base Chapel. I probably looked desperate when I stopped and ran into the Chapel. I asked the first chaplain I saw for help. I needed someone to meet with my survivors and say a prayer for our recent loss. The chaplain turned and followed me immediately and jumped into my truck. When we reached the flight line, it was blocked "NO ENTRY" but the Security Police waved me through this time.

Meanwhile, back at my unit, everyone was anticipating bad news. When they saw the chaplain, there was a calm among the men. I had no idea what to say or do, but it was soon obvious I had made a good choice. After more than an hour of prayers and meditation, we had bonded with the chaplain and from that moment on he was a member of my unit, visiting often and always on Sundays.

Nerves were raw the day after the attack, when seemingly out of nowhere, a *Chicago Tribune* reporter appeared and I overheard him asking my airmen, "How would you like to be the last man killed in Vietnam?" I threatened to shoot him if he didn't leave my men alone and leave the base! The reporter turned red in the face, cursed me with something like "You can't do this" and just as suddenly as he had appeared, the reporter was gone, probably back to the Continental Palace Hotel in Saigon. My troops were still grieving over the attack, but the news reporter irritation was gone.

We believed that many war reporters and their bán nhiều lắms (sweethearts), spent the war at the air-conditioned Continental Palace hotel in Saigon, drinking and writing "fake news" about American soldiers and the "War."

My other casualty was wounded and had been admitted to the Army Hospital in Saigon. I went to visit and was happy to see he had lots of flowers and a family of Vietnamese friends already visiting. Because of the unknown nature of shrapnel wounds the surgeon had cut him open from his throat to his navel as a precaution. No shrapnel allowed. Soon-there-after, he was airlifted back to USA.

As commander, I first acted as Summary Court Officer to resolve any outstanding debts or obligations of my casualty. Then I was responsible for sending personal effects and a letter of condolences to the Next of Kin (NOK) and surviving family. For this daunting and unpleasant task I sought help from Air Force documents and attempted to find a Commander's Guide while seeking advice from other local commanders. in this regard, I soon learned that I was on my own.

I had to find out who was the actual NOK. I went to my TSgt's barracks and worked backwards to fill his "foot locker" a small wooden box (about 18" x 30" x 18") with his personal belongings. I packed a clean uniform and a flight suit. He was so proud of his flying status. Then I neatly placed his helmet that he had not been wearing during the attack. Then I gathered all of his letters, photographs, and Vietnam souvenirs that he had saved. I used the letters to confirm his last known address.

Wait a minute: there were many letters from his family in one state and a few more recent letters from California. Taking no chances of up-setting a family already upset, I reopened and read the California letters, carefully putting them in order. Once completed, I removed them and mailed them separately in a manila envelope back to the writer with a separate letter of condolences, which included advice to please contact me directly and not to contact any NOK. This was based on the words I had read and to my relief, no reply was ever received. I then took the box over to Mortuary Affairs to be sealed and shipped to accompany his flag draped aluminum coffin back to the USA.

I also submitted both soldiers for the Purple Heart and the Bronze Star Medal awards.

ROBERT CERTAIN

US Air Force — Navigator
Dates of Military Service (Active and Reserve): 1969 to 1999
Unit Served with in Vietnam: Strategic Air Command
Dates in Vietnam: 100 missions in 1971 and 1972, POW Dec 1972 to Mar 1973
Highest Rank Held: Colonel
Place and Date of Birth: Savannah, GA — 1947

ROTATION DAY

My Blytheville AFB (AR) B-52G crew flew its last combat mission of our temporary duty tour to Andersen AFB, Guam on Thursday, December 14, 1972, and were packing for our return to our homes and families the next Monday. I had flown 50 combat missions out of U-Tapao Royal Thai Navy Airbase in Thailand in 1971 and had completed 49 missions from Guam — one short of the coveted 100 missions. Since arriving on Guam for Operation Bullet Shot in July (six weeks after marrying my wife), our copilot had been reassigned to AC-47s, and our aircraft commander had resigned his commission and left the crew. Our replacement pilot had been sent from Blytheville to complete the tour; and the new copilot was from a different AFB back in the States. Since his tour was not complete, he was replaced by a Blytheville copilot at the end of his tour.

The next day, we were notified that all missions scheduled for Saturday and Sunday were cancelled, all crews were to enter crew rest, and all rotations were suspended. Our first thought (and hope) was that the war was over, and we were being held on Guam to bring all the planes back to the US. Why else would the weekend sorties be cancelled? We managed to check out a crew truck and take a tour of the flight line on Saturday morning. What we saw dashed all hope for the end of the war. All BUFFs (nickname for B-52 bombers) were being refueled and loaded with bombs. Some enormously important and probably dangerous mission was clearly in the works. We suspected we were going "downtown" to Hanoi.

The flight schedule was posted on Sunday. Twenty-seven B-52s on the base were going to war on Monday morning, and we were scheduled to fly as Charcoal 3 (three aircraft flying in formation were known by various colors), the twelfth G-model in a train of eighteen behind nine D models. Twenty-one D models out of U-Tapao Royal Thai Airbase in Thailand would precede us. President Nixon was finally going to unleash a war-ending strike against the heart of North Vietnam, Hanoi.

I was elated at being able to be part of an effort that was designed to force the government of North Vietnam to sign a treaty and to release the POWs. Finally, B-52s would be used for their original purpose, and we were going to be a part of history. Since 1970, I had seen our primary goal in SAC as bringing the war to a conclusion and to free the prisoners. Now I was going to be part of that long-sought event. We would give all those fighter pilots in the prisons of Hanoi a reason to respect the courage of the crews of the Stratofortress and an appreciation for their soon-to-be-granted freedom.

Arriving at Stratofortress number 58-0201, we carefully completed the preflight checks on the entire aircraft and its systems and then sat in the shade under the wing for another fifteen to twenty minutes engaging in quiet chatter and reflection. My mood was a mixture of somber, anxious, and excited anticipation as I focused on the challenges and the danger that might await us in the next seventeen hours and twenty minutes. We had flown a lot of missions, but almost all of them had been over the friendlier skies of South Vietnam and Cambodia where there were no enemy air defenses capable of reaching our 30-35,000 foot bombing altitude.

A few seconds into the roll, Charlie Tower called to say two aircraft ahead of us had been moved to the rear, moving us into the Charcoal 1 position. We were now in the lead where we normally flew, and I felt a lot more comfortable being there rather than back in the line. Charcoal 1 was the lead aircraft in the third wave (a cell was three B-52s; a wave was three cells) and would be the nineteenth B-52 over the target, a railroad marshalling yard in the northeast corner of Hanoi.

For the next eight hours we would be making our way across the Pacific Ocean, heading to war. As the navigator, I was responsible for crew coordination, position fixes, rendezvous with a KC-135 Stratotanker, and most critically, absolute time control in order to arrive at the bomb release point exactly on time, on altitude, and on target. As we turned eastbound out of Laos to enter North Vietnam for the bomb run, we were all focused on making this the best, most accurate mission we had ever flown. We would be within lethal range of surface-to-air missiles (SAMs) for about twenty minutes, and we couldn't be distracted by the threats.

The bombardier and I turned off our exterior radios so we could concentrate only on our checklists and crew coordination. I became all business, super-organized, and aware of where we were and what we needed to do in the next several minutes. Any fear I had felt earlier was now gone. We were headed downtown to break the back of enemy transportation and warehousing and there was no doubt in my mind that we could do that.

As we made our turn at the initial aiming point, about seventy-five nautical miles from the target, the radar easily found the target and all four offset aiming points (OAPs). The crosshairs were steady, with no drift. Our initial heading was 147 degrees, with a dogleg turn to 152 for the final run. The wind was a quartering tailwind, giving us 7 degrees of drift compensation. I calculated the time to target and confirmed that our bombs would reach the ground at exactly 2014:00 local time, or 1314:00 Greenwich Mean Time. We were on time, on target, and with the best bombing system possible.

With our outside radios off and the crew maintaining only checklist and bombing instructions on the intercom, the radar and I were able to concentrate on this critical offensive phase of the mission. By this time I was aware of no emotion other than dogged determination, no words other than checklist items, and few thoughts other than mission accomplishment. Thirty seconds before bombs away, Dick opened the doors, and twenty seconds later I was to restart my stopwatch as a backup to the drop should anything go wrong. Just short of ten seconds to go, time seemed to stand still and speed ahead, all at the same time.

At 1313 GMT, the radar screens went blank and other instruments lost power and the aircraft shuddered and yawed slightly left. We had been struck by the first of two SAMs, mortally wounding the pilot and gunner, knocking out engines and electronics, and setting fires. Since we were over the target, the bombardier dropped the bombs, and the members of the crew who could began a controlled ejection. The pilot, co-pilot, and gunner all died that night, while the radar navigator (bombardier), navigator, and electronic warfare officer made it safely to the ground and were captured in short order by groups of civilians and militiamen. We were transported within a few hours into the capital city and arrived at the Hoa Lo prison, the infamous Hanoi Hilton.

For the next eighteen hours I was interrogated, threatened with death, and required to stand, tightly roped, blindfolded, barefoot, and clad only in skivvies on the cold concrete floor. After I identified myself by name, rank, service number, and date of birth, I was pressed for the home base of our airplane, the organization at Andersen, the number of planes stationed there, the type and model of aircraft, numbers of sorties that could be flown in a day, names of other crew members, ingress/egress routes, targets, and a myriad of technical questions about the airplane. My standard response was, "I'm a celestial navigator. I guide planes over water using the sun and stars. When we arrive over land the radar navigator takes over. I know nothing about anything else." When they challenged me I replied that I didn't know—I just flew.

About noon on Tuesday, the guards delivered the magenta and gray pajamas of the POW, along with "Ho Chi Minh sandals." I was told to put them on. A little later, Tom and I were again badly blindfolded, taken from our interrogation rooms, securely handcuffed, and put back into the jeep. We were driven through the streets of Hanoi to what appeared to be a hotel. We were individually led into a large room crammed with people with still and moving cameras and instructed to walk to a bank of microphones. The international press corps was busy snapping photos and taking videos. I walked to the microphones, looked forward, left, and right, but said nothing. In the crowd I spotted an American folk singer: Joan Baez.

Questions were being shouted, but I ignored them. Somebody's photograph of me would be on the wire services soon. After that, I was home free. I had no intention of compromising my integrity and sense of duty by speaking to these people. More threats would mean nothing, because the North Vietnamese would never allow an identified POW to be killed or lost.

For the next ten days, Hanoi was again the target of about 100 B-52 strikes each night, as the USA relentlessly and noisily made the point—sign the treaty and release the POWs. When the bombing stopped after December 29, we were placed in a larger cell with seven men, then moved from the Hanoi Hilton to the Ku Loc (Zoo) prison. When we first arrived at the Zoo, it was a mess. Ceilings had collapsed and there was shrapnel from exploded bombs all over the ground. We were told that fighter planes had dropped high-drag bombs just outside the walls, causing the damage. A few days before Christmas, the locals had tried to break in, necessitating an evacuation of the prisoners. The Vietnamese said the locals wanted to kill the Americans, but we figured they were trying to get in where it was safe. They knew the Americans would never drop a bomb into a known POW camp.

For the next two weeks, life in prison was routine and boring. We discovered that everyone in the Zoo had been down for eighteen months or less, with longer-held prisoners and the December wounded in the larger Hilton. We assumed the reorganization was to facilitate our releases. On the first of February, that assumption was confirmed. The Paris Peace Accords had been signed a few days earlier, and POWs on both sides would be gradually repatriated over the next sixty days. As the last captured, we would be the last to leave North Vietnam, on March 29, 1973.

On that day, we were driven to the Gia Lam airport, where we climbed out of the bus and formed two lines in order of our shootdown dates. When our names were called we marched across the tarmac to a USAF Colonel, saluted, and reported in, "Sir, Captain Certain reporting for duty." From there I turned right and walked toward the ramp of the C-141 known by now as the Hanoi Taxi, where more officers and

NCOs greeted me, and then up the ramp into the aircraft to be greeted by doctors, nurses, and a masseur. We settled into our seats and buckled in so we could get underway. Quite soon the ramp came up, the doors closed, and the Hanoi Taxi moved out smartly to the runway. Although it seemed like hours, in a few minutes we were airborne. Up to that point we were all quite solemn, talking in soft tones or praying. I was anxious and nervous that something might yet go wrong.

When the wheels left the ground and began their retraction, a spontaneous cheer arose. Fifteen or twenty minutes later, the aircraft commander, announced we were "feet wet" (over water) and then that we were clear of North Vietnamese airspace. Each announcement brought another cheer and the swift ebbing of the fear and anxiety of prison life. The last one resulted in seatbelts being unbuckled, cigars lit, and general movement, hubbub, and frivolity. U.S. magazines were plentiful. The masseur busied himself relieving the tension in our bodies that had amassed over the previous hundred days. Our rotation home had finally arrived. I had flown my 100th combat mission, spent 100 days in prison, and would be in the arms of my wife in 100 hours.

Editor's Note: Robert Certain's book Unchained Eagle is available from www.deedspublishing.com and from Amazon.com

VIETNAMIZATION ADVISORY PHASE

(1973-1975)

January 27, 1973—Draft comes to an end.

February 12, 1973—POWs return to United States. First C-141 flight returns first American POWs to US as part of Operation Homecoming; an additional 53 flights returned over 2,000 servicemen by April 4, 1975.

March 29, 1973—Last combat troops leave Vietnam.

August 8, 1974—President Nixon resigns as President of the US In the wake of the Watergate scandal. Gerald Ford sworn in as 39th President.

April 30, 1975—North Vietnamese troops enter Saigon, ending the Vietnam war.

Source: www.vvmf.org/VietnamWar/Timeline

DAN HOLTZ

U.S. Air Force – Healthcare Administration

Dates of Military Service (Active Duty and Reserves Combined): 25 years

Unit Served with in Vietnam: HQ MACV/CORDS, Military Provincial Health Assistance Program, Ninh Thanh Province (Phan Rang, RVN)

Dates you were in Vietnam: Nov 1969 to Nov 1970

Highest Rank Held: Colonel

Place of Birth and Year: Indianapolis, IN – 1943

ESCORT OFFICER FOR A PRISONER OF WAR RETURNEE

In March of 1973, I had the honor and privilege to serve as an escort officer for a newly promoted Air Force major, James Cutter. Jim had been an F-105 pilot who was shot down over North Vietnam in late December 1971. He was taken to the Hanoi Hilton and released as one of the last POW's to come back to the States following the end of hostilities.

Jim was originally from Oklahoma and his father was a coach at Oklahoma State University. Jim and I were matched up for the "homecoming" at Sheppard Air Force Base (SAFB) in Wichita Falls, Texas. SAFB was the closest Air Force base with a regional hospital that was a designated site for POW returnees to meet with their families. I was stationed at SAFB at the School of Health Care Sciences, which is how our paths crossed. However, the first question Jim and I had was, how did a non-combatant medical service corps officer get matched to an F-105 pilot, more later.

My wife Linda and our two children, Laura (then four years old) and Jack (then a year and a half) lived in base housing. Jim's wife, Ginny, her two sons, and Jim's parents all came to our house the morning of Jim's return so we could get acquainted and ready for Jim's arrival that afternoon. The bottom line is, we made lasting friendships with them that have lasted to this day.

When it came time to go to Base Operations, all of us piled into three cars and drove across the base and parked in the reserved lot. We went into the operations center lobby, where we joined about five to seven other families and escort officers who were there waiting for the landing of the airplane carrying the precious cargo. After wonderful and in many cases tearful reunions, the returnees were taken to the USAF Regional Hospital Sheppard for medical examinations and debriefings, which lasted three to five days, depending on each returnee's needs. During this period, family members were permitted to visit with their loved ones each day.

Following the debriefings, the returnees were permitted to do interviews with the news media and then released to their families to go on leave before returning to duty. The media interviews were mandatory if the returnees wanted to do later interviews with local press in their hometowns. Jim and Ginny decided to stay at SAFB for the weekend and depart for Oklahoma on Monday, so Linda and I invited them to go to church with us Sunday morning at our church, Christ United Presbyterian Church on the Northside of Wichita Falls. It was at this point we figured out how Jim and I were matched, he too was a Presbyterian. That seemed to be the common thread.

We got to the church in time for the 11:00 AM worship service. Linda and I were adult leaders for the youth group, so I asked Jim if he would be willing to speak with our youth group about his experiences as a POW and he said he would. I then told our pastor, Rev. Herman Boles, about what was going to be our activity at the youth group meeting that evening. Herman was an old radio preacher from west Texas and took it upon himself to announce to the congregation about the special guest who was with us at the worship service and who had graciously agreed to speak at the youth group meeting that evening. The congregation gave a standing ovation to Jim and a special prayer was offered for his wellbeing and thanks for his safe return to the States. Following the worship service, the members of the congregation were all taking the opportunity to meet Jim and Ginny and welcome Jim back home.

That evening was an even bigger surprise for everyone. The youth group usually met in the small fellowship hall of the church, or sometimes in the front pews of the sanctuary. Well, that evening we decided to meet in the sanctuary, since Rev. Boles had invited anyone interested to come to the youth group meeting that evening. That decision was a great one because we had not only the youth group members attending the gathering, but I believe most of the entire congregation, largely made up of local Texans with some members from the base, who showed up. The sanctuary was filled with every seat taken and quite a few folks standing. Jim did an informal opening with a few remarks and then spent almost two hours answering questions about who, what, when, where, how, why, etc. The kids in the youth group were thrilled, with many asking questions. The adults probably learned more about what had happened during the war and especially during Jim's imprisonment than they had ever expected to learn.

Looking back on the evening and subsequent events over the years, I believe this gathering was the first of the telling of the truth about the Vietnam War by someone who had actually been there and seen a side of it; thankfully, few others had seen.

Jim went back to flying for the Air Force and was stationed in Germany. One day while on a routine mission over the Black Forest, his aircraft experienced a malfunction and he had to eject from the plane. Fortunately, he was not injured, his plane was destroyed and the U.S. government had to pay the German government money for the destruction to the forest caused by the plane. Following that incident, Jim decided he had enough of piloting for the Air Force and that flying a desk was the better part of continuing a long and healthy life. He subsequently retired from the Air Force and settled in the western part of the U.S., where he and Ginny lived most of the time until his death about five years ago in 2014—2015. Ginny still lives there and the boys have gone on to have families of their own.

Our contact with Ginny today is limited to exchanging Christmas cards every year. However, we look forward to them and are thankful each year when we get them. This experience for me started out as a challenge and became a labor of love for which I am most thankful.

BILL HACKETT

USMC – Combat Engineer Officer

Dates of Military Service (Active and Reserve): 1968 to 1988

Units Served with in Vietnam: 1st Engineer Bn, 1st Marine Div; 9th MAB, 1st Marine Div.

Date in Vietnam: Apr 1969 to May 1970; April to May 1975

Highest Rank Held: Lt. Colonel

Place of Birth and Year: Griffin, GA – 1946

OPERATION FREQUENT WIND – THE END

I reported to RVN in April 1969. My tour was as a Combat Engineer Officer with 1st Marine Division south of DaNang. As you know, Marines spent 13 months of duty. About 30 days before my RTD, my unit transferred me to a unit being sent home. My job was commanding officer of a medical unit. Eighteen long days on a big grey boat on the high seas… Coming home by ship did give us time to decompress and had a nice welcome at San Diego.

Forty-eight months later, I found myself back overseas in Okinawa. Things were heating up in Vietnam in April 1975. My Battalion was tasked to provide a 72 man Ships Detachment for deployment aboard Merchant shipping tasked to assist picking up boat people/refugees leaving country. The specific reason I was given command of the Detachment Hotel (one of ten our Division provided) was I was the only Captain who had a previous RVN tour.

We boarded a C-130 and flew to Cubi Point in the Philippines and boarded the USS Dubuque, an amphib with five helo spots and a ramp for amphibs. Our unit was designated the 9th Marine Amphibious Brigade (9th MAB).

We sailed up and down the coast of RVN, preparing and planning for the evacuation of Saigon. This became known as Operation FREQUENT WIND. I will never forget being in the CIC of the Dubuque on April 30, 1975 and observing the SAM missile sites plotted along the helo evac routes. CH 46's flying almost 24 hours straight picking up refugees. I observed many helo's tossed overboard with no space for them.

Later we debarked the Dubuque and boarded USNS merchant ship TRANSCOLORADO and gathered off coast and transported 4,250 refuges and families to ANDERSON AFB in Guam.

We prepared for numerous operational exigencies such as sailing in a Mike Boat up the Vung Tau river to bring refuges out, but never had to actually set foot back in country. This phase of my RVN experience will never be forgotten.

Semper Fi,

Bill Hackett

JOHN W. PATTON

USAF -C-130E pilot

Dates of Military Service: 1973 to 2003

Unit Served with in Vietnam: 21st Tactical Airlift Squadron

Dates in Vietnam: Jun 1974 to Apr 1975

Highest Rank Held: Colonel

Place of Birth and Year: Shreveport, LA 1950

THE EVACUATION OF SAIGON – A PERSONAL REMEMBRANCE

None of us knew quite what to expect. We knew what was happening but we still did not know how we would fit into the puzzle. We had been told to report to the wing theater at Clark AB, Republic of the Philippines, for our briefing. My crew consisted of Capt. Bill Lundberg, Aircraft Commander; 1Lt. Paul Boudreau, Navigator; MSgt. John Kays, Flight Engineer; TSgt Steve Tkach, Loadmaster; and 1Lt. Frank Jershe, Scanner and relief pilot. I was the Co-Pilot. There were many other crews similar to ours at the briefing, all wondering the same things. When were we going? Where exactly were we going? How were we going? What were we going to do?

All this happened over forty-four years ago but much of it is still fresh and clear in my mind. I guess that is how fear and apprehension can affect a person. I know that there was plenty of fear and apprehension in this 24-year-old First Lieutenant.

South Vietnam was collapsing. The North Vietnamese Army was moving through the countryside just like the tide rolling over a beach. And it was all happening so fast! I could hardly believe that as recently as March 21st I had been in Hanoi. Each week we had a mission to transport the different parties in the peace talks back and forth between Saigon and Hanoi. Only a month ago I had been on one of these trips and now in April, just a month later, it looked as if Saigon would be overrun at any time. Ban Me Thout had fallen on March 10 and the South Vietnamese government by mid- March had abandoned Pleiku. When would Saigon take its turn?

"Room Attention" came the command. As we all got to our feet, the 374th Tactical Airlift Wing Commander, Colonel James Baginski, entered the theater for the briefing. Now perhaps some of our questions would be answered.

It was called Operation Frequent Wind. We would be flying into Tan Son Nhut AB at Saigon to evacuate Americans, their dependents, and those South Vietnamese people who would be in grave danger after the communists took over. We were to load these people into our aircraft and fly them to safety back at Clark AB.

Once the details of the briefing were completed, we were released to crew rest until we were needed. As the operation called for flights to depart every thirty minutes, some crews left immediately while others would see a couple of days pass before they got the call.

When our turn to fly came, we once again had a briefing to update us on the latest developments. This done, we set about to flight plan, draw our weapons, ammunition, survival vests, and flak jackets before proceeding to our aircraft. Normally we weren't issued weapons, much less survival vests and flak jackets for our flights, so this was a little unnerving. By the time we made all the necessary stops to pick up equipment and arrived at the aircraft, I felt like I was carrying an extra fifty pounds of gear.

Once our pre-flight checks were done, we were finally ready to go. The missions were initially set up so that the C-141s flew by day and the C-130s at night. But now that things had begun to heat up, only the C-130s were allowed to fly into Tan Son Nhut. We began our take-off roll just as the sun was going down. In about four hours we would be in Saigon.

A four-hour flight over water at night gives one a lot of time to think. I do not remember much conversation inbound to Vietnam; we each had too much to think about. Our Flight Engineer, John Kays, and our Loadmaster, Steve Tkach, had each served in Vietnam before, when the war was really hot. They knew what to expect. Our Aircraft Commander, Bill Lundberg, had flown EC-47s earlier over Vietnam and knew what was ahead. The rest of us, all first lieutenants, were rookies. We had no way of knowing what to expect and we had no way of knowing how we would react. But old hands or rookies, we all had our thoughts and emotions to deal with.

Somewhere inbound to Saigon I began to ponder all the equipment I had been issued — flak jacket, survival vest, and .38 pistol and ammunition. Our C-130s had armored seats — panels of thick material on all sides of the crew seats — all sides except the bottom, that is. I decided that this would be a very good place for all of that equipment I had been issued. So I wore the survival vest and .38 pistol, and slid the flak jacket under my seat in the probably foolish notion that it afforded me some degree of protection. It was well that I did all this over the water because I got so busy later that I never thought about any personal preparation.

Eventually the coastline of Vietnam began to show up, first on radar, then after a while we could see the lights from the villages along the coast. I still remember the taste in my mouth. Now there was no more time for private thoughts; we began to get busy. There were many radio calls to make and answer; often I was listening to three radio frequencies at the same time. Frank Jershe, the other co-pilot, assigned as a scanner for our flight was able to help with the radios until we got overhead Saigon, but then he had to return to his duties in the cargo compartment with the loadmaster. As busy as I was, Bill Lundberg, our aircraft commander, clearly had the most difficult job. In addition to flying the plane, he had to coordinate with each one of us to make sure that things went as smoothly as possible.

We had been briefed to fly overhead Saigon—right over the city at twenty thousand feet. We were all wearing parachutes and our flight helmets and oxygen, and we were to depressurize the aircraft and open the paratroop doors on each side in the back of the cargo compartment as we descended in circles. The loadmaster and the scanner each hooked up safety harnesses and stood in the open doors with flare pistols loaded with phosphorus flares. If they saw SA-7 shoulder fired missile launches, they were to fire these flares at the oncoming missiles in the hope that the heat from the flares would decoy the missiles away from us. Just offshore we had turned off all our navigation and position lights and our red rotating beacon so that we were completely dark.

Now we were over the city, flying blacked out, depressurized, with two crewmembers strapped at the open doors ready to launch flares at any missiles as we spiraled down in the dark. All the way down we had to look out for missiles and small arms fire, find Tan Son Nhut AB (it was blacked out too), and hope to avoid any other aircraft in the area. We could hear other crews calling out their altitudes over the radio as they spiraled up or down over the city just like we were doing. We all hoped that they were not too near. I remember straining my eyes looking through the dark at the ground as we descended, trying to look out for shoulder fired missiles, and wondering if the twinkling lights I saw below were the result of ground lights shining and flickering through the trees in the city or small arms fire.

Finally, as we got lower and flew more towards the northwest part of the city, we located the blacked out patch of ground called Tan Son Nhut AB. Our final spirals in the dark brought us in for a landing over the dark runway. All of the lights on the runways and taxiways had been turned off lest they provide illumination and range for any enemy soldiers in the area. We made our final approach as steep as possible, completely blacked out onto a blacked out runway. I let out an audible sigh of relief as the landing gear made contact with the pavement and Bill threw the propellers into reverse. We had made it!

It only took a few minutes to taxi to the part of the ramp where we were to load our passengers; by the time we got there I realized how tired and drained I was. But we weren't through. We had to run this gauntlet again on our way out of Vietnam!

Some crews experienced lengthy delays on the ground at Tan Son Nhut waiting for their passengers. This never happened to us. Both times we went into Saigon our passengers were ready and we loaded them immediately. Certainly on one occasion we left the engines running, loaded our people and took off again all in the space of a few minutes. I do not remember much from our ground time in Saigon. I guess we filed a "round-robin" flight plan before we left Clark AB, because I do not remember filing one or even leaving the aircraft in Saigon.

The C-130 is designed to hold about 75 troops in full combat gear or about 90 regular passengers. Normal maximum take-off weight is 155,000 pounds. I have no idea how much we weighed on these flights—none of us knew. I am sure that we did not weigh more than the limit, but we sure carried a lot more than 90 people. We loaded our passengers from the ramp door at the tail of the aircraft and just kept putting them on until the airplane would not hold any more. I can remember our loadmaster losing count after 200. This was not anything unusual or isolated; all the other crews were doing the same things. We had been given a job to do and we were going to do it.

Finally, we took off and climbed up over the city. Once again we were totally blacked out, spiraling up, calling out altitudes, hearing altitude calls in reply, and looking out for small arms fire and missiles. I prayed that we wouldn't see any, but I prayed even harder that if we did see missiles we would see them in time.

All the way from take-off to coast out, really all the way from coast in to coast out, I was filled with both apprehension and uncertainty. What a sense of relief when we were safely off the coast of South Vietnam headed back to the Philippines!

Once we were over the water and safely on our way back to Clark

AB, I got out of my seat and went to the steps leading to the cargo compartment. I guess I felt that I had to see what it was like in the back. The cargo space was packed with people, all of them sitting on empty cargo pallets on the aircraft floor. It was so crowded that no one could move anywhere. None of us in the cockpit could go to the back, and neither our loadmaster nor our scanner could come up front. This airborne sea of humanity was very subdued. There were no smiling faces in spite of being airlifted out of a war, only faces that reflected fear and uncertainty. These people were leaving behind their homes and in many cases, their families. What they must have been feeling, I can only imagine. It was a pretty quiet flight home, as we each had a lot to think about on our way out of Vietnam.

While my flights evacuating Saigon were uneventful, some of my fellow crews from the 21st Tactical Airlift Squadron had different experiences. Our ground times were very short, but some of the other crews spent many hours waiting for their passengers to be processed before they could be boarded onto the airplanes. I can only imagine what that must have felt like, spending hours in the dark on the ramp at Tan Son Nhut, wondering when the airfield would be attacked.

Tan Son Nhut was attacked at least twice while C-130 crews were on the ground. One attack came from South Vietnamese A-37 aircraft either captured in the fall of Pleiku or flown by disgruntled South Vietnamese airmen. One of my friends, First Lieutenant Fritz Pingel, was the aircraft commander of a C-130 on the ground when the field was bombed by these A-37s. He took off as quickly as he could, only to be chased by one of the A-37s. Fritz did about the only thing you can do in a C-130 in a situation like this; he flew as fast and as low as he could and headed for the coast. He also tried to fly into any cloud he could find in the hopes of losing his pursuer. Only when he got to the coast did the A-37 turn away. He told me later that he could never figure out why they did not get shot up that day unless the A-37 was out of ammunition.

Shortly after this incident, on April 29, the field came under rocket attack. One of the rockets hit the wing of a parked C-130 and it immediately burst into flames. Captain Greg Chase and his crew were parked nearby and watched in horror as the airplane burned. Fortunately the plane had not yet been loaded with passengers and all the crewmembers managed to escape the wreckage and scramble into Greg's airplane. They immediately took off for Clark AB. I believe that this was the last American C-130 to leave Tan Son Nhut.

My crew made two trips in and out of Saigon that April. On April 29th, about the time Tan Son Nhut was under rocket attack, we were inbound to Saigon for our third time when we got the call from the Command Post. "Come on back to Clark, it's over." The following day, April 30, 1975, Saigon fell to the North Vietnamese Army. The war was truly over.

Over forty-four years have passed since then, and while I am sure that I have forgotten or confused some of the details of those trips, one thing is sure. I will never forget the faces of those people, mostly women and children, on my airplane that April in 1975.

POST-VIETNAM EXPERIENCES
AND WRAP-UP

JIM COCHRAN

USAF — C-141 Pilot

Dates of Military Service: 1970 to 1977

Unit Served with in Vietnam: 7th MAS (Military Airlift Squadron)

Dates in Vietnam: Sep 1971 to Mar 1974 (various times)

Highest Rank Held: Captain

Place of Birth and Year: Zanesville, Ohio — 1947

BRINGING HOME 12 FALLEN SERVICEMEN

On March 14, 1977, over three years after the Vietnam War had ended, I was one of three pilots selected to fly the C141 Starlifter jet transport into Gia Lam Airport, Hanoi, North Vietnam. The mission was to fly President Carter's delegation into North Vietnam for talks and possibly bringing back remains of our brave men who had been killed serving in the war.

This mission was a DV Code 2, reserved for the Vice President of the United States. There were no weather minimums, no crew duty day limits, nor any other normal operating restrictions. We were to do whatever it took to get this mission done.

Major Comstock, Major Abernathy, and I had flown from Travis AFB, CA to Hawaii, and then on to Clark AB, in the Philippines. We arrived two days prior to meet the presidential delegation. The delegation was brought to Clark AB in a presidential aircraft (VC-137) which the North Vietnamese did not want on their airport ramp.

On March 18th, we went out to our aircraft, tail number 38075. The delegation had already been flown into Hanoi two days prior to this by Major Comstock and Major Abernathy. I remained at Clark AB to monitor the first trip into Hanoi.

Since this was the second trip in, Major Abernathy told me I would fly the aircraft into Gia Lam Airport, Hanoi, North Vietnam, since I had stayed back on the first trip.

This second trip went as planned except when we got close to the airport, burning rice fields made visibility very limited. We had no precision approach plates, as expected. We did find a crude one which had been used to bring back our POWs. I flew the approach based off a radio station signal. When we broke out of the smoke, I was right of the runway, but maneuvered to a successful landing. Having a presidential delegation onboard, you can imagine how nervous I was.

In Hanoi, we taxied to our parking spot behind a "follow me" vehicle. Then we were taken to an area where we were given tea, as well as gifts of vases and hand-painted plates.

We then returned to the aircraft, there were 12 small boxes placed on tables behind the aircraft. Some of these boxes had dog tags, some had military ID's, and some had both on top of them. After a few minutes, we (the aircrew) placed U.S. flags folded into the normal triangle on top of each one. The delegation arrived and stood silently while we carried these remains on board the aircraft.

We departed with the delegation and flew them to Vientiane, Laos where they departed for more official talks.

After Laos, we flew to Bangkok, Thailand and refueled; then around the southern end of South Vietnam back to Clark AB.

At Clark, the mortuary team came onboard and placed the remains into normal deceased transfer cases. Now the folded U.S. flags were unfolded and placed full length on top of each service member's case. They were secured to our aircraft floor for safe transport home.

After this was completed and we had refueled again, we flew on to Anderson AFB, Guam. There another crew was standing by to relieve us. They flew on to Hickam AFB, Hawaii where the remains would be positively identified.

This was an extremely and unusual 27-hour "on duty" day consisting of over 12 hours of flight time. We were very tired, but honored to have been part of bringing these servicemen back to their families.

THEY HAD GIVEN THEIR LIVES FOR THE UNITED STATES OF AMERICA!

ROBERT O. "BOB" BABCOCK

US Army—Infantry Officer

Dates of Military Service (Active Duty and Reserves Combined): 1965 to 1974

Unit Served With in Vietnam: Bravo Company, 1st Bn, 22nd Inf Regt, 4th Inf. Div.

Dates Served in Vietnam: July 1966 to July 1967

Highest Rank Held: Captain

Place of Birth: Heavener, OK—1943

VIETNAM 3-MAN STATUE DEDICATION, 1984

January, 1983. Sleet and snow pelted Atlanta, Georgia, bringing the city to a screeching halt. The IBM meeting I was running for salesmen from the Midwest was temporarily suspended. A festive atmosphere permeated the group as we congregated into one apartment and partied while we weathered the storm.

Many changes had come to my life since I left Vietnam. My wife and I had gotten on with our life, had two children, gotten divorced, and were both remarried. I had been working for IBM for fifteen years.

Though never totally out of my mind, Vietnam was still not a popular topic of conversation. Most of my friends and fellow workers had either missed serving in the war or were too young to have participated.

I am not sure why the subject of Vietnam came up that night. But when it did, two of us started relating our stories. The circle of people became quiet as they listened with intense interest. It was as if they were curious about Vietnam, did not know how to ask about it, but were very interested when they found someone who would talk of their experiences. I found it was not an unusual phenomenon, especially in bars after a few drinks.

As we talked, I found the other vet, Hal Reynolds, was also a veteran of my division, the 4th Infantry Division. He had served a year after me and had the distinction of having been in a tank battle between American and Russian tanks, having been wounded when his tank was hit. Hal and I took turns telling tales of our wartime exploits.

As the evening wore on and the conversation shifted, Hal and I continued our discussion. He had stayed in tune with Vietnam. He told me of the 4th Infantry Division Association and the Vietnam Veterans of America, organizations of which I have since become a life member. And, he had attended the dedication of the Vietnam Memorial Wall in Washington, DC in 1982.

After I returned to my home in Kansas City and Hal to Oklahoma City, we stayed in touch via phone calls. During one of those calls in the summer of 1984, he asked, "Babcock, do you want to go to Washington, DC with me on Veterans Day? They're dedicating a Statue of three fighting men at the Vietnam Memorial Wall." Without hesitation, I committed to be there.

When I told my wife, Jan, about my plans, I was surprised when she said she wanted to go. She had never shown much interest in Vietnam, had not experienced it with me, and had never asked much about it.

It was with much anticipation, and some trepidation, that I boarded the plane that day. I did not know what to expect as we arrived on Friday before the Sunday dedication ceremony. We stayed in the hotel where the 4th Infantry Division Association was making their headquarters. My hopes were high that I would meet some of the men I had served with. Since returning from Vietnam, I had only been in touch with two men. Hal was not scheduled to arrive until later, so Jan and I were on our own as we went to the hospitality suite.

As I scanned the room, none of the faces looked even familiar. Many of the men seemed to have similar uncertainties as mine. After giving up on seeing a familiar face, I decided to wade in and start talking. As I approached each individual or group, I always asked the same question, "What unit were you in and when were you there?"

Most of the men had served later than I but I did find one man from our battalion who had gone over on the troop ship with us from Fort Lewis, WA. Although we did not know each other, we knew some of the same people and remembered many of the same experiences.

I was talking to a man, Swede Ekstrom, who introduced himself as a helicopter crew chief and had served during the same time period as I. Another man joined us and pointed out a cluster of men at the other end of the room. In that group was a thin fellow he identified as a recipient of the Medal of Honor, First Sergeant David H. McNerney. We immediately moved down the room to meet him, shake his hand, and listen to what he had to say.

I learned he had earned the nation's highest honor in the spring of 1967 while with the 1st Battalion, 8th Infantry Regiment. His action was in the same area where our battalion had been. He was an unassuming man. As we continued to ask questions, he pulled a small, laminated copy of the Medal of Honor citation from his pocket. Squinting to read the small print, Swede and I eagerly read it. (To save space, I won't include it here—you can find it at www.4thinfantry.org on the Medal of Honor page).

Wow! I was impressed and awe struck standing next to this hero. However, the next few minutes' conversation showed that some of his biggest feats of bravery were not included in the citation. Swede, my new friend, started talking to him about the evacuation of the wounded. It turned out he had, in fact, been the crew chief on a helicopter that came in to take out the wounded! David McNerney's eyes lit up brightly as he and Swede jointly told the remainder of the story. It took as much courage, albeit of a different kind, as what he had done earlier under enemy fire.

The chopper had been flying news reporters to a fire base when they heard the call for a "dust-off" chopper. Since they were close, the pilot diverted and headed for the landing zone. As the chopper landed, McNerney and his men brought the wounded out to load on the chopper. Seeing that the helicopter was partially loaded with reporters, McNerney ordered them off.

When the reporters refused to get off, he, along with Swede, physically removed them to make room for more wounded. The helicopter left with the wounded while the reporters scurried for the relative safety of the company's defensive position. As more "dust-off" choppers came in to evacuate the wounded, McNerney steadfastly refused to let the reporters back on until all his wounded men had been evacuated. Swede and his chopper made several trips with wounded before making a final trip to pick up the reporters.

This was the first of many memorable events that Jan and I experienced that weekend. Hal Reynolds soon arrived and we decided to visit other hotels that were hosting hospitality suites. The same thing was happening at each of the reunions. Men from all walks of life, dressed in everything from jungle fatigues to business suits, were standing around reliving that most memorable time of their lives. Everywhere we went, we could feel the emotion of the moment.

Soon it was 2:00 AM. Several men were talking about making the several block trip to the shadow of the Lincoln memorial where the Vietnam Wall stands. Since we had a car, we volunteered to drive. Later, reflecting back on the events, it was a strange happening. Here we were, with my wife, in a strange, crime infested city, inviting several men we had never met to get into the car with us in the middle of the night. Danger never once crossed my mind. These were not potential assailants; they were brothers in arms, men who had lived the same experiences I had.

We were surprised to find a large number of visitors at the Wall at this early hour in the morning. With the exception of a few wives, the vast majority of the visitors were men of a similar age as me, paying homage to their comrades in arms who had paid the ultimate price in defense of our country. Each man was looking for a name or names that brought back a flood of memories and pain.

One of the men who had ridden with us kept running his hand over a portion of the Wall. He obviously was deeply moved so I asked him if he wanted to talk about it. The story he told sent chills down my spine. Twenty-two of his comrades had been killed in an ambush. He was the only survivor. He had played dead while the NVA searched the bodies and killed the wounded.

We walked the short distance from the Wall to the Statue of the Three Soldiers that was to be dedicated that weekend. We were struck by the authenticity of the figures and the emotion they evoked in us as they stood there in silence, gazing toward the Wall which held the names of over 58,000 fallen comrades. As we studied the figures in detail, we all found one minor flaw—there were no vent holes in their jungle boots to let the rice paddy water out.

The next afternoon's activities were centered on the Mall. A crowd had gathered in front of the temporary stage. After some warm up singers, Chris Noel, the voice I remembered so well from Armed Forces Radio came on stage. Although eighteen years had passed since I had ogled her picture in "Stars & Stripes," she had taken good care of herself and was still a beautiful lady. And, her sultry way of opening and closing her radio show with a "Hi, love" and "Bye, love" had not diminished over the years.

To wrap up the show, Chris introduced Lee Greenwood to sing his relatively new song, "God Bless the USA." There wasn't a dry eye on the Mall as Lee sang the song over and over again.

We awoke to a gray, misty, cold morning on Sunday. But the rain did not keep us indoors. Nothing was going to keep me from being at the Wall; the reason we had congregated there that weekend.

The area around the Vietnam Memorial Wall was a sea of humanity as we waited for the helicopter to make the short flight from the White House with President Reagan. I do not remember what he said that afternoon nor do I totally remember how I felt. I do remember the emotions I saw on veterans faces as they caressed the Wall, saw old buddies for the first time in years, and struck up conversations with total strangers. The feeling of camaraderie and belonging was everywhere. The greeting among the vets was an emotional, "Welcome Home, Brother!"

That weekend changed me. From being a passive Vietnam vet who fought the urge to share my experiences, I welcomed the chance to share my views on the Vietnam experience. I wanted Americans to understand we Vietnam vets were not bad guys, that we were good Americans doing what our country had asked us to do, and had done it very well (thank you very much).

I was not standing on soap boxes yet, but I was starting to get involved. It also started the spark to smolder in me to write my book, *What Now, Lieutenant?* covering my Vietnam experiences.

A year later, on the tenth anniversary of the fall of Saigon, the national conscience seemed to change. It was then that the American public started paying attention to us, listening to what we had to say, and recognizing us with long overdue welcome home parades. Five years later, Jan gave me a twelve inch tall replica of the Three Fighting Men Statue which sits in a place of honor in our home. Seldom does a day go by that I don't look at the Statue and think of the experiences of the men and women who fought the Vietnam War.

Note from Bob: For all veterans and Family members of Vietnam veterans, I highly encourage you to visit the Wall if you have not been there. It truly is a Wall that Heals. Memorial Day and Veterans Day are great times to be there, with large crowds. But any day or night is good. Be sure to wear something showing you are a veteran and the unit you served with—that helps meet others from your unit. I always wear my 4th Infantry Division cap and my Combat Infantryman's Badge to identify me as a vet.

Almost two years later, I attended the Chicago Welcome Home Parade, here are my memories of that. Both the above and this memory, along with many others, are recorded in my What Now, Lieutenant? *book.*

VIETNAM VETERAN'S PARADE, CHICAGO, 1986

May 8, 1985—This page one article caught my eye as I read the newspaper prior to leaving home for work, "New Yorkers Roar Thanks to Veterans." Reading further, the article said: "Vietnam veterans accepted New York City's thanks yesterday at a bittersweet ticker-tape parade through the canyons of lower Manhattan.

"The belated welcome home from a war that ended 10 years ago began Monday night when Mayor Koch lighted the Vietnam Veterans Memorial, adjacent to 55 Water Street. It continued yesterday when a thunderously appreciative crowd, which the police estimated at one million, lined the parade route from Cadman Plaza in Brooklyn to the Battery…"

As I finished the article, a sick, hollow feeling tore through the pit of my stomach. Pictures of the World War II homecoming parades had always been vivid in my mind. As I served my tour in Vietnam, I had often thought about a real homecoming—and a New York City ticker tape parade was right at the top of the list.

It made me even sicker knowing I could have been there that day. On Friday of the previous week, I had decided against going to an IBM meeting in New York that would have put me right on the parade route. Instead, I had sent one of the men who worked for me. If only I had known about the parade, wild horses could not have kept me from New York.

I said nothing and went on to work. For the next several weeks, I scoured the newspapers and weekly news magazines to read about the event. It had been a very emotional experience for those in attendance. Approximately 25,000 Vietnam veterans marched in the parade, thousands more watched from the sidewalks. Men found friends they had not seen in years. The American public showed a real appreciation for the contributions the Vietnam vet had made. For the first time, the Vietnam vet seemed to be truly accepted for the sacrifices he had made for his country.

I went on about my life but, often, in moments of reflection, I felt sad I had missed that once in a lifetime experience. Like all veterans, I felt I had earned a parade. The only appreciation I had ever felt was at the Vietnam Memorial Statue dedication the previous November. That was from other vets, not from the American public.

A year later, in the spring of 1986, when I was participating in the Moving Wall ceremony in Atlanta, I heard about an upcoming Vietnam veteran's parade to be held in Chicago in June. Like in New York, it was to be a ticker tape parade. Over 50,000 veterans were expected to participate.

Immediately, my mind started debating with myself about whether or not I could go to the Chicago parade. There was no question I wanted to go, being able to afford the time and money was the question. It had never been my nature to spend money on myself, and I traveled so much in my job I felt guilty to leave my family for what, to them, would appear to be such a frivolous reason.

A few days later, I mentioned it to my wife, Jan. Much as I expected, she listened and told me to do whatever I thought was right. If anything, she encouraged me to go ahead and do it. I muddled the problem around in my mind for a few more days. I had concluded I could justify the one day away from my family to fly up there and back. I was still having a hard time justifying the $400 airplane ticket.

Finally, it hit me. A few years earlier, I had spent almost $400 one night to take our family to see Michael Jackson, the famous singer, on the first night of his world tour at Arrowhead Stadium when we lived in Kansas City. This had to be a more important event to me than Michael Jackson was to our kids. If they could go see Michael Jackson, I could go participate in a Vietnam veteran's parade—and not feel guilty about doing it! My mind was made up; I called Delta Airlines and made my plane reservation.

It was with mixed emotions that I boarded the first Delta flight from Atlanta to Chicago on the morning of June 13, 1986. Would I find any of the guys I had served with? How would I react to the parade? Would it be what I expected or would it be a disappointment? A flood of memories coursed through my brain on the flight to Chicago.

By 7:30 am, I was in a cab en route from O'Hare airport to the Navy pier, marshalling area for the parade. As I paid the cab driver, I saw hundreds of fatigue clad vets milling around the pier. I scanned them all, looking for a familiar face or a 4th Infantry Division patch.

For the next two hours, I roamed through the crowd, my eyes constantly scanning around me. Not since Vietnam had I been so observant of my surroundings. I found a few 4th Infantry Division patches, but no one I knew.

As the parade began to form, I moved to the group standing behind the banner with the 4th Infantry Division "Ivy Leaf" patch proudly emblazoned on it. Still, no familiar faces. Then suddenly, I saw one. General

William Westmoreland, commander of all troops in Vietnam and the grand marshal of this parade, approached our group.

Not being bashful, I was the first in our group to step forward and shake his hand. It was around the time he was in the middle of a lawsuit with CBS news over their reporting of his actions during the war. Many of our emotions were mixed about the job he did in Vietnam, but he was our commander, and he deserved our respect. I was proud to have the opportunity to shake his hand.

He said one thing which still stands out in my mind. "You men in the 4th Infantry Division had one of the toughest jobs in Vietnam. No one else spent as much time in the jungle as you. You humped the hills for months on end without getting a break. And, the press seldom made it out to report on your accomplishments. You did a great job in the Central Highlands — I thank you for it."

He did not have to say that. I think he believed what he said — and I agree with him. The 4th Infantry Division was not the darlings of the press like the 1st Cavalry Division, the Marines, or the 173rd Airborne Brigade, but, by God, we did a job to be proud of.

My emotions and unit pride were starting to build. I never wanted to take anything from another unit; I just wanted our accomplishments to be recognized. I could sense this was going to be a pivotal day in my life.

Several things during the course of the parade swept me with emotion. First, the parade was led by Medal of Honor recipients from World War II, Korea, and Vietnam. There, in a group, were men who had earned the ultimate honor in defense of our country. Just like my experience in Washington the previous fall, I was in awe of Medal of Honor recipients (and still am).

Behind the Medal of Honor men were the disabled vets. My heart went out to the men who had given one, two, three, or even all four limbs and more in doing what their country had asked of them. Some pushed themselves along in wheelchairs or on four wheeled platforms; others were pushed by other vets. Some paralyzed vets were pushed on gurneys. You could see the emotion in each of these men, and feel it in yourself.

As we progressed down LaSalle street, I was overwhelmed by the reaction of the public. Men in business suits, women, children, poor people, rich people, middle class people, old men, veterans in fatigue jackets—all walks of life were represented along the parade route. The shouts of encouragement and thanks they gave us will live forever in my heart and mind. Tears streamed down my cheeks as I marched proudly along closely behind our 4th Infantry Division banner.

And, before long, I found myself walking through knee deep ticker tape. This parade was everything I had ever dreamed it would be, and more. Never have I felt as proud of my service to my country as I did that day parading through the streets of Chicago.

As our unit came to the end of the parade, I peeled off and walked back through the crowd to add to the cheers and encouragement for subsequent units coming down the street. Standing on the sideline watching the emotions in other vets passing by was as gripping as the parade had been. It struck me that men from all walks of life, many who had not had a significant break in all their life, were getting an experience they had earned and would never forget. I stood and shouted, "Welcome Home!" and "Thanks!" as each unit passed. Sheer emotion showed on each of their faces.

This was an important day to me, and I have been fortunate to receive many breaks in my life—it had to be even more overwhelming to those who had struggled all their lives, only to be snubbed for their most important accomplishment. Finally, the American public was expressing their appreciation to us.

As the parade wound down, people wandered toward Grant Park, on the shores of Lake Michigan. There, a sea of fatigue clad veterans was everywhere the eye could see. Again, my eyes swept over every person as I tried to find even one long lost friend and comrade. I walked and I wandered, I went to the amphitheater where Lee Greenwood was entertaining with his "God Bless the USA" and other songs.

I am not a fatigue wearing veteran. In fact, I wore my "IBM casual" uniform—khaki pants, blue button down shirt, and navy blue sports jacket. The only thing which designated me as a Vietnam vet was the metal 4th Infantry Division patch I wore on my jacket pocket. (Well, I did buy a "Vietnam Veteran and Proud of It" cap). Still no old friends were to be found. But new friends were everywhere.

That was the day I learned about the common bond that exists among all Vietnam vets. Never have I been hugged so genuinely, so often, and with so much emotion as by the men I ran into that day. And each hug was accompanied by what has become the slogan of the Vietnam vet, "Welcome Home, Brother." Black, white, officer, enlisted, Army, Navy, Marine, Air Force, Coast Guard, we were all the same that day—brothers.

For the rest of the day, I wandered aimlessly around Grant Park. Finally, as time approached for me to return to the airport and the flight home, I stopped by the hotel where a reception was being held for the Medal of Honor recipients. I was in the company of the largest assembly ever of recipients of our nation's highest honor. Eighty of the 247 surviving Medal of Honor men from all wars were expected to be in attendance. I wandered through the hotel lobby to pay homage to those heroes and reflect on how great it is to be an American.

Reluctantly, I hailed a cab and started my trip home. I did not want to leave but I had accomplished what I had come for. Despite not finding any old friends, I had an emotional experience of a lifetime. It was on this day I took on a new avocation—standing up for, helping, and advancing the cause of the Vietnam veteran. I forever quit trying to ignore my Vietnam experience. I started trying to understand its importance to me, and to others. What had started at the Vietnam Wall almost two years earlier was bonded into my being that one day in Chicago.

It continues as I preserve the stories of so many Vietnam veterans in this book you are reading now.

GEORGE MURRAY

US Army — Armor Officer / Aviator

Dates of Military Service: 1964 to 1968

Unit Served with in Vietnam: A/82d Airmobile Light, re-designated 335th
Assault Helicopter Company

Dates in Vietnam: Feb 1966 to Feb 1967

Highest Rank Held: Captain

Place of Birth and Year: Grenada, MS — 1942

THE WALL

On a much-anticipated trip to Washington, D.C. to visit The Vietnam Wall again, my wife and I went with a group from the Atlanta Vietnam Veterans Business Association (AVVBA).

On March 27, 2018, we arrived at The Wall at about 1930 hours (7:30 p.m.). The place was mobbed. It was spring break and a large crowd of people stretched the length of The Wall. It was dark, but The Wall is lighted by subdued lights that allow viewing of the names on The Wall.

My mission was to visit Panel 9E82 to honor Warrant Officer Joe Sampson, a member of my flight class in 1965. He was the aircraft commander of a UH1D Huey Helicopter which crashed on July 27, 1966. His helicopter was No. 8 in a formation of 10 moving troops of the 173d Airborne Brigade. I was the Aircraft Commander of No. 7.

On the night visit to The Wall, I accidentally went to Panel 19, as opposed to Panel 9. As I searched, a young lady of about 13 asked if she could help me. I told her I was looking for Sampson. Suddenly, I had a group of teenagers all helping me. They asked about Sampson and I related the accident in which the rotor head flight controls failed and his crew of pilot Jamie Rutherford, crew chief Jim Collins, and gunner Harold Reinbott, and six paratroopers of the 173rd Airborne perished in a fall from 1500 feet.

As they helped me search, I realized I was searching Panel 19 rather than Panel 9. I apologized to my helpers, but they excitedly followed me to Panel 9. They quickly located Joe Sampson and the members of his crew memorialized on The Wall. Thanking them for their help gave me a feeling that the USA is in very good hands with the future generation.

Our AVVBA group moved over to the Three Soldiers statue immediately south of The Wall. Maybe because of our age, or because some of the guys with me had on AVVBA hats, we were suddenly surrounded by teenagers shaking our hands. They thanked us for our service and wanted their picture taken with us. We learned from some of the students that they were from Dalton High School in Dalton, Georgia. They displayed a lot of knowledge of Vietnam War history, much more than I expected.

These young people related to us that they had been to Arlington Cemetery earlier in the day. They viewed the grave of a classmate's Grandpa, who was killed in Korea after service in World War II.

The large number of people viewing The Wall was a surprise, many of the people there had a very specific mission of viewing a loved one or simply someone they had heard about. The patriotism was outstanding. Our great country will be served by young people who will follow us proudly.

RICK LESTER

US Army - Aviator, Armor Officer
Dates of Military Service: 1967 - 1994
Unit Served with in Vietnam: 10th Combat Aviation Battalion (1st tour); 48th Assault Helicopter Company (2nd tour)
Dates in Vietnam: 1969, 1970 to 1971
Highest Rank Held: Lt. Colonel
Place of Birth and Year: Marietta, GA - 1948

A PROUD LEGACY

I was going to college when Dad went to Vietnam, but I soon left school to join the Army where I attended the Warrant Officer Rotary Wing Flight training program. When I finished training, I went home on leave prior to deploying to Vietnam. My Dad talked with me about his experience in combat and how he wished he could recall specifics about events and people he met. He talked about "my brothers" and I was always asking him questions about his time in service. He told us that he couldn't imagine what his life would be like if it weren't for the men and women he met through his military service. He only wished that he had made some record of what they did together and how to contact them after the war.

He lifted a box from a chair near his desk and placed it in front of me and told me to open it. It contained twenty-four "any year diaries." He handed me one of the books and said, "My last mandate to you as you leave my household is to maintain a journal. You'll find that when you are in combat your time won't be your own and dates and places will be blurred. In war you will be witness to many things you will have a difficult time comprehending, and an even more difficult time expressing to the laymen. You will have a tough personal choice, you can let these things effect you in a negative way and hang over you like a dark cloud hampering everything you attempt in life, or you can let them effect you in a positive way and temper you like good steel, making you stronger for enduring them. It will be your choice, no one else's. Because I know you, I know you will persevere and I know you will help those in your charge do so as well. Your credibility will always be tied to your integrity and your character will not only be forged by your actions, but your word as well, don't ever rationalize those traits away."

"I want you to promise me that at some time during your day, you will take time for yourself, stay grounded in all you have been taught, close your eyes and think of your God, your family, your standards, and what you ascertain to be your true, important priorities. Say our family prayer, order your thoughts, and make time to purge them into your journal. One day you will realize the importance of this. This journal will nurture your memory, helping you by keeping things in it in order. Through time you will find that others will be unable to recall even important events. They will, at first, joke about it, then be confused and then be angry. I'm not a doctor and can only tell you what I've witnessed."

I left for Vietnam a short time later and was soon caught in that blur of strange smells, strange surroundings, the true realization of danger, fatigue, and anxiety that every "NEWBIE" encounters. While trying hard to not look like a newbie, I moved from the replacement center to another place where I waited in the transient barracks for my orders to another in-processing center. At some point, someone handed me a postcard to send my family.

"Hello_____ I have arrived safely in country and will write you as soon as I have in- processed with my permanent unit and have a unit mailing address for you."

After nine days of traveling and finally getting to my assigned unit, I sat down to write to my mom and dad. I knew the first question my dad was going to ask was, "Are you maintaining your journal?" So, I reached into my bag and took out one of the two "any year diaries" I had brought with me. I sat there for a few minutes trying to remember: What day did I actually depart CONUS? What was the name of that contract air line? Was the replacement center at Binh Long or was it Long Binh? My dad was right! I couldn't even put the first nine days of my tour in chronological order, much less provide details. I promised not to let that happen again, then started my letter home with; "Well I have arrived in my new unit safely and started working on my journal..."

I have used my journal through the years to present facts related to investigations, award recommendations, Officer and NCO professional develop programs, and Staff College reports. After I retired from the Army, I continued to record in my journal, but primarily as a periodic summary rather than a daily log. I recorded events like when my son and I were present at the bombing in Centennial Olympic Park and details of the historical development of the Atlanta Vietnam Veterans Business Association.

I also recorded the details of how I was able to acquire an actual AH-1 helicopter rotor blade from Vietnam and get it to Atlanta for a memorial honoring Frances McDowell in the atrium of the Atlanta Hartsfield-Jackson Airport. I made a record of how I worked to gather soil from every site around the world where soldiers from our country fought, bled, and died for the cause of freedom. And then noted how, in

2013, I used the soil in a "Sacred Soil Ceremony" to commemorate the opening of the remodeled Veteran's Park at the Atlanta History Center. I made notes about working to present a unique design for our memorial at Atlantic Station.

I also used my journal to provide details to the Joint Personnel Accounting Center in Hawaii which led to the recovery of remains for two pilots from my unit who had been missing since 1971.

After their recovery, I used my journal notes to honor them at their funeral at Arlington National Cemetery in 2000. The eulogy not only served to honor those two men, but highlights my thoughts about our role in the Vietnam War. I think it fits with our efforts to promote a counter to the media's perception of Vietnam Veterans and would like to present it here.

Through the fog of controversy surrounding our country's involvement in Vietnam, some may ask why, but let no one ever question the intentions of these honorable men. There are so many good things we can say about them. In the finest traditions of our country, they answered their call and went forth with only the best intentions. What more could our country or anyone ask?

Though some wars have been necessary, there has never been a good one and true warriors know there never will be. The cost of war is great. It robs nations of their most precious resource, their young. It also has a way of bringing out the best in men. War strips men to their most basic moral standards, facades are quickly torn away, and you are judged as your true self, good or bad. Those of us who knew these men saw them in that light and can tell you they were truly dedicated, strong, and courageous.

Those of us who served with them also came to know their heart. In the quiet times, we heard of their love for their families, shared their laughs, and listened to their stories of life before Vietnam. We found pleasure in simple things such as mail from home, hot food, cold beer, a periodic hot shower, and time shared in the "Club." We grew close.

In the violence of that war, we also shared our fear and frustration, endured physical pain, and the bitter experience of losing friends. We came to know indefinable fatigue from seemingly endless hours of flying in the most demanding and dangerous conditions, yet if we weren't flying, we were not happy. Though for the most part we dealt with the confusion, complexity, and violence of battle in our own way, it was understood there was no shame in showing your emotion. We were only human. We endured and became stronger for it. We were sometimes hard on each other, but it was with purpose and we knew we could turn to each other for anything. We grew closer.

We may have been sent in harm's way with a broken sword, but we stood as one. Our shield was our pride and the respect we had for each other and our duty was to carry out the mission. We were in this thing together and our strength would come from our commitment to each other and to our unit. We learned a special trust common only to those who have learned to hide their fear and willingly place their lives at risk, not just for the cause, but for those with whom they served. The common theme was a bond of mutual respect and unspoken love and friendship forged and tempered through the trials of battle.

You realized, once you had fought for them, that freedom and life are indeed very special. You no longer took things for granted. You noticed for the first time how really intense and beautiful a sunrise can be and how nice it was to once again feel the warmth of the sun on your face after the monsoons had passed.

You no longer said prayers. You spoke with God. You now knew the fragility of life and therefore it became more intense. Through all of this we quickly realized what an honor it was to know and serve with men like these and how truly blessed we are to have had them in our lives.

We will remember them always and to our absent companions we now say, "Catch the wind, good friends, take the lead, and soar high into that warm light of God and on your wing, keep watch for us."

Rick Lester
"JOKER 94"

HISTORY OF THE ATLANTA VIETNAM VETERANS BUSINESS ASSOCIATION (AVVBA)

It was Veteran's Day 1982 and Joe Harrison was in his office working hard to expand his business in the growing commercial and residential rental property market in Atlanta, but on this special day he had a desire to reach out to others who knew it's significance.

He had served in the Navy aboard the USS Hancock during two tours in the South China Sea supporting operations in Vietnam. He had met a few veterans of the Vietnam War through his real estate connections and decided to reach out to them in hopes of getting out of the office early to enjoy a little camaraderie. He called the office of a commercial real estate appraiser and spoke with Mike Turner who had served in Vietnam with the U. S. Army's 75th Rangers in 1970-71 and Don Purdue who served in Vietnam with the Army's First Infantry Division in 1968.

Joe explained his desire to get together with other veterans and asked for their help. They decided on a site and time for their "link-up" and began calling everyone they knew in the local business community who had served in Vietnam and asked those contacts to do the same. Their efforts proved successful and their first gathering at a local bar named "Penrod's" included; Joe Harrison, Mike Turner and his sister, Don Plunkett, Don Pardue and friend, Mal Garland, Walter Stroman, and Al Roberts.

The event was a hit and soon plans were underway to compile a contact list of Vietnam veterans and schedule another social. Though the group lacked a formal structure, they stayed in touch and continued to get together as their business schedules allowed. There were socials at "Harrison's," "Otto's," and other popular bars, primarily on Memorial

Day and Veteran's Day, but in 1984 Walter Stroman and Mal Garland coordinated a meeting at the Holiday Inn in Marietta, specifically to gather as many Vietnam veterans from the business community as possible.

Walter's wife Dancy collected the names and contact info for all attendees. These contacts formed the base list for future events. One of the attendees who made his first meeting with the group was Rick Lester who had been invited by Duke Doubleday. Rick was still serving on active duty with the Army and was stationed at Fort McPherson. He had been introduced to Duke by Georgia's Secretary of State, Max Cleland. Rick had been a patient at Walter Reed Army Medical Center after being medically evacuated from Vietnam and had met Max there.

Rick suggested setting up a regular event to progressively grow the organization and volunteered to set up a monthly luncheon and coordinate with local media to recruit veterans and inform prospective members about the scheduled luncheons. I volunteered to write a story about our efforts and record our initial history. Rod Knowles and Duke Doubleday had contacts with the Atlanta Business Chronicle so we decided to ask them to publish a story about our group. Rod and Walter proposed trying to coordinate a social and invite the staff from the Atlanta Business Chronicle. Soon this idea became one of challenging the staff to a softball game to provide them a different perspective of Vietnam veterans. Rick set up the game to be played at the recreational facility at Fort McPherson. This game became an annual event which not only involved softball, but included "combat bowling" at the Ft Mac bowling alley and dinner at their grill!

In November, 1984, the first of our monthly luncheons was held at the Vietnamese restaurant "Cha Gio" at the corner of Tenth and Peachtree. Larry Taylor, Rick Lester, and his brother ate at this restaurant regularly and thought a Vietnamese Restaurant was a perfect fit for the group. There were twelve attendees at the first luncheon, nine at the second and seventeen at the third. The number of monthly attendees grew progressively until we outgrew that facility and moved to the 57th Fighter Squadron Restaurant near the Peachtree Dekalb Airport. (Note:

as the organization formally developed and grew, it soon outgrew the 57th Fighter Squadron Restaurant and moved to the Asian "Happy Valley Restaurant" on Buford Highway, and then followed that restaurant's move to Jimmy Carter Boulevard for a while, but after members complained about the location being so far from the center of Atlanta, the luncheon site was soon moved to The Retreat At Dunwoody. The size of our group and new ownership of that facility finally led us to our current location at the Dunwoody United Methodist Church.)

The monthly luncheons helped us to develop as an organization. Walter Stroman formalized our group by drafting our charter as a 501 C19 organization. Steve Martin filed the charter through his law firm and Mal Garland stepped up to be our first President. Our group supported a veteran's outreach program for a while, but looked to do something else to show a positive side of the Vietnam veteran.

Mal Garland had been associated with a youth camp in North Carolina and at one of our luncheons, he told the group a story about one of the young camp counselors who had become a favorite with all the kids who attended the camp. This young man left to join the Army and after his training, came back to visit prior to going to Vietnam. The kids and their families had a big going away party for him. They were devastated when they learned he had been killed in action shortly after arriving in Vietnam. Mal said the kids were pretty emotional about the tragic loss of their friend and were having a tough time coping so he, the staff, and many of the families decided to honor their friend by naming the camp's boat house, where he had served as counselor, lifeguard, and canoeing instructor, in his honor. He related how this had such a positive impact on all the people associated with the camp and suggested that we look into the possibility of honoring someone from the Atlanta area.

Walter Stroman had a classmate from the Citadel who was from Atlanta and had been killed in Vietnam, so a decision was made to honor this man, Lt John Fuller, at the Galleria Complex in 1987, as our first memorial as an organization. These memorials became an annual tradition and were emplaced around Atlanta in high visibility sites where people lived, worked, and played. The organization took great pride in the fact that these ceremonies were planned, financed, and conducted by Vietnam veterans to honor the sacrifice of their fellow Vietnam veterans to the highest possible standard. The Atlanta Vietnam Veterans Business Association soon gained recognition as one of the premier veterans' groups in the state.

Through the years, the memorials were our central focus and our membership took great pride in our efforts and readily contributed financially to the success of our memorial each year. The memorials were initially done by the same small group of members, primarily due to business commitments of most of the membership. The initial memorial committee members encouraged other members to get involved and bring new ideas as they also highlighted how rewarding it was to see how much our actions meant to the family members. Soon the memorial committee grew from the original six or seven member team to more than twenty members and many more volunteers on the day of the memorial dedication. It was truly inspiring to witness the efforts of so many of our group, from different military branches and civilian business professions pull together to execute a mission they felt to be so important.

Our charter, "To promote a positive image of the Vietnam veteran through our actions in the community and to honor those from the Atlanta area lost in Vietnam," was now a living testament to the true heart of all Vietnam veterans. This was acknowledged on numerous occasions by the news media in the area, and even by the dean of a prominent private school in a 'left handed compliment' after our memorial there. The dean approached Rick Lester with whom he had been working to address every facet of the memorial and said, "I just want to take a minute to tell you how much our staff was impressed with the organizational skill, professionalism, and unbelievable efficiency of your group. When you set up the initial meeting you arrived thirty minutes early, 'wearing suits' and bringing a 'read ahead' notebook and video which highlighted your past memorials. When I stated that I wasn't certain I would be able to clearly articulate a description of your idea for the memorial in order to get approval from our board of director's at a meeting the next morning you asked, 'Would a scale model of the memorial help you do that?' I asked if you had one and you said you could make one by the next morning and then brought the most beautiful architectural model to me an hour prior to our board meeting the next morning.

"You told us the AVVBA memorial committee had arranged for a guest speaker who had served with the person we were honoring. The committee had arranged for accommodations for all the honoree's family members at no expense to them and coordinated a reception the night before so that the committee, my staff, the guest speaker, and family could interact prior to the ceremony. When we told you we intended to keep the students in their classrooms during the ceremony so they wouldn't be a disruption, your team insisted they be allowed to assemble around the memorial site so they could witness the event and also asked that our Chamber Orchestra be allowed to participate with the Army band. The committee even had bagpipers present to open and close the ceremony and then coordinated a precision flyby of military aircraft. After the ceremony, your association had a catered reception for the family, my staff, and your team and invited me to participate in presenting replicas of the memorial to each family member."

"My staff and faculty were in awe of the AVVBA's attention to detail and unbelievable ability to orchestrate every facet of such a complex, multidimensional event in such close proximity to the Atlanta Airport. To a person we were totally impressed and just found it difficult to understand, with the immense talent of your group and the organization skills your group exhibited, how you lost that war."

In 1996, when the terrorist's bomb exploded in Atlanta's Centennial Olympic Park, Rick Lester and his son were twenty five paces away. They were soon working with first responders to assist fifteen critically injured victims and continued to assist them until 4:00 a.m. when they were taken to the security headquarters to provide their accounts of what they witnessed. Rick contacted Rod Knowles who was serving as a managing director for The Atlanta Committee for The Olympic Games to tell him what he had been involved in and to offer his assistance if further needed.

Rod informed the senior leadership at NBC news and soon Rick was asked to talk about what he and his son did to show a positive side of the events that morning. Tom Brokow gave a lot of attention to Rick's actions and his background as a two-tour Vietnam helicopter pilot. In the days following the bombing, Rick and his son were asked to attend receptions with various Olympic sponsors. At a reception with Coca Cola, Rick met many of the Trustees and members of the Board of Directors from the Atlanta History Center and took the opportunity to tell them about the AVVBA. In 1997, 1998, and 1999, Rick and his son delivered invitations for the AVVBA memorial ceremonies to the Atlanta History Center President.

In 1999, Rick Beard, the AHC president contacted Rick, who was then serving as the President of the AVVBA, to tell him that the Atlanta History Center staff had attended two of the AVVBA memorial ceremonies and they were so impressed that they had decided to invite the AVVBA to emplace a memorial on the AHC property. They said it would be the first "war memorial" at the AHC. Rick Lester was a native of Marietta and the son a career military veteran who had served in WWII, the Berlin Airlift, Korea, and Vietnam. He had attempted to get support for a veteran's park in Cobb County, but with the invitation from the Atlanta History Center and the knowledge that there wasn't a veteran's memorial in Atlanta, he approached Rick Beard to discuss making the area at the corner of Slayton and West Paces Ferry Road a veteran's park rather than just a site for the AVVBA memorial. Rick submitted his design for the park to many of the AHC trustees and members of the Board of Directors that he and his son had met and received unanimous approval with the stipulation that he represent the AHC as the project manager with Post properties to complete the park.

Rick saw the park as the perfect site and opportunity to honor ALL Atlantans lost in Vietnam. This would allow the AVVBA to complete its mission to honor all those from the Atlanta area who were lost and also give the AVVBA an "anchor memorial" that, once the AVVBA terminated the annual memorial ceremonies, would be able to gather on Memorial Day and Veteran's Day each year to remember the sacrifice of their fellow veterans and those members of the AVVBA who had passed away over the years.

In 1998, one of our members introduced us to a gentleman named Delous Yancey who had attended one of our memorials and was so impressed by our efforts that he graciously offered the AVVBA a donation of $25,000.00 to be disbursed to the association over a five-year period. He encouraged us to form a foundation which would enable him to solicit support from his fellow business associates to raise the funds which would give us a financial base from which we could continue to support our memorials. We established a foundation and used Mr. Yancey's donation to offset memorial expenses and build 'seed' money for a scholarship program.

The memorials were becoming more difficult to conduct to the standard we had established due to numerous circumstances, such as trying to locate immediate family members of those we wanted to honor, the availability of sites in the developing downtown area, and a loss of support from the local military community with the closing of Fort McPherson and Fort Gillem. In 2001, the AVVBA president, Pete Mitchell, selected Rick Lester, Max Torrence, and Steve Martin as a strategic planning committee to evaluate our association's future missions, giving consideration to our charter, financial status, and feedback from our membership. The recommendations were to look at a point in the future to terminate the annual memorials and develop a scholarship program for veterans and/or family members of veterans.

Events following the terrorists attack on the World Trade Center Towers expanded our goals as members Bob Hopkins and Mark Walker recommended that we include a priority to support the USO. Max Torrence, Mark Steele, Bob Babcock, along with John and Judy Vail stepped up to coordinate the AVVBA volunteers to provide unfaltering support to the soldiers moving through Atlanta's Hartsfield-Jackson Airport USO. During the peak of the wars in Iraq and Afghanistan, AVVBA USO volunteers served 300 active duty military personnel twice a month, greeting and feeding them at the USO, as they passed in and out of the combat zones, and providing them with much needed rest and relaxation at home before returning to the war. AVVBA continues this mission as this book goes to press, with a smaller volume of military personnel coming through the airport, never missing a single one of their twice monthly volunteer opportunities since 2004.

The AVVBA membership stepped up to support our scholarship program and expand the scholarship opportunities to several schools in the community. The scholarship program and continued support of the USO are the primary focus of the AVVBA presently. We continue to meet for our monthly luncheons and proudly serve our community.

This recorded AVVBA history is attributed to Joe Harrison, Ed Hines, (from an article prepared for the Atlanta Business Chronicle) and Rick Lester, Past President and Historian, Atlanta Vietnam Veterans Business Association

Reviewed and approved — Don Plunkett, Maj Gen Larry Taylor, Rod Knowles.

To learn more about AVVBA, go to the web page at www.avvba. org — all Vietnam vets with a valid DD-214 are welcome to join our association.

Atlanta Vietnam Veterans Business Association
Proud ★ Professional ★ Patriotic

FINAL THOUGHTS

I never dreamed when I started this project that we would end up with a book over 600 pages in length. I am honored and humbled that the veterans and wives of the Atlanta Vietnam Veterans Business Association responded so strongly to my pleas to get their stories to me.

For you other Vietnam veterans reading this, remember what I said up front—people want to hear about your experiences. You owe it to your Family, the unit you served with, and American history to tell your story. As we are closer to the sunset than the dawn of our lives, don't put it off any longer to preserve your stories for posterity. There are many ways to do it, among the best is the Veterans History Project, a law passed by Congress in 2002 and managed by the Library of Congress. Google that and you can learn where the closest volunteer to preserve your story is located.

In Atlanta, the Atlanta History Center is a partner of the Veterans History Project and a repository of interviews and does more interviews each month. Other cities have similar places. And, for other ideas, you can go to my web site at www.americansremembered.org to get some of my ideas on how to preserve your memories for posterity, by video, written form, audio form, there are lots of options. I am a Founding Official Partner of the Veterans History Project through my American's Remembered non-profit organization.

For you Cold War, Desert Storm, Iraq, Afghanistan, and other younger veterans reading this, don't put it off like far too many of your WWII, Korea, Cold War, and Vietnam forefathers did and take your stories to your grave, never to be read or heard again. Go ahead and start taking the actions now to preserve your memories for your Family, your unit, and America.

And to all Veterans and their Families—thanks for your service to America— Welcome Home!

I will finish as I started this book...

Steadfast and Loyal—Deeds not Words!—Proud, Professional, Patriotic

ROBERT O. "BOB" BABCOCK

Bob grew up in the small railroad town of Heavener, Oklahoma. He served in Vietnam with B/1-22 Infantry, 4th Infantry Division as an Infantry officer in 1966-1967. From 1968 to 2002, Bob was a Sales/Marketing Executive with IBM. Bob is a founding official partner of the Veterans History Project, part of the Library of Congress, preserving memories of America's veterans.

Bob recently ended eight years as president of the National 4th Infantry Division Association. He continues as their historian. He served for ten years as president of the 22nd Infantry Regiment Society.

Bob is author of ten books and is founder/CEO of Deeds Publishing LLC and of Americans Remembered Inc. Bob has published over 300 books for established and aspiring authors.

Bob is a member of multiple national and local veterans organizations, including Atlanta Vietnam Veterans Business Association and leads the Veterans ministry at Mt. Bethel United Methodist Church. Bob and his wife, Jan, have four grown children and five grandchildren.

BOOKS BY ROBERT O. BABCOCK

War Stories: Utah Beach to the Liberation of Paris
War Stories: Paris to VE Day
War Stories: Vietnam 1966 to 1970
What Now, Lieutenant?
You Don't Know Jack… or Jerry
Operation Iraqi Freedom I: A Year in the Sunni Triangle
Operation Iraqi Freedom 07-09: Dispatches from Baghdad
World War II WAC — Helen Denton as told to Bob Babcock
With Honor We Served — Veterans of Mt. Bethel UMC
I'm Ready to Talk — Vietnam Veterans Preserve Their Stories

These and many other military books published by Deeds Publishing can be found at www.deedspublishing.com

CPSIA information can be obtained
at www.ICGtesting.com
Printed in the USA
FFHW010603221119
56054099-62049FF